HIGH-ENERGY RADIATION BACKGROUND IN SPACE

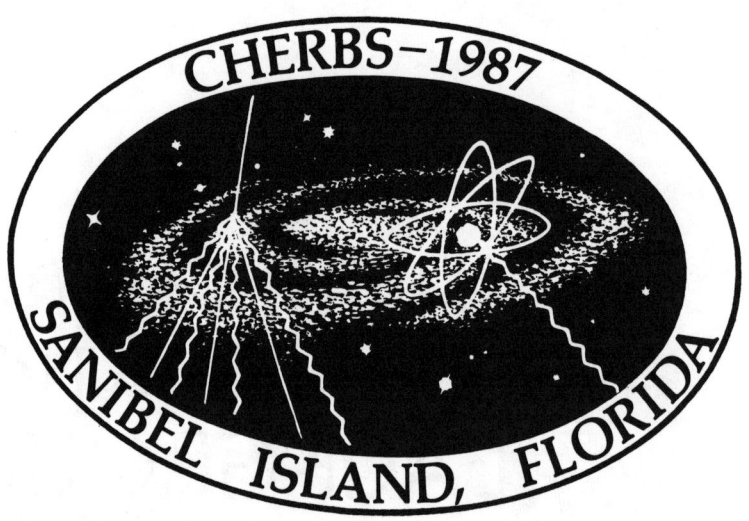

AIP CONFERENCE PROCEEDINGS 186

RITA G. LERNER
SERIES EDITOR

HIGH-ENERGY RADIATION BACKGROUND IN SPACE
SANIBEL ISLAND, FL 1987

EDITORS:
A. C. RESTER, JR.
UNIVERSITY OF FLORIDA

J. I. TROMBKA
GODDARD SPACE FLIGHT CENTER

AMERICAN INSTITUTE OF PHYSICS   NEW YORK 1989

Authorization to photocopy items for internal or personal use, beyond the free copying permitted under the 1978 US Copyright Law (see statement below), is granted by the American Insitute of Physics for users registered with the Copyright Clearance Center (CCC) Transactional Reporting Service, provided that the base fee of $3.00 per copy is paid directly to CCC, 27 Congress St., Salem, MA 01970. For those organizations that have been granted a photocopy license by CCC, a separate system of payment has been arranged. The fee code for users of the Transactional Reporting Service is: 0094-243X/87 $3.00.

Copyright 1989 American Institute of Physics.

Individual readers of this volume and non-profit libraries, acting for them, are permitted to make fair use of the material in it, such as copying an article for use in teaching or research. Permission is granted to quote from this volume in scientific work with the customary acknowledgment of the source. To reprint a figure, table or other excerpt requires the consent of one of the original authors and notification to AIP. Republication or systematic or multiple reproduction of any material in this volume is permitted only under license from AIP. Address inquiries to Series Editor, AIP Conference Proceedings, AIP, 335 E. 45th St., New York, NY 10017.

L.C. Catalog Card No. 89-083833
ISBN 0-88318-386-2
DOE CONF 8711149

Printed in the United States of America.

Contents

Preface ix

I. THE RADIATION ENVIRONMENT

Charged Particle Radiation Exposure of Geocentric
Satellites 3
    E. G. Stassinopoulos
Transient X-Rays, Gamma-Rays, and Neutrons in Space 64
    R. E. Lingenfelter
Approximate Angular Distributions and Spectra
for Geomagnetically Trapped Protons in Low Earth Orbit 75
    J. W. Watts, T. A. Parness, and H. H. Hechman
Solar Particle Composition, Spectra, and Frequency
of Occurrence 86
    D. V. Reames

II. PARTICLE INTERACTIONS AND PROPAGATION: DYNAMIC MODELING

Products from Cosmic-Ray Interactions in Extraterrestrial
Matter: What They Tell Us About Radiation
Background in Space 97
    M. C. Reedy
Bremsstrahlung Production by Electrons: Cross Sections and
Electron–Photon Transport Calculations 103
    S. M. Seltzer
Interactions of Multi-MeV Gamma Rays with Matter 125
    R. L. Coldwell, F. E. Dunnam, M. Katoot, and P. S. Haskins
Three-Dimensional Monte-Carlo Simulation of Gamma-Ray
Scattering and Production in the Atmosphere 132
    D. J. Morris
High-Energy Radiation Environment During Manned
Space Flights 146
    R. Silberberg, C. H. Tsao, J. H. Adams, Jr., and J. R. Letaw
High-Energy Outer Radiation and Belt Dynamic Modeling 159
    Y. T. Chiu, R. W. Nightingale, and M. A. Rinaldi

III. DATA BASES

Nuclear Cross Sections for Estimating Secondary Radiations
Produced in Spacecraft 177
    L. W. Townsend and J. W. Wilson
Nucleon Interaction Data Bases for Background Estimates 192
    J. W. Wilson and L. W. Townsend

Reference Nuclear Data for Space Applications 203
    S. Pearlstein
ENVIRONET: An Interactive Space-Environment
Information Resource 210
    A. L. Vampola, W. Hall, and Michael Lauriente
SEL Monitoring of the Earth's Energetic Particle
Radiation Environment 216
    H. Sauer

IV. INSTRUMENT BACKGROUND AND DOSIMETRY

Gamma Radiation Background Measurements from Spacelab 2 225
    W. S. Paciesas, J. C. Gregory, and G. J. Fishman
Radioactivity Observed in Scintillation Counters
During the HEAO-1 Mission 232
    D. E. Gruber, G. V. Jung, and J. L. Matteson
Background Observations on the SMM High-Energy Monitor
at Energies > 10 MeV 243
    D. J. Forrest
Long-Term Variations in the Gamma-Ray Spectrometers
on OSO-7 and SMM Spacecraft 250
    J. D. Kurfess, G. H. Share, R. L. Kinzer, W. N. Johnson,
    J. H. Adams, Jr., E. L. Chupp, D. J. Forrest, and C. Reppin
Comparison of Backgrounds in OSO-7 and SMM Spectrometers
and Short-Term Activation in SMM 259
    P. P. Dunphy, D. J. Forrest, E. L. Chupp, and G. L. Share
Instrumental and Atmospheric Background Lines
Observed by the SMM Gamma-Ray Spectrometer 266
    G. H. Share, R. L. Kinzer, M. S. Strickman, J. R. Letaw, E. L. Chupp,
    D. J. Forrest, and E. Reiger
Radioactivity Induced in Gamma Ray Spectrometers 278
    C. S. Dyer, P. R. Truscott, N. D. A. Hammond, and C. Comber
Space Radiation Shielding Analysis and Dosimetry
for the Space Shuttle Program 289
    W. Atwell, E. R. Beever, A. C. Hardy, R. G. Richmond, and B. L. Cash
The Radiation in a Molniya-Type Orbit 297
    J. B. Blake and J. E. Cox
The HEAO-3 Background: Spectrum Observed by a Large
Germanium Spectrometer in a Low-Earth Orbit 304
    W. A. Wheaton, J. C. Ling, W. A. Mahoney, L. S. Varnell,
    and A. S. Jacobson
On-Orbit Observations of Single-Event Upsets
in Harris HM-6508 RAMs: An Update 323
    J. B. Blake and R. Mandel

V. DETECTORS AND EXPERIMENTAL PROGRESS

The Space Radiation Environment at 840 km 329
    E. G. Mullen, M. S. Gussenhoven, and D. A. Hardy

The Cosmic Radiation Effects and Activation Monitor 343
    C. S. Dyer, A. J. Sims, R. J. Hutchings, D. Mapper, J. H. Stephen,
    and J. Farren

Characterization of Space Radiation Environment in Terms
of the Energy Deposition in Functionally
Important Volumes 350
    L. A. Braby, N. F. Metting, W. E. Wilson, and C. A. Ratcliffe

The GRAD High-Altitude Balloon Flight Over Antarctica 359
    G. Eichhorn, R. L. Coldwell, F. E. Dunham, A. C. Rester, J. I. Trombka,
    R. Starr, and G. P. Lasche

VI. BIOLOGICAL EFFECTS

Biophysical Aspects of Heavy Ion Interaction in Matter 369
    W. Schimmerling, M. Wong, B. Ludewigt, M. Phillips, E. L. Alpen,
    P. Powers-Risius, R. J. Guzman, L. W. Townsend, and J. W. Wilson

Delayed Effects of Proton Irradiation in Macaca
Mulatta Primates (22-Year Summary) 381
    D. H. Wood, K. A. Hardy, A. B. Cox, Y. L. Salmon, M. G. Yochmowitz,
    and R. E. Cordts

Responses of Carausius Morosus to Spaceflight Environment 393
    G. Reitz, H. Bücker, R. Facius, G. Horneck, W. Ruther, R. Beaujean,
    and W. Heinrich

The Protons of Space and Brain Tumors I: Clinical
and Dosimetric Considerations 407
    G. V. Dalrymple, W. A. Nagle, A. J. Moss, Jr., L. A. Cavin,
    J. R. Broadwater, E. L. McGuire, C. S. Eason, J. C. Mitchell,
    K. A. Hardy, D. H. Wood, Y. A. Salmon, and M. G. Yochmowitz

The Protons of Space and Brain Tumors II: Cellular
and Molecular Considerations 412
    W. A. Nagle, A. J. Moss, Jr., G. V. Dalrymple, A. B. Cox, J. F. Wigle,
    and J. C. Mitchell

New Astronaut Radiation Exposure Limits and Implications
of Proposed Changes in Quality Factors 432
    D. S. Nachtwey and R. J. M. Frey

Promotion of a New Radioprotective Antioxidative Agent 434
    J. Matsubara, A. Ikeda, and T. Kinoshita

VII. FUTURE NEEDS AND STRATEGIES

Space Station: Infrastructure for Radiation Measurements
in Low Earth Orbit 445
    B. D. Meredith

Scientific Considerations in the Design of the Mars Observer Gamma-Ray Spectrometer 453
    J. R. Arnold, W. V. Boynton, P. Englert, W. C. Feldman, A. E. Metzger,
    R. C. Reedy, S. W. Squyres, J. I. Trombka, and H. Wanke

Particle Background Effects for the Hubble Space Telescope (HST) and the Lyman Far Ultraviolet Spectroscopic Explorer 468
    B. E. Woodgate and W. B. Fowler

Radiation Environment Evaluation for ESA Projects 483
    E. J. Daly

Preface

The interaction of human beings and their plants, animals, and machines with the nuclear radiation background of space has been of interest to researchers from the earliest days of space exploration. As we enter into the era of ultra-high altitude airliners, permanently manned space stations, a new generation of space probes, and manned interplanetary voyages, the exact characterization of that background is becoming increasingly important in the development of shielding strategies for sensitive instrumentation and biological systems. This is also the age of nuclear proliferation; agencies around the globe responsible for monitoring the compliance of signatory nations with non-proliferation treaties have need of the best possible understanding of the nuclear radiation background as they push against the limits of technology with their instrumentation. We are also witnessing the beginning of the era of high-energy astronomy, a promising new field which can for the most part be pursued only above the earth's atmosphere. The extraction of weak signals from a highly varying background which interacts with the instruments themselves to produce even more background is a complex problem whose solution is not trivial. It was our purpose in organizing the Conference on the High-Energy Radiation Background in Space (CHERBS) to bring together workers from these and related fields to share useful but often unpublished data, examine their common needs, and develop a consensus on the types of measurements and model studies which would be of the most benefit in the future.

Scheduling of the conference for November 3–5, 1987 proved to be a very timely decision. The recent spate of disasters in the American and European space programs had provided the community of space researchers a window of time to contemplate this and other important issues before becoming caught up again in the rush to meet one deadline after another in preparation for space missions. It also came at a time when there had begun to develop a feeling in the astronomical community that, unless a solid body of high-quality observations on the pristine nuclear radiation background is accumulated before man-made pollution causes it to be permanently altered, high-energy astronomy from earth orbit will become not just difficult, but impossible.

At the height of preparations for the conference there occurred one of those great events which change everything; in this case, Supernova 1987a burst upon the scene to make history. The whole world of astrophysics and astronomy suddenly became consumed by a passion to measure the emanations from that body with any and every type of instrument that could be pointed in its direction. We confess that we, too, succumbed to a raging case of supernova fever. Shortly after the close of the conference we were winging our way to Antarctica to launch a gamma-ray telescope heavenward on a balloon. Obtaining supernova data from such an exotic locale was a challenging enough task; determining exactly what it was that we had observed proved to be at least as formidable. During the ensuing year of intense data analysis, we found ourselves wrestling again and again with the background problem. Data presented at the conference proved to be most valuable in helping us to untangle our few supernova events from the complex jumble of instrumental background lines that we had gathered in such abundance. Having thus already enjoyed some of the fruits of the present proceedings, we now present them with the hope that others may find them at least as useful as we have.

We thank the program advisory committee members for helpful advice on setting up the conference program and for actively promoting the meeting in their respective communities.

We are grateful to the sponsoring agencies for their support of the CHERBS. These include the University of Florida Division of Sponsored Research, the Oak Ridge Associated Universities, the NASA Goddard Space Flight Center, the Santa Fe Community College Endowment Corporation, and the Committee on Interagency Radiation Research and Policy Coordination. Mr. Robert Tanner served faithfully and cheerfully as our editorial assistant; Mrs. Harriet Bennetts, Miss Stephanie Sanchez, Miss Chandra Burrows, Ms. Corinne Etchison, and Mrs. Cathy Lawson worked diligently as secretaries; and Mrs. Patti Perkins served as treasurer.

The success of a conference is ultimately dependent on the work of the participants. In the present case, 81 scientists presented 57 excellent papers on every aspect of the high-energy radiation background problem from astrophysics to biophysics to cosmochemistry and beyond. These proceedings are the result of their dedication.

A. C. Rester
J. I. Trombka
February, 1989

I. The Radiation Environment

CHARGED PARTICLE RADIATION EXPOSURE OF GEOCENTRIC SATELLITES

E. G. Stassinopoulos
Radiation Physics Office
NASA-Goddard Space Flight Center
Greenbelt, MD 20771

Introduction

The space radiation environment is an important consideration in the development of spacecraft electronics. In this paper, the earth's space radiation environment is reviewed in terms of trapped and transient charged particles. The nature and magnitude of spatial and temporal variations in the trapped radiation environment are presented. Transient cosmic rays of galactic and solar origin are described, and their interaction with the earth's magnetic field are considered in terms of rigidity and geomagnetic shielding effects on missions in earth orbit. The critical to electronic devices, internal radiation environment of a spacecraft is described in terms of shielding considerations and transport calculations.

1. The Magnetosphere

The geomagnetic cavity is formed by the earth's magnetic field in the solar wind as it sweeps by the earth. The cavity is hemispherical on the day side, with a boundary at ~10-12 earth radii (R_e). On the night side, it is cylindrical ~40 R_e in diameter and because of the sweeping action of the solar wind, it extends over several hundred R_e in the antisolar direction. The geomagnetic cavity is illustrated in **Figure 1**. The main particle trapping region is the crosshatched area labeled plasmasphere.

The total magnetic field of the magnetosphere is defined in terms of two interacting and superimposed sources of internal and external origin. The internal field of the earth is thought to be caused by convective motion in the molten nickel-iron core of the planet and by a residual permanent magnetism in the earth's crust. The external field is comprised of the sum-total effect of currents and electric fields set up in the magnetosphere by the solar wind.

The advent of the space age and the satellite era has stimulated the proliferation of quantitative geomagnetic field descriptions in the last two decades. This has resulted in an abundance of numerical models

[Barraclough et al., 1975; IAGA Division I et al., 1986]. The internal field component of the earth's magnetic field exhibits gradual changes with time modeled as secular variations. These temporal effects are also observed in the shrinking value of the Earth's dipole moment and the drift in the location of the boreal (north) and austral (south) magnetic poles.

In addition, superimposed on these slow internal changes are external cyclic variations whose magnitudes depend on the degree of perturbation experience by the magnetosphere. Specifically, strong perturbations of the geomagnetic field are present in the outer magnetosphere. These perturbations depend on local time (diurnal effects) and season (tilt effects) as well as solar wind conditions [Stassinopoulos et al., 1984]. All of these, in turn, affect magnetosphere current systems. The aggregate effects of these cyclic variations are usually represented by external source models.

A characteristic of the geomagnetic field, of particular significance to space radiation effects in electronics, is the Brazilian or South Atlantic Anomaly (SAA). This is primarily caused by the fact that the dipole term of the geomagnetic field is offset by approximately 11 degrees from the Earth's axis of rotation and displaced by about 500 km toward the Western Pacific, causing an apparent depression of the magnetic field over the coast of Brazil with significantly reduced field strength. There, the Van Allen belts reach lower altitudes extending deep down into the atmosphere. The SAA is responsible for nearly all the trapped radiation encountered in low earth orbits (LEO). In contrast, the SE-Asian Anomaly displays correspondingly stronger field values and the trapped particle belts are located farther away from the earth, at higher altitudes.

2 The Radiation Environment

A. Trapped Radiation

I. Domains

The earth's magnetic field above the densest part of the atmosphere is populated with trapped electrons, protons, and small amounts of low energy heavy ions. These particles gyrate around and bounce along magnetic field lines and are reflected back and forth between pairs of conjugate mirror points (regions of maximum magnetic field strength values along their trajectories) in opposite hemispheres. At the same time, because of their charge, electrons drift eastward around the earth while protons and heavy ions drift westward. **Figure 2** [Spjeldvik and Rothwell, 1983] illustrates the "spiral", "bounce", and "drift" motion of the trapped particles.

Figure 3 is a map of the magnetosphere which is divided into five regions or domains, indicating the particle species populating or visiting these regions. The strong dependence of trapped particle fluxes on altitude and latitude is expressed in terms of the McIlwain L parameter [McIlwain, 1961], where L is approximately equal to the geocentric distance of a field line in the geomagnetic equator, commonly used as a dimensionless ratio of the earth's radius. *Note that the geometric interpretation and physical* *significance of L becomes increasingly invalid for equatorial distances greater than 4 Re* because of the more complex particle motion in the real geomagnetic field and the distortion of the geomagnetic cavity by the solar wind interaction effects. An arrangement corresponding to that, but in R-Λ space, is also shown in **Figure 3**. The dipole fieldline equation ($R = L \cos^2 \Lambda$) is used to map the domains, where R is the radial distance and Λ is the invariant latitude.

The indicated domain boundaries in either notation should be considered only transition areas, not actual lines. These boundaries are assumed for modeling purposes and, additionally, are used here to convey a qualitative picture of the charged particle distribution. "Real" boundaries are diffused areas fluctuating in their L position due to several factors, such as magnetic perturbations (storm and substorm effects), local time effects (diurnal variation), solar cycle variations (minimum and maximum activity phases), individual solar events, etc. The boundaries also vary with particle energy.

Electrons. Energetic Van Allen belt electrons are distinguished into "inner zone" and "outer zone" populations occupying, respectively, volumes of space extending at the equator to about 2.4 earth radii (R_e), and from 2.8 to about 12 R_e. These domains are indicated, correspondingly, by regions 1 and 2-3-4 in **Figure 3**. The L = 2.8 line is used to separate the inner and outer zone domains, while the termination of the outer zone at L = 12 is intended only to delineate the maximum outward extent of stable or pseudo electron trapping. The region between L = 2.4 and 2.8 is called the "slot". During magnetospherically quiet times it is occupied by electrons of very low intensities. However, during magnetic storms, the electron fluxes in that domain may rise by several orders of magnitude.

The inner zone electrons are more benign as compared to the more severe outer zone electrons. Specifically, the outer zone has peak fluxes exceeding those of the inner zone by about an order of magnitude. Also, the outer zone spectra extend to much higher energies (~7 MeV) than the inner zone spectra (<5 MeV).

Protons. In contrast to the electrons, the energetic trapped protons (E>1 MeV) occupy a volume of space which varies inversely and monotonically with their energy, as shown in **Figure 4**, and, consequently, these particles *cannot* be assigned to "inner" and "outer" zones. Shown in **Figure 3** are protons with energies E>10 MeV populating regions 1 and 2 with an approximate trapping domain boundary placed at L = 3.8. This has the effect that in low altitude earth orbits, the most intense and penetrating radiation is encountered in the form of protons in the South Atlantic Anomaly (SAA). **Figure 5** shows the proton flux intensities as a function of radial distance and energy.

II. Variations

The trapped particle fluxes respond to changes in the geomagnetic field induced by solar activity, and therefore, exhibit a strong dynamic behavior, especially in the outer belts. Satellite measurements in geosynchronous

(GEO) equatorial orbits have revealed a complicated temporal pattern consisting of a superposition of several *cyclical* variations in conjunction with *sporadic* fluctuations [Lanzerotti et al., 1967; Lin and Anderson, 1966; O'Brien, 1963]. The main *periodic* variations include a diurnal cycle which in GEO is characterized by order-of-magnitude flux changes [Lanzerotti et al., 1967], and the 11-year solar activity cycle.

Sporadic magnetic storms in GEO can produce a modulation of the electron flux above .05 MeV by an order of magnitude within a period of less than 10 minutes [Lin and Anderson, 1966] and with a corresponding decay in days. Substorms, which are a common feature of the midnight to dawn sector of a GEO orbit, result in the injection of electrons with energies between 50 and 150 KeV from the magnetospheric tail region. The electron flux above 200 KeV remains constant or actually decreases.

However, for electrons, *short term* variations in flux levels can occur over most regions of magnetospheric space (L>2). Depending on particle energy and the type and intensity of the causative event, these excursions from average flux values can be as much as a factor of 10^2-10^3. The effect of short term magnetic perturbations on the "slot" region of the trapping domain (L=2.4-2.8) has already been discussed.

The *diurnal* or *local time* variation of electron fluxes in the outer belt results from the interaction of the solar wind with the magnetosphere, which produces the modulations of the local magnetic field and, hence, of the trapping strength. Significant in this process is the compression of the geomagnetic cavity in the solar direction (local noon) and the elongation in the antisolar direction (local midnight). The local time variation for electron fluxes above 1.9 MeV is illustrated in **Figure 6** [Singley and Vette, 1972] for L values from 3 to 10. For example, at L=7.0 the ratio of maximum to minimum flux is about 16, with the extrema separated by about 12 hours. In general, this ratio increases with increasing L and energy.

Another important *solar activity induced modulation* of the trapped particle population, *particularly protons*, occurs in the *low altitude* regime of the magnetosphere. Here, during the active phase of the solar cycle, the increased energy output from the sun causes the atmosphere to expand, thereby raising the density of the atmospheric constituents normally encountered at heights between 200 and 1000 kilometers. This global increase in atmospheric density depletes, through coulomb scattering, the populations of those trapped particles that have their mirror points at these low altitudes, with significant effects on the radiation exposure of satellites orbiting in that domain.

The solar cycle variations observed in some areas of the trapped particle domain, are functions of energy and magnetic parameter L. They generally have opposite effects on each particle specie, particularly in the low altitude regime:

	Solar Min	Solar Max
Electron Intensities	lower	higher
Proton Intensities	higher	lower

No solar cycle changes of consequence have been measured in the heart of the proton trapping domain. And *no significant long-term variations, that can be modeled, occur in the electron populations at geostationary altitudes.* However, in the *atmospheric cutoff regions*, electron and proton variations may range from a factor of 1 to 5.

III. Artificial

A severe hazard for space missions could be introduced by a high altitude *nuclear explosion*. Such an event would result in the injection into the magnetosphere of energetic electrons from the beta decay of fission fragments and nuclear decay, with subsequent trapping in the earth's magnetic field [Teague and Stassinopoulos, 1972]. This could produce an enhancement of the trapped electron population by many orders of magnitude.

The principal hazard would be to missions in low earth orbits, mainly because of an expected very stable trapping with lifetimes up to *eight years* [Teague and Stassinopoulos, 1972], as shown in **Figure 7** for the STARFISH exoatmospheric nuclear explosion of July 1962 over Johnston Island in the Pacific. However, depending on the location of the explosion (latitude *and* altitude), the injection could also produce a temporary large enhancement of the electron environment at geostationary orbits. Here the trapping would be less stable, with exponential decay periods of between 10 and 20 days. The apparent longevity, or conversely the decay rate, of such fission electrons depends to a large extent on injection latitude and altitude, that is, it is a function of magnetic dipole shell parameter L and to a lesser degree of field strength B [Stassinopoulos and Verzariu, 1964].

IV. Models

Available radiation measurements from space form the basis for models of the trapped electron and proton environments that have been developed by the U.S. National Space Science Data Center (NSSDC) at NASA's Goddard Space Flight Center. All models are constructed with several dozen data sets from a corresponding number of satellites, providing a wide *spatial* and a long *temporal* coverage.

The most recent of these models, AP8 for protons [Sawyer and Vette, 1976] and AE8 for electrons [to be published], permit long term average predictions of trapped particle fluxes encountered in any orbit and currently constitute the best estimates for the trapped radiation belts. However, they do not provide any statistics associated with random fluctuations, and cyclical variations have been averaged out. The exception is the solar cycle dependence, which is reflected in the solar minimum and solar maximum versions of the models, that represent *average* conditions for these solar activity phases.

It should also be noted that these models have a serious *deficiency* in the low altitude regime (<1000 km), that is, in the atmospheric cutoff region. In that domain they predict fluxes that are too high by factors of ~2 (at 800-1000km) to ~50 (at 200-500 km), *if used with current or projected (future) field strength values.*

This happens because the geomagnetic field is changing with time (secular variations). In this process, the dipole moment is decreasing, pulling heavily populated field lines down into the denser regions of the atmosphere, where the models have no mechanism to account for the corresponding significant particle losses, mostly due to coulomb scattering.

To illustrate the predictions of these models for low earth missions of interest to NASA and DoD projects, **Tables 1 and 2** contain integral proton and electron spectra for solar minimum conditions and **Figures 8 and 9** give the integral *electron* and *proton* spectra, respectively, in a circular 500 km, 60° inclination orbit, for both solar minimum and maximum conditions. The relative hardness of the proton spectrum should be noted, which between 50 and 500 MeV declines by only a factor of 4. By comparison, the *proton* spectrum in *geosynchronous* orbits is very soft and essentially is totally depleted above 1.75 MeV; thus, trapped protons in this orbit are stopped by very small material thicknesses (~ .005 cm Al) and are not of concern to electronic systems within a satellite.

The *geosynchronous* integral *electron* spectrum is given in **Table 3** and is ploted in **Figure 10**, based on the AE8-MAX model. *Worst* and *best* cases are shown, corresponding to "parking" longitudes of 160°W (L=6.6) and 70°W (L=7.0), respectively, with a flux ratio of about 1.8 for energies E>1 MeV and 2.3 for energies E>2 MeV. This happens because of the relative inclination of the geomagnetic to the geographic equatorial plane (~11°). Thus, parking positions along the geostationary great circle in the geographic equator may occupy L values ranging from 6.6 to 7.0 [Stassinopoulos, 1980].

The *reverse* is true for galactic or solar *cosmic ray* exposures, because of geomagnetic shielding effects. That is, the 160°W (L=6.6) position is less exposed while the 70°W (L=7.0) is more vulnerable to the *low energy* part of the cosmic ray spectra. The table below gives the corresponding approximate cutoff energies E_c and rigidity R:

		160°W	70°W
Protons (Z=1)	E_c	60 MeV	48 MeV
Heavy Ions (Z≥1)	E_c	15 MeV/n	12 MeV/n
Rigidity	R	0.304 GV	0.342 GV

V. Flux-Free Time

As mentioned in a previous section, the South Atlantic Anomaly (SAA) is a region with high intensities of trapped particle radiation close to the earth. Hence, for low altitude, low inclination orbits, the SAA is the most important factor in determining the level of radiation exposure of spacecraft or of astronauts in extra-vehicular activities (EVAs). For low earth orbits (LEO) with higher inclinations (i > 40°), the protrusions of the outer zone electron belts (the electron "horns") in the midlatitude regions must also be considered. **Figure 11** shows the electron flux profile for a circular 900 km, 99° inclination orbit during its worst pass through the SAA.

Note in **Figure 11** that even in a worst case orbit there are time periods during which instantaneous electron fluxes above .5 MeV are *below 1 particle* per square centimeter per second. The same is true for protons above 5 MeV. These time periods are called *flux free time* (FFT) intervals. They are an important feature of some orbital configurations. They may occur over short orbit segments (partial FFT per period) or over the entire length of a revolution (total FFT per period). In terms of geomagnetic geometry, the FFTs establish the duration for which the trajectory lies outside the trapping domain of the corresponding particle species, evaluated at the given energies. Or conversely, they are a measure of the degree to which the trajectory is exposed to the charged particle trapping domains.

The number of consecutive flux-free orbits of *circular* trajectories is primarily a function of altitude and inclination and to a lesser degree a function of particle energy. Generally, higher energies will yield longer FFTs because the more energetic particles occupy a smaller volume of space, particularly in the case of protons. For an orbit configuration similar to the one illustrated in **Figure 11** and for protons with energies E > 5 MeV and electrons of E > .5 MeV, no completely flux-free orbits exist. In this trajectory, the total FFT is entirely composed of contributions from partially exposed revolutions. It can be summarized as follows, in percent of total mission duration:

	Protons (E_p > 5 MeV)	Electrons (E_e > .5 MeV)
Solar Minimum	81%	33%
Solar Maximum	83%	53%

For a 500 km, 30° inclination LEO trajectory the FFT includes *six completely flux free orbits per day*, that is, orbits which do not pass through the SAA or the electron "horn" regions. The total FFT can be summarized (in percent of total mission duration) as follows:

	Protons (E_p > 5 MeV)	Electrons (E_e > .5 MeV)
Solar Minimum	90%	89%
Solar Maximum	92%	88%

B. Transient Radiation

    I. Solar Cosmic Rays

        a. Solar Flare Protons

Disturbed regions on the sun sporadically emit bursts of energetic charged particles into interplanetary space. These solar particle events (or solar flares) are composed primarily of protons, with a minor constituent of alpha particles (5-10%), heavy ions (see next section on Cosmic Rays), and electrons, and can last as long as several days.

The *time history* of solar flare particles at 1 astronomical unit (AU)* *after* the occurrence of the parent flare is of great interest: 1) they arrive in tens of minutes to several hours (depending on their energy), 2) they peak within two hours to one day, and 3) they decay within a few days to one week.

It is important to remember that the most energetic protons arrive at the earth in about 10-30 minutes.

Solar flare phenomenology distinguishes between ordinary (OR) events and anomalously-large (AL) events. AL events are quite rare (e.g. 3 occurred during the 19th solar cycle, 1 during the 20th cycle, and none in the 21st cycle, as seen in **Figure 12** [Goswami et al., 1988]); they occur mostly near the first and last year of the solar maximum phase. Their prediction was initially based on an empirical model [Stassinopoulos and King, 1974] and later on a probabilistic treatment involving modified Poisson statistics [King, 1974]. A simple statistical predictive model for solar flares is provided by SOLPRO [Stassinopoulos, 1975] which is based exclusively on satellite spectral measurements covering nearly the entire 20th solar cycle. This model predicts, for a given mission duration and a specified confidence level, the mission integrated proton fluence spectrum from OR events, and the number of AL events to be expected with their event-integrated fluence spectra.

Since AL events are rare, small-sample statistics are the only appropriate prediction technique. This implies that for *unmanned satellites* with mission durations of $\tau \geq 1$ year, OR event fluences are *not significant* because probabilistic theory predicts the possible occurrence of at least one AL event, even for the lowest allowable confidence level (Q = 80%).

b. Solar Heavy Ions

Cosmic ray emission by the sun occurs sporadically, during solar flares. For ordinary solar events, the relative abundance of the He ions in the emitted particle fluxes is usually between 5 and 10%, while the fluxes of heavier ions relative to protons are very small, and significantly below the galactic background. However, during major solar events, *the abundance of some heavy ions may increase* rapidly by 3 or 4 orders of magnitude above the galactic background, for periods of several hours to days, presenting a very serious hazard to missions exposed to these high cosmic ray intensities, particularly in terms of increased Single Event Upset (SEU) occurrence in electronic devices or human exposure to these HZE particles.

---

*AU: mean distance from sun to earth (given in km or miles)

II. Galactic Cosmic Rays

The region outside the solar system in our part of the galaxy is believed to be filled uniformly with cosmic rays. These consist of about 85% protons, about 14% alpha particles, and about 1% heavier nuclei, ranging in energy to above 10 GeV/nucleon. **Figure 13** shows the spectral distributions for hydrogen, helium, carbon and oxygen ions. The differential energy spectra of hydrogen and heavier ions near the earth tend to peak around 1 GeV/nucleon. Toward lower energies the spectral shape is depressed by interaction with the solar wind and the interplanetary magnetic field; this reduction in flux becomes more pronounced during the active phase of the solar cycle (solar modulation). The total flux of cosmic ray particles seen outside the magnetosphere at 1 AU is ~4 cm^{-2}·sec^{-1} (primarily composed of protons) and is, for all practical purposes, *omnidirectional*. **Figure 14** shows the relative abundances of the galactic cosmic ray ions. A model for these particles is available [Adams et al., 1981].

III. Geomagnetic Shielding

Low altitude and latitude earth orbits are essentially shielded from solar or galactic cosmic rays by the geomagnetic field up to inclinations of about 45°. The earth's field acts as an energy filter preventing particles with less than a given momentum value from penetrating to certain altitude-latitude combinations. **Figures 15 and 16** show the total ion energy required to penetrate the magnetosphere in terms of the dipole parameter L. **Table 4** and **Figure 17** shows the effect of geomagnetic shielding on solar flare protons for high inclination (> 60°) LEO orbits and **Figure 18** shows the effect of shielding on cosmic ray silicon ions for LEO orbits. **Figure 19** shows the magnetospheric attenuation dependence of the galactic cosmic ray iron spectrum on energy and L.

For *geostationary* orbits, however, magnetic shielding is relatively ineffective and such orbits will be exposed to *galactic cosmic ray hydrogen* of all energies above approximately 60 MeV and *heavier ions* above 15 MeV/nucleon. This applies also to *solar flare protons*. as seen in **Table 5** Otherwise, GEO missions are not affected by satellite position (or parking longitude): all positions have the *same exposure* for particles with energies greater than stated above.

For all practical purposes, the cosmic ray fluxes can be considered *omnidirectional*, except for very low altitude orbits where the solid angle subtended by the earth defines a region inaccessible to these particles.

For economy reasons, geomagnetic shielding effects on geocentric missions are usually being evaluated from simple momentum considerations which determine how deep the cosmic rays can penetrate into the magnetosphere. Such approximations are reasonable for obtaining long-term averages, but may not be adequate for short term estimates because of substantial diurnal variations in the cutoff latitudes, associated with geomagnetic tail effects (2-4 degrees), and storm-induced changes (>4 degrees).

C. Secondary Radiation

In interacting with spacecraft materials, the electrons and protons from the trapped radiation belts, the solar flare protons, as well as the cosmic ray fluxes, produce various secondary radiations. These can extend the penetrability of the primary radiation and lead to an increase in the dose deposition or the SEU rate over that of the primary particle alone.

The most significant secondary radiation contributing to the total dose damage is the bremsstrahlung or "braking radiation" produced in the deceleration of electrons penetrating the spacecraft. This is a continuous X-ray spectrum emitted roughly in the direction of electron penetration, with a mean X-ray photon energy of about 1/3 the initial electron energy. The bremsstrahlung intensity depends linearly on the atomic number of the target material and on the square of the initial electron energy. *Bremsstrahlung* from energetic electrons populating the radiation belts is *very penetrating* and thus difficult to attenuate, especially with the low-atomic number materials prevalent on spacecraft (e.g. aluminum). On the other hand, these materials also tend to produce *less* bremsstrahlung.

Nuclear reactions induced by protons and heavier ions with energies above an effective threshold of a few MeV/nucleon provide another source of secondary radiations, both prompt and delayed, of considerable importance to SEU sensitive electronic components. Above several 100 MeV/nucleon nuclear reactions surpass atomic ionization as the main attenuation mechanism in a target. At higher energies the interaction of the incident particle tends to occur primarily with individual nucleons in the target nucleus, and can lead to the ejection of several energetic protons and neutrons. This "spallation" process leaves the product nucleus highly excited, the de-excitation occurring through the "evaporation" of additional nucleons and the emission of gamma rays. For 400 MeV protons incident on aluminum, the total nucleon emission on the average, is 4.8, including 2.8 spallation nucleons with an average energy of 120 MeV [Haffner, 1967]. The process can generate a rich variety of residual nuclei especially in heavier elements, as a result of the multiplicity of statistically possible reaction paths (i.e. specific number of protons and neutrons emitted). These product nuclei are frequently radioisotopes decaying by beta-ray emission, with a variety of life-times.

3. Transport, Shielding, and Doses

A. Emerging Radiation Spectra

I. Electrons and Bremsstrahlung:

As the curves in **Figure 20a** clearly indicate, the trapped electrons are very effectively attenuated by aluminum shields and are nearly all stopped by thicknesses $s>3$ gm/cm^2 even at the highest energies predicted for the trapped radiation environment.

The same, however, does not hold for the secondary photons. The bremsstrahlung flux levels for energies above .04 MeV are not significantly affected by any of the aluminum shields from 0.1 to 10.0 gm/cm^2 (**Figure 20b**). The bremsstrahlung intensities show little attenuation over the evaluated depth range. However, the important fact to remember is that above .1 MeV, the photon fluxes are, on the average, over 3 orders of magnitude lower than the incident electrons at corresponding energy levels.

II. Trapped Protons

Figure 20c shows that the incident and emerging spectral curves bunch together for energies greater than about 30 MeV for all investigated shield thicknesses. Considering the large range of material depths evaluated here (from .1 to 10. gm/cm^2 of aluminum), this clustering is indicative of the very small effect that shielding has on protons with energies above 30 MeV. It is also characteristic of the very "hard" incident proton spectra between 30 and 500 MeV.

The consequence of this, in terms of radiation exposure, is that proton incidence on a sensitive volume is only a weak function of shield thickness as it shows very little attenuation over the evaluated depth range from about 1 to 10 gm/cm^2. Thus, in order to get an appreciable reduction in the number of energetic protons reaching the target, say by one order of magnitude, a nearly 20-fold increase in shield thickness is necessary. By the way, the corresponding ranges and LET values of the emerging protons are given in the following table for a silicon target:

ENERGY (MEV)	RANGE (mg/cm^2)	LET		
		(MeV/cm)	(MeV.cm^2/g)	(pc/cm)
.1	0.198	1311	565	26
.5	1.295	622	268	12
1	3.623	415	179	8
5	49.04	139	60	2.7
10	160.45	84	36	1.6
50	2762	23	10	.45
100	9504	14	6	.27

III. Solar Flare Protons

The materially attenuated emerging spectra reflect the shielding effect on the distribution of these particles. It can be seen from **Figure 20d** that the fluxes in the .1 to 10 MeV range emerging behind spherical aluminum shields of thicknesses .3-5 gm/cm^2 are substantial.

IV. Galactic Cosmic Rays

Figure 20e shows for silicon cosmic ray ions the unattenuated interplanetary spectra, the magnetospherically attenuated orbit-integrated spectra (incident on the surface of the spacecraft), and, derived from this latter, the shielded spectra of emerging particles behind selected thicknesses of spherical aluminum geometries. Also plotted in **Figure 20e**, is the integral LET spectrum obtained for the geomagnetically and materially shielded cosmic ray spectrum. These values provide useful information for Single Event Upset (SEU) evaluations, particularly for soft error analyses.

The most important (but predictable and expected) features of the data in **Figure 20e** are:

1. the substantial attenuation by the earth's magnetic field of all particles in the investigated energy domain (10-10,000 MeV/n);

2. the insignificant effect of material shielding in the energy range from about 90 to 10,000 MeV/n: *no substantial decrease in fluxes even for 10 gm/cm^2 aluminum (approximately 1.5 inches);*

3. the unavoidable shielding side-effect of *a significant increase in the low energy (.8-50 MeV/n) fluxes for shield thicknesses >.1 gm/cm^2 aluminum.*

Although not scientifically correct, the cosmic ray incidence may be assumed omnidirectional for practical application purposes. The error introduced by this simplification lies, for most cases, well within the large intrinsic uncertainties associated with these type of estimates and calculations.

B. <u>Variables Affecting Dose Evaluations</u>

The external radiation environment is filtered and attenuated by the spacecraft structure. The resulting internal radiation field defines the dose level received by various radiation-sensitive parts and systems. Obtaining estimates of the dose on a given target in space is a complex process involving several variables that directly affect the results of dose calculations. **Table 6** lists some of these variables.

Four areas stand out that are of particular concern to shielding and transport evaluations and which are completely independent from and unrelated to the definition of the vehicle-encountered radiation background. These are: a) *shield geometry*, b) *shield composition*, c) *target composition*, and d) *dose units*. Each of these offers a multiplicity of choices and conditions, as shown in **Table 7**, that *need to be clearly identified and defined*, whenever calculations are performed and results presented. Otherwise, the comparison of dose data compiled by several independent sources, *although derived from the same spacecraft surface incident spectrum*, becomes meaningless and futile. In such cases, disagreements by factors up to 20 have been known to occur.

At this point it should be emphasized that the physical dose-unit of radiation, the "rad", measures the energy deposition per unit mass of absorber material (1 rad = 10 erg/gm)[*] *without reference to effects in the target.*

---

[*] 100 rad = 1 Gy

The calculation of radiation penetration and dose deposition in principle is well understood (with the possible exception of intra-nuclear cascades) and usually can be carried out to adequate accuracy with a variety of available radiation transport codes. Analysis of the internal radiation environment in specific space system configurations, while complex, is possible and has been performed using ray tracing techniques, solid angle sectoring, and Monte Carlo modeling [Jordan, 1982].

Generally, though, space radiation transport and dose calculations use idealized shielding configurations such as solid or hollow spheres, semi-infinite slabs, and cylinders, usually with aluminum as a reference material. This readily permits parametric analysis of dose attenuation and exploration of the consequences of environmental uncertainties, and identifies the shielding required for a given space system. In comparing results from different geometries, it should be noted that for omnidirectional isotropic flux incidence, spherical shields yield dose results roughly 2 to 6 times higher than 4π exposure of slab shields with centered dose points, while cylindrical shields yield intermediate results. Specific differences, however, depend on particle species, energy spectrum, shield thickness, and particularly target composition.

C. <u>Ionizing Radiation Dose</u>

The exposure of a satellite to the magnetospherically attenuated cosmic rays and solar flare protons, to the trapped protons and electrons, and to the secondary bremsstrahlung photons, as well as the interaction of these types of radiation with the spacecraft in terms of equivalent aluminum shields, are usually evaluated for a single and simple geometry. The materially attenuated doses and fluxes, presented in this document, including Linear-Energy-Transfer (LET) values emerging behind selected shield thicknesses, were calculated with state-of-the-art transport codes [Jordan, 1982; Seltzer, 1980].

Tables 8 through **13** contain daily doses for low Earth orbits (LEO) at 300 and 500 km altitudes and inclinations of 28.5°, 60° and 90° for solar minimum conditions based on the trapped electron and proton models. Daily silicon doses in a low earth orbit (LEO) at 500 km altitude and 30° inclination, for solar minimum and maximum are shown in **Figure 21** for a two-sided exposure of aluminum slab shields and for a solid spherical shield, as an average over 15 orbits. The electron dose includes the bremsstrahlung contribution.

As discussed in previous sections, the SAA is the primary contributor to the doses accumulated by spacecraft in LEO. **Figure 22** shows contours of total dose for an altitude of 500 km for a spherical shield thickness of 2 gm/cm^2 of aluminum. Superimposed on the world maps are worst case passes through the SAA for 28.5°, 57°, and 90° inclination orbits. For low inclination orbits (<45°), there are periods during the day when complete revolutions do not experience any dose accumulations. These time periods are especially important when considering EVA activities.

The corresponding electron-plus-bremsstrahlung daily dose for an aluminum shield of solid sphere geometry in the geostationary (GEO) orbit at the parking longitude with the lowest average flux (70°W), is illustrated in **Figure 23** in the form of a dose-depth curve. For the parking longitude with the largest average flux (160°W), the dose behind a 2 g/cm^2 shielding thickness is a factor of about 1.7 higher, regardless of geometry.

These daily-dose values, however, are inferred from long-term averages, and can be exceeded substantially (by as much as two orders of magnitude) in the short term, especially during the local morning to mid-afternoon.

Particle fluxes from solar flares are heavily attenuated by the geomagnetic field, which prevents their penetration to low orbital altitudes and inclinations. For a 500 km, 30° inclination orbit the attenuation is nearly total, while in a 500 km 57° inclination orbit, some penetration occurs. In contrast, a polar orbit (any altitude) experiences a substantial degree of exposure. *Altitude dependence is very small!*

In GEO the geomagnetic shielding is relatively ineffective. Even so, the *average yearly dose* from OR events behind a 2 g/cm^2 spherical aluminum shield is quite small: ~18 rad/year in a silicon target, as shown in **Figure 24**. In comparison, the event-integrated dose from an AL flare at 70°W would be ~600 rad/event for the same shield and target (**Table 14**). Tripling the shield thickness to 6 gm/cm^2 would still result in ~300 rad/event. Overall, the *total dose contributions* from solar flares (OR or AL) in the GEO domain *are not significant in terms of damage to and degradation of* electronic parts or materials, for long mission durations (5-10 years). *However, these particles may have a devastating impact upon electronic devices susceptible to SEUs.* Interplanetary integral LET spectra from 1 AL event, emerging behind two aluminum shield thicknesses: .01 and 10.0 gm/cm^2, are given in **Figure 25**.

REFERENCES

Adams, J.H., R. Silberberg, and C.H. Tsao, "Cosmic Ray Effects on Microelectronics, Part I: The Near-Earth Particle Environment", Naval Research Laboratory, NRL Memorandum Report 4506, August 1981.

AE8 Trapped Electron Model: to be published.

Barraclough, D. R., R. M. Harwood, B. R. Leaton, and S. R. C. Malin, "A Model of the Geomagnetic Field at Epoch 1975", Geophys. J. R. Astron. Soc., 43, 645, 1975.

Bethe, H., "Theory of the Passage of Fast Corpuscular Rays Through Matter", Ann. Physik, Series 5, 5-325, 1920.

Goswami, J. N., R. E. McGuire, R. C. Reedy, D. Lal, and R. Jha, "Solar Flare Protons and Alpha Particles During the Last Three Solar Cycles", submitted to J. Geophys. R. (Los Alamos Preprint LA-UR-87-1176).

Haffner, J. W., "Radiation and Shielding in Space", Academic Press 1967.

IAGA Division I, Working Group 1, "International Geomagnetic Reference Field, Revision 1985", EOS, Vol. 67, 523-524, June 1986.

Jordan, T.M., "Adjoint Monte Carlo Electron Shielding Calculations", ANS Transactions, Vol. 41, June 1982.

King, J. H., "Solar Proton Fluences for 1977-1983 Space Missions", J. Spacecraft and Rockets, 11: 401-408, 1974.

Lanzerotti, L.J., C.S. Roberts, and W.L. Brown, "Temporal Variations in the Electron Flux at Synchronous Altitude", J. Geophys. Res., 72, No. 23, 5893-5902; December 1967.

Lin, R.P and K.A. Anderson, "Periodic Modulation of the Energetic Electron Fluxes in the Distant Radiation Zone", J. Geophys. Res., 71, No. 7, 9172-1835; April 1966.

Littmark, U. and J. F. Ziegler, "Handbook of Range Distributions for Energetic Ions in all Elements", Pergamon Press, 1980.

McIlwain, C. E., "Coordinates for Mapping the Distribution of Magnetically Trapped Particles", J. Geophys. Res., 66, No. 11, 3681-3691, 1961.

O'Brien, B.J., "A Large Diurnal Variation of the Geomagnetically Trapped Radiation", J. Geophys. Res., 68, No. 4, 989-995, February 1963.

Sawyer, D. M. and J. I. Vette, "AP8 Trapped Proton Environment for Solar Maximum and Solar Minimum", NSSDC 76-06, National Space Science Data Center, Greenbelt, Maryland, December 1976.

Seltzer, S., "SHIELDOSE: A Computer Code for Space Shielding Radiation Dose Calculations", U.S. Department of Commerce, National Bureau of Standards, NBS Technical Note 1116, May 1980.

Singley, G.W. and J.I. Vette, "A Model Environment for Outer Zone Electrons", NSSDC 72-13, National Space Science Data Center, Greenbelt, MD., December 1972.

Spjeldvik, W. N. and P. L. Rothwell, "The Earth's Radiation Belts", AFGL-TR-83-0240, Air Force Geophysics Laboratory, Hanscom AFB, Massachusetts, September 20, 1983.

Stassinopoulos, E.G., "The Geostationary Radiation Environment", J. Spacecraft and Rockets, Vol. 17, No. 2, pp.145, March-April 1980.

Stassinopoulos, E. G., "SOLPRO: A Computer Code to Calculate Probabilistic Energetic Solar Proton Fluences", NSSDC 75-11, National Space Science Data Center, Greenbelt, Maryland, April 1975.

Stassinopoulos, E. G. and J. H. King, "Empirical Solar Proton Model for Orbiting Spacecraft Applications", IEEE Trans. AES, Vol. AES-10, No.4, July 1974.

Stassinopoulos, E. G., L. J. Lanzerotti, and T. Rosenberg, "Temporal Variations in the Siple Station Conjugate Area", J. Geophys. R., 89, No. A7, 5655-5659, July 1984.

Stassinopoulos, E. G. and P. Verzariu, "General Formula of Decay Lifetimes of Starfish Electrons", J. Geophys, R., 76, No. 7, 1841-1844, March 1971.

Teague, M.J. and E. G. Stassinopoulos, "A Model of the Starfish Flux in the Inner Radiation Zone", NASA/GSFC Report X-601-72-487, December 1972.

Figures:

1. Geomagnetic Cavity
2. Motions of Trapped Particles
3. Charged Particle Distribution in the Magnetosphere
4. Trapped Proton Population Domain as a Function of Energy
5. Equatorial Radial Profiles of Proton Fluxes
6. Local Time Variation of Outer Zone Electrons
7. Isochronal Contours for STARFISH Electrons: Longevity
8. Low Earth Orbit (LEO) Proton Fluxes: 60°/500km
9. Low Earth Orbit (LEO) Electron Fluxes: 60°/500km
10. Geostationary (GEO) Electron Spectra: 160° W + 70° W
11. Positional Electron Flux Profile for Polar Orbit
12. Solar Events of Cycles 19, 20, 21
13. Cosmic Ray Spectra
14. Cosmic Ray Abundances
15. Schematic of Magnetospheric Cosmic Ray Attenuation
16. Penetration Energy of Cosmic Rays into the Magnetosphere
17. Magnetospheric Attenuation of Solar Flare Protons
18. Magnetospheric Attenuation of Cosmic Rays
19. Galactic Cosmic Ray Attenuation vs L + E
20a. Emerging Differential Spectra: Electrons (e)
20b. Emerging Differential Spectra: Bremsstrahlung (B)
20c. Emerging Differential Spectra: Trapped Protons (p)
20d. Emerging Differential Spectra: Solar Flare Protons (SFP)
20e. Emerging Differential Spectra: Galactic Cosmic Rays (GCR)
21. Low Earth Orbit (LEO) Doses: 60°/500km
22. World Map of Constant Dose Contours at 500 km Altitude
23. Geostationary (GEO) Electron and Bremsstrahlung Doses
24. Unattenuated and Magnetospherically Attenuated Solar Flare Proton (SFP) Doses (AL&OR events)
25 Solar Flare Proton (SFP) LET Spectra

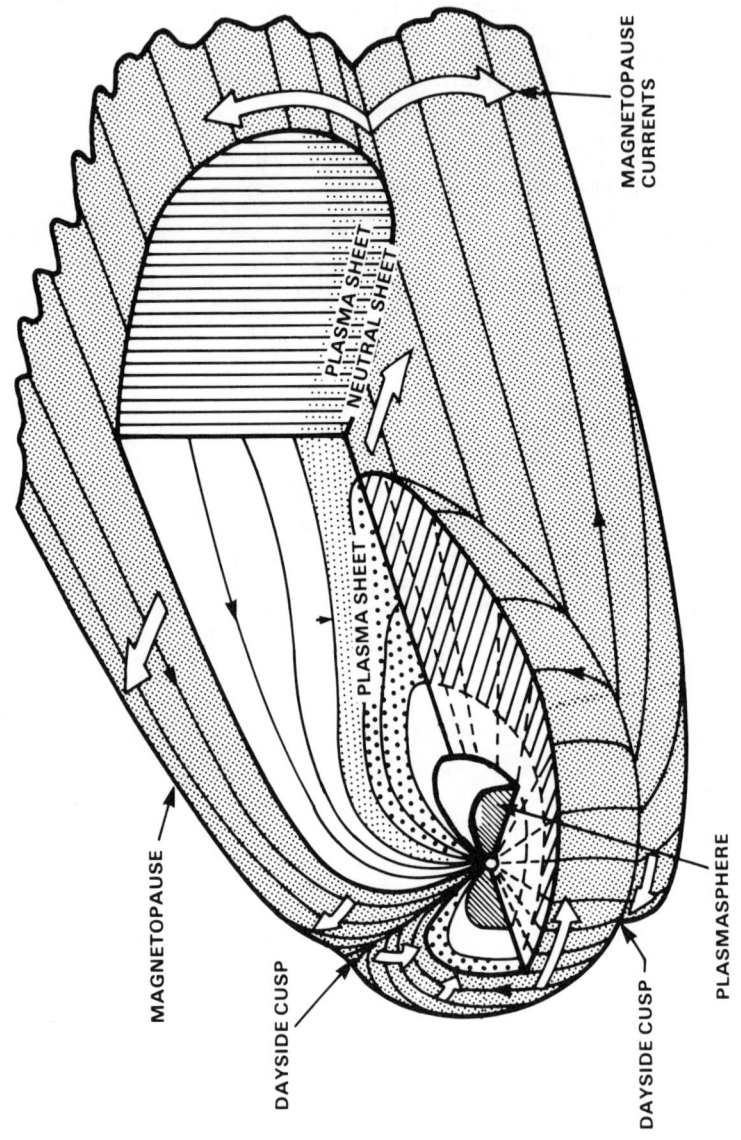

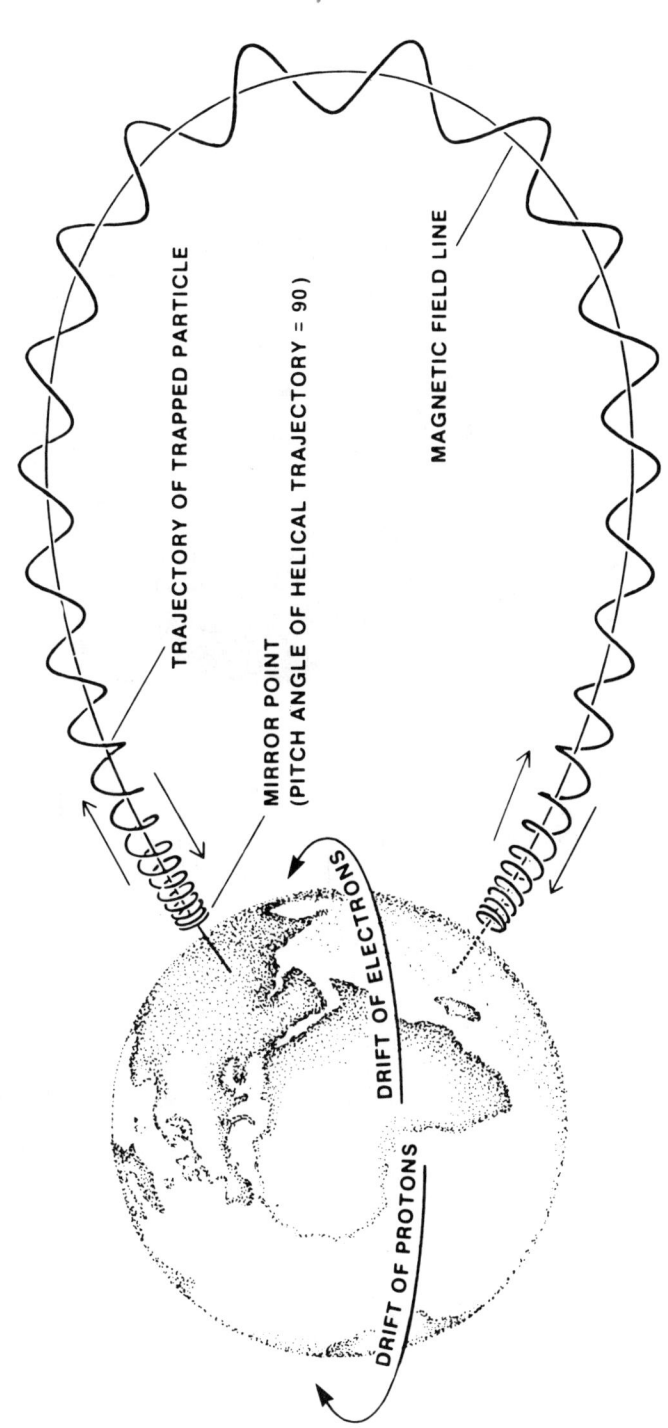

CHARGED PARTICLE DISTRIBUTION IN THE MAGNETOSPHERE

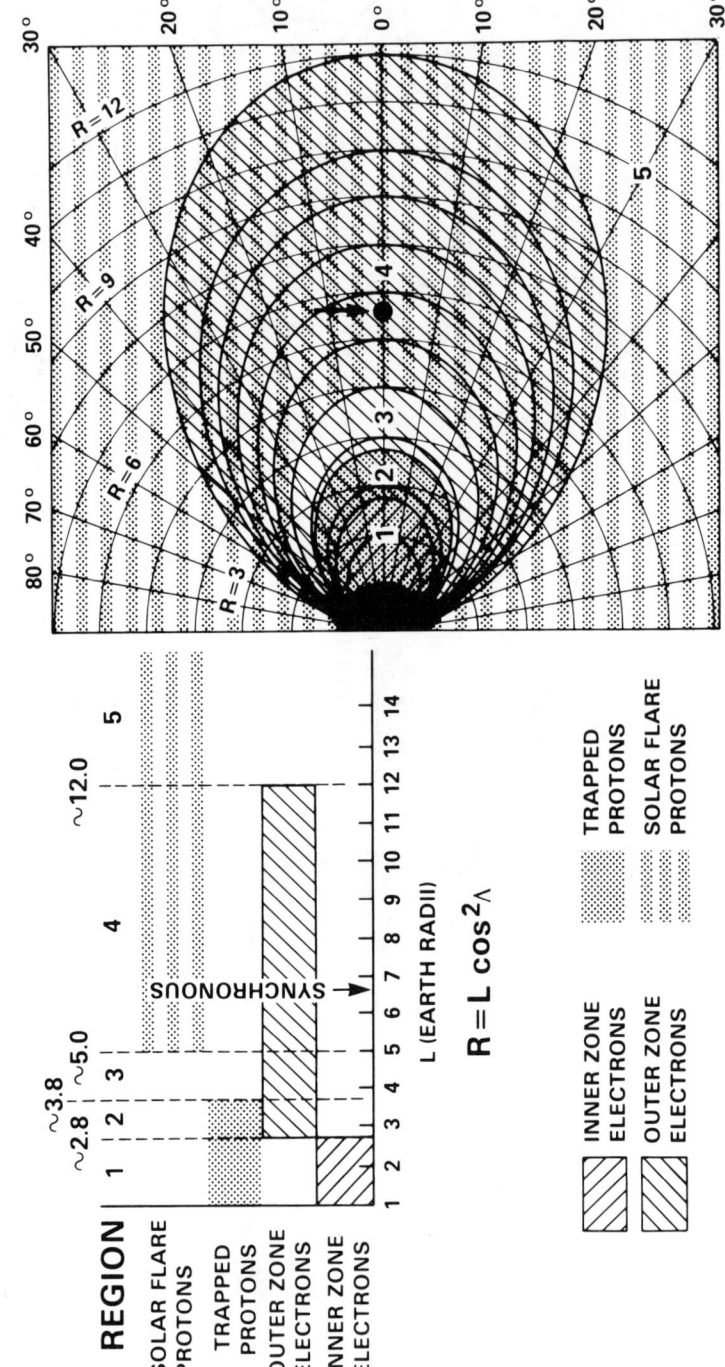

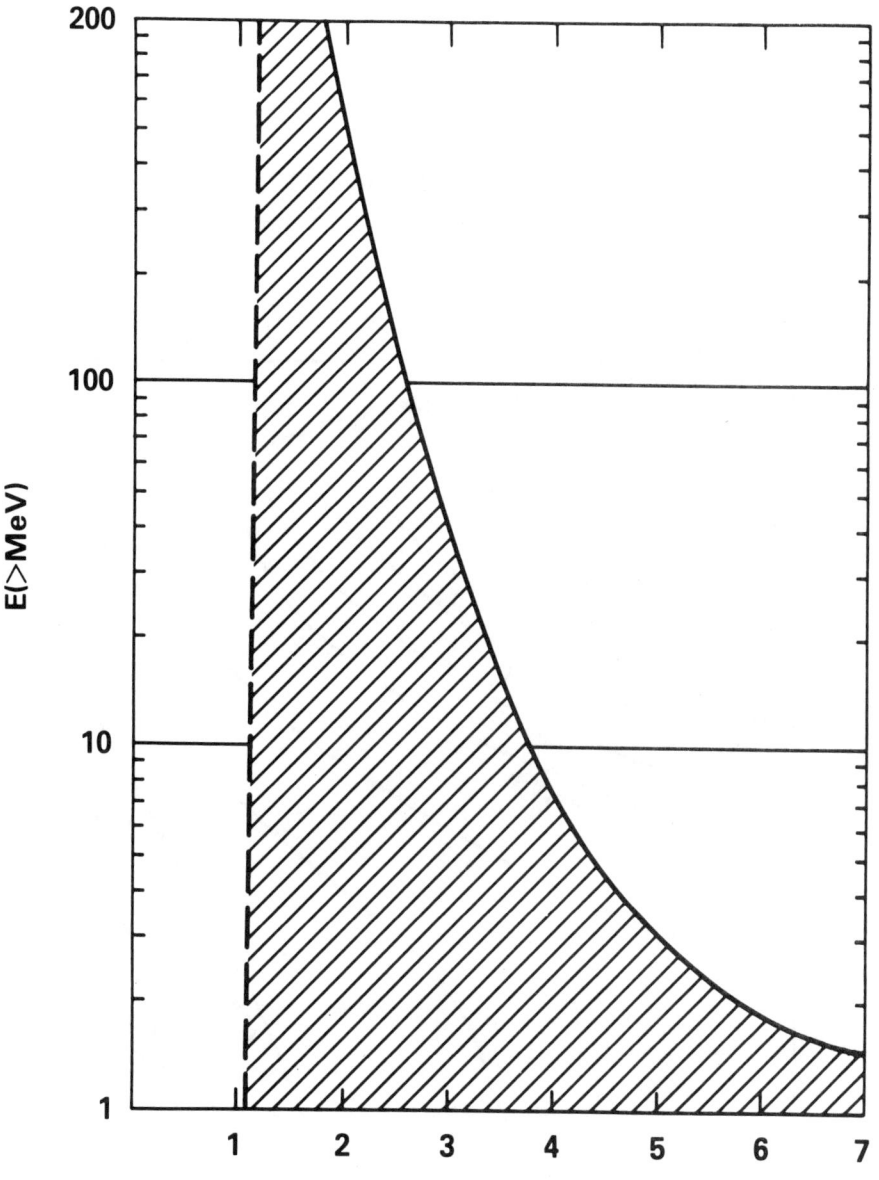

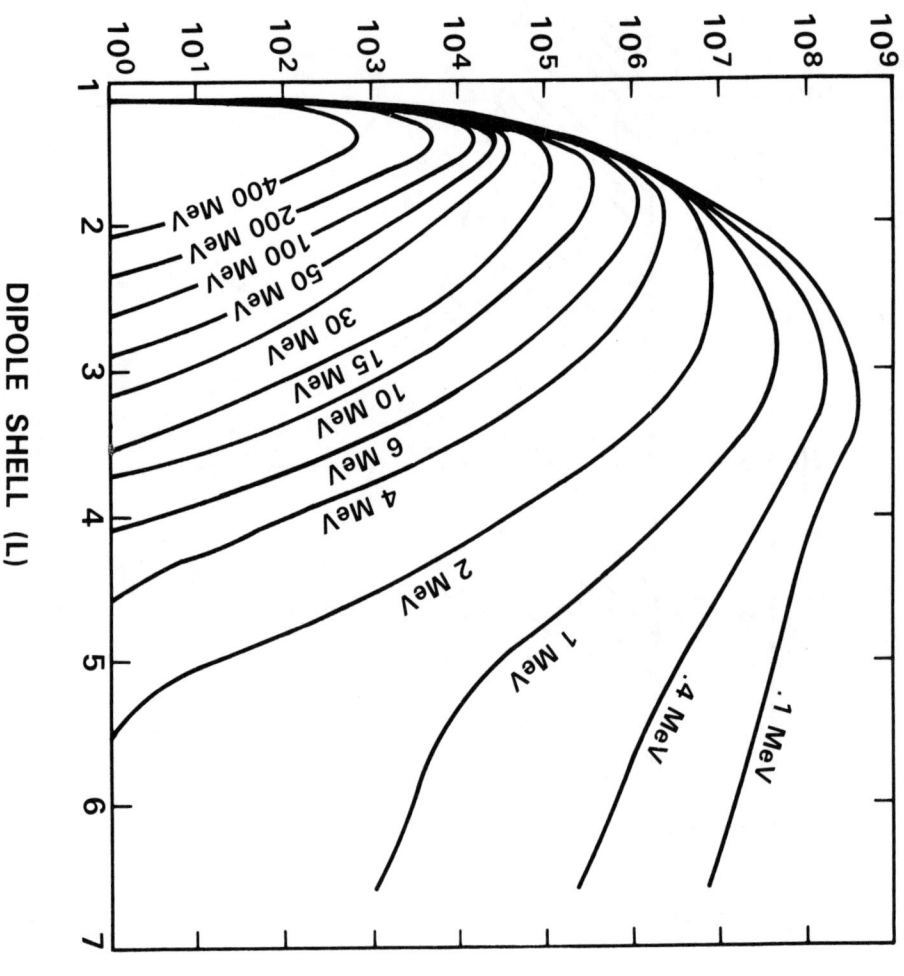

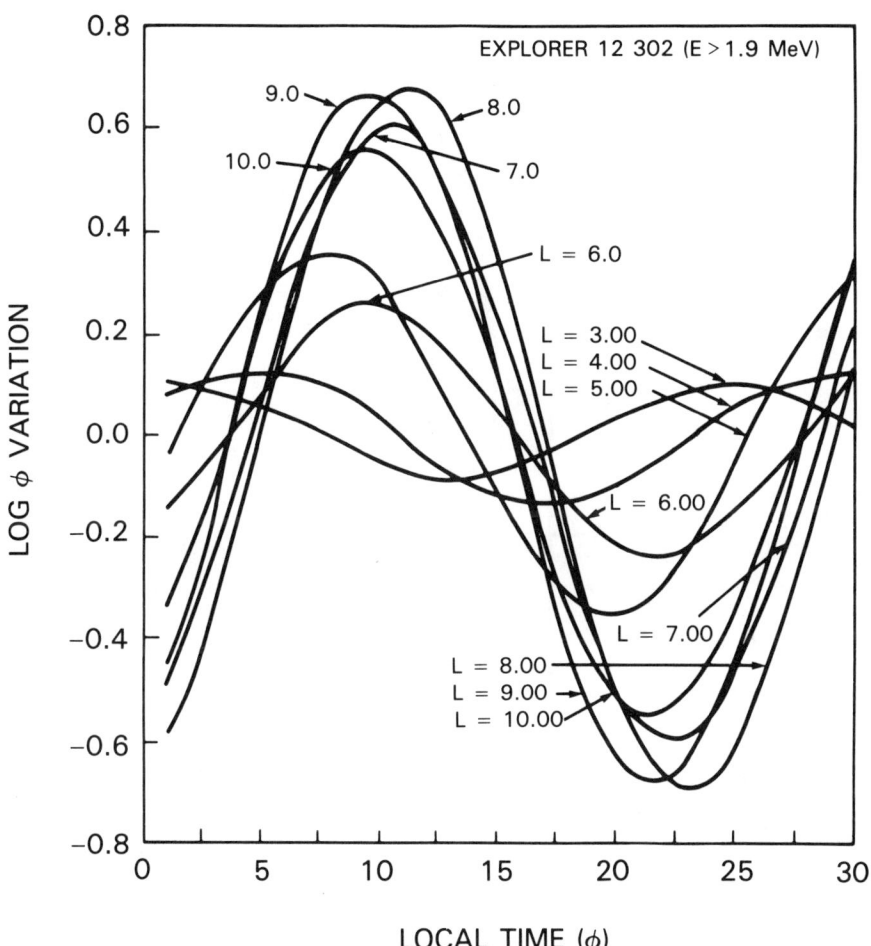

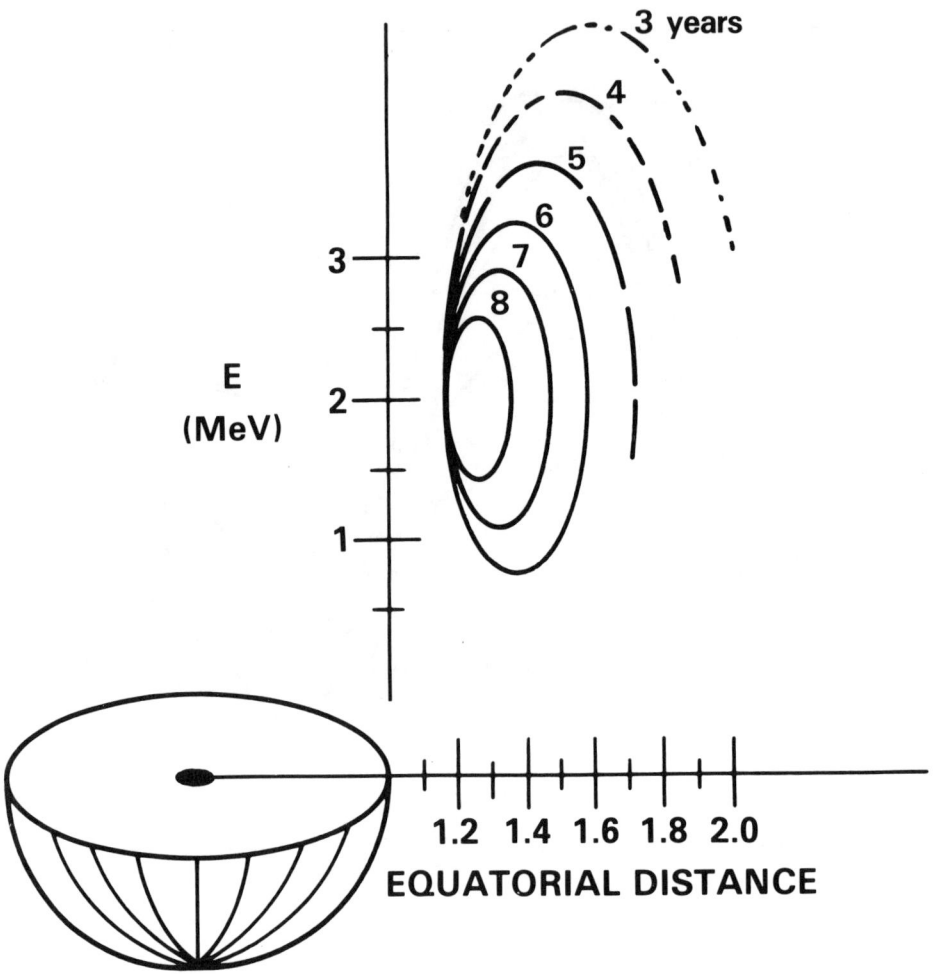

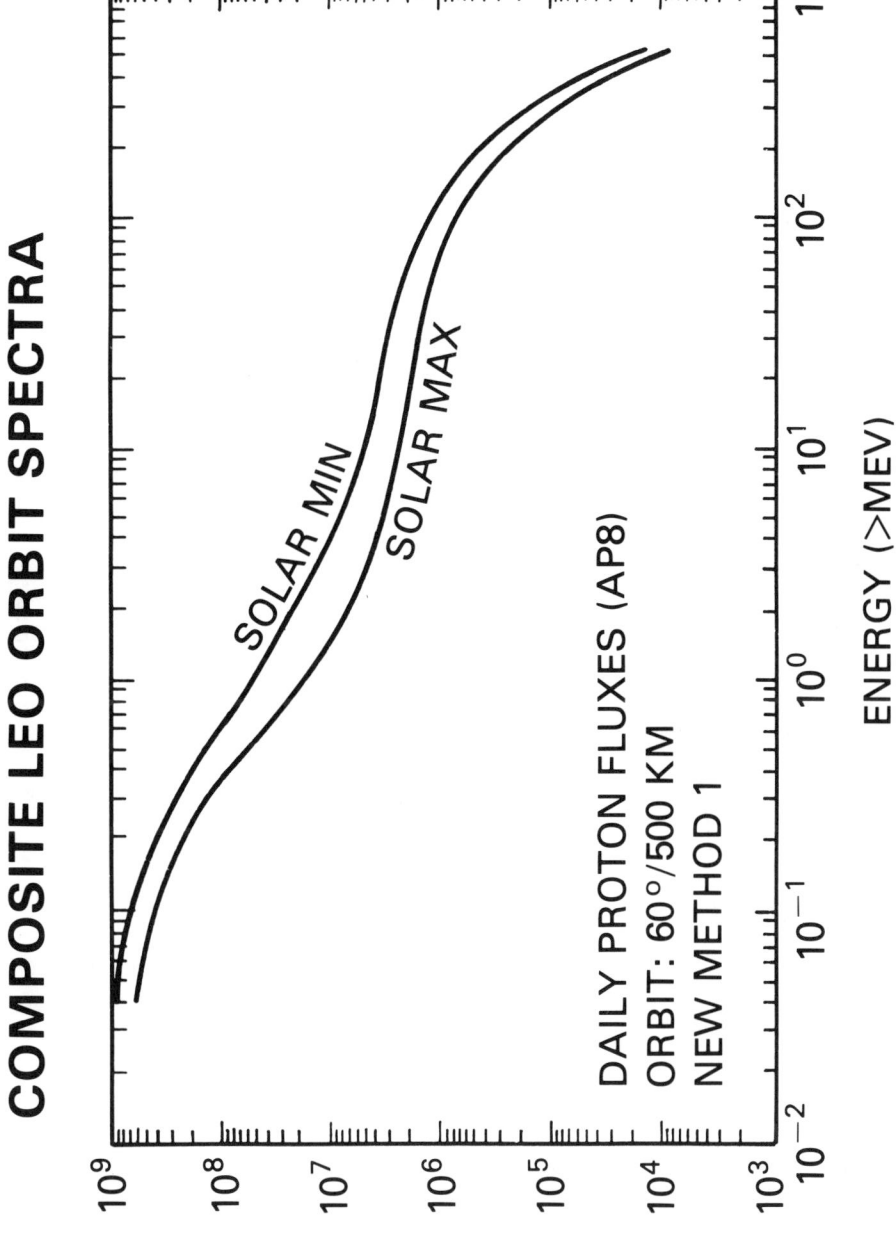

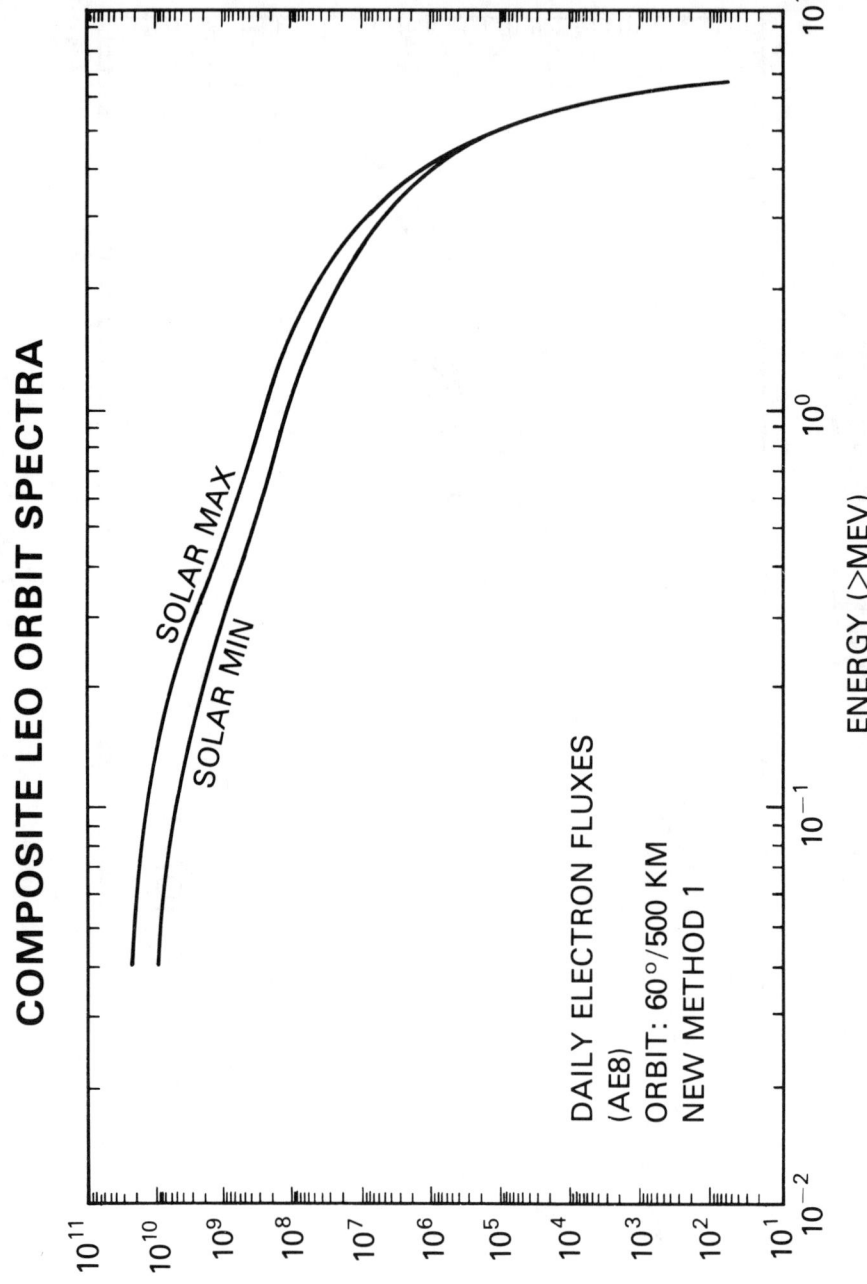

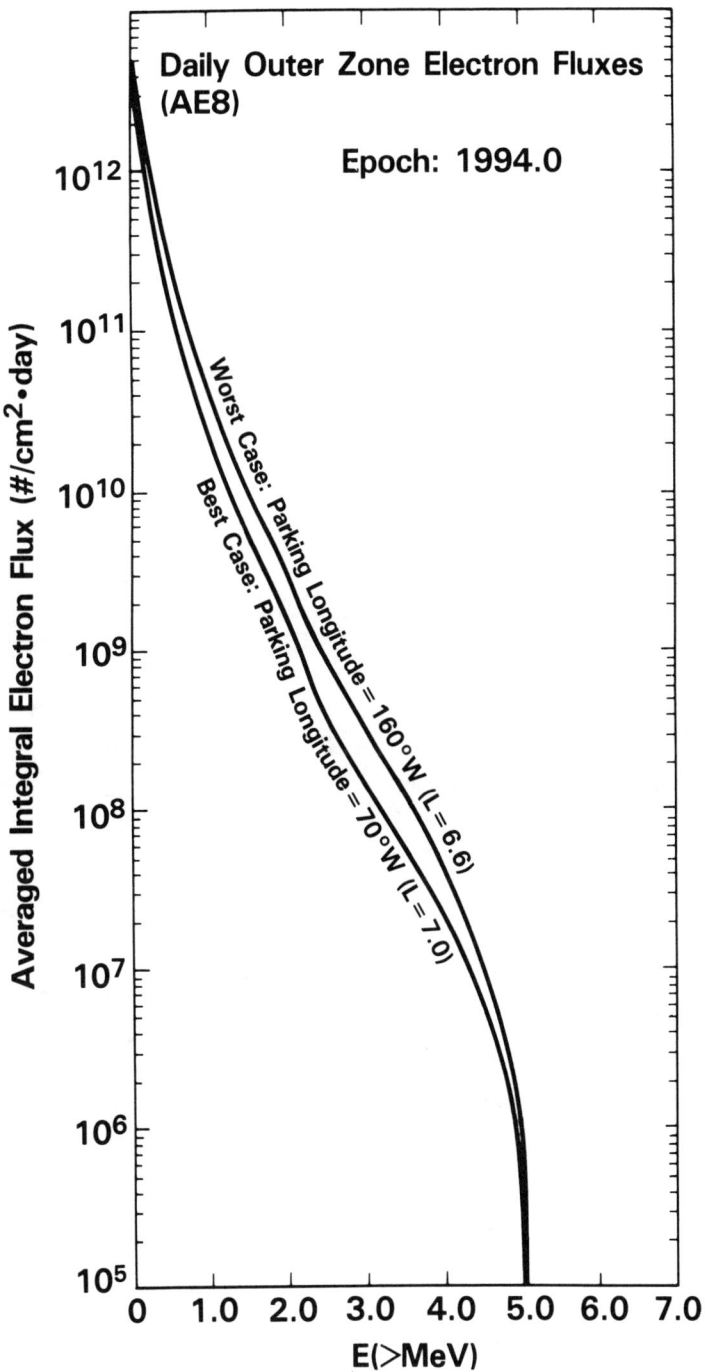

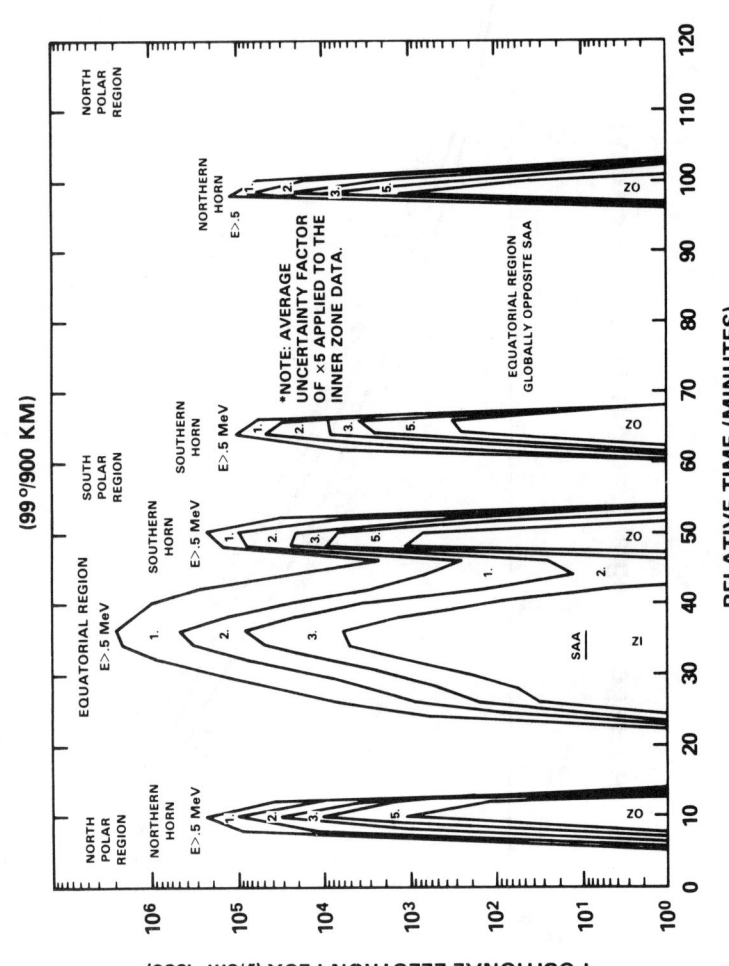

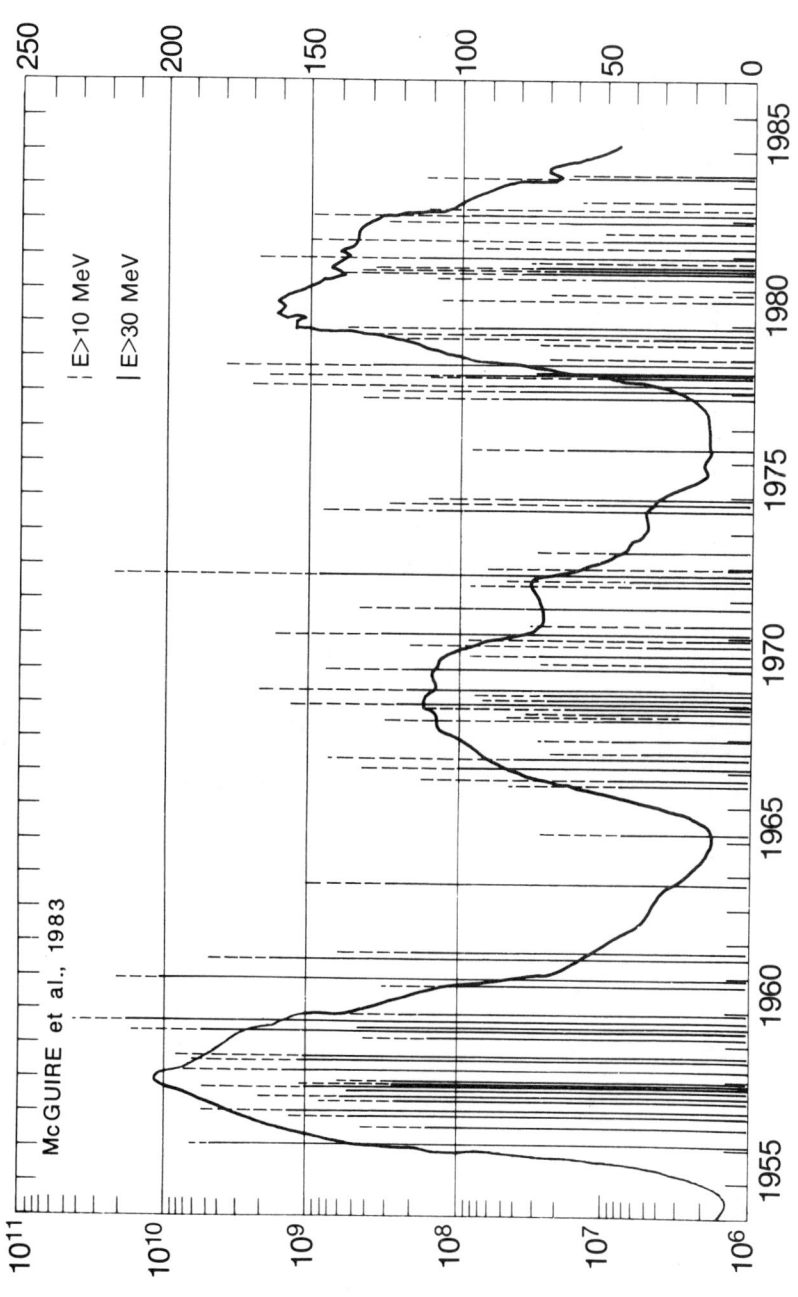

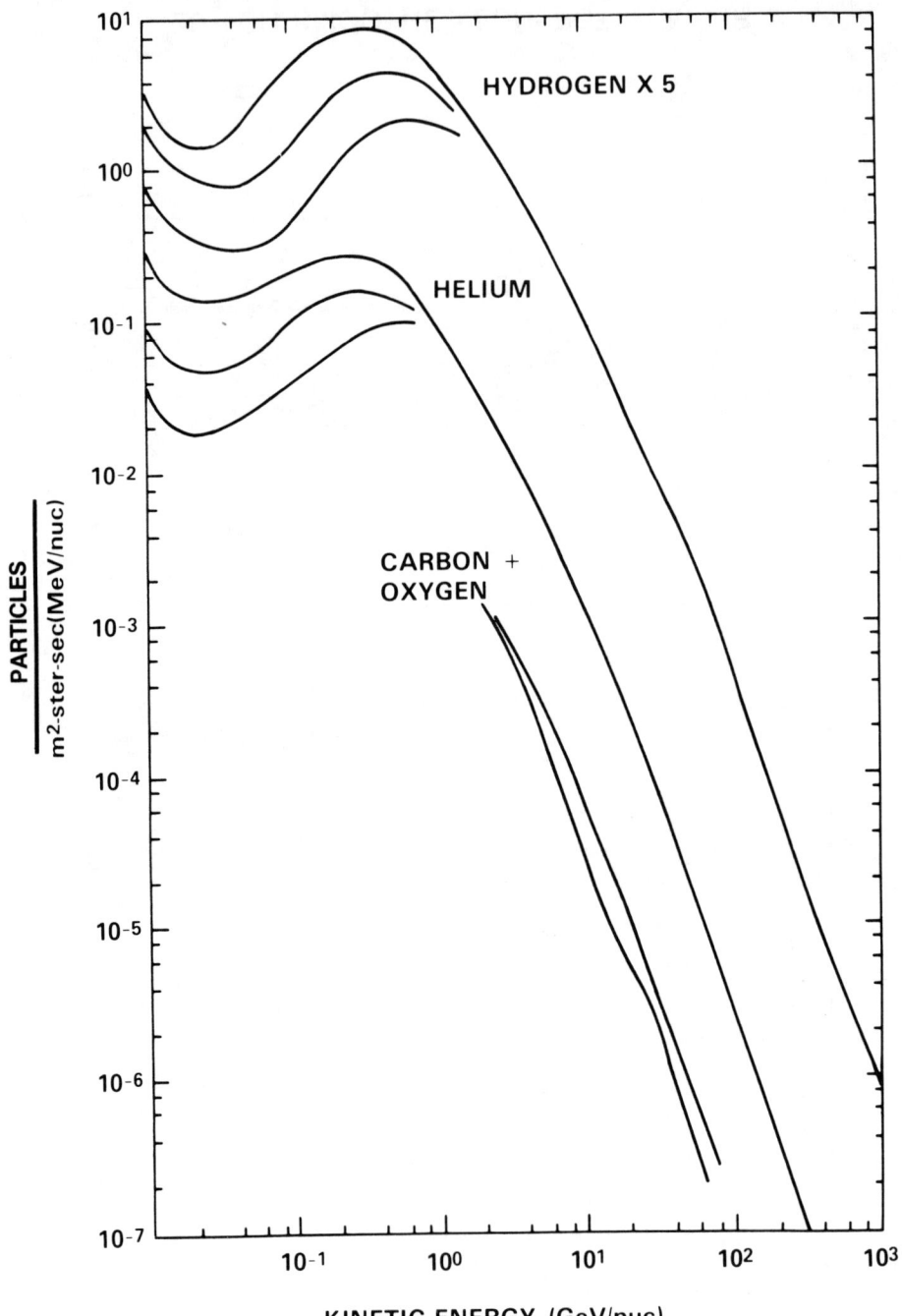

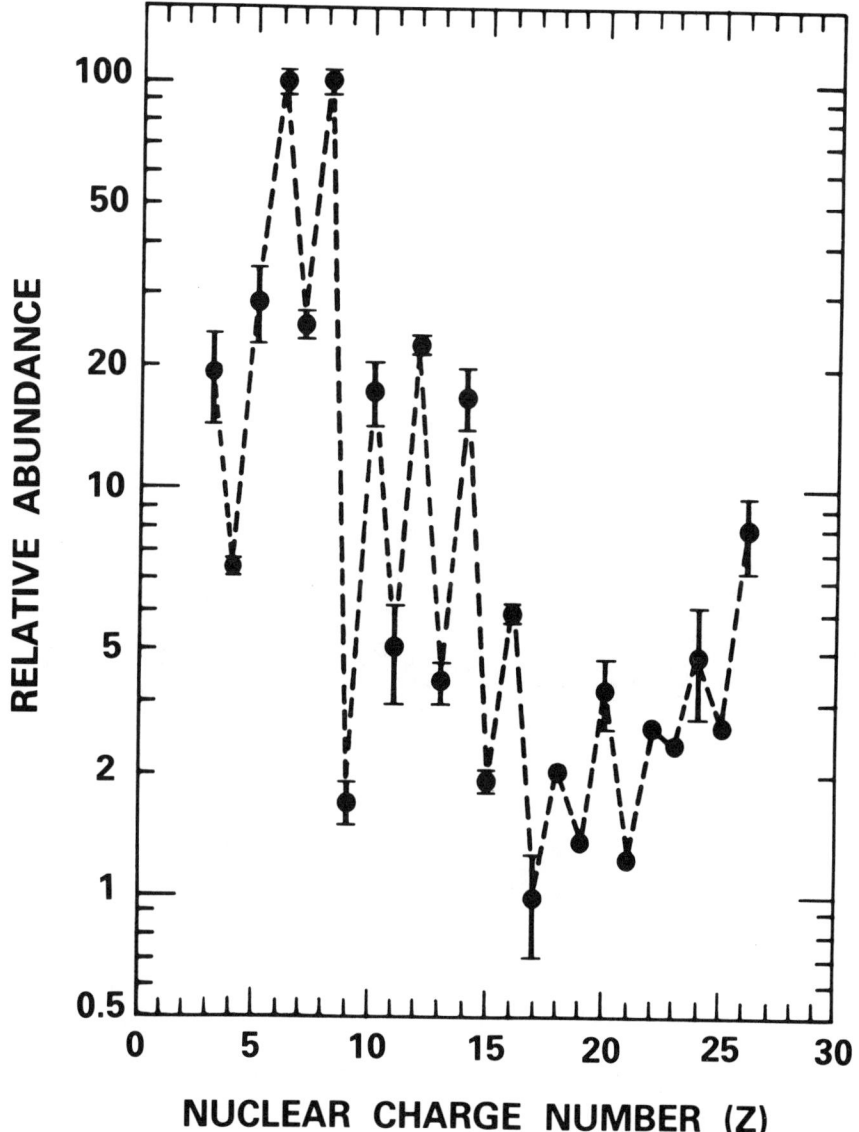

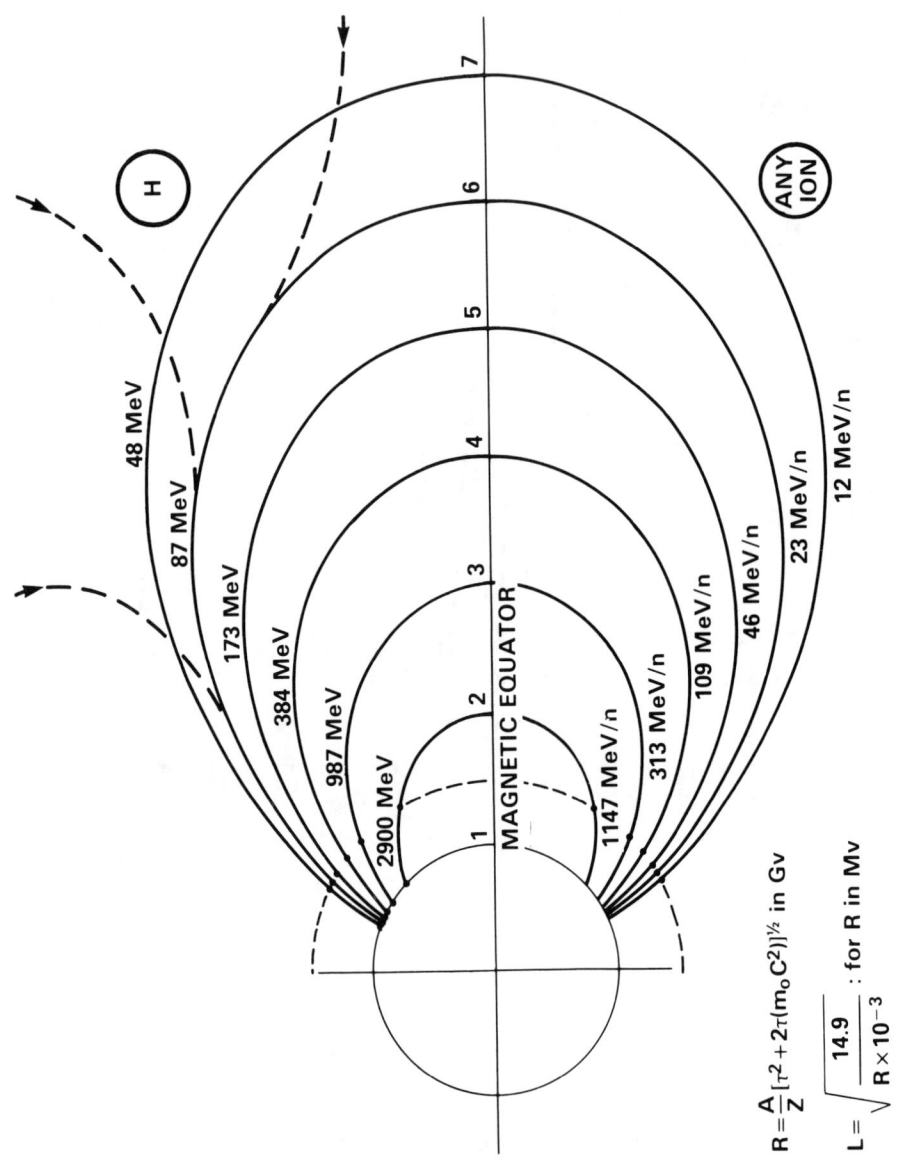

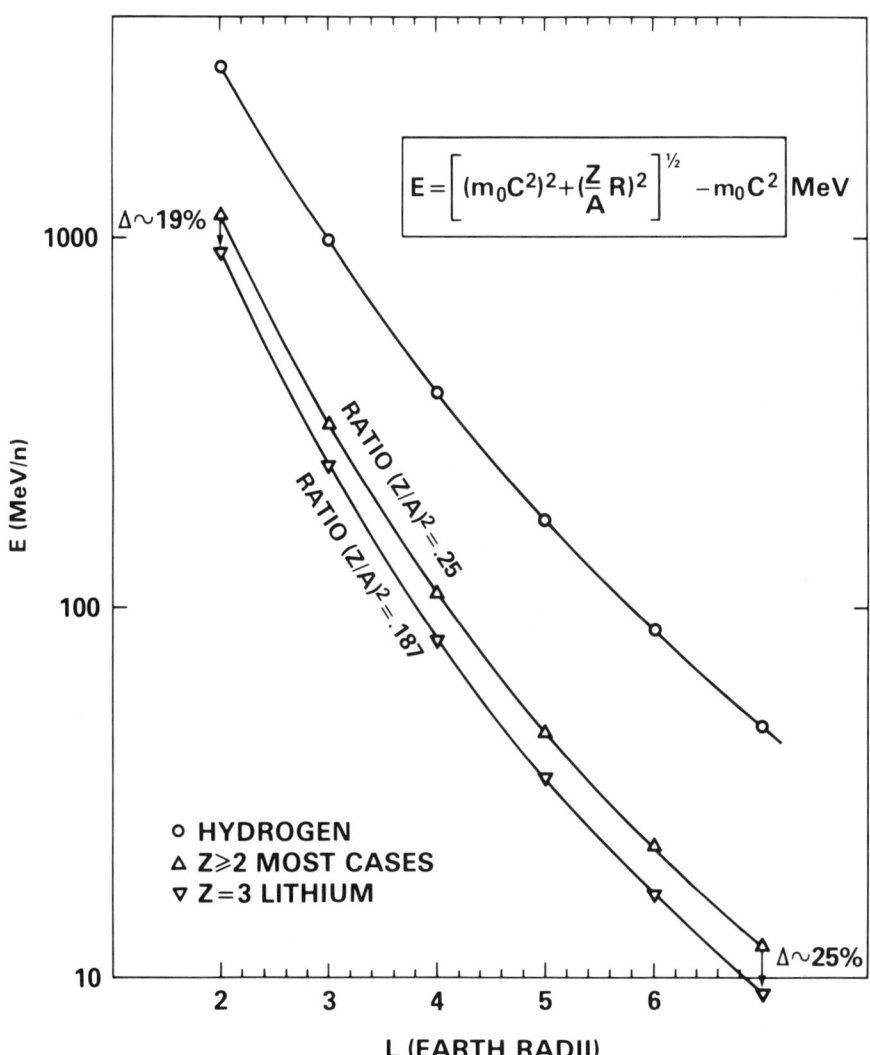

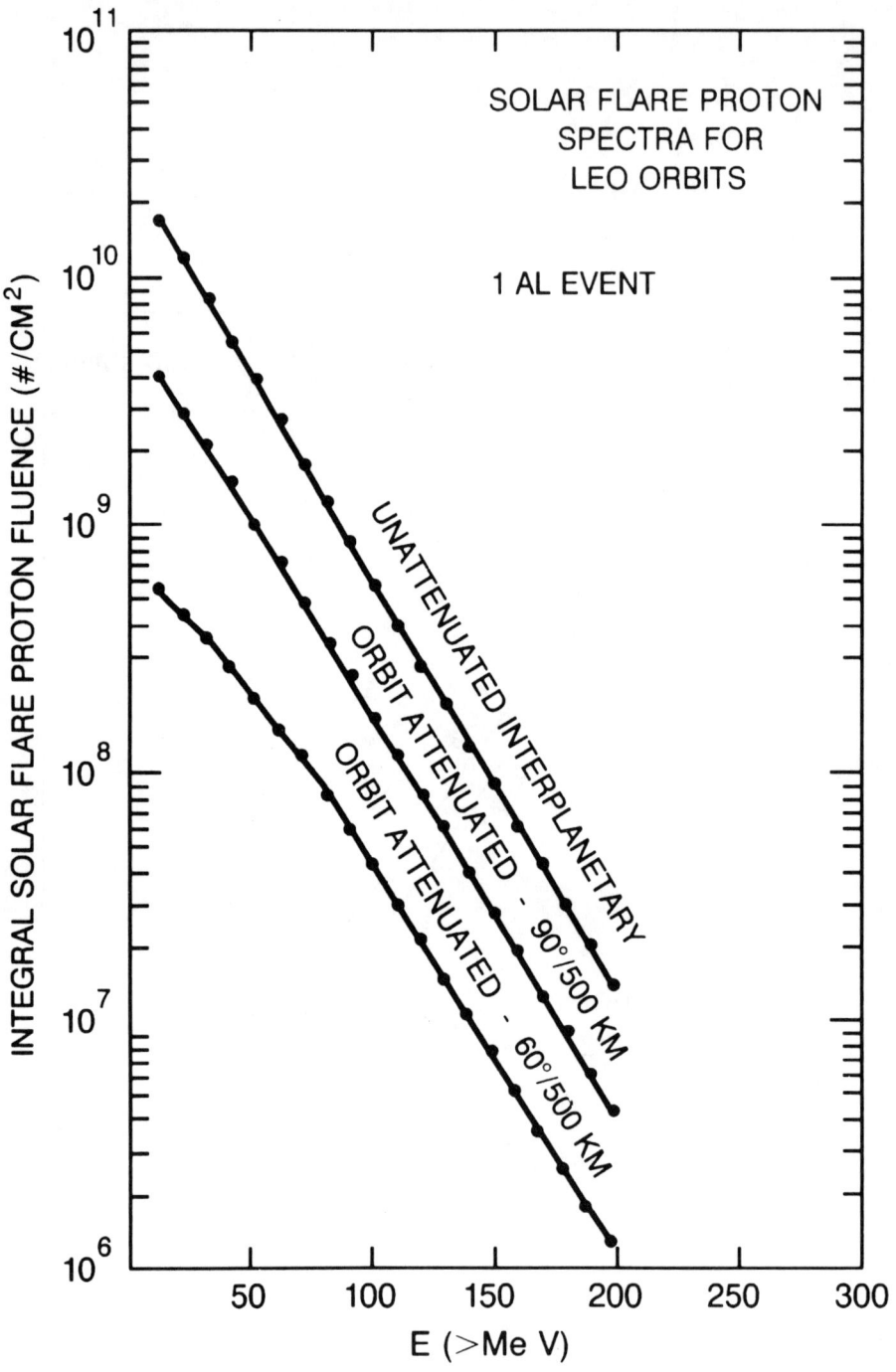

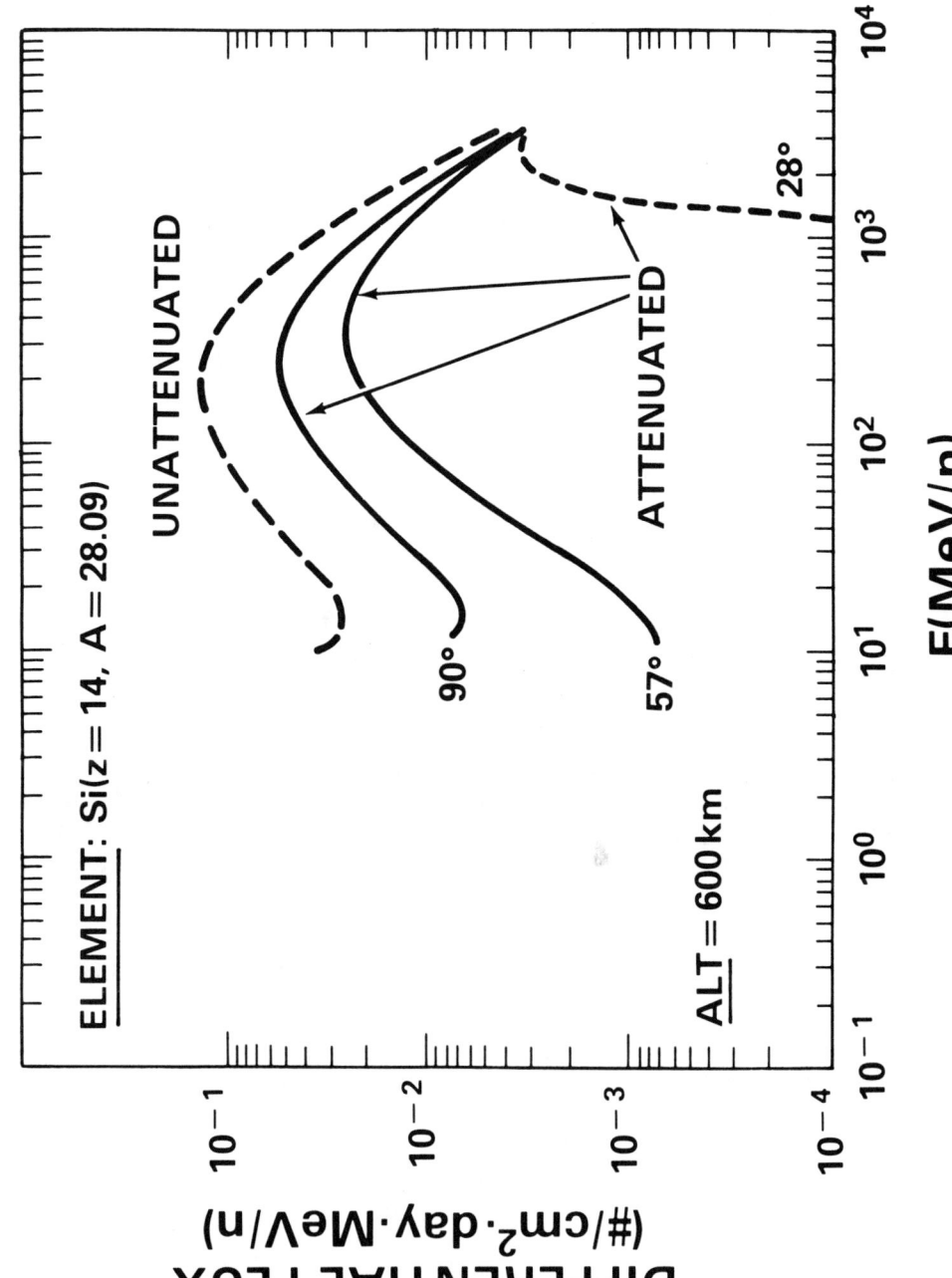

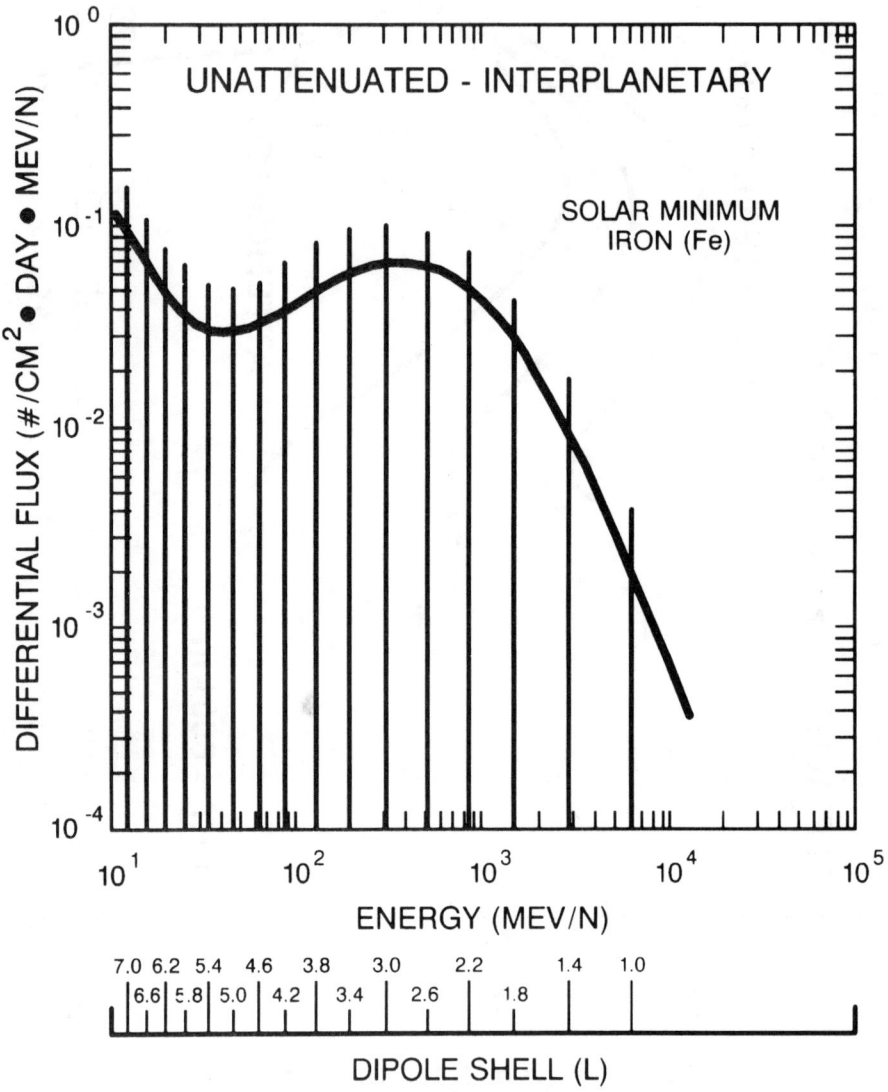

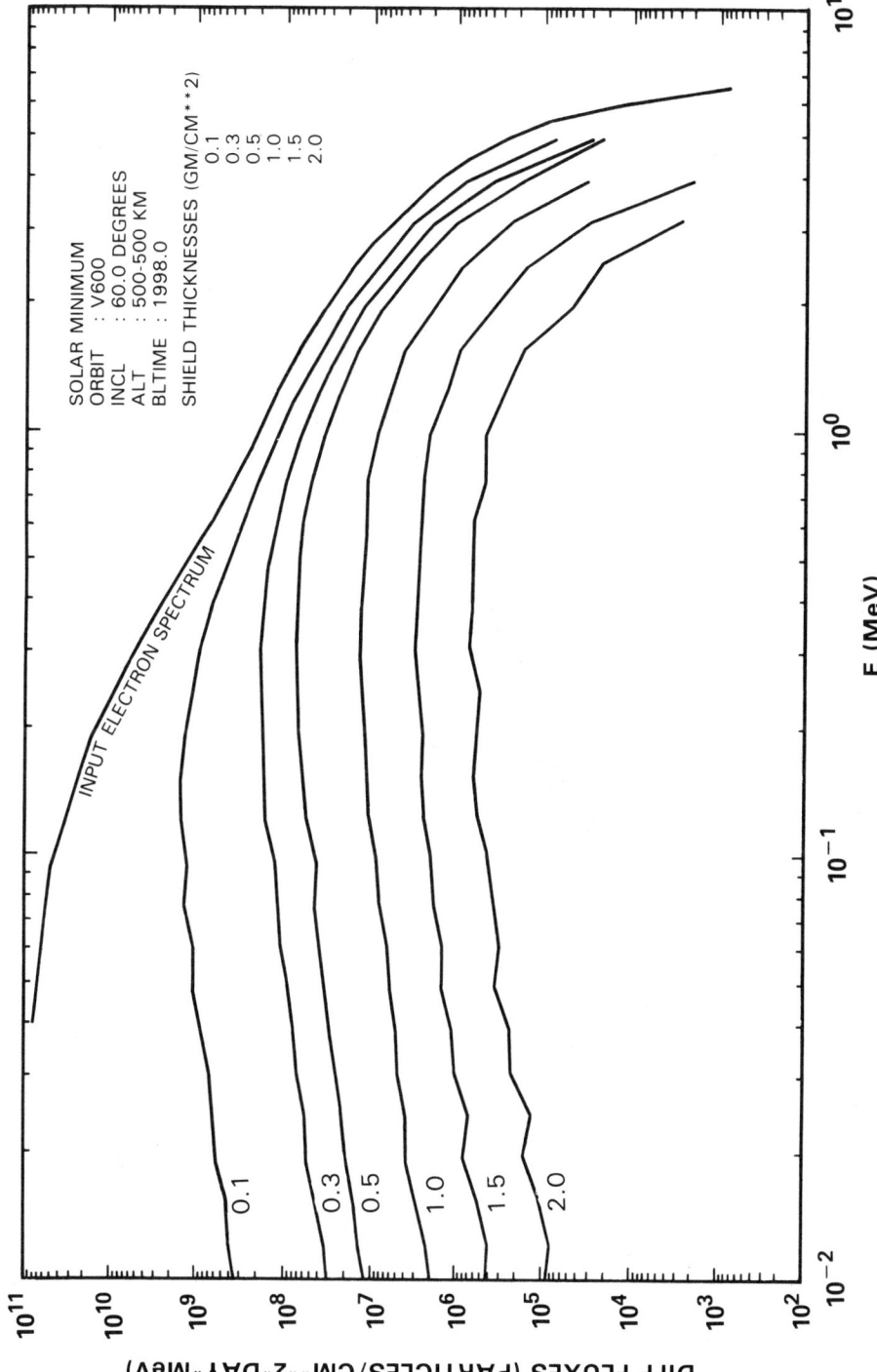

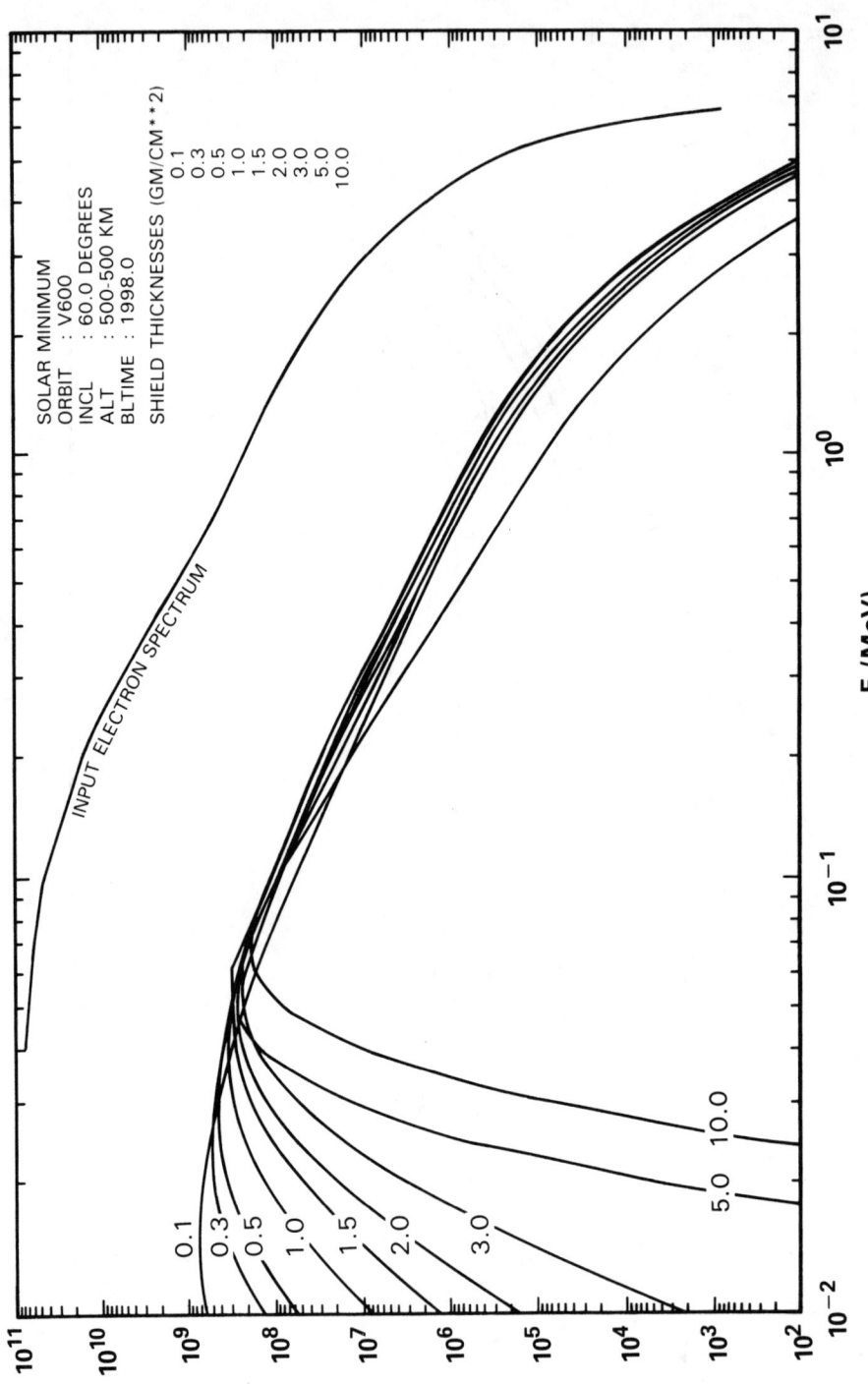

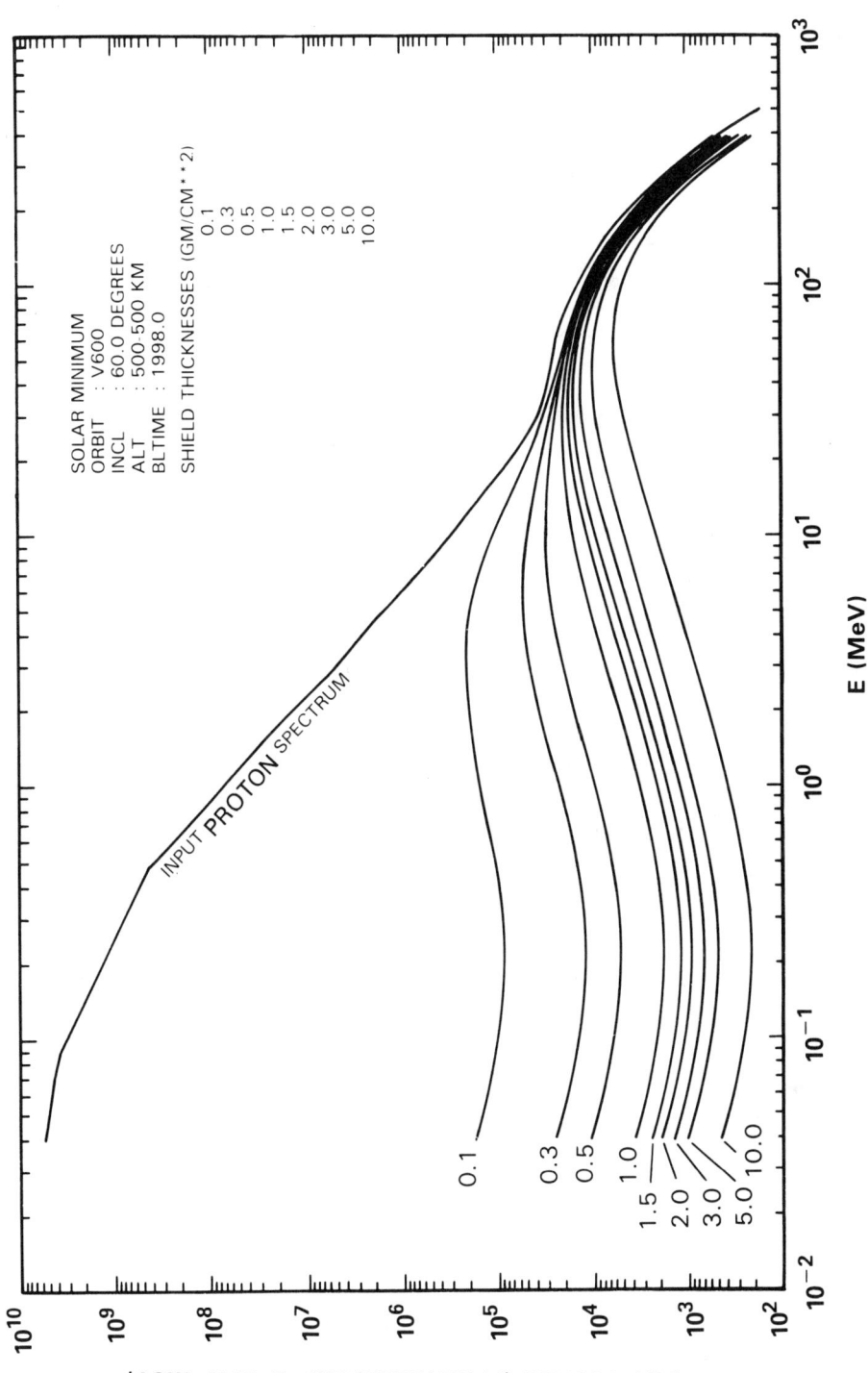

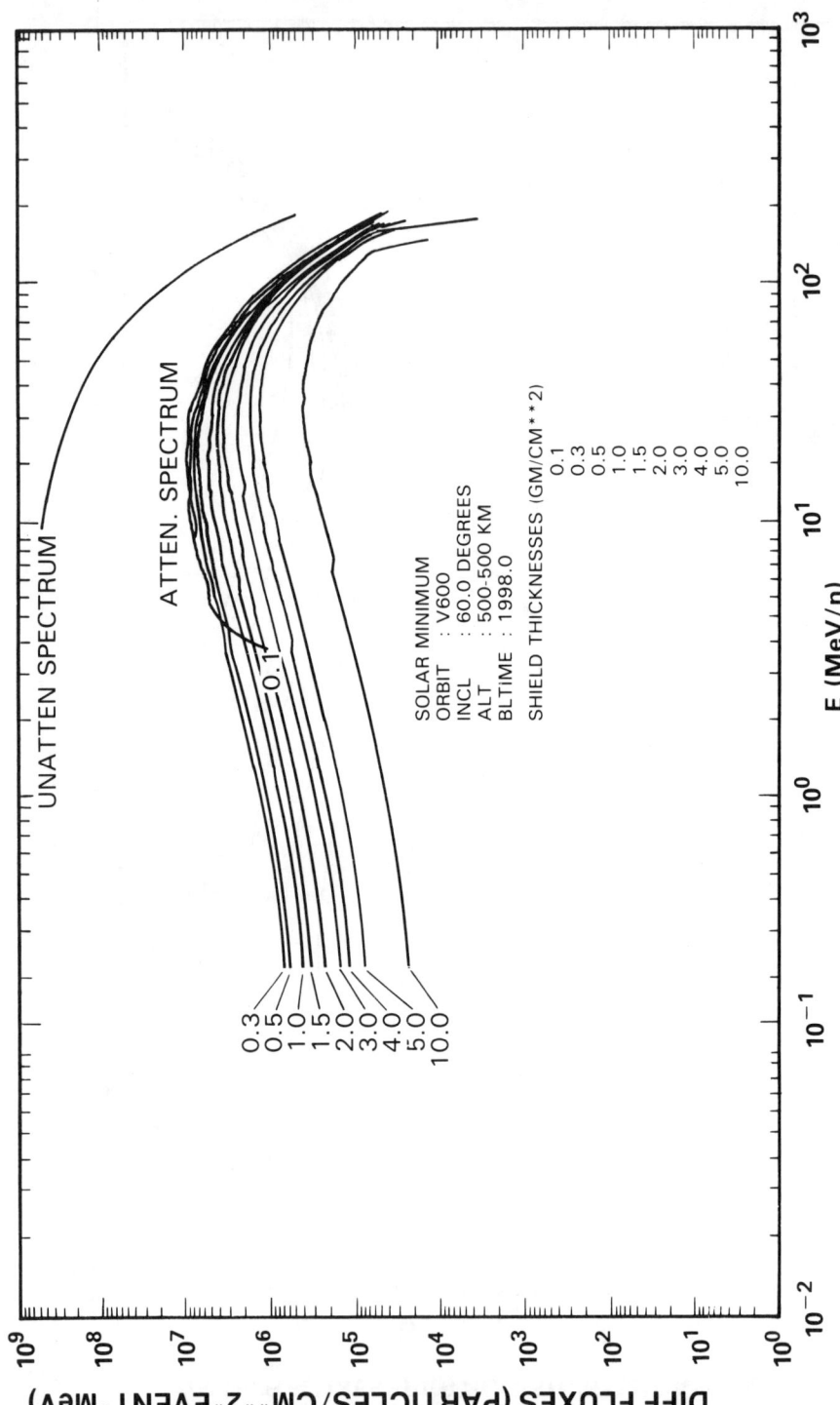

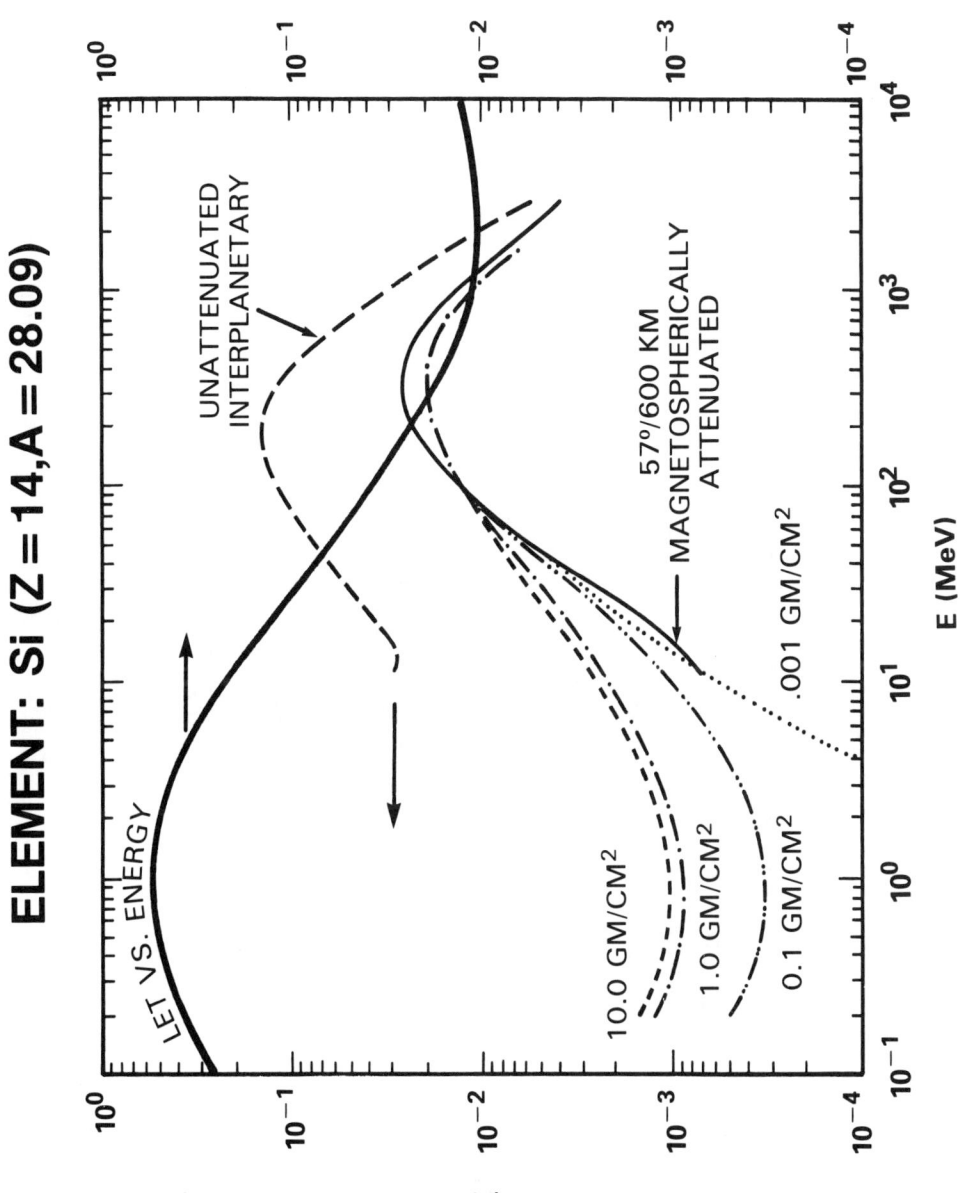

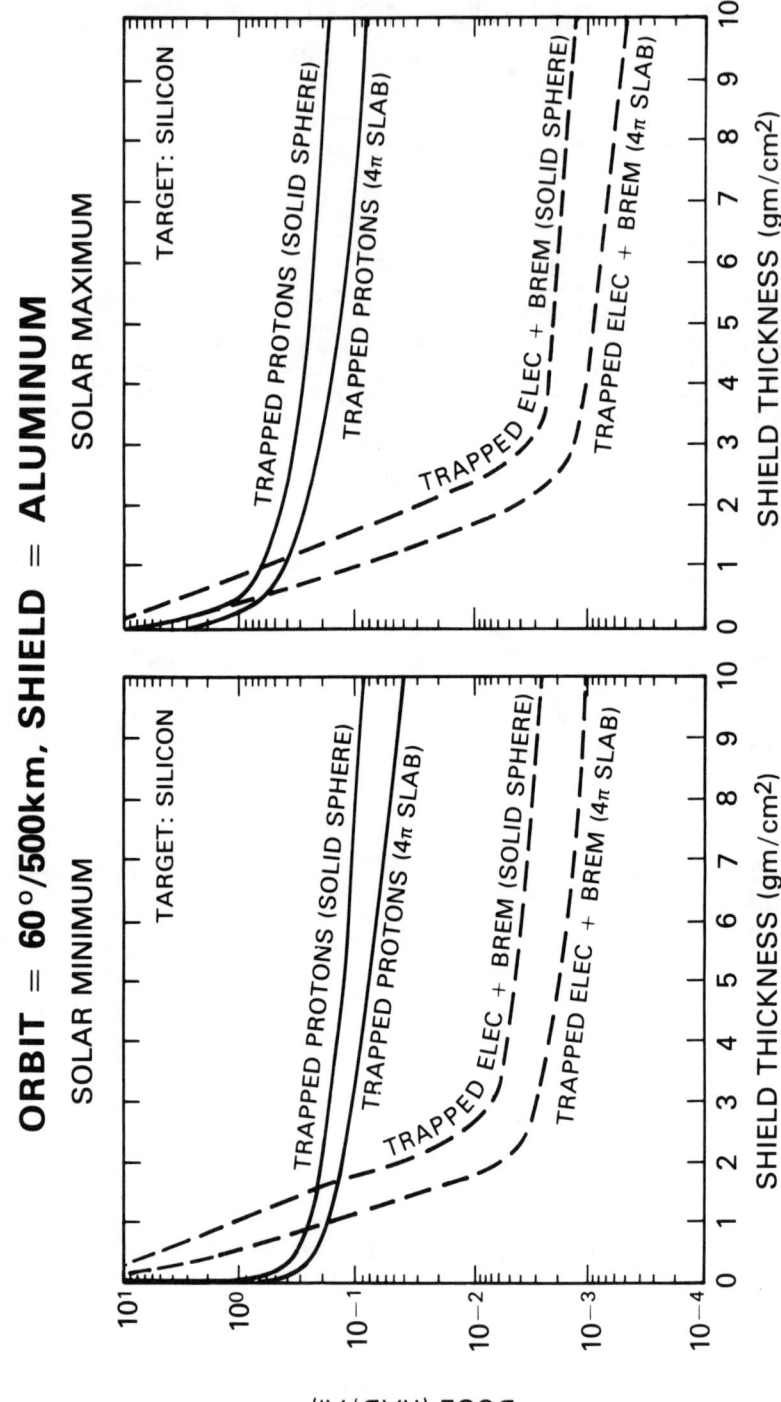

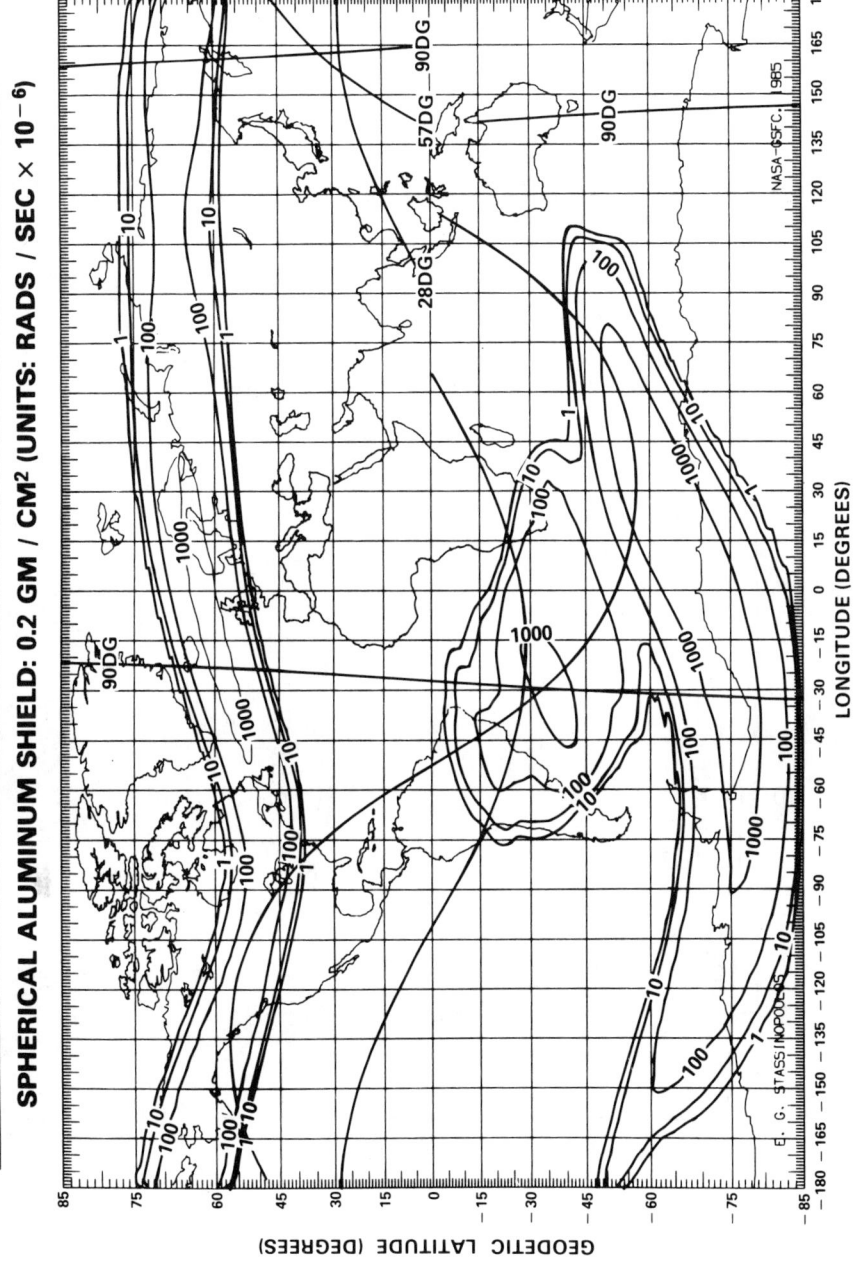

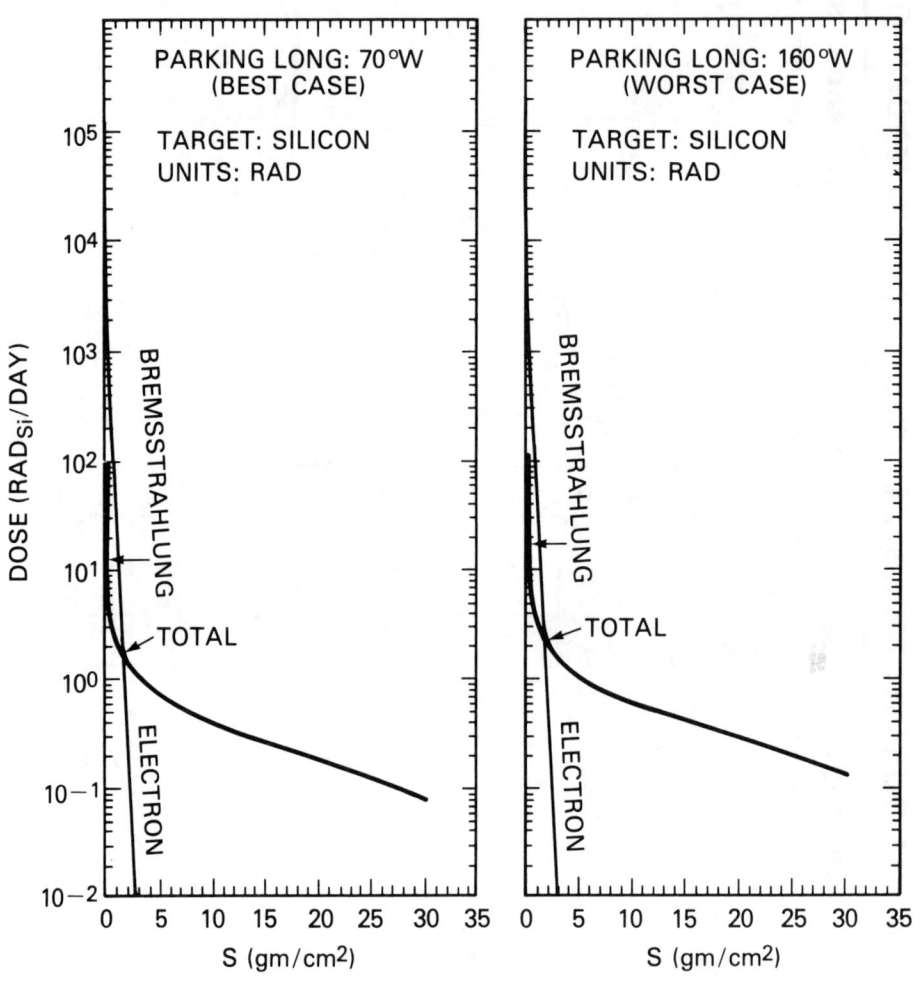

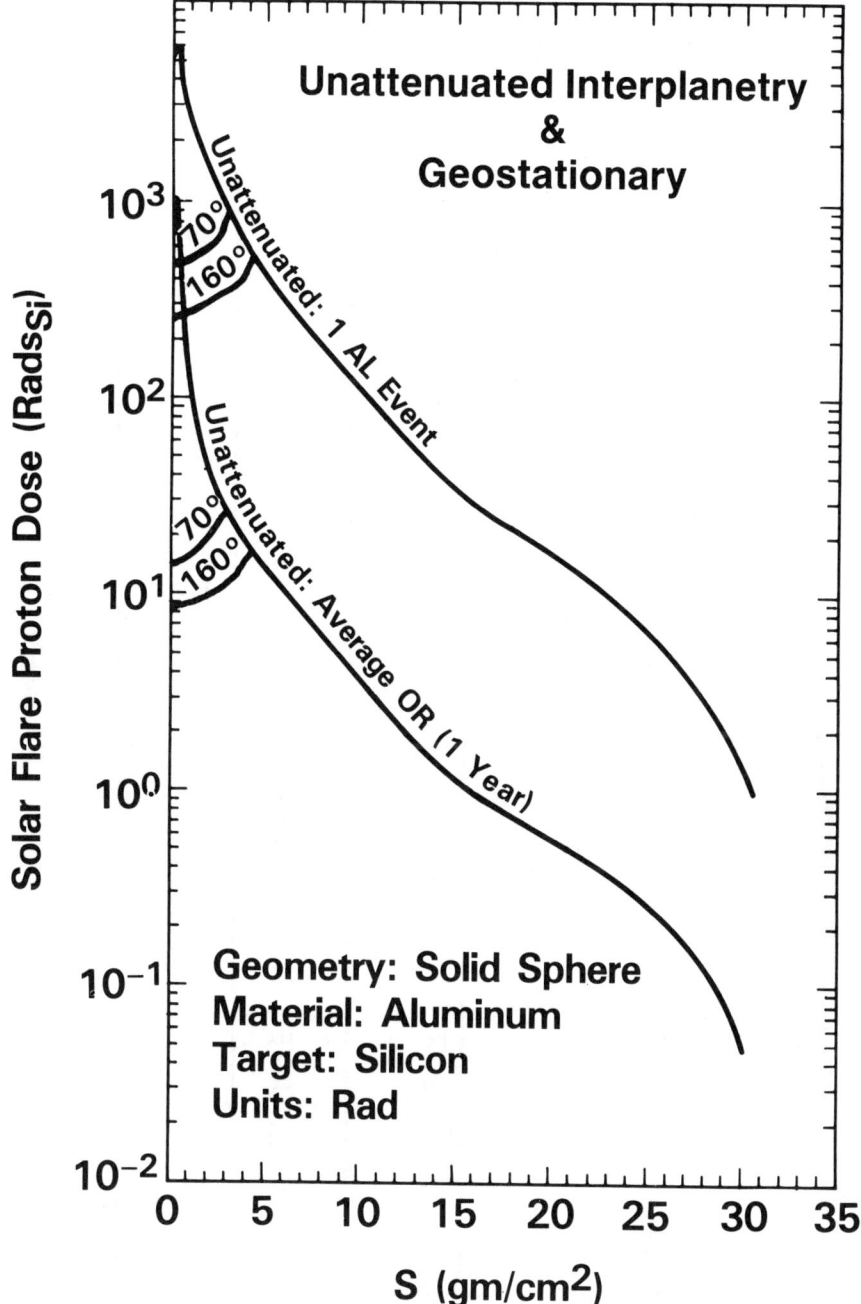

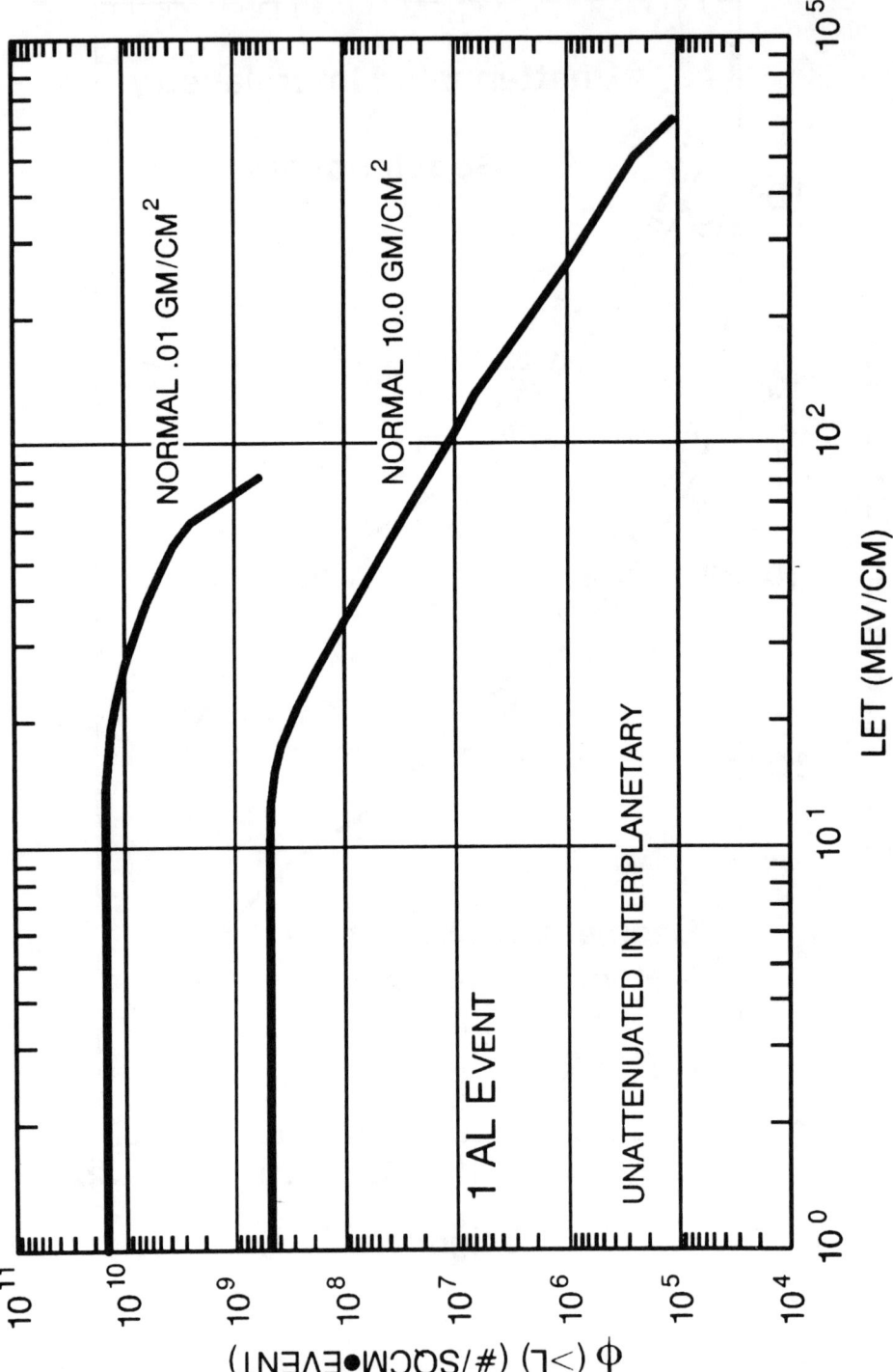

Tables:

1. Proton Fluxes: 28.5°-60°-90°/300-500 km (min)
2. Electron Fluxes: 28.5°-60°-90°/300-500 km (min)
3. GEO Electrons: 70° + 160°
4. SFP Unatt. + Atten (AL): 28.5°-60°-90°/500
5. SFP Unatt. + Atten (AL): 70°-160°
6. Variables Affecting Dose
7. Areas of Concern
8. Doses (e, B, p, T): 28.5°/300 (min)
9. Doses (e, B, p, T): 60°/300 (min)
10. Doses (e, B, p, T): 90°/300 (min)
11. Doses (e, B, p, T): 28.5°/500 (min)
12. Doses (e, B, p, T): 60°/500 (min)
13. Doses (e, B, p, T): 90°/500 (min)
14. SFP doses (AL): 70° + 160°

TRAPPED PROTONS
AVERAGED ORBIT INTEGRATED FLUXES
(#/SQCM*DAY)

SOLAR MINIMUM

E(>MEV)	300 KM			500 KM		
	INCLINATION			INCLINATION		
	28.5 DEG	60 DEG	90 DEG	28.5 DEG	60 DEG	90 DEG
0.04	5.420E+05	3.038E+08	1.509E+08	6.730E+06	9.595E+08	5.013E+08
0.07	5.399E+05	2.577E+08	1.297E+08	6.714E+06	7.947E+08	4.176E+08
0.10	5.379E+05	2.198E+08	1.121E+08	6.699E+06	6.620E+08	3.512E+08
0.50	5.202E+05	5.145E+07	3.055E+07	6.550E+06	1.329E+08	8.003E+07
1.00	5.028E+05	2.056E+07	1.386E+07	6.411E+06	5.126E+07	3.384E+07
2.00	4.945E+05	8.707E+06	6.445E+06	6.305E+06	2.246E+07	1.598E+07
3.00	4.890E+05	5.687E+06	4.265E+06	6.208E+06	1.539E+07	1.124E+07
4.00	4.835E+05	3.895E+06	2.949E+06	6.113E+06	1.123E+07	8.390E+06
5.00	4.781E+05	2.792E+06	2.129E+06	6.020E+06	8.679E+06	6.606E+06
6.00	4.728E+05	2.092E+06	1.606E+06	5.929E+06	7.054E+06	5.453E+06
8.00	4.613E+05	1.535E+06	1.184E+06	5.739E+06	5.663E+06	4.448E+06
10.00	4.501E+05	1.191E+06	9.242E+05	5.556E+06	4.774E+06	3.795E+06
15.00	4.348E+05	9.010E+05	7.073E+05	5.234E+06	3.947E+06	3.169E+06
20.00	4.203E+05	7.359E+05	5.827E+05	4.936E+06	3.422E+06	2.760E+06
25.00	4.064E+05	6.609E+05	5.241E+05	4.720E+06	3.169E+06	2.556E+06
30.00	3.930E+05	6.026E+05	4.779E+05	4.517E+06	2.958E+06	2.384E+06
35.00	3.770E+05	5.588E+05	4.433E+05	4.313E+06	2.780E+06	2.248E+06
40.00	3.616E+05	5.201E+05	4.129E+05	4.119E+06	2.617E+06	2.123E+06
45.00	3.470E+05	4.857E+05	3.857E+05	3.935E+06	2.468E+06	2.009E+06
50.00	3.331E+05	4.548E+05	3.613E+05	3.761E+06	2.330E+06	1.902E+06
60.00	2.999E+05	3.917E+05	3.118E+05	3.382E+06	2.055E+06	1.681E+06
80.00	2.441E+05	2.959E+05	2.363E+05	2.748E+06	1.613E+06	1.324E+06
100.00	1.997E+05	2.276E+05	1.823E+05	2.243E+06	1.279E+06	1.053E+06
150.00	1.018E+05	1.055E+05	8.646E+04	1.279E+06	6.951E+05	5.742E+05
200.00	5.303E+04	5.103E+04	4.278E+04	7.439E+05	3.896E+05	3.226E+05
250.00	2.684E+04	2.526E+04	2.144E+04	4.334E+05	2.246E+05	1.856E+05
300.00	1.377E+04	1.281E+04	1.100E+04	2.547E+05	1.313E+05	1.082E+05
350.00	6.940E+03	6.559E+03	5.680E+03	1.506E+05	7.733E+04	6.359E+04
400.00	3.219E+03	3.139E+03	2.714E+03	8.914E+04	4.594E+04	3.753E+04
500.00	4.961E+02	7.257E+02	5.937E+02	3.108E+04	1.618E+04	1.328E+04

TABLE 1

ELECTRONS
AVERAGED ORBIT INTEGRATED FLUXES
(#/SQCM*DAY)

SOLAR MINIMUM

E(>MEV)	300 KM INCLINATION			500 KM INCLINATION		
	28.5 DEG	60 DEG	90 DEG	28.5 DEG	60 DEG	90 DEG
0.04	2.973E+08	3.203E+09	2.971E+09	5.153E+09	9.171E+09	7.876E+09
0.07	2.351E+08	2.391E+09	2.257E+09	4.082E+09	7.007E+09	6.066E+09
0.10	1.861E+08	1.795E+09	1.730E+09	3.236E+09	5.382E+09	4.712E+09
0.20	5.629E+07	6.779E+08	7.424E+08	9.975E+08	1.908E+09	1.816E+09
0.30	2.227E+07	3.631E+08	4.262E+08	3.969E+08	9.484E+08	9.514E+08
0.40	1.144E+07	2.384E+08	2.849E+08	2.017E+08	5.895E+08	6.029E+08
0.50	5.897E+06	1.616E+08	1.950E+08	1.030E+08	3.807E+08	3.944E+08
0.60	3.985E+06	1.283E+08	1.526E+08	6.850E+07	2.917E+08	3.007E+08
0.70	2.701E+06	1.027E+08	1.204E+08	4.574E+07	2.258E+08	2.315E+08
0.80	1.948E+06	8.399E+07	9.744E+07	3.268E+07	1.813E+08	1.845E+08
0.90	1.494E+06	7.001E+07	8.051E+07	2.494E+07	1.504E+08	1.515E+08
1.00	1.147E+06	5.850E+07	6.669E+07	1.904E+07	1.252E+08	1.248E+08
1.25	7.213E+05	3.857E+07	4.262E+07	1.179E+07	8.119E+07	7.931E+07
1.50	4.549E+05	2.554E+07	2.742E+07	7.310E+06	5.292E+07	5.076E+07
1.75	3.051E+05	1.747E+07	1.828E+07	4.870E+06	3.561E+07	3.371E+07
2.00	2.053E+05	1.199E+07	1.224E+07	3.250E+06	2.407E+07	2.250E+07
2.25	1.392E+05	8.275E+06	8.289E+06	2.194E+06	1.644E+07	1.516E+07
2.50	9.419E+04	5.725E+06	5.637E+06	1.484E+06	1.127E+07	1.026E+07
2.75	3.788E+04	3.899E+06	3.751E+06	5.934E+05	7.373E+06	6.610E+06
3.00	1.521E+04	2.695E+06	2.529E+06	2.405E+05	4.956E+06	4.361E+06
3.25	4.850E+03	1.856E+06	1.695E+06	7.591E+04	3.324E+06	2.862E+06
3.50	1.357E+03	1.292E+06	1.148E+06	2.394E+04	2.274E+06	1.914E+06
3.75	3.874E+02	8.495E+05	7.316E+05	7.263E+03	1.474E+06	1.206E+06
4.00	0.000E+00	5.650E+05	4.726E+05	8.860E+02	9.693E+05	7.712E+05
4.50	0.000E+00	2.066E+05	1.643E+05	0.000E+00	3.493E+05	2.633E+05
5.00	0.000E+00	6.828E+04	5.129E+04	0.000E+00	1.143E+05	7.979E+04
5.50	0.000E+00	1.572E+04	1.188E+04	0.000E+00	2.659E+04	1.751E+04
6.00	0.000E+00	2.858E+03	1.970E+03	0.000E+00	3.923E+03	2.470E+03
6.50	0.000E+00	0.000E+00	0.000E+00	0.000E+00	1.235E+02	6.052E+01
7.00	0.000E+00	0.000E+00	0.000E+00	0.000E+00	0.000E+00	0.000E+00

TABLE 2

GEOSTATIONARY ELECTRONS
AVERAGED ORBIT INTEGRATED FLUXES
#/SQCM*SEC

E(>MEV)	70 DEG W	160 DEG W
0.04	3.775E+07	4.643E+07
0.07	3.023E+07	3.847E+07
0.10	2.421E+07	3.188E+07
0.20	1.145E+07	1.587E+07
0.30	5.944E+06	8.575E+06
0.40	3.383E+06	5.044E+06
0.50	1.925E+06	2.967E+06
0.60	1.224E+06	2.048E+06
0.70	7.788E+05	1.414E+06
0.80	5.290E+05	9.879E+05
0.90	3.838E+05	6.983E+05
1.00	2.784E+05	4.935E+05
1.25	1.338E+05	2.475E+05
1.50	6.435E+04	1.242E+05
1.75	3.497E+04	7.171E+04
2.00	1.900E+04	4.142E+04
2.25	9.313E+03	2.128E+04
2.50	4.653E+03	1.093E+04
2.75	2.816E+03	6.494E+03
3.00	1.737E+03	3.856E+03
3.25	1.118E+03	2.484E+03
3.50	7.196E+02	1.600E+03
3.75	4.260E+02	8.527E+02
4.00	2.522E+02	4.546E+02
4.50	6.825E+01	1.187E+02
5.00	1.673E+00	4.519E+00
5.50	0.000E+00	0.000E+00
6.00	0.000E+00	0.000E+00
6.50	0.000E+00	0.000E+00
7.00	0.000E+00	0.000E+00

TABLE 3

SOLAR FLARE PROTON FLUENCES
1 ANOMALOUSLY LARGE EVENT
ALTITUDE = 500 KM

ENERGY (>MEV)	1 AL UNATTEN #/SQCM*EVENT	28.5 DEG #/SQCM*EVENT	60 DEG #/SQCM*EVENT	90 DEG #/SQCM*EVENT
10.0	1.680E+10	0	5.314E+08	3.837E+09
20.0	1.152E+10	-	4.429E+08	2.761E+09
30.0	7.900E+09	-	3.502E+08	1.960E+09
40.0	5.417E+09	-	2.654E+08	1.386E+09
50.0	3.714E+09	-	1.996E+08	9.735E+08
60.0	2.547E+09	-	1.461E+08	6.819E+08
70.0	1.746E+09	-	1.062E+08	4.762E+08
80.0	1.197E+09	-	7.717E+07	3.322E+08
90.0	8.210E+08	-	5.516E+07	2.316E+08
100.0	5.629E+08	-	3.903E+07	1.608E+08
110.0	3.860E+08	-	2.762E+07	1.117E+08
120.0	2.646E+08	-	1.976E+07	7.766E+07
130.0	1.815E+08	-	1.417E+07	5.389E+07
140.0	1.244E+08	-	1.011E+07	3.723E+07
150.0	8.531E+07	-	7.225E+06	2.571E+07
160.0	5.850E+07	-	5.177E+06	1.779E+07
170.0	4.011E+07	-	3.696E+06	1.231E+07
180.0	2.750E+07	-	2.610E+06	8.508E+06
190.0	1.886E+07	-	1.827E+06	5.874E+06
200.0	1.293E+07	0	1.276E+06	4.056E+06

TABLE 4

SOLAR FLARE PROTON FLUENCES FOR GEOSTATIONARY ORBITS
1 ANOMALOUSLY LARGE EVENT

ENERGY (>MEV)	1 AL UNATTEN #/SQCM*EVENT	ENERGY (>MEV)	1 AL 160 W #/SQCM*EVENT	ENERGY (>MEV)	1 AL 70 W #/SQCM*EVENT
10.0	1.680E+10	NOT ACCESSIBLE		NOT ACCESSIBLE	
20.0	1.152E+10		0		0
30.0	7.900E+09		0		0
40.0	5.417E+09	48.1	3.990E+09		0
50.0	3.714E+09	50.0	3.714E+09		0
60.0	2.547E+09	60.0	2.547E+09	60.4	2.509E+09
70.0	1.746E+09	70.0	1.746E+09	70.0	1.746E+09
80.0	1.197E+09	80.0	1.197E+09	80.0	1.197E+09
90.0	8.210E+08	90.0	8.210E+08	90.0	8.210E+08
100.0	5.629E+08	100.0	5.629E+08	100.0	5.629E+08
110.0	3.860E+08	110.0	3.860E+08	110.0	3.860E+08
120.0	2.646E+08	120.0	2.646E+08	120.0	2.646E+08
130.0	1.815E+08	130.0	1.815E+08	130.0	1.815E+08
140.0	1.244E+08	140.0	1.244E+08	140.0	1.244E+08
150.0	8.531E+07	150.0	8.531E+07	150.0	8.531E+07
160.0	5.850E+07	160.0	5.850E+07	160.0	5.850E+07
170.0	4.011E+07	170.0	4.011E+07	170.0	4.011E+07
180.0	2.750E+07	180.0	2.750E+07	180.0	2.750E+07
190.0	1.886E+07	190.0	1.886E+07	190.0	1.886E+07
200.0	1.293E+07	200.0	1.293E+07	200.0	1.293E+07

TABLE 5

- PRIMARY ENVIRONMENT DEFINITION
- SECONDARY CONTRIBUTIONS
- DESCRIPTION OF INPUT SPECTRA
- SHIELD GEOMETRY
- SHIELD COMPOSITION
- SHIELD EVALUATION METHOD
- TARGET COMPOSITION
- ASSUMPTIONS & APPROXIMATIONS

TABLE 6: Variables Affecting Dose Calculations

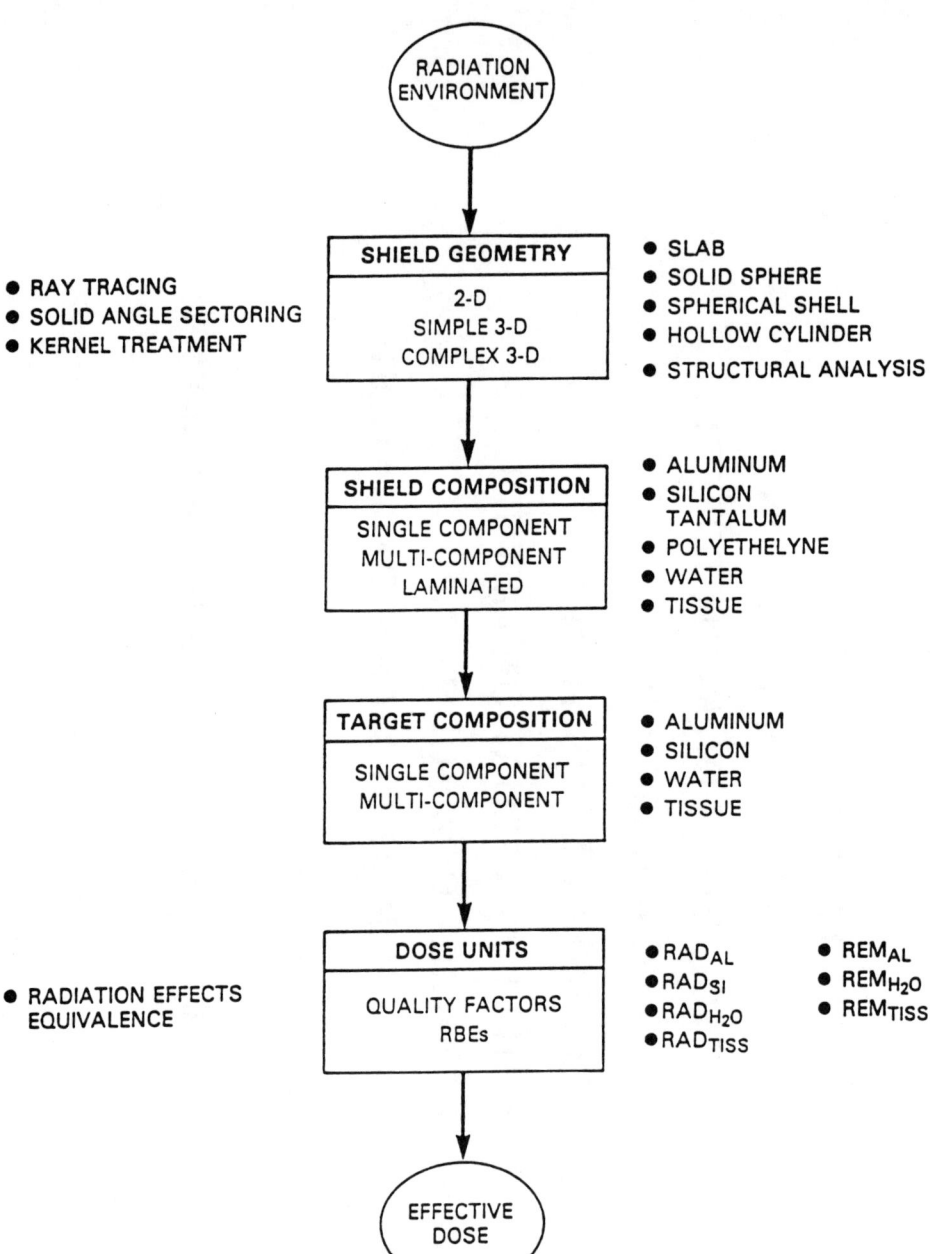

TABLE 7: AREAS OF CONCERN FOR SHIELDING AND TRANSPORT EVALUATION

DAILY DOSE AT CENTER OF ALUMINUM SPHERES FOR ALUMINUM TARGETS
28.5DEG/300KM

SOLAR MINIMUM

S GM/SQCM	T MM	T MILS	ELEC RADS-AL	BREM RADS-AL	PROTON RADS-AL	TOTAL RADS-AL
0.01	0.04	1.	1.438E+01	5.482E-03	1.228E-01	1.451E+01
0.02	0.07	3.	8.107E+00	3.863E-03	1.079E-01	8.219E+00
0.03	0.11	4.	5.060E+00	2.942E-03	1.022E-01	5.165E+00
0.04	0.15	6.	3.405E+00	2.372E-03	9.838E-02	3.506E+00
0.05	0.19	7.	2.409E+00	1.980E-03	9.482E-02	2.506E+00
0.06	0.22	9.	1.773E+00	1.683E-03	9.271E-02	1.868E+00
0.07	0.26	10.	1.346E+00	1.451E-03	9.041E-02	1.438E+00
0.08	0.30	12.	1.047E+00	1.266E-03	8.833E-02	1.137E+00
0.09	0.33	13.	8.334E-01	1.119E-03	8.641E-02	9.210E-01
0.10	0.37	15.	6.745E-01	9.978E-04	8.436E-02	7.599E-01
0.20	0.74	29.	1.610E-01	4.940E-04	7.192E-02	2.335E-01
0.30	1.11	44.	7.063E-02	3.425E-04	6.600E-02	1.370E-01
0.40	1.48	58.	4.008E-02	2.648E-04	6.263E-02	1.030E-01
0.50	1.85	73.	2.600E-02	2.163E-04	6.008E-02	8.630E-02
0.60	2.22	87.	1.795E-02	1.835E-04	5.852E-02	7.665E-02
0.80	2.96	117.	9.375E-03	1.419E-04	5.627E-02	6.579E-02
1.00	3.70	146.	4.956E-03	1.165E-04	5.416E-02	5.924E-02
1.25	4.63	182.	1.809E-03	9.573E-05	5.227E-02	5.418E-02
1.50	5.56	219.	3.940E-04	8.170E-05	5.000E-02	5.048E-02
1.75	6.48	255.	5.627E-05	7.159E-05	4.827E-02	4.840E-02
2.00	7.41	292.	3.438E-06	6.395E-05	4.679E-02	4.686E-02
2.50	9.26	365.	0.000E+00	5.340E-05	4.411E-02	4.416E-02
3.00	11.11	437.	0.000E+00	4.649E-05	4.189E-02	4.194E-02
3.50	12.96	510.	0.000E+00	4.151E-05	3.937E-02	3.941E-02
4.00	14.81	583.	0.000E+00	3.764E-05	3.732E-02	3.735E-02
4.50	16.67	656.	0.000E+00	3.452E-05	3.551E-02	3.554E-02
5.00	18.52	729.	0.000E+00	3.192E-05	3.367E-02	3.370E-02
6.00	22.22	875.	0.000E+00	2.775E-05	3.058E-02	3.060E-02
8.00	29.63	1167.	0.000E+00	2.212E-05	2.586E-02	2.588E-02
10.00	37.04	1458.	0.000E+00	1.819E-05	2.207E-02	2.209E-02

TABLE 8

DAILY DOSE AT CENTER OF ALUMINUM SPHERES FOR ALUMINUM TARGETS   SOLAR MINIMUM
60 DEG/300KM

S GM/SQCM	T MM	T MILS	ELEC RADS-AL	BREM RADS-AL	PROTON RADS-AL	TOTAL RADS-AL
0.01	0.04	1.	1.477E+02	5.825E-02	2.636E+01	1.741E+02
0.02	0.07	3.	8.142E+01	4.093E-02	1.036E+01	9.183E+01
0.03	0.11	4.	5.238E+01	3.205E-02	6.345E+00	5.876E+01
0.04	0.15	6.	3.715E+01	2.675E-02	4.318E+00	4.150E+01
0.05	0.19	7.	2.808E+01	2.314E-02	3.077E+00	3.110E+01
0.06	0.22	9.	2.227E+01	2.042E-02	2.412E+00	2.470E+01
0.07	0.26	10.	1.827E+01	1.816E-02	1.939E+00	2.023E+01
0.08	0.30	12.	1.538E+01	1.655E-02	1.618E+00	1.701E+01
0.09	0.33	13.	1.320E+01	1.514E-02	1.389E+00	1.460E+01
0.10	0.37	15.	1.148E+01	1.396E-02	1.202E+00	1.270E+01
0.20	0.74	29.	4.732E+00	8.540E-03	5.255E-01	5.266E+00
0.30	1.11	44.	2.844E+00	6.564E-03	3.386E-01	3.189E+00
0.40	1.48	58.	1.909E+00	5.400E-03	2.497E-01	2.164E+00
0.50	1.85	73.	1.338E+00	4.575E-03	1.984E-01	1.540E+00
0.60	2.22	87.	9.592E-01	3.959E-03	1.707E-01	1.134E+00
0.80	2.96	117.	5.159E-01	3.118E-03	1.380E-01	6.570E-01
1.00	3.70	146.	2.882E-01	2.592E-03	1.172E-01	4.080E-01
1.25	4.63	182.	1.425E-01	2.166E-03	1.032E-01	2.479E-01
1.50	5.56	219.	7.123E-02	1.877E-03	9.290E-02	1.660E-01
1.75	6.48	255.	3.441E-02	1.667E-03	8.616E-02	1.222E-01
2.00	7.41	292.	1.636E-02	1.508E-03	8.104E-02	9.891E-02
2.50	9.26	365.	2.205E-03	1.290E-03	7.252E-02	7.602E-02
3.00	11.11	437.	2.262E-04	1.150E-03	6.614E-02	6.751E-02
3.50	12.96	510.	9.704E-07	1.051E-03	6.014E-02	6.119E-02
4.00	14.81	583.	0.000E+00	9.756E-04	5.559E-02	5.656E-02
4.50	16.67	656.	0.000E+00	9.137E-04	5.175E-02	5.267E-02
5.00	18.52	729.	0.000E+00	8.605E-04	4.808E-02	4.894E-02
6.00	22.22	875.	0.000E+00	7.715E-04	4.208E-02	4.285E-02
8.00	29.63	1167.	0.000E+00	6.433E-04	3.348E-02	3.412E-02
10.00	37.04	1458.	0.000E+00	5.493E-04	2.725E-02	2.780E-02

TABLE 9

DAILY DOSE AT CENTER OF ALUMINUM SPHERES FOR ALUMINUM TARGETS
90DEG/300KM
SOLAR MINIMUM

S GMSQCM	T MM	T MILS	ELEC RADS-AL	BREM RADS-AL	PROTON RADS-AL	TOTAL RADS-AL
0.01	0.04	1.	1.370E+02	5.460E-02	1.799E+01	1.551E+02
0.02	0.07	3.	7.918E+01	3.995E-02	7.540E+00	8.676E+01
0.03	0.11	4.	5.321E+01	3.216E-02	4.707E+00	5.794E+01
0.04	0.15	6.	3.918E+01	2.740E-02	3.230E+00	4.244E+01
0.05	0.19	7.	3.058E+01	2.412E-02	2.314E+00	3.291E+01
0.06	0.22	9.	2.486E+01	2.159E-02	1.820E+00	2.670E+01
0.07	0.26	10.	2.080E+01	1.956E-02	1.466E+00	2.229E+01
0.08	0.30	12.	1.778E+01	1.788E-02	1.225E+00	1.902E+01
0.09	0.33	13.	1.544E+01	1.649E-02	1.053E+00	1.651E+01
0.10	0.37	15.	1.356E+01	1.530E-02	9.123E-01	1.449E+01
0.20	0.74	29.	5.650E+00	9.501E-03	4.011E-01	6.068E+00
0.30	1.11	44.	3.321E+00	7.277E-03	2.599E-01	3.588E+00
0.40	1.48	58.	2.179E+00	5.951E-03	1.929E-01	2.378E+00
0.50	1.85	73.	1.494E+00	5.011E-03	1.542E-01	1.653E+00
0.60	2.22	87.	1.047E+00	4.310E-03	1.334E-01	1.185E+00
0.80	2.96	117.	5.408E-01	3.373E-03	1.086E-01	6.528E-01
1.00	3.70	146.	2.923E-01	2.795E-03	9.247E-02	3.876E-01
1.25	4.63	182.	1.392E-01	2.333E-03	8.153E-02	2.230E-01
1.50	5.56	219.	6.663E-02	2.023E-03	7.337E-02	1.420E-01
1.75	6.48	255.	3.068E-02	1.799E-03	6.805E-02	1.005E-01
2.00	7.41	292.	1.393E-02	1.631E-03	6.403E-02	7.958E-02
2.50	9.26	365.	1.749E-03	1.399E-03	5.729E-02	6.044E-02
3.00	11.11	437.	1.715E-04	1.249E-03	5.225E-02	5.367E-02
3.50	12.96	510.	7.036E-07	1.143E-03	4.753E-02	4.868E-02
4.00	14.81	583.	0.000E+00	1.061E-03	4.397E-02	4.503E-02
4.50	16.67	656.	0.000E+00	9.934E-04	4.096E-02	4.195E-02
5.00	18.52	729.	0.000E+00	9.356E-04	3.805E-02	3.099E-02
6.00	22.22	875.	0.000E+00	8.389E-04	3.334E-02	3.418E-02
8.00	29.63	1167.	0.000E+00	6.997E-04	2.656E-02	2.726E-02
10.00	37.04	1458.	0.000E+00	5.975E-04	2.168E-02	2.227E-02

TABLE 10

DAILY DOSE AT CENTER OF ALUMINUM SPHERES FOR ALUMINUM TARGETS
28.5DEG/500KM
SOLAR MINIMUM

S GM/SQCM	T MM	T MILS	ELEC RADS-AL	BREM RADS-AL	PROTON RADS-AL	TOTAL RADS-AL
0.01	0.04	1.00	2.494E+02	9.518E-02	1.805E+00	2.513E+02
0.02	0.07	3.00	1.417E+02	6.745E-02	1.629E+00	1.434E+02
0.03	0.11	4.00	8.896E+01	5.151E-02	1.540E+00	9.055E+01
0.04	0.15	6.00	6.008E+01	4.162E-02	1.472E+00	6.160E+01
0.05	0.19	7.00	4.266E+01	3.479E-02	1.411E+00	4.410E+01
0.06	0.22	9.00	3.142E+01	2.959E-02	1.373E+00	3.283E+01
0.07	0.26	10.00	2.387E+01	2.552E-02	1.334E+00	2.523E+01
0.08	0.30	12.00	1.859E+01	2.227E-02	1.300E+00	1.991E+01
0.09	0.33	13.00	1.477E+01	1.967E-02	1.268E+00	1.606E+01
0.10	0.37	15.00	1.195E+01	1.753E-02	1.236E+00	1.320E+01
0.20	0.74	29.00	2.781E+00	8.622E-03	1.032E+00	3.821E+00
0.30	1.11	44.00	1.191E+00	5.951E-03	9.200E-01	2.117E+00
0.40	1.48	58.00	6.660E-01	4.589E-03	8.468E-01	1.517E+00
0.50	1.85	73.00	4.268E-01	3.742E-03	7.885E-01	1.219E+00
0.60	2.22	87.00	2.915E-01	3.170E-03	7.501E-01	1.045E+00
0.80	2.96	117.00	1.497E-01	2.451E-03	6.962E-01	8.483E-01
1.00	3.70	146.00	7.071E-02	2.012E-03	6.534E-01	7.341E-01
1.25	4.63	182.00	2.060E-02	1.654E-03	6.189E-01	6.492E-01
1.50	5.56	219.00	6.463E-03	1.412E-03	5.844E-01	5.923E-01
1.75	6.48	255.00	1.053E-03	1.238E-03	5.595E-01	5.617E-01
2.00	7.41	292.00	1.032E-04	1.105E-03	5.386E-01	5.398E-01
2.50	9.26	365.00	0.000E+00	9.225E-04	5.008E-01	5.017E-01
3.00	11.11	437.00	0.000E+00	8.022E-04	4.704E-01	4.712E-01
3.50	12.96	510.00	0.000E+00	7.156E-04	4.386E-01	4.393E-01
4.00	14.81	583.00	0.000E+00	6.485E-04	4.140E-01	4.146E-01
4.50	16.67	656.00	0.000E+00	5.942E-04	3.923E-01	3.929E-01
5.00	18.52	729.00	0.000E+00	5.490E-04	3.710E-01	3.715E-01
6.00	22.22	875.00	0.000E+00	4.770E-04	3.345E-01	3.350E-01
8.00	29.63	1167.00	0.000E+00	3.795E-04	2.797E-01	2.801E-01
10.00	37.04	1458.00	0.000E+00	3.118E-04	2.381E-01	2.384E-01

TABLE 11

DAILY DOSE AT CENTER OF ALUMINUM SPHERES FOR ALUMINUM TARGETS 60DEG/500KM SOLAR MINIMUM

S GM/SQCM	T MM	T MILS	ELEC RADS-AL	BREM RADS-AL	PROTON RADS-AL	TOTAL RADS-AL
0.01	0.04	1.00	4.304E+02	1.679E-01	6.493E+01	4.955E+02
0.02	0.07	3.00	2.406E+02	1.187E-01	2.498E+01	2.657E+02
0.03	0.11	4.00	1.539E+02	9.230E-02	1.545E+01	1.694E+02
0.04	0.15	6.00	1.076E+02	7.633E-02	1.071E+01	1.184E+02
0.05	0.19	7.00	7.992E+01	6.540E-02	7.814E+00	8.780E+01
0.06	0.22	9.00	6.208E+01	5.710E-02	6.268E+00	6.841E+01
0.07	0.26	10.00	4.989E+01	5.060E-02	5.159E+00	5.510E+01
0.08	0.30	12.00	4.115E+01	4.537E-02	4.395E+00	4.559E+01
0.09	0.33	13.00	3.468E+01	4.110E-02	3.844E+00	3.857E+01
0.10	0.37	15.00	2.967E+01	3.756E-02	3.389E+00	3.310E+01
0.20	0.74	29.00	1.090E+01	2.170E-02	1.675E+00	1.260E+01
0.30	1.11	44.00	6.208E+00	1.625E-02	1.164E+00	7.389E+00
0.40	1.48	58.00	4.088E+00	1.316E-02	9.099E-01	5.011E+00
0.50	1.85	73.00	2.833E+00	1.104E-02	7.575E-01	3.601E+00
0.60	2.22	87.00	2.008E+00	9.501E-03	6.737E-01	2.691E+00
0.80	2.96	117.00	1.056E+00	7.438E-03	5.723E-01	1.636E+00
1.00	3.70	146.00	5.778E-01	6.156E-03	5.052E-01	1.089E+00
1.25	4.63	182.00	2.756E-01	5.123E-03	4.589E-01	7.396E-01
1.50	5.56	219.00	1.309E-01	4.425E-03	4.216E-01	5.570E-01
1.75	6.48	255.00	6.178E-02	3.921E-03	3.964E-01	4.621E-01
2.00	7.41	292.00	2.811E-02	3.540E-03	3.764E-01	4.081E-01
2.50	9.26	365.00	4.293E-03	3.016E-03	3.411E-01	3.484E-01
3.00	11.11	437.00	4.175E-04	2.677E-03	3.140E-01	3.171E-01
3.50	12.96	510.00	8.088E-06	2.436E-03	2.882E-01	2.907E-01
4.00	14.81	583.00	0.000E+00	2.251E-03	2.689E-01	2.711E-01
4.50	16.67	656.00	0.000E+00	2.099E-03	2.524E-01	2.545E-01
5.00	18.52	729.00	0.000E+00	1.970E-03	2.364E-01	2.384E-01
6.00	22.22	875.00	0.000E+00	1.756E-03	2.098E-01	2.116E-01
8.00	29.63	1167.00	0.000E+00	1.453E-03	1.706E-01	1.721E-01
10.00	37.04	1458.00	0.000E+00	1.233E-03	1.421E-01	1.433E-01

TABLE 12

DAILY DOSE AT CENTER OF ALUMINUM SPHERES FOR ALUMINUM TARGETS SOLAR MINIMUM
90DEG/500KM

S GM/SQCM	T MM	T MILS	ELEC RADS-AL	BREM RADS-AL	PROTON RADS-AL	TOTAL RADS-AL
0.01	0.04	1.00	3.693E+02	1.449E-01	4.279E+01	4.123E+02
0.02	0.07	3.00	2.123E+02	1.049E-01	1.725E+01	2.297E+02
0.03	0.11	4.00	1.395E+02	8.307E-02	1.090E+01	1.504E+02
0.04	0.15	6.00	9.986E+01	6.959E-02	7.660E+00	1.076E+02
0.05	0.19	7.00	7.573E+01	6.027E-02	5.652E+00	8.144E+01
0.06	0.22	9.00	5.986E+01	5.315E-02	4.573E+00	6.449E+01
0.07	0.26	10.00	4.885E+01	4.748E-02	3.789E+00	5.269E+01
0.08	0.30	12.00	4.079E+01	4.288E-02	3.247E+00	4.408E+01
0.09	0.33	13.00	3.471E+01	3.907E-02	2.855E+00	3.761E+01
0.10	0.37	15.00	2.997E+01	3.586E-02	2.529E+00	3.254E+01
0.20	0.74	29.00	1.127E+01	2.103E-02	1.287E+00	1.258E+01
0.30	1.11	44.00	6.340E+00	1.573E-02	9.079E-01	7.263E+00
0.40	1.48	58.00	4.096E+00	1.270E-02	7.170E-01	4.826E+00
0.50	1.85	73.00	2.789E+00	1.062E-02	6.008E-01	3.400E+00
0.60	2.22	87.00	1.948E+00	9.110E-03	5.356E-01	2.493E+00
0.80	2.96	117.00	1.001E+00	7.104E-03	4.559E-01	1.464E+00
1.00	3.70	146.00	5.370E-01	5.874E-03	4.019E-01	9.448E-01
1.25	4.63	182.00	2.496E-01	4.890E-03	3.644E-01	6.189E-01
1.50	5.56	219.00	1.148E-01	4.227E-03	3.351E-01	4.541E-01
1.75	6.48	255.00	5.203E-02	3.751E-03	3.153E-01	3.711E-01
2.00	7.41	292.00	2.264E-02	3.392E-03	3.003E-01	3.263E-01
2.50	9.26	365.00	3.159E-03	2.896E-03	2.740E-01	2.800E-01
3.00	11.11	437.00	2.841E-04	2.573E-03	2.538E-01	2.567E-01
3.50	12.96	510.00	4.912E-06	2.344E-03	2.340E-01	2.363E-01
4.00	14.81	583.00	0.000E+00	2.168E-03	2.187E-01	2.209E-01
4.50	16.67	656.00	0.000E+00	2.022E-03	2.056E-01	2.076E-01
5.00	18.52	729.00	0.000E+00	1.899E-03	1.927E-01	1.946E-01
6.00	22.22	875.00	0.000E+00	1.694E-03	1.713E-01	1.730E-01
8.00	29.63	1167.00	0.000E+00	1.402E-03	1.396E-01	1.410E-01
10.00	37.04	1458.00	0.000E+00	1.191E-03	1.165E-01	1.177E-01

TABLE 13

DOSE FOR 1 ANOMALOUSLY LARGE SOLAR FLARE PROTON EVENT IN GEOSTATIONARY ORBITS
(ALUMINUM SHIELDS AND TARGETS)

S (GM/SQCM)	MAGNETOSPHERICALLY UNATTENUATED		MAGNETOSPHERICALLY ATTENUATED TO 70DEG W		MAGNETOSPHERICALLY ATTENUATED TO 160DEG W	
	SLAB(4PI) (RADS-AL)	SOLID SPHERE (RADS-AL)	SLAB(4PI) (RADS-AL)	SOLID SPHERE (RADS-AL)	SLAB(4PI) (RADS-AL)	SOLID SPHERE (RADS-AL)
0.01	4.582E+03	4.490E+03	4.674E+02	4.661E+02	2.612E+02	2.607E+02
0.02	4.654E+03	4.536E+03	4.688E+02	4.660E+02	2.618E+02	2.606E+02
0.03	4.704E+03	4.579E+03	4.702E+02	4.661E+02	2.622E+02	2.606E+02
0.04	4.738E+03	4.627E+03	4.716E+02	4.663E+02	2.628E+02	2.607E+02
0.05	4.760E+03	4.677E+03	4.728E+02	4.665E+02	2.634E+02	2.607E+02
0.06	4.770E+03	4.738E+03	4.740E+02	4.667E+02	2.638E+02	2.608E+02
0.08	4.764E+03	4.836E+03	4.764E+02	4.673E+02	2.648E+02	2.610E+02
0.10	4.722E+03	5.102E+03	4.786E+02	4.680E+02	2.658E+02	2.612E+02
0.20	3.918E+03	5.751E+03	4.874E+02	4.727E+02	2.698E+02	2.628E+02
0.30	3.132E+03	5.009E+03	4.936E+02	4.778E+02	2.730E+02	2.647E+02
0.40	2.620E+03	4.334E+03	4.982E+02	4.825E+02	2.754E+02	2.668E+02
0.50	2.248E+03	3.872E+03	5.016E+02	4.871E+02	2.774E+02	2.687E+02
0.60	1.961E+03	3.476E+03	5.040E+02	4.923E+02	2.790E+02	2.705E+02
0.70	1.735E+03	3.146E+03	5.056E+02	4.980E+02	2.802E+02	2.723E+02
0.80	1.551E+03	2.901E+03	5.062E+02	5.039E+02	2.812E+02	2.743E+02
0.90	1.396E+03	2.667E+03	5.062E+02	5.100E+02	2.820E+02	2.765E+02
1.00	1.267E+03	2.456E+03	5.054E+02	5.164E+02	2.826E+02	2.787E+02
1.25	1.014E+03	2.087E+03	5.006E+02	5.350E+02	2.828E+02	2.847E+02
1.50	8.312E+02	1.755E+03	4.918E+02	5.547E+02	2.818E+02	2.912E+02
1.75	6.958E+02	1.556E+03	4.792E+02	5.853E+02	2.798E+02	2.984E+02
2.00	5.842E+02	1.371E+03	4.624E+02	6.000E+02	2.766E+02	3.064E+02
2.50	4.328E+02	1.013E+03	4.144E+02	7.641E+02	2.670E+02	3.250E+02
3.00	3.346E+02	8.570E+02	3.328E+02	8.348E+02	2.530E+02	3.535E+02
3.50	2.592E+02	6.861E+02	2.592E+02	6.829E+02	2.336E+02	3.831E+02
4.00	2.086E+02	5.585E+02	2.086E+02	5.592E+02	2.066E+02	4.853E+02
4.50	1.693E+02	4.844E+02	1.693E+02	4.843E+02	1.693E+02	4.891E+02
5.00	1.385E+02	4.057E+02	1.385E+02	4.057E+02	1.385E+02	4.044E+02
6.00	9.630E+01	3.000E+02	9.630E+01	3.000E+02	9.630E+01	3.003E+02
8.00	4.998E+01	1.658E+02	4.998E+01	1.658E+02	4.998E+01	1.658E+02
10.00	2.802E+01	1.082E+02	2.802E+01	1.082E+02	2.802E+01	1.082E+02
12.00	1.626E+01	6.634E+01	1.626E+01	6.634E+01	1.626E+01	6.634E+01
13.50	1.125E+01	4.717E+01	1.125E+01	4.717E+01	1.125E+01	4.717E+01
14.00	1.001E+01	4.240E+01	1.001E+01	4.240E+01	1.001E+01	4.240E+01

TABLE 14

TRANSIENT X-RAYS, GAMMA-RAYS AND NEUTRONS IN SPACE

R. E. Lingenfelter
Center for Astrophysics and Space Sciences
University of California, San Diego, CA 92093

ASTRACT

We survey the various astrophysical sources of transient x-ray, gamma-ray and neutron emission which can produce significant increases in the high energy (> 1 keV) radiation and particle background flux in space. These sources include solar flares, gamma-ray bursts, supernovae and other astrophysical transients. We review the observations of these sources and then discuss the expected rate of occurrence of their transient emission as a function of intensity and energy.

INTRODUCTION

One of the peculiarities of high energy astrophysics is that most of the known sources of hard x-rays, gamma-rays and neutrons are highly variable emitters, often seen only by their transient emission. Even the sun is observable at high energies solely in its transient emission during flares. The only significant steady x-ray and gamma ray emission is the diffuse cosmic and galactic background. Transient fluxes have exceeded this background by at least four orders of magnitude. The observations and theoretical studies of these transient sources, including solar flares, gamma-ray bursts, supernovae and a variety of other objects, have been the subject of several recent reviews[1-5]. Here we discuss only the contribution of these sources to the high energy radiation background in nearby space. In particular we review measurements of the fluxes, durations, spectra and frequency of transient sources. Based on these measurements, we discuss the expected rate of occurrence of such transients as a function of their intensity and energy. For comparison we also review measurements of the steady, or at least slowly varying, background flux from persistent sources. Lastly, we discuss the need for the future monitoring and study of these transients and what we can hope to learn from such observations.

X-RAY AND GAMMA-RAY TRANSIENTS

Solar Flares - The Sun is the most intense source of transient x-ray flux at energies below about 30 keV, and it is the most intense source of persistent, extraterrestrial emission below several keV. Solar emission in the 1.5 to 12.5 keV has been measured almost continuously since 1975 by the SMS and GOES satellites. These measurements[6] show that the monthly-average background energy flux has varied from around 3×10^{-5} ergs/cm^{2}s during solar minumum to around 2×10^{-3} ergs/cm^{2}s during solar maximum. This emission is thermal bremsstrahlung from hot plasma at temperatures of the order of a few times 10^6 K. Measurements[7-8] of the energy spectrum of the background energy flux can be approximated by an exponential in photon energy with an e-folding energy of 0.3 keV during both solar minimum and maximum.

Superimposed on this background are flares lasting from about 10^2 to 10^4 s with peak energy fluxes extending up to at least 3×10^{-1} ergs/cm²s. The integral size-frequency distribution of the peak energy flux, P, in the 1.5 to 12.5 keV band from the flares observed[6] over roughly half a solar cycle from January 1977 through September 1981 can be approximated by $N(>P) = 0.25\, P^{-1.3}$ flares/yr with peak energy fluxes greater than P in ergs/cm²s above 1.5 keV. This distribution is shown in Figure 1, together with the statistical uncertainty of the current measurements.

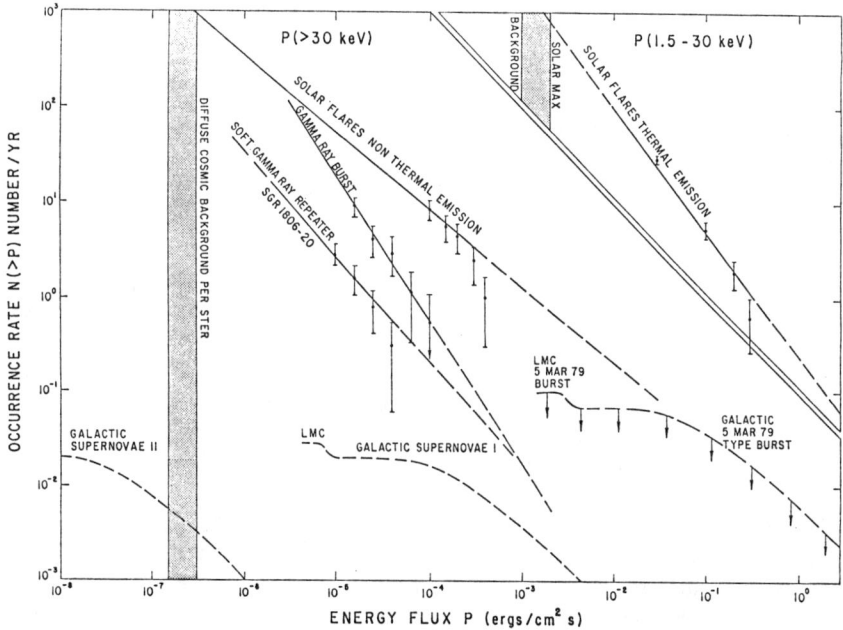

Figure 1. The occurrence rate of x-ray and gamma-ray transients with peak energy fluxes greater than P ergs/cm²s above photon energies of 1.5 and 30 keV. The solid lines represent analytical approximations to the measured rates over the range of the observations and the dashed lines are extrapolations. Data points show the measured values for the larger transients together with their statistical uncertainty. The sources of the data are given in the text. The persistent background flux is also indicated.

As can be seen, the distribution is determined only up to a peak flux of 3×10^{-1} ergs/cm²s, corresponding to a rate of 1 per year. An extrapolation of this distribution to more energetic flares would suggest that peak energy fluxes of 2 ergs/cm²s might occur once every 10 years and about 10 ergs/cm²s once every 10^2 years. But the actual observations already fall slightly below the simple power law distribution at 3×10^{-1} ergs/cm²s so that such an extrapolation may significantly overestimate the occurrence rate of larger flares. Analysis of data from the second half of the last solar cycle, which doubles the period of observation, should provide a better statistical sample of large flares, but continued measurements are required.

The flare emission in this energy range is also thermal bremsstrahlung from hot plasmas, but at much higher temperatures which reach as much as a few times 10^7

K. Measurements[8] of the spectrum of the peak energy flux of the more intense flares can also be approximated by exponentials in photon energy with an e-folding energy of about 3 keV. Nominal flare spectra of differing peak energy fluxes are shown in Figure 2 together with their expected occurrence rates and the underlying solar background flux at solar maximum and minimum described above. The differential energy flux shown here is equal to E^2 times the photon flux and has the units of ergs/cm^{2}s lnE. A constant value corresponds to equal an energy flux per logarithmic energy interval. As can be seen, this thermal emission dominates the transient fluxes up to about 30 keV.

At energies above 30 keV these same flares are also a major source of transient flux with predominantly nonthermal emission extending up to energies of more than 100 MeV, and generally lasting for 10^2 to 10^3s. This emission is primarily bremsstrahlung from flare accelerated electrons and the spectra can be approximated by simple inverse power laws in photon energy[9-10]. The spectral index of the peak flux is typically about -4, but the indices in different flares range from around -2 to as much as -10. In addition, however, the spectra of many of the larger flares also show contributions from thermal bremsstrahlung in superhot (several times 10^7 °K) plasmas and from nuclear line emission, including nuclear deexcitation, radioactive decay, positron annihilation, neutron capture and pion decay, all resulting from the interactions of accelerated ions.

The most extensive observations[10-11] in this energy range are those of the solar photon count rate above 30 keV by the Hard X-Ray Burst Spectrometer (HXRBS) on the SMM satellite from February 1980 to the present. Over 7000 flares were observed during the first five years of observation covering the last half of the last solar cycle. The differential size-frequency distribution of these flares follows almost exactly an inverse power law in size with an index of -1.8 for at least three decades in size. This same index was found previously[9,12] from smaller samples of flares which included microflares an order of magnitude less intense. Converting the peak count rates to peak energy fluxes, assuming an effective area[13] of 36 cm^2 for the detector and an average spectral index[10] of -4 for the peak flux, we find that the integral size-frequency distribution is approximately $N(>P) = 5.5\times10^{-3}\, P^{-0.8}$ flares/yr with peak energy fluxes greater than P in ergs/cm^{2}s above 30 keV.

This distribution is shown in Figure 1 together with the statistical uncertainty of the current measurements[11]. This distribution is much flatter than that for the thermal emission below 30 keV. Moreover, since it falls less rapidly than P^{-1}, it means that, if the time-integrated energy fluxes, or fluences, from these flares are proportional to the peak flux, then the time-integrated energy flux from just the few largest flares exceeds that from all of the other flares combined. If so, such a distribution obviously cannot extend to arbitarily large fluxes. As can be seen in Figure 1, the observed occurrence rate of the largest flares already seems to be falling below that expected from the power law distribution for peak energy fluxes $> 3\times10^{-4}$ ergs/cm^{2}s. Thus although an extrapolation of the power law would suggest that flares with peak fluxes of about 1.5×10^{-3} ergs/cm^{2}s might occur once a year and about 3×10^{-2} ergs/cm^{2}s every 10 years, these rates may be much too high. Clearly further monitoring of the largest flares over a longer period of time is needed in order to determine exactly where the distribution begins to break away from the well-defined power law, how it behaves above that point, and what the maximum hard x-ray luminosity may be.

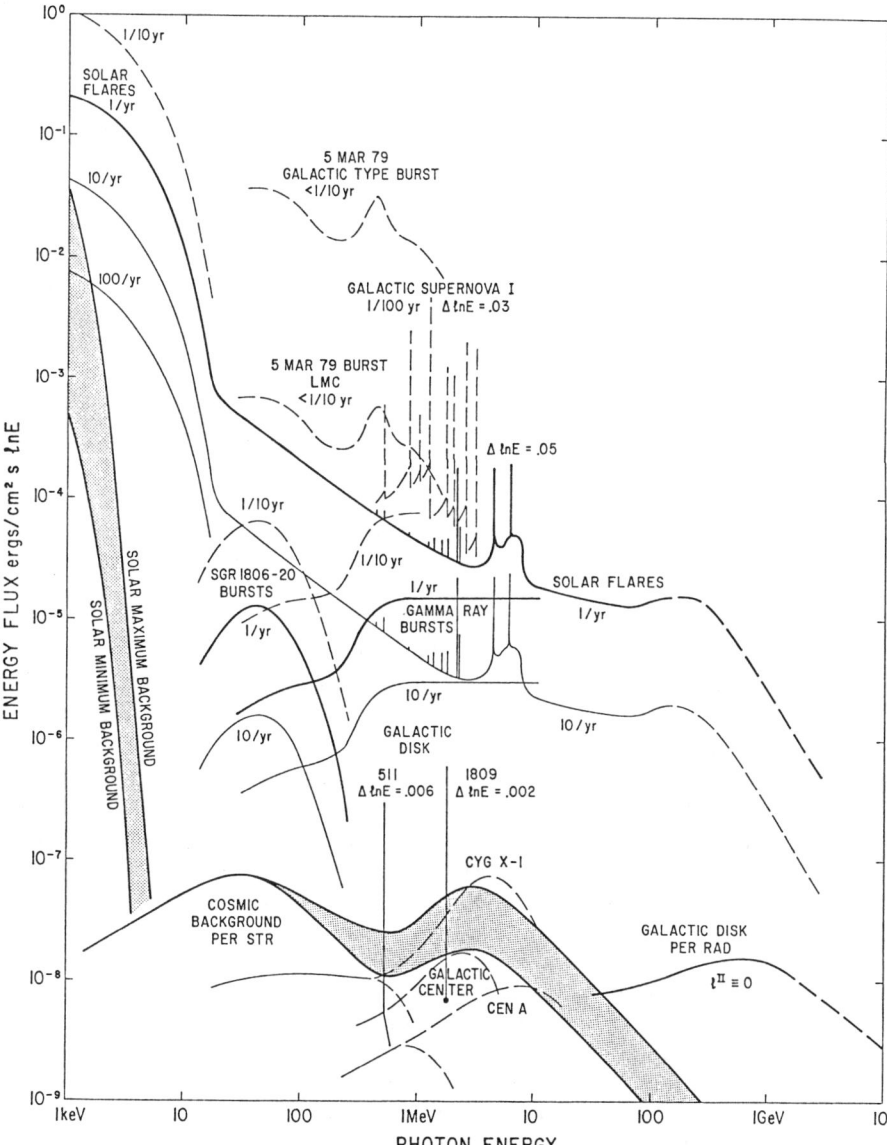

Figure 2. Nominal spectra for x-ray and gamma-ray transients corresponding to the peak energy fluxes expected for different frequencies of occurrence. The differential energy flux with units of ergs/cm²s lnE is equal to E^2 times the photon flux. The spectra of the principal sources of persistent background flux are also shown.

It is also important to recognize that the size-frequency distribution shown in Figure 1 is essentially the solar-cycle averaged rate. The occurrence rate on shorter times scales can vary by more than an order of magnitude from solar minimum to solar maximum and also shows significant variation over a newly discovered[10] period of 152 to 158 days.

The nominal energy spectra of these flares are also shown in Figure 2 for differing energy fluxes together with their expected occurrence rates. The spectrum for 1/yr corresponds to that of the flare of 3 June 1982 which had a peak energy flux of about 7×10^{-4} ergs/cm^{2}s. This is a conservative estimate in that this is the peak flux actually observed at that occurrence rate, and is only half of that expected from an extrapolation of the size-frequency distribution. The power law spectral index of -2.7 for the photon flux was determined from the combined HXRBS measurements[11] > 30 keV and the Gamma Ray Spectrometer (GRS) measurements[14] > 300 keV. The gamma ray lines fluxes, such as the prominent 4.44 and 6.13 MeV lines from ^{12}C* and ^{16}O* deexcitation, the 2.2 MeV line from neutron capture on H, the 0.511 MeV line from positron annihilation and the broad pion decay feature above about 50 MeV are also taken from the GRS measurements[15]. However, the deexcitation line widths ($\Delta \ln E$ = 0.05), which were narrower than the GRS energy resolution, and the extension of the pion decay spectrum above the measured values at 150 MeV are taken from theoretical calculations[5]. The 0.511 and 2.2 MeV lines are expected to be even much narrower than the deexcitation lines, but for simplicity they are shown here with the same relative widths. The weaker deexcitation lines from N, Ne, Mg, Si and Fe are also taken from theoretical calculations. Nominal spectra for other values of the energy flux are simply scaled to this, using the measured size-frequency distribution, although the typical spectra of the less intense flares many be significantly softer.

Gamma Ray Bursts - Gamma-ray bursts are the next most important source of transient gamma-ray flux. Several hundred bursts at energies > 30 keV have been observed[16] over the past 20 years, reaching peak energy fluxes of > 10^{-3} ergs/cm^{2}s. These bursts appear to belong to at least two distinct spectral and temporal classes: the soft repeating bursts with very short ($\sim$ 0.1 s) durations and very soft ($\sim$ 50 keV) spectra, and the more common, hard bursts which last from 1 sec to > 100 s, often with complex time dependences and spectra that may extend to > 10 MeV. Source positions have been determined for over 100 bursts, yet, with only one exception (the supernova remnant N49 in the Large Magellanic Cloud), none of these positions coincide with any known source. A neutron star origin, however, is suggested by apparent absorption and emission features, which have been interpreted as cyclotron absorption in 10^{12} gauss magnetic fields and gravitationally redshifted positron annihilation radiation, respectively. A variety of energy sources have been proposed, including sporadic accretion onto the surface of a neutron star, thermonuclear runaway of the accreted matter, and neutron star quakes, perhaps driven by accretion. A comparable variety of emission processes have also been suggested, including gyrosynchrotron emission for the soft repeating bursts and the low energy (< 300 keV) component of the hard burst spectra, which also generally have an exponential form, and Compton scattering and positron annihilation radiation for the higher energy component of the spectra, which have a power law form.

Although gamma-ray bursts have been observed by a large number of spacecraft experiments, the only published[17-18] distribution of peak energy flux of the hard bursts is that obtained by the KONUS experiments on VENERA 11 and 12 between September 1978 and February 1980. When spectral selection biases against the detection of less intense bursts are taken into account[19], these measurements are consistent with a cumulative distribution, $N(> P) = 6 \times 10^{-7} P^{-1.5}$ bursts/yr, as shown in Figure 1. Such a $P^{-1.5}$ power law would be expected from a uniform spatial distribution of sources. The observed[17-18] isotropy of source positions on the sky also requires such a distribution. This is consistent with the sources being relatively nearby neutron stars in the galactic disk, lying within a scale height of a few 10^2 pc to perhaps a few kpc. If so, the size-frequency distribution should flatten at lower intensities to a P^{-1} power law and directions should become anisotropic with a concentration along the galactic equatorial plane as more distant bursts are observed. Measurements of bursts with peak energy fluxes $> 4 \times 10^{-7}$ ergs/cm^{2}s with balloon-borne detectors already suggest[20] such a break in the distribution, but more extensive observations are required to confirm it and determine its form.

From Figure 1 we see that a burst with a peak energy flux of about 6×10^{-5} ergs/cm^{2}s has been observed once a year. Extrapolating the power law size frequency distribution to more intense bursts, we also see that a burst of about 3×10^{-4} ergs/cm^{2}s would be expected once every 10 years and one of about 1.5×10^{-3} ergs/cm^{2}s every 10^2 years. These occurrence rates may be quite realistic because the observed rates of the larger bursts show no indication yet of deviating from the power law distribution.

Nominal energy flux spectra for these bursts are shown in Figure 2 for a two component spectrum, such as that observed[21] in the 25 March 1978 burst in which the bulk of the energy flux is in the hard power-law component > 300 keV. The power-law index of the differential photon spectrum from that burst was -2, typical of the values observed[22] in other bursts. As can be seen, observations are needed at higher energies in order to determine where and how the spectra start to steepen.

All of the reported soft repeating bursts come from just a few sources and their size-frequency distribution is at the present completely dominated by that of a single source, known as SGB 1806-20. This burster was only recently recognized[23-24] to be the source of > 100 repeating bursts observed between August 1978 and June 1986. Most of these bursts, however, were clustered in four groups during 1983-1984. The separation time between the bursts ranges from seconds to years, but it is typically less than a day. No correlation has been found between the intensities of the bursts and their separations in time.

The integral size-frequency distribution of peak energy flux for this source can be estimated from the observed[24] distribution of peak count rates, normalized to the energy flux > 30 keV for the most intense bursts and an effective observing time of 4.5 yr. This distribution can be approximated by $N(> P) = 9 \times 10^{-6} P^{-1.1}$ bursts/yr. As can be seen in Figure 1, on the average a burst with a peak energy flux of about 2×10^{-5} ergs/cm^{2}s was observed once a year, and a simple extrapolation of the power-law would suggest that a burst of about 2×10^{-4} ergs/cm^{2}s would be expected once every 10 years and one another order of magnitude larger every 100 years. Nominal energy flux spectra for these bursts are again shown in Figure 2. But any estimate of expected rates for this source is highly uncertain because of the highly nonrandom distribution of the burst in time.

Similar soft repeating bursts were observed[25] from the source GBS 0526-66, which appears to lie in the supernova remnant N49 in the Large Magellanic Cloud at a distance of 52 kpc. The typical fluxes are also an order of magnitude lower than those from SGR

1806-20, which would not be unexpected if the source was in the nearby galaxy. But GBS 0526-66 was also the source of the most intense burst yet observed[25], that of 5 March 1979. That burst had a peak energy flux of 3×10^{-3} ergs/cm^{2}s lasting 0.15 s, followed by weaker pulsed emission with an 8 s period. The peak flux of this burst was 10^3 times greater than that of the 15 later repeaters, all of which had peak fluxes within a factor of 3 of one another. The 5 March 1979 burst, therefore appears[26] to have been a unique event, caused by a different, much more powerful mechanism, possibly a neutron star quake[27]. Moreover, if the source is, in fact, in the Large Magellanic Cloud, then a similar burst on a source, such as SGR 1806-20, in our own galaxy could typically have a peak flux of the order of 0.1 ergs/cm^{2}s.

Only a crude upper limit can be set on the occurrence rate of such superbursts, however, since so far only one has been seen in roughly 10 years of observation by various satellite and spacecraft detection systems. The estimated limit for a galactic superburst, shown in Figure 1, is based on the assumption that the relative rates in the Large Magellanic Cloud and our galaxy are equal to the relative number of neutron stars, which are in turn equal to the relative rates of supernova explosions (0.4:1). The energy flux spectrum of the 5 March 1979 burst is shown in Figure 2, together with that expected from a similar burst in our galaxy.

Supernovae - Galactic supernovae, although rare, are the only other known source of potentially intense transient gamma-ray emission. This emission is expected to be primarily in nuclear line emission from the decay of radioactive ^{56}Co which has a mean life of 0.31 yr. The ^{56}Co results in turn from the decay of much shorter lived ^{56}Ni which is the principal end product of explosive nuclear burning. The peak energy flux from a supernova depends not only on its distance but on the mass of ^{56}Ni produced and on the mass of overlying ejecta which can greatly attenuate the gamma ray emission at early times. The principal lines from ^{56}Co decay are at 0.511, 0.847, 1.038, 1.238, 1.771, 2.015, 2.035, 2.599, 3.202 and 3.254 MeV. The relative line fluxes depend on the branching ratios of the decay and the differential attenuation by the ejecta.

The peak energy flux in the lines from Type I supernovae, which are generally thought to be exploding white dwarfs with about 0.5 M$_\odot$ of ^{56}Ni in about 1 M$_\odot$ of ejecta, is thus expected[28] to be much greater than that from Type II supernovae, which are thought to be explosions of much more massive (typically 10 to 20 M$_\odot$ stars forming ~ 1.5 M$_\odot$ neutron stars from most of their Ni-Fe cores and ejecting only about 0.1 M$_\odot$ of ^{56}Ni under all of the rest of the stellar mass. The expected[29] peak line energy fluxes from a Type I supernova at the nominal distance (~ 7 kpc) of the galactic center is shown in Figure 2, based on the current SNI model[30]. The line widths ($\Delta \ln E = 0.03$) reflect a mean expansion of 5000 km/s. The integrated energy flux in these lines is 4×10^{-4} ergs/cm^{2}s and the occurrence rate of a flux of this intensity or greater is roughly once per 100 years, assuming a total galactic supernova rate of one per 25 years, equally divided between Type I and II supernovae[31].

The peak energy fluxes of both Type I and II supernovae as a function of occurrence rate for a simple uniform distribution of supernovae in the galactic disk are shown in Figure 1. Also shown is the contribution of supernovae in the nearby Large Magellanic Cloud, which is about 0.4 of that in our galaxy. The expected[29] peak energy flux from Type II supernovae is based on a typical 15 M$_\odot$ star model[32]. As can be seen, the Type II fluxes are only about 3×10^{-4} of those expected from Type I supernovae, and the recent Type II Supernova 1987A in the Large Magellanic Cloud is likely to have a peak flux of $< 10^{-8}$ ergs/cm^{2}s.

It should be noted that since the peak flux from a Type I supernova occurring once every 100 years exceed the diffuse background flux by a factor of about 10^3 and lasts for a time comparable to the ^{56}Co decay mean life of 0.3 years, supernovae should be the dominant source of the long-term, time integrated flux, or fluence.

Other Transients and Background - Several other types of x-ray and gamma-ray transients of varying duration and intensity have been observed[1,2,4] from X-ray bursters, novae, pulsars, and accreting compact objects. But all have peak energy fluxes > 30 keV that are negligible compared to those of solar flares and gamma-ray bursts. Nonetheless, a few of the more intense, accreting black hole candidates, Cyg X-1, the Galactic Center and Cen A, each show[4] flux variations that can double the background flux above an Mev for periods of months to years, as can be seen in Figure 2.

The relative importance of such sources is not surprising, however, since it has been suggested[2,33-34] that most of the diffuse cosmic background may be just the superposition of a large number of similar, but as yet unresolved extragalactic sources, most likely Seyfert galaxies, BL Lac objects and quasars, known collectively as active galactic nuclei. The energy flux spectrum of the diffuse cosmic background has been measured[1-2] from a few keV to over 100 MeV. As can be seen in Figure 2, this spectrum consists of two distinct components with peak energy fluxes at a few tens of keV and a few MeV, suggesting two seperate classes of sources or distinct evloutionary phases. There is also considerable uncertainty in the fluxes above 100 keV which is be due in part to the difficulty of resolving discrete sources and subtracting their contributions. But for the present purpose of determining the overall background flux this is not an important problem.

The only other significant source of steady gamma-ray emission is the diffuse galactic emission, also shown in Figure 2. This emission is thought[2] to be produced primarily by cosmic ray interactions with both gas and photons in the interstellar medium. The cosmic ray protons produce pion-decay gamma rays and electrons produce bremsstrahlung and Compton-scattered gamma rays. The magnitude of this flux dependes strongly on galactic longitude and latitude, with the most of the emission coming from within $\pm 5°$ latitude of the galactic equator and $\pm 30°$ longitude of the galactic center. In addition there is diffuse gamma-ray line emission at 0.511 MeV from positron annihilation and at 1.809 Mev from the decay of ^{26}Al, both presumably[4] resulting from sources of galactic nucleosynthesis.

NEUTRON TRANSIENTS

Solar flares are the only detected source of extraterrestrial neutrons in space and the relatively short (918 s) neutron lifetime makes it highly unlikely that any more distant sources could be important. These neutrons are produced in flares by nuclear interactions of accelerated ions in the solar atmosphere. At the present, solar flare neutrons have now been observed[3,5] directly from at least three flares with detectors on SMM and with ground-based, cosmic-ray neutron monitors, and indirectly with instruments on IMP-8 and ISEE-3 which detected the protons from their decay. The peak energy flux spectrum from the largest observed flare, that of 3 June 1982, is shown in Figure 3, together with the cosmic-ray produced background leakage flux from the earth's atmosphere. The background neutron flux is based on balloon measurements[35] at a geomagnetic latitude of 40° during solar maximum and scaled by theoretical models[36–37] to give global average, polar and equatorial values for both solar maximum and minimum.

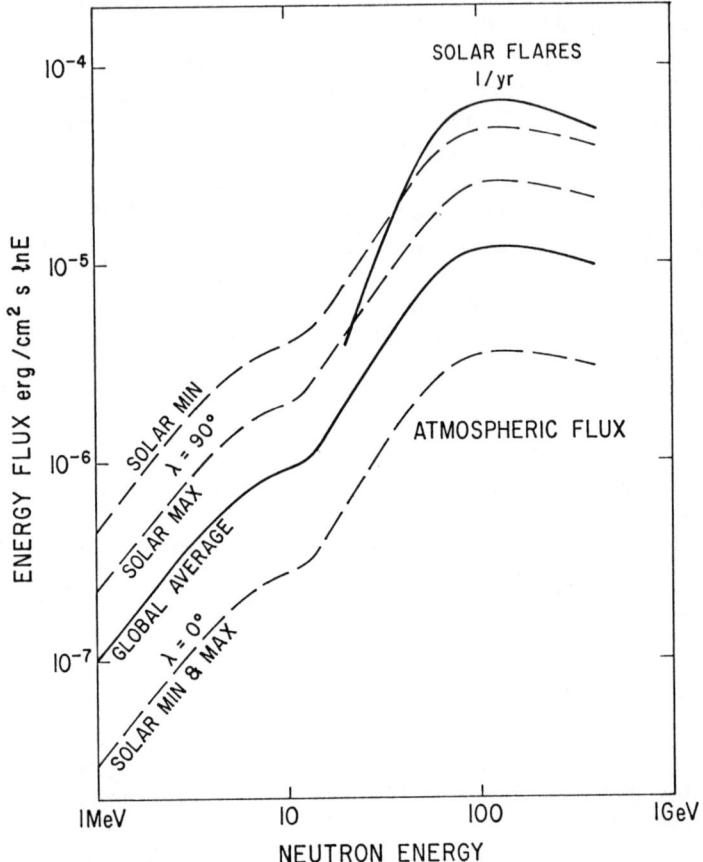

Figure 3. Nominal solar flare neutron peak energy spectrum together with the cosmic-ray produced background leakage flux from the earth's atmosphere.

As can be seen, this flux is only slightly larger than that background. Because of the poor statistics it is difficult to estimate the relative occurrence rate of flare neutron transients of larger or smaller magnitude. However, since the solar flare neutron fluxes should be roughly proportional to those of the solar flare gamma rays, and the 3 June 1982 flare was typical of the magnitude expected at least once a year, a neutron flux of this magnitude should also be expected with the same frequency. Thus the size frequency distribution for less intense flares should also scale proportionally, suggesting that the integral distribution $N(>P) \approx 6 \times 10^{-4} \, P^{-0.8}$ flares/yr with peak energy fluxes greater than P in ergs/cm²s. Extrapolating the power law to larger neutron transients, however, is likely to overestimate the occurrence rate because the observed occurrence rate of such flares in gamma rays seems to be falling below that expected from the power law distribution. Clearly further monitoring of solar flares with ground based neutron monitor and satellite experiments is needed.

CONCLUSIONS

The expected transient fluxes of x-rays, gamma rays and neutrons from the most important astrophysical sources, together with their spectra and their occurrence rates, are summarized in the figures. But these estimates should be considered only as preliminary, because much further work is still required. Not only are further, systematic observations needed to continuously monitor and measure transients in the range of energies, fluxes and durations already observed in order to obtain a better data base for analysis, but also new experiments are needed to extend observations to higher energies and longer time scales where we know virtually nothing of transient emission. In addition, we need to more thoroughly analyze existing data bases, which can already greatly improve our statistical sample. Finally, we need to further develop the theoretical models in order to make more informed prediction from the data gathered.

ACKNOWLEDGEMENTS

I would like to thank Brian Dennis, Richard Donnelly, Gary Heckman, Hugh Hudson, Reuven Ramaty, and Richard Rothschild for valuable discussions and NASA for financial support under grants NSG 7541 and NAG 5-945.

REFERENCES

1. S. S. Holt and R. McCray, *Ann. Rev. Astron. Astrophys.*, **20**, 323 (1982).
2. R. Ramaty and R. E. Lingenfelter, *Ann. Rev. Nucl. Part. Sci.*, **32**, 235 (1982).
3. E. L. Chupp, *Ann. Rev. Astron. Astrophys.*, **22**, 359 (1984).
4. R. E. Lingenfelter, *20th Internat. Cosmic Ray Conf. Papers*, **7**, in press (1988).
5. R. Ramaty and R. J. Murphy, *Space Sci. Rev.*, in press (1988).
6. S. D. Bouwer et al., *NOAA Tech. Mem. REL SEL-62* (1982).
7. J. A. Bowles et al., *Planet. Space Sci.*, **15**, 931 (1967).
8. R. W. Kreplin et al., in *The Solar Output and Its Variation*, ed. O. R. White (Colo. Assoc. Univ. Press, Boulder, 1977) p. 287.
9. D. W. Datlowe, M. Elcan and H. S. Hudson, *Solar Phys.*, **39**, 155.
10. B. R. Dennis, *Solar Phys.*, **100**, 465 (1985).
11. B. R. Dennis et al., *NASA Tech. Mem. 86236* (1985).
12. R. P. Lin et al., *Astrophys. J.*, **283**, 421 (1984).
13. B. R. Dennis, personal communication (1987).
14. W. T. Vestrand et al., *Astrophys. J.*, **322**, 1010 (1987).
15. D. J. Forrest et al., *19th Internat. Cosmic Ray Conf. Papers*, **4**, 126 (1985).
16. S. E. Woosley, ed. *High Energy Transients in Astrophysics*, (Am. Inst. Phys., N.Y., 1984) 715pp.
17. E. P. Mazets and S. V. Golenetskii, *Astrophys. Space Phys. Rev.*, **1**, 205, (1981).
18. E. P. Mazets, *19th Internat. Cosmic Ray Conf. Papers*, **9**, 415 (1986).
19. J. C. Higdon and R. E. Lingenfelter, *Astrophys. J.*, **307**, 197 (1986).
20. C. A. Meegan, G. J. Fishman and R. B. Wilson, in *High Energy Transients in*

Astrophysics, ed. S. E. Woosley, (Am. Inst. Phys., N.Y., 1984) p.422.
21. G. J. Hueter, Unpublished Dissertation, Univ. of California, San Diego (1987).
22. S. M. Matz, Unpublished Dissertation, Univ. of New Hampshire (1986).
23. J. L. Atteia et al. *Astrophys. J.*, **320**, L105 (1987).
24. J. G. Laros et al. *Astrophys. J.*, **320**, L111 (1987).
25. S. V. Golenetskii, V. N. Ilyinskii, and E. P. Mazets, *Nature*, **307**, 41 (1984).
26. T. L. Cline, in *Gamma Ray Transients and Related Astrophysical Phenomena*, eds. R. E. Lingenfelter et al., (Am. Inst. Phys., N.Y., 1982) p. 17.
27. R. Ramaty et al., *Nature*, **287**, 122 (1980).
28. S. E. Woosley and T. A. Weaver, *Proc. Tenth Texas Symposium on Relativistic Astrophysics, N.Y. Acad. Sci.* **375**, 357 (1981).
29. K. W. Chan and R. E. Lingenfelter, in preparation (1988).
30. K. Nomoto, H. K. Thielemann, and K. Yokoi, *Astrophys. J.*, **286**, 644 (1984).
31. V. Trimble, *Rev. Mod. Phys.*, **54**, 1183 (1982).
32. T. A. Weaver and S. E. Woosley, *Proc. Ninth Texas Symposium on Relativistic Astrophysics,*
    N.Y. Acad. Sci. **336**, 335 (1980).
33. R. Giacconi and G. Zamorini, *Astrophys. J.*, **313**, 20 (1987).
34. C. E. Fichtel, in *Supermassive Black Holes*, ed. M. Kafatos, (Am. Inst. Phys., N.Y., in press 1988).
35. A. M. Prezler, G. M. Simnett and R. S. White, *Phys. Rev. Letters*, **28**,982 (1972).
36. R. E. Lingenfelter, *J. Geophys. Res.*, **68**, 5633 (1963).
37. R. E. Lingenfelter, in *Spallation Nuclear Reactions and Their Applications*, eds. B. S. P. Shen and M. Merker, (Reidel, Dordrecht, 1976), p. 193.

APPROXIMATE ANGULAR DISTRIBUTION AND SPECTRA FOR GEOMAGNETICALLY TRAPPED PROTONS IN LOW-EARTH ORBIT

J. W. Watts and T. A. Parnell
ES62, NASA/Marshall Space Flight Center, AL 35812

H. H. Heckman
Lawrence Radiation Laboratory, University of California, Berkeley, CA 94720

ABSTRACT

The highly anisotropic nature of the radiation in the low-Earth orbit has been ignored for most spacecraft shielding calculations made to date because the standard environmental models describe the omnidirectional flux only, because the varying attitude of the spacecraft in the environment is assumed to average out the effect and because of the added complexity of the calculation. The Space Station is planned to be stabilized with respect to the velocity vector and local vertical. Thus it will pass through the South Atlantic Anomaly where most of the radiation flux is encountered in much the same attitude on each pass. Any calculation including a complex shielding geometry should thus consider the angular distribution of the incident radiation. An approximate trapped proton angular distribution is presented which includes both the "pan caked" distribution relative to the magnetic field direction and the east-west effect which is energy dependent. This distribution is then used with a planar shielding geometry to obtain an estimate of the effect of the anisotropy on radiation dose rates in spacecraft.

INTRODUCTION

The Space Station is planned for a 28.5° inclination circular orbit with an altitude in the range from 300 to 500 km. It will be attitude stabilized and oriented so that one end of the habitation modules always points along the velocity vector. For this orbital altitude and inclination two components contribute most of the penetrating charge particle radiation encountered—the galactic cosmic rays and the geomagnetically trapped Van Allen protons. Both sources are strongly modulated by the Earth's magnetic field. Here we confine our consideration to the trapped protons. Almost all the proton flux will be encountered in the region called the South Atlantic Anomaly produced because the Earth's magnetic field, though approximately dipolar, is not centered on the Earth. The protons follow a helical path about a magnetic field line as shown in Figure 1. As the field intensity increases, both the diameter and the pitch of the helix decrease until the pitch becomes zero. The point with zero pitch angle is called the mirror point and the center of the helical path is called the guiding center. From here the helix reverses direction and protons travel up the field line toward decreasing field intensity and away from the Earth. In the South Atlantic Anomaly almost all the protons observed are near their mirror points.

© 1989 American Institute of Physics

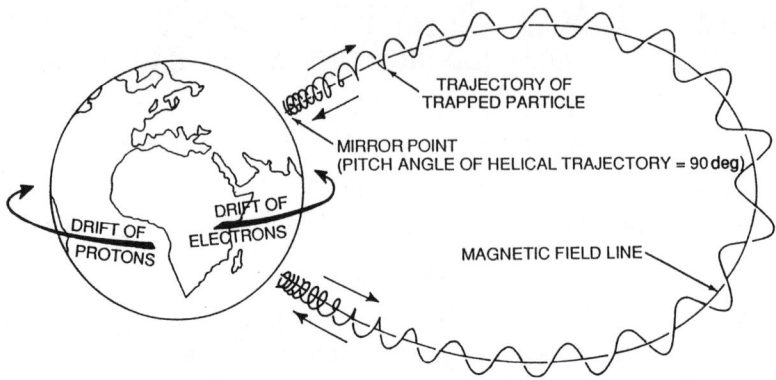

Fig 1. Path of trapped charged particles in the geomagnetic field.

Thus the flux is anisotropic with most of the flux arriving from a narrow band perpendicular to the local geomagnetic field direction. Any protons there which are not nearly mirroring will travel deep into the atmosphere and be scattered or stopped by atmospheric interactions.

Atmospheric interactions also affect the proton angular distribution in another fashion. Protons that are observed traveling eastward are following guiding centers above the observation point and protons traveling westward are following guiding centers below the observation point. The gyroradius for energtic protons in the Anomaly is on the same order as the atmospheric density scale height. Thus westward traveling protons encounter a significantly denser atmosphere and are more likely to suffer atmospheric interactions and be lost. The resulting energy-dependent anisotopy is called the east-west effect.

Most flux and dose analysis of the low-altitude radiation environment have ignored the anisotropic angular distribution of the flux. Although the anisotropy was well documented, Henley[1] first noted its greater importance for Space Station because of its stabilized orbital attitude. The current proton environment model in use is the "Vette" model[2] together with the associated magnetic field models[3]. Given an orbital trajectory the field models yield the magnetic field coordinates B and L. Given the B and L along the trajectory the Vette model yields the omnidirection flux as a function of proton energy along the trajectory. In the work reported here we calculate an angular distribution to associate with the omnidirection flux using the atmospheric density profile and local magnetic field direction.

METHODS

Assuming that the mirror point density along a field line is proportional to the atmospheric density along the field line, Heckman and Nakano[4] showed that

$$f(\theta)d\theta \propto \rho(\theta)^{-1}d\theta \tag{1}$$

where $f(\theta)$ is the pitch angle distribution and $\rho(\theta)$ is the atmospheric density expressed in terms of the pitch angle. For an exponential atmosphere the density is given by

$$\rho \propto \exp\left(-\frac{h_m}{h_o}\right) \tag{2}$$

where h_m is the mirror point altitude and h_o is the scale height. Since most protons in the South Atlantic Anomaly are very near their mirror points h_m is well approximated by

$$h_m \approx h - l\sin I \tag{3}$$

where l is the distance along the field line from the observation height h to h_m and I is the magnetic dip angle. For a dipole field Heckman and Nakano[4] then showed that

$$l \approx K\theta^2/2 \tag{4}$$

where

$$K = \frac{4/3R}{\sin I(2 + \cos^2 I)} \tag{5}$$

and R is the dipolar radius. Thus

$$f(\theta)d\theta \propto \exp\left(-\frac{K\sin I}{2h_o}\theta^2\right)d\theta \tag{6}$$

The pitch angle distribution is Gaussian with a standard deviation given by

$$\sigma = \sqrt{\frac{h_o}{K\sin I}} \tag{7}$$

For typical values of the magnetic field parameters and atmospheric scale height at low altitudes in the Anomaly σ is about 8°. Heckman and Nakano[5] made oriented emulsion measurements at an altitude of 364 km during passes through the Anomaly that are well described by a Gaussian distribution. (The measurements also detected the east-west effect.) The measured σ was 7.77±0.15° for proton flux normal to the magnetic meridian and 13.00±0.23° in the plane of the magnetic meridian. The observed distribution is wider than the actual pitch angle distribution because the magntic field direction changed relative to the emulsion plate during the passes. Fisher et al.[6] measured proton pitch angle distributions at near-equatorial latitudes as a function of L and B/B_o. At the lowest L value, 1.15, where they thought atmospheric interactions were the dominant loss mechanism, the full width at half maximum of the pitch angle distribution was 17°.

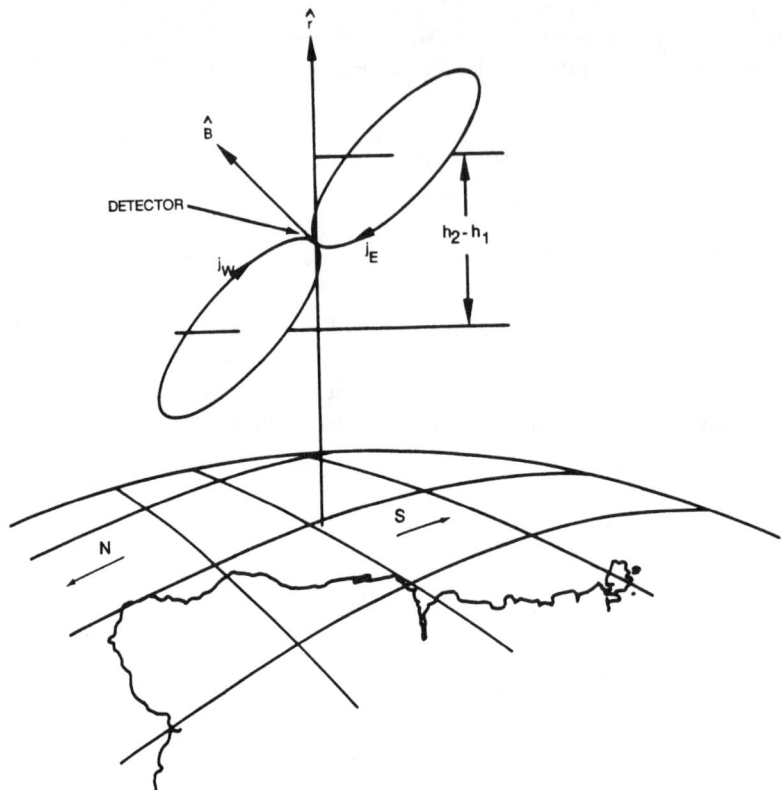

Fig 2. East-west effect coordinates.

Lenchek and Singer[7] give an expression for the east-west effect assuming, as observed, that the proton flux rises exponentially with altitude. That is

$$j \propto \exp\left(\frac{h}{h_o}\right) \tag{8}$$

where h_o is the scale height and j is the flux. Then

$$\frac{j_2}{j_1} = \exp\left(\frac{h_2 - h_1}{h_o}\right) \tag{9}$$

and

$$h_2 - h_1 = (r \cos \beta_2 - r \cos \beta_1) \cos I \tag{10}$$

where h_1 and h_2 are the heights of the guiding centers of proton observed traveling in directions β_1 and β_2 (see Figure 2). The angles are measured relative to magnetic east. r is the proton gyroradius given by

$$r = \frac{10^4 p}{cqB} \tag{11}$$

where r is in kilometers, p is momentum in MeV/c, c is the speed of light in km/s, q is the particle charge in electron charges, and B is the magnetic field intensity in Gauss. Then

$$\frac{j_2}{j_1} = \exp\left(\frac{r \cos I(\cos \beta_2 - \cos \beta_1)}{h_o}\right) \quad (12)$$

For typical values of the magnetic field parameters and atmospheric scale height in the Anomaly the gyroradius for 50 MeV protons is 52 km and the ratio of the eastward traveling ($\beta_2 = 0°$) to the westward traveling ($\beta_1 = 180°$) flux is 3. At 400 MeV the values are 145 km and 22. Heckman and Nakano[5] quote observed ratios at 60 MeV of 1.57±0.16 and at 116 MeV of 5.49±0.77. The cosine difference term for their observations was considerably below 2. Reagan and Imhof[8] report a ratio of 7 for the energy band 23.5 to 80 MeV at lower altitudes that was in good agreement with atmospheric model calculations. Heckman and Nakano[9] show good agreement with atmospheric scale height data for altitudes between 340 and 500 km and suggest that particle data are useful monitors of average properties of the atmosphere at these altitudes.

Given the omnidirectional flux from the Vette model[2], the vector magnetic field[3], appropriate atmospheric scale height data, the pitch angle distribution from equation (6), and the east-west distribution from equation (12), an approximate vector proton flux distribution can be derived. Selecting a coordinate system with the z-axis point along $\vec{B}$ and its y-axis pointing toward magnetic east from the definition of the pitch angle distribution

$$f(\theta)d\theta = \int_0^{2\pi} \frac{j(E,\theta',\phi)}{j_o(E)} d\phi d\cos\theta' \quad (13)$$

where $j(E,\theta',\phi)$ is the vector flux in direction (θ',ϕ) at energy E, $j_o(E)$ is the omnidirectional flux, and $\theta = \pi/2 - \theta'$. Applying the normalization condition to equation (6) and using the relationship between θ and θ'

$$\int_0^{2\pi} \frac{j(E,\theta',\phi)}{j_o(E)} d\phi = \frac{\exp\left(\frac{(\pi/2-\theta')^2}{2\sigma^2}\right)}{\sin\theta'\sqrt{2\pi}\sigma \operatorname{erf}\left(\frac{\pi}{2\sqrt{2}\sigma}\right)} \quad (14)$$

This equation provides the normalization for a ring in constant θ'. The distribution around that ring is provided by equation (12) together with the relationship

$$\cos\beta = \sin\theta' \sin\phi \quad (15)$$

Thus

$$\frac{j(E,\theta',\phi)}{j_o(E)} = \int_0^{2\pi} \frac{j(E,\theta',\phi)}{j_o(E)} d\phi \cdot \frac{\exp\left(\frac{r \cos I \cos \beta(\theta',\phi)}{h_o}\right)}{\int_0^{2\pi} \exp\left(\frac{r \cos I \cos \beta(\theta',\phi)}{h_o}\right) d\phi} \quad (16)$$

and

$$j(E,\theta',\phi) = j_\circ(E)\frac{\exp\left(\frac{(\pi/2-\theta')^2}{2\sigma^2}\right)}{\sin\theta'\sqrt{2\pi}\sigma\,\mathrm{erf}\left(\frac{\pi}{2\sqrt{2}\sigma}\right)}\frac{\exp\left(\frac{r\cos I\cos\beta(\theta',\phi)}{h_\circ}\right)}{\int_0^{2\pi}\exp\left(\frac{r\cos I\cos\beta(\theta',\phi)}{h_\circ}\right)d\phi} \quad (17)$$

The right integral must be performed numerically. Equation (17) yields the flux distribution in coordinates fixed to the magnetic field. Given the attitude of a spacecraft relative to the magnetic field the distribution is then transformed by a rotation to the spacecraft coordinates.

RESULTS

Avoiding the confounding variable of spacecraft attitude, Figure 3 shows the proton spectra average over many orbits for particles traveling in the direction of magnetic east and magnetic west. The omnidirectional proton environment used was AP8MIN[2] and the magnetic field model was the IGRF 1965.0 80-term[3] projected to 1964, the epoch of the environmental model. The atmospheric scale height was for solar minimun[10] with corrections for proton longitudinal drift and mirror point oscillation.[9] The scale height values used were 73 km at an altitude of 300 km and 123 km at an altitude of 500 km. In Figure 4 the directional model spectra are displayed for protons traveling toward the geographic east and west. The fluxes are lower than those for the equivalent magnetic direction because magnetic north deviates by about 24° to the west of geographic north in the Anomaly placing the geographic east and west out of the mirror plane.

Figures 5 and 6 show dose rates for a simple shielding geometry using vector fluxes. The dose point is behind a plane alumimun shield 5 g/cm^2 thick and has infinite material behind it. Directions about the detector were segmented into 36 ϕ and 20 $\cos\theta'$ uniformly spaced bins yielding a total of 720 directions. In Figure 5 the plane was oriented relative to the magnetic direction. As expected one sees the highest dose rates to the west and the lowest when looking parallel to the field. The results looking parallel to the field have a questionable uncertainty because most of the flux arrives at large angles to the mirror plane where the small angle approximation used to derive the pitch angle distribution is not applicable. One improvement to the method would be the introduction of a pitch angle cutoff. For Figure 6 the plane was perpendicular to the geographic directions. Again the highest dose rates are seen when looking to the west. An initial surprise is that dose rates looking north are about as low as rates looking east. The cause of this is the westward deviation of magnetic north in the Anomaly.

The ratio of the dose rate looking west to that looking south is about 1.5 and the ratio for west versus east is 3. The plane shielding geometry with an infinite backing shield probably enhances the directional effect relative to a realistic spacecraft shielding situation. However, in cases where most of the flux arrives from a few thinly shielded areas, the effect could be considerably larger.

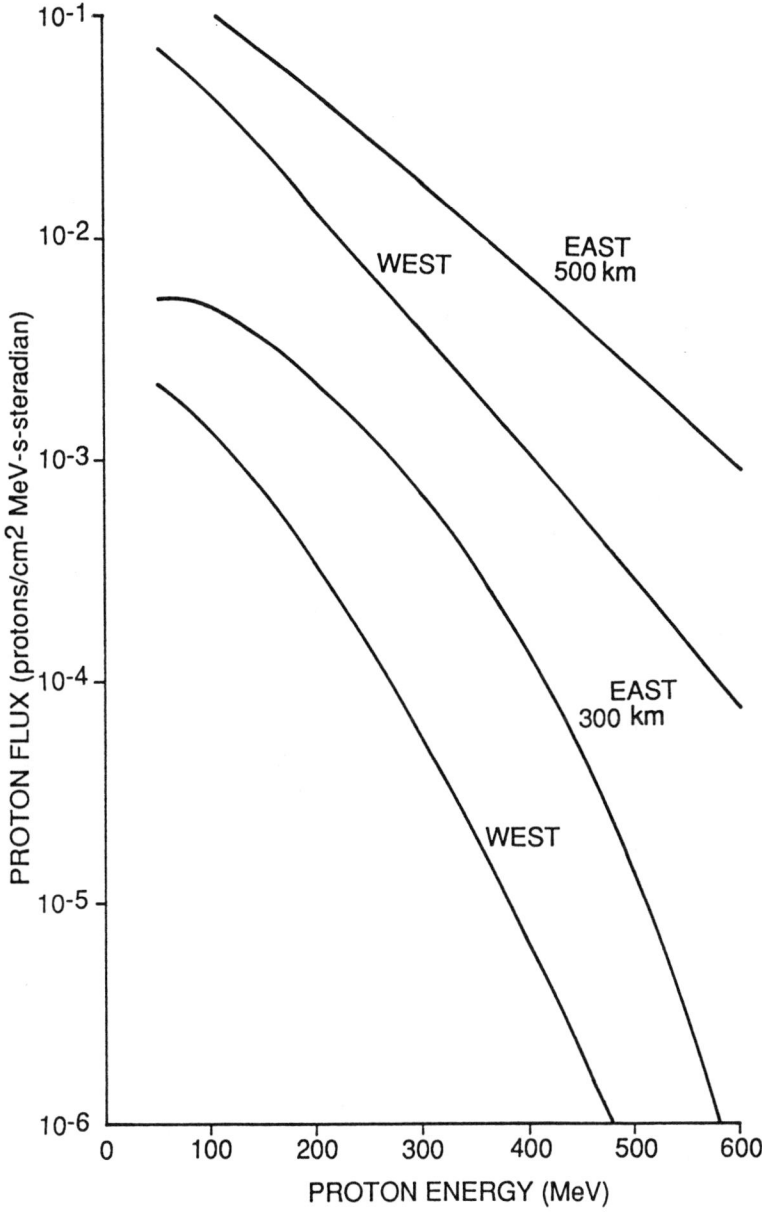

Fig 3. Average proton energy spectra on a 28.5° orbit for particle traveling in the directions of magnetic east and west from the directional model.

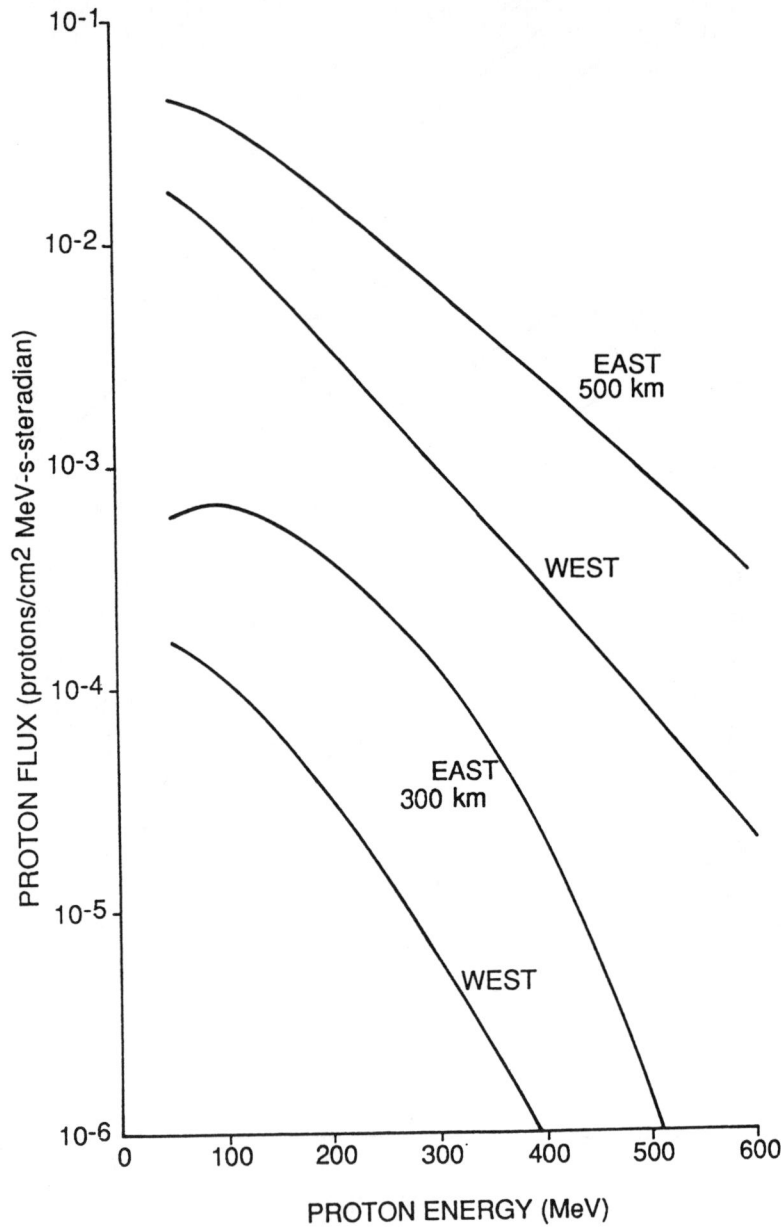

Fig 4. Average proton spectra on a 28.5° orbit for particles traveling in the directions of geographic east and west from the directional model.

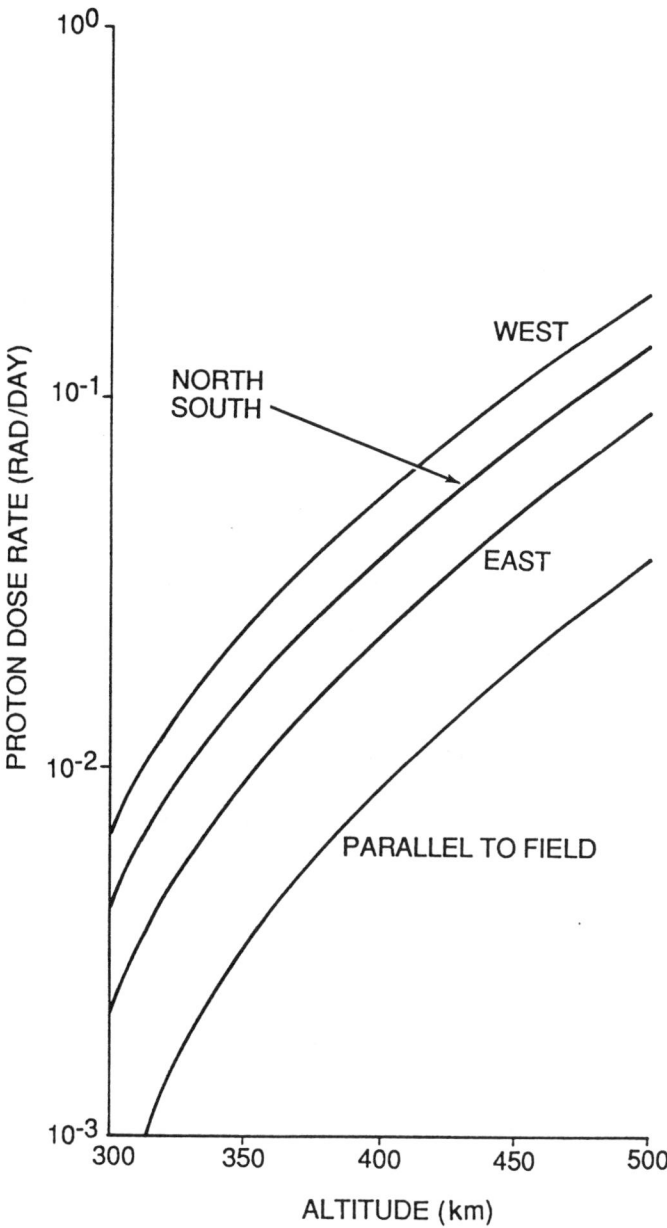

Fig 5. Average proton dose rate on a 28.5° orbit for a detector behind a 5 g/cm² aluminum shield with infinite backing with the plane facing the specified directions relative to the magnetic field.

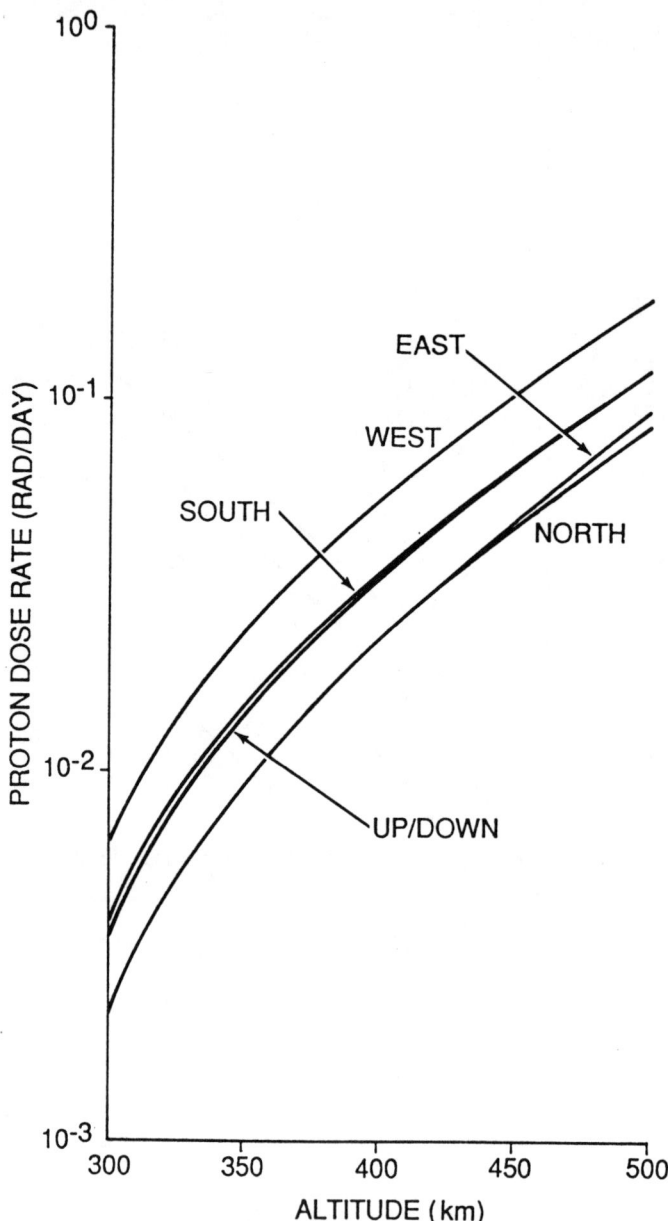

Fig 6. Average proton dose rates for 28.5° orbit for a detector behind a 5 g/cm² aluminum shield with infinite backing with the plane facing in the specified directions relative to the geographic coordinates.

On spacecraft that maintained no long-term stability relative to the magnetic field such as Skylab and Shuttle, doses varied by about a factor of 2 about the spacecraft due to shielding differences. Due to the effect described here, the variation will be more extreme on stabilized spacecraft. We note that the anisotropy of the trapped flux was observed in ion chamber dose measurements on early manned missions such as Gemini IV[11].

CONCLUSIONS

For spacecraft stabilized in Earth or magnetic coordinates in low-altitude orbits the anisotropic nature of the flux distribution is important if complex shielding geometries are being modeled. Using an omnidirectional flux distribution would introduce significant errors in the predictions at a specific location. Although the results from the presented model have the large uncertainty (a factor of 2 in values of flux) of the omnidirectional model used, the relative results for various look directions are a useful guide in consideration of shielding and instrumentation.

REFERENCES

1. Mark W. Henley, J. Spacecraft Roc., 23, 108 (1986).
2. Donald M. Sawyer and James I. Vette, AP-8 Trapped Proton Environment for Solar Maximun and Solar Minimun (National Science Data Center, Goddard Space Flight Center, NSSDC/WDC-A-R&S 76-06, 1976).
3. E. G. Stassinopoulos and Gilbert D. Mead, ALLMAG, GDALMG, LINTRA: Computer Programs for Geomagnetic Field and Field-Line Calculations (National Space Science Data Center, Goddard Space Flight Center, NSSDC 72-12, 1972).
4. H. H. Heckman and G. H. Nakano, J. Geophys. Res., 74, 3575 (1969).
5. H. H. Heckman and G. H. Nakano, J. Geophys. Res., 68, 2117 (1963).
6. Harald M. Fisher, Volker W. Auschrat, and Gerd Wibberenz, J. Geophys. Res., 82, 537 (1977).
7. A. M. Lenchek and S. F. Singer, J. Geophys. Res., 67, 4073 (1962).
8. J. B. Reagan and W. L. Imhof, Observations of the East-West Asymmetry of Protons Trapped at Low Altitudes, Adv. Space Res., 10, 853 (1970).
9. H. H. Heckman and G. H. Nakano, Direct Observations of Mirroring Protons in the South Atlantic Anomaly, Adv. Space Res., 5, 329 (1965).
10. Dale L. Johnson and Robert E. Smith, The MSFC/J70 Orbital Atmosphere Model and the Data Bases for the MSFC Solar Activity Prediction Technique, NASA TM-86522 (1965).
11. M. Schneider and J. Janni, Aerospace Medicine, 40(12), 1535 (1969).

SOLAR PARTICLE COMPOSITION, SPECTRA, AND FREQUENCY OF OCCURRENCE

D. V. Reames
NASA/Goddard Space Flight Center, Greenbelt, MD 20771

ABSTRACT

The radiation background from solar-particle events can assume different forms as the energy and composition of the particles changes from event to event. It has recently become clear that different particle composition and spectra arise from different classes of events at the sun and we have learned how to associate the properties of the particles with the radio, X-ray and optical observations of the parent flares or with the related interplanetary shock.

INTRODUCTION

Solar particle events present a significant hazard to a variety of human enterprises. On earth, they can disrupt communications, elevate the radiation background for high-altitude aircraft and enhance radio-isotopes like ^{14}C. Outside the earth's magnetosphere, they can endanger the lives of astronauts, disrupt electronic circuits and damage exposed surfaces, coatings and sensors.

Perhaps the most threatening aspect of the events is their sudden occurrence that remains essentially unpredictable. Intensities of energetic particles can increase by many orders of magnitude in a matter of minutes and a single large event can produce more particles than all of the other events in an 11-year solar cycle. The energy source for the solar particles comes from rearrangements in the solar magnetic fields that are generated by turbulence in the solar atmosphere[1]. Such turbulence is akin to terrestrial weather but it occurs in magnetized, ionized plasma at an inconvenient distance for measurement.

Large solar flares are the source of the most penetrating radiation, predominantly protons, at energies as high as 100 MeV or more. About 30 events with substantial intensities of protons above 100 MeV have been seen at earth during the last 11 years. The unique statistics of large events have been studied by King[2] and recent summaries of the events themselves are available elsewhere[3,4].

Not all of the potential damage is caused by 100 MeV protons, however. Heavy ions with their high rate of ionization can cause single-event upsets in circuits and cause severe biological damage. Proton from 1 to 100 MeV contribute to polar-cap absorption affecting communications. Lower energy particles are much more numerous in a given event and they are accelerated with greater ease in a wide variety of events.

The last few years have brought a significant advancement in our understanding of the relationships between the particles and

© 1989 American Institute of Physics

other solar phenomena such as radio emission, hard and soft X-rays, gamma rays and coronal mass ejections (CME's). Different modes of particle acceleration at the sun result in different populations of particles in space. It is important to distinguish these populations since they contribute unequally to different particle species and energy regimes and hence, alter the nature of the radiation background.

Two classes of solar event have been distinguished that have the properties shown in table I [5,6,7,8]. The mechanisms leading to the two classes of phenomena are occasionally triggered in the same event and, in fact, they were once thought to be phases of the acceleration process that were required to occur in sequence.

Table I Classes of Solar Particle Events.

	Class I	Class II
Radio type:	III, V	II, IV
Soft X-rays:	Impulsive Low, Compact	Long Duration (LDE) High, Extended
Coronagraph:	no CME's	CME's
Particles:	Electron-rich ^{3}He-rich High He/H High Fe/O	Protons Normal Coronal Abundances
Examples:	^{3}He-rich Events	Disappearing- Filament Events

Particles accelerated at the sun are constrained to spiral outward along the interplanetary magnetic field (IMF) lines. Because of solar rotation, the IMF is drawn into a spiral pattern by the solar wind so that the earth is best connected to a region about 40° to 60° west of central meridian on the sun. CME's and coronal shocks accelerate particles over an extended area of the sun, however, so that solar particles can be observed from flares that occur from about E30° through about W140° (*i.e.* 50° behind the west limb of the sun)[9].

Yet another population of particles is accelerated by the interplanetary shock waves that are produced by large CME's. The high-energy particles from these events usually reach maximum intensity after the passage of the shock, 1 to 3 days after the event at the sun. Particle events near central meridian on the sun usually have a strong component from the interplanetary shock and the particles from flares east of E30° are almost entirely shock associated [9]. It is important to distinguish this particle population since it is, in principle, predictable.

The properties of the various solar-particle populations are best observed in small events where a single component dominates. Such studies allow us to understand the relationships of the properties of the particles to those of the acceleration mechanism. In large events all mechanisms can occur, leading to "big flare syndrome"[10] that causes false correlations among observations. In subsequent sections I discuss each particle population and its contribution to the radiation background in space.

CLASS I EVENTS

These events were first identified by their anomalously high abundance of electrons relative to protons[7]. Since these two species usually have different spectra it is difficult to quantify the enhancement, but the e/p ratio is typically a factor 100 larger than that in class II events at energies of a few MeV. Abundant electrons at lower energies generate the metric and kilometric type III radio emission as they stream outward from the event. The events are associated with the flash phase of solar flares; they have impulsive X-ray profiles and the larger events are accompanied by gamma-ray emission[10]. In fact, more of the energy in class I events may appear as photons because most of the particles are trapped in low coronal loops until they plunge into the chromosphere and interact to produce photons.

Striking examples of events in this class are found in the ^{3}He-rich events. These events show the characteristic electron richness[11] and the impulsive X-ray profiles[12] but they are most strongly characterized by enhancements in the ^{3}He/^{4}He ratio by several orders of magnitude. According to Fisk[13], the enhancement occurs because of preferential heating of the ^{3}He by the absorption of electrostatic ion-cyclotron waves.

From the standpoint of radiation background in space, it is not the enhancements in ^{3}He that are significant, however, but the related enhancements in heavy nuclei. Figure 1 shows the relative abundances of elements in four ^{3}He-rich events. Events can be found in which there are nearly as many Fe nuclei as He nuclei. A more extensive summary of element abundances in these events is given by Mason et al.[14].

Recently it has been found that heavy-element enhancements correlate with the soft X-ray temperature in the flare[15]. This occurs because the equilibrium charge state of the elements like Fe depend upon temperature, and the charge state affects their gyrofrequency in the magnetic field and determines the resonant absorption of waves[13]. Direct charge-state measurements first suggested this higher temperature in ^{3}He-rich events[16,17].

Even though class I events can occur frequently, often occurring in clusters of several events per day, they are usually small and are observed only at low energies. Occasionally, extremely large events of this kind occur, and when they do, large fluxes of energetic Fe nuclei can be present.

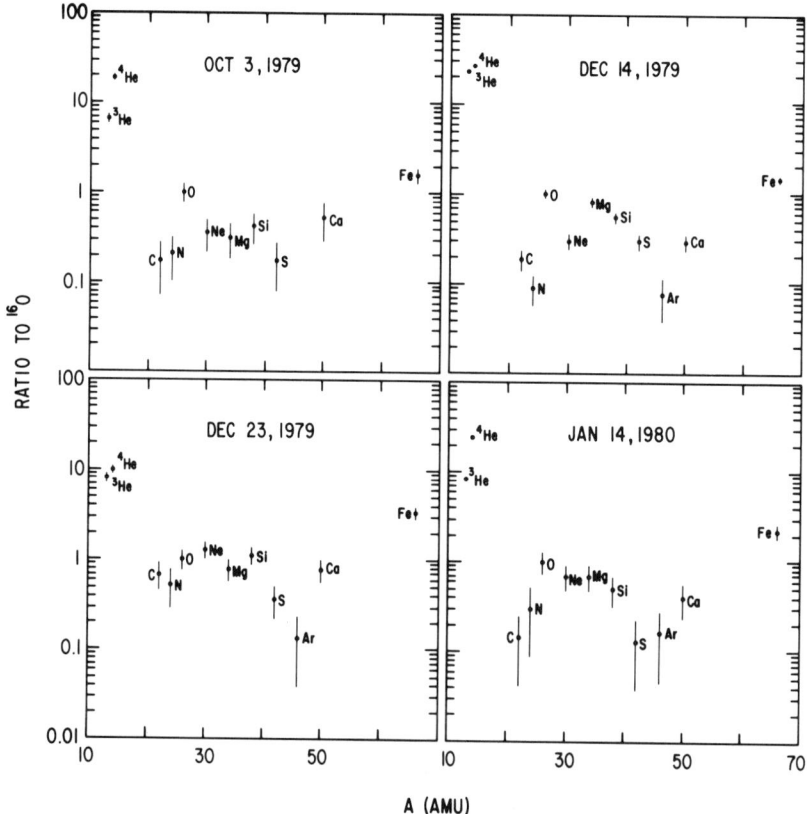

Fig. 1. Abundances of elements in four ^{3}He-rich events

CLASS II EVENTS

Most of the large solar-proton events seen at earth are class II events. At the sun, a class II event is dominated by a CME[18] and a coronal shock that generates type II radio emission as it moves outward. The events occur high in the corona[6] over a large spatial region and most of the accelerated particles can escape.

As the shock wave propagates away from the flare site, most of the material it traverses has not been heated by the flare. In the absence of any preferential heating, the shock accelerates the ambient coronal material in an unbiased way so that the abundances of elements in the energetic particle population reflects those of the local coronal material.

Extreme examples of class II particle events are found in the solar filament eruptions (also called disappearing-filament events)[19,20]. These events are characterized by the complete absence of an impulsive phase and they frequently occur outside active regions. However, they show the characteristic CME and the long-duration soft X-ray event.

In addition to the prompt particles, class II events frequently produce an interplanetary shock, probably the bow shock of the CME, that accelerates particles throughout the interplanetary medium. This latter particle population reaches maximum intensity 1 to 3 days after the solar event[9].

OBSERVATIONS

In this section I have collected typical observations that compare the relative contributions of the different particle populations to different types of background radiation.

Figure 2 shows the intensity of energetic protons for a period of several years during the last solar cycle. The dense

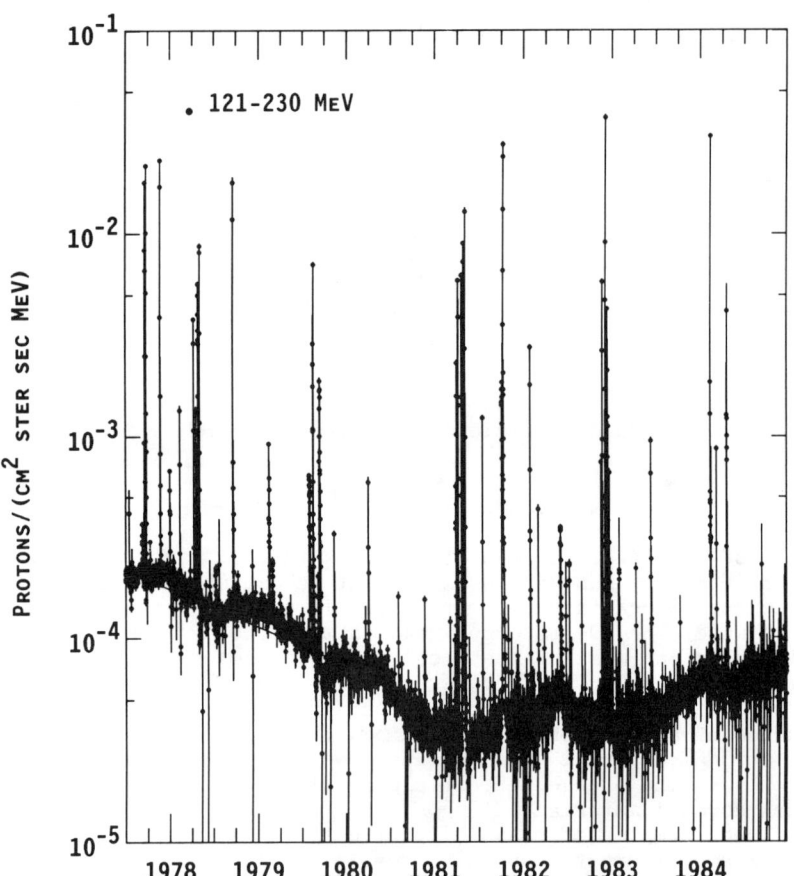

Fig. 2. Intensity of energetic protons showing solar events rising from galactic cosmic-ray background.

black region in the figure traces the intensity of the galactic cosmic-ray background. The galactic background is high during solar minimum and then decreases during solar maximum. Solar events project upward from the background by as much as 3 orders of magnitude during this cycle. Each point in the figure is one 8-hr. averaged period so the duration of an increase can be estimated from the number of points in it. (The points below the galactic intensity result from poor statistics caused by data gaps; they have large error bars as shown.)

Figures 3 and 4 compare the intensities of lower-energy protons with those of the 121-230 MeV protons for a given time

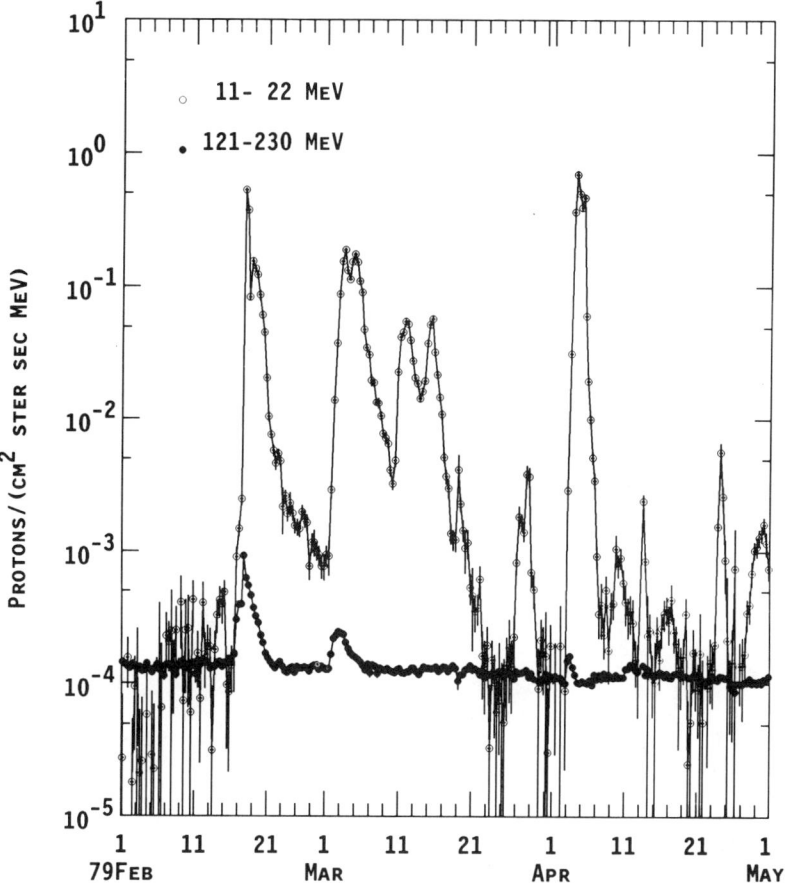

Fig. 3. A comparison of proton intensity variations at 11-22 MeV with those at 121-230 MeV for a sample time period.

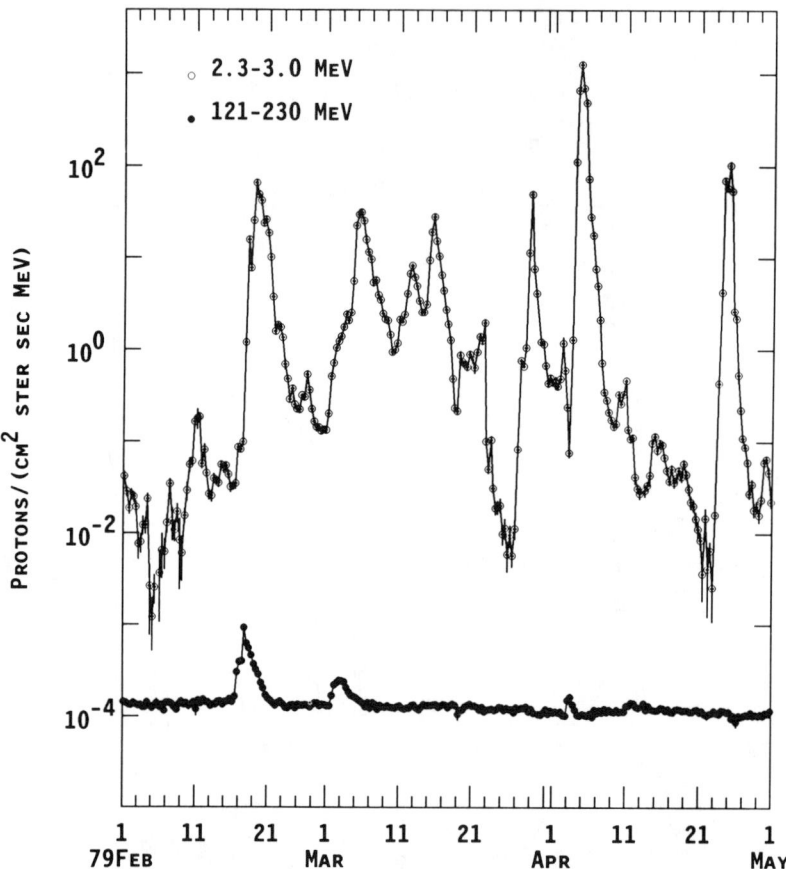

Fig. 4. A comparison of proton intensity variations at 2-3 MeV with those at 121-230 MeV for the same time period as in Fig 3.

period. Substantial increases occur much more often at lower energies and the intensities are orders of magnitude larger than at high energies. In fact, many of the large peaks in the lower energy intensities have no high-energy counterpart. Two of the six largest peaks in the 2-3 MeV data in figure 4 are caused by disappearing-filament events at the sun.

An example of the variation in composition in particle events is shown in figure 5. The Fe/O ratio can be seen to vary from about 1.0 to 0.01 from event to event. This is typical of the range of the variation seen over a 7-year period of ISEE-3 observations. A summary of the range of variation in the abundances of other elements is given by Mason et al.[21] for an earlier 4-year period.

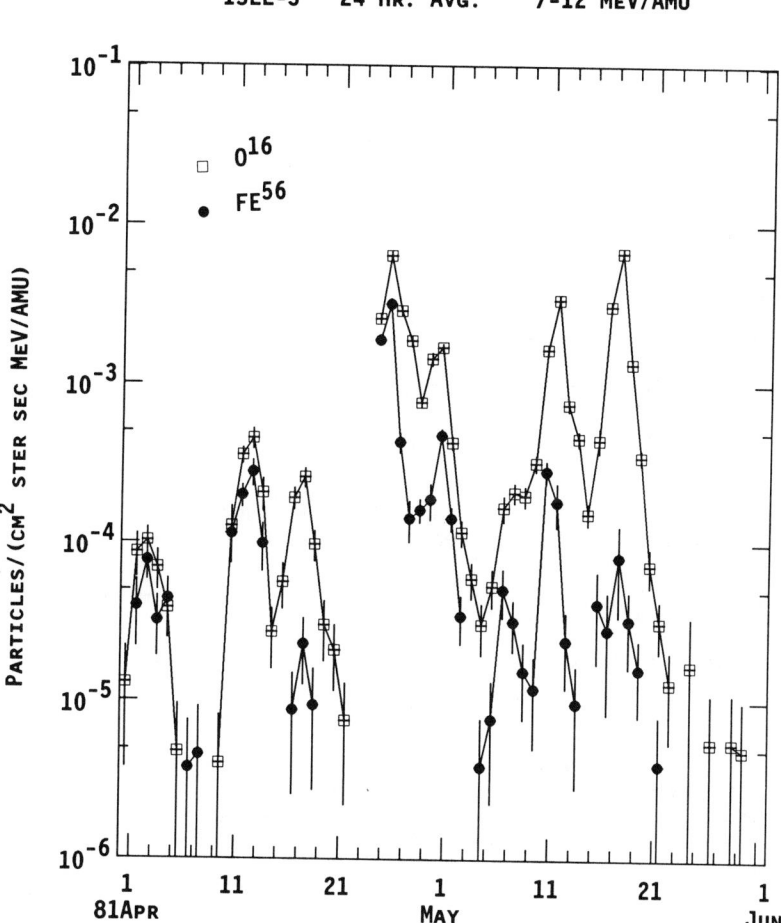

Fig. 5. Daily-averaged intensities of oxygen and iron during a sample time period.

The largest solar-particle event seen in high-energy particles during the last solar cycle occurred on June 3 1982. The event occurred at a solar longitude of E72° so that few particles were able to reach earth, however, the event was well connected to the Helios 1 spacecraft at 0.57 AU off the east limb of the sun. Intensities of protons above 120 MeV were about an order of magnitude larger than the largest event shown in figure 2 [22]. The interesting features of this event are the large enhancement of Fe (Fe/O≈2.5), the large electron/proton ratio, and the presence of ^{3}He [23]. This large event clearly has a strong flash phase as well as a strong shock phase.

I would like to thank H. V. Cane for her comments on this manuscript.

REFERENCES

1. E. Parker, Physics Today, July 1987, p 36
2. J. H. King, *J. Spacecraft Rockets* **11**, 401 (1974).
3. J. N. Goswami, R. E. McGuire, R. C. Reedy, D. Lal and R. Jha, *J. Geophys. Res.* (1988, to be published).
4. R. E. McGuire, this conference (1988).
5. J. P. Wild, S. F. Smerd and A. A. Weiss, *Ann. Rev. Astr. Ap.* **1**, 291 (1963).
6. R. Pallivicini, S. Serio and G. Vaiana, *Ap. J.*, **216**, 108 (1977)
7. H. V. Cane, R. E. McGuire, and T. T. von Rosenvinge, *Ap. J.*, **301**, 448 (1987).
8. D. V. Reames and R. G. Stone, *Ap. J.*, **308**, 902 (1986).
9. H. V. Cane, D. V. Reames and T. T. von Rosenvinge (to be published, 1988).
10. E. L. Chupp, *Ann. Rev. Astr. Ap.* **22**, 239 (1984).
11. D. V. Reames, T. T. von Rosenvinge, and R. P. Lin, *Ap. J.* **292**, 716 (1985).
12. D. V. Reames, B. R. Dennis, R. G. Stone, and R. P. Lin, *Ap. J.* (in press, April 1988).
13. L. A. Fisk, *Ap. J.* **224**, 1028 (1978).
14. G. M. Mason, D. V. Reames, B. Klecker, D. Hovestadt, and T. T. von Rosenvinge, *Ap. J.* **303**, 849 (1986).
15. D. V. Reames, *Ap. J. (Letters)* **325**, L53 (1988).
16. A. Luhn, B. Klecker, D. Hovestadt, and E. Möbius, *Proc. 19th Internat. Cosmic Ray Conf.* (La Jolla) **4**, 285 (1985).
17. R. P. Lin, *Revs. Geophys.* **25**, 676 (1987).
18. S. W. Kahler, N. R. Sheeley, Jr., R. A. Howard, M. J. Koomen, D. J. Michels, R. E. McGuire, T. T. von Rosenvinge and D. V. Reames, *J. Geophys. Rev.* **12**, 209 (1984).
19. S. W. Kahler, E. W. Cliver, H. V. Cane, R. E. McGuire, R. G. Stone and N. R. Sheeley, Jr. *Ap. J.* **302** (1986).
20. H. V. Cane, S. W. Kahler, N. R. Sheeley, Jr. *J. Geophys. Res.* **91**, 13321 (1986).
21. G. M. Mason, L. A. Fisk, D. Hovestadt and G. Gloeckler, *Ap. J.* **239**, 1070 (1980).
22. F. B. McDonald and M. A. I. Van Hollebeke, *Ap. J. (Letters)* **290**, L67.
23. M. A. I. Van Hollebeke, F. B. McDonald and J. H. Trainor, *Proc. 19th Internat. Cosmic Ray Conf.* (La Jolla) **4**, 209 (1985)

II. Particle Interactions and Propagation: Dynamic Modeling

PRODUCTS FROM COSMIC-RAY INTERACTIONS IN EXTRATERRESTRIAL MATTER: WHAT THEY TELL US ABOUT RADIATION BACKGROUNDS IN SPACE*

Robert C. Reedy
Los Alamos National Laboratory, Los Alamos, NM 87545

ABSTRACT

The nuclides and the heavy-nuclei "tracks" made by the interactions of solar and galactic cosmic-ray particles with meteorites, lunar samples, and the Earth have been extensively studied, simulated, and modelled. Most research involves the use of these cosmogenic products to study the history of the "targets" or of the cosmic rays. However, much work has also been done in understanding these interactions and in predicting their rates as a function of the target's size and shape and of the location inside the target. These studies apply to any object exposed to cosmic rays. The fluxes as a function of depth for cosmic-ray primary and secondary particles vary greatly with particle energy and type. The variations of the fluxes of these cosmic rays in the past have been studied. Energetic solar particles are unpredictable and are the greatest potential radiation hazard in space.

INTRODUCTION

A wide variety of "cosmogenic" products from cosmic-ray interactions have been measured in lunar samples and meteorites. These products include radiation damage tracks (produced in certain minerals by nuclei with $Z \geq 20$) and rare nuclides that are made by spallation or neutron-capture reactions.[1,2] They are usually used to study the history of the "target" (such as the time period that it was exposed to cosmic-ray particles), but they often have been used to determine the fluxes and composition of cosmic-ray particles in the past.[1,2] These products can also be used to investigate the nature of cosmic-ray interactions with matter in space, complementing studies of the interactions of high-energy particles with matter done at accelerators[3] or with theoretical models.[4-8] Products made by both the high-energy ($\sim$GeV) galactic cosmic rays and the energetic ($\sim$ 1–100 MeV) particles emitted from the Sun have been extensively studied in meteorites and lunar samples.[1,2] Studies of cosmogenic products in natural extraterrestrial matter can usually be directly applied to spacecraft and other artificial materials in space, especially far from the Earth.

COSMIC-RAY PARTICLES AND THEIR INTERACTION PRODUCTS

The major particles in space that have been studied in extraterrestrial matter are the galactic cosmic rays (GCR), energetic ($\sim$1 to >100 MeV) particles from the Sun (hereafter called solar cosmic rays, SCR), and the solar wind. The low-energy ions in the solar wind have been observed implanted in the outer $\sim$50 nm of lunar samples. As the solar wind contributes very little to radiation

---

* This work was supported by NASA and done under the auspices of the U.S. Department of Energy.

backgrounds in space, it won't be discussed any further here. The nature of the galactic and solar cosmic rays and their interactions will be discussed in detail. The galactic cosmic rays have fairly low fluxes ≈ 3 cm^{-2} s^{-1} and high energies ($\sim$ GeV) whereas the particles emitted by solar flares have high fluxes (up to $\sim 10^6$ cm^{-2} s^{-1} at the peak of an event and a long-term average flux of $\sim$100 cm^{-2} s^{-1}) with fairly low energies (mostly $\sim$1–100 MeV).[1,2,4,5,9,10] A summary of the energies, mean fluxes, and interaction depths in solid (density $\sim$3 g cm^{-3}) matter of the GCR and SCR particles is given in Table I.

GCR fluxes are fairly constant, with solar activity being the dominant source of variation. The lowest and highest GCR fluxes are during periods of the 11-year solar cycle with the maximum and minimum solar activity, respectively.[4] Higher fluxes of GCR (by factors of a few times) are present in the local interstellar space beyond the heliosphere and could be present in the inner solar system during prolonged periods of unusually low solar activity, such as occurred during 1645–1715 (the Maunder minimum).[1,4] While heavy nuclei in the GCR are mainly stop by ionization energy losses in the outer few centimeters before they can react, most GCR protons and α particles react and produce a cascade of secondary particles, including many pions and neutrons. These secondary particles, especially the penetrating neutrons, induced all types of nuclear reactions down to depths of meters in large objects exposed to GCR particles.[5]

SCR fluxes vary from essentially nothing for most of the time to $\sim 10^6$ cm^{-2} s^{-1} at the peak of the 4 August 1972 flare. Observations of event-integrated solar particle fluxes over the last three solar cycles (1954–1986) have been summarized.[9,10] From 1956 to 1986, there have been $\approx$116 events with total event-averaged omnidirectional fluences above 10 MeV of $> 10^7$ protons cm^{-2} s^{-1}.[10] Average fluxes of solar protons over periods of $\sim 10^4$ to 5 $\times 10^6$ years have been determined from measurements of radionuclides in lunar samples[1] and are $\sim$100 protons cm^{-2} s^{-1}. Almost all SCR nuclei heavier than protons and most solar protons are stopped by ionization energy losses in the outer $\sim$0.1–1 cm of matter in space.[1,5] The few reactions induced by SCR particles are low-energy ones that emitted few secondaries and produce residual nuclei close in mass to that of the target nucleus.

Table I. Energies, mean fluxes, and interaction depths of galactic and solar cosmic-ray particles.

Radiation	Energy[a] (MeV nucleon^{-1})	Mean flux (cm^{-2} s^{-1})	Effective depth[b] (cm)
Galactic cosmic rays			
Protons & α particles	$\sim$100–3000	$\approx$3	$\approx$0–100
VH, VVH nuclei (Z$\geq$20)	$\sim$100–3000	$\approx$0.03	$\approx$0–10
Solar cosmic rays			
Protons & α particles	$\sim$5–100	$\sim$100[c]	$\approx$0–2
VH, VVH nuclei (Z$\geq$20)	$\sim$1–50	$\sim$0.03[c]	$\approx$0–0.1

[a] Typical energies, actual energies range to lower and much higher values.
[b] Assuming typical lunar rock or meteorite densities ($\sim$3 g cm^{-3}).
[c] Long-term averages, actual fluxes vary from zero to much high values.

PRODUCTS FROM COSMIC-RAY INTERACTIONS

There are two major types of cosmic-ray products that can be detected in extraterrestrial matter as having resulted from cosmic-ray interactions: rare nuclei and "tracks." The cosmogenic nuclei that can readily be identified as having been produced by cosmic-ray-induced reactions are radionuclides (like ^{10}Be) and the minor isotopes of the noble gas (like ^{21}Ne) that are normally not present in matter.[1,2] GCR particles can produce almost any nucleus lighter in mass than the target,[5] and the types of reactions vary from high-energy spallation reactions, such as ^{56}Fe(p,X)^{10}Be (where X can be any of a great variety of nucleon and particle combinations) to low-energy reactions induced by low-energy secondary neutrons, such as ^{24}Mg(n,α)^{21}Ne. The low-energy protons and α particles in the SCR mainly induce reactions that produce residual nuclei close in mass to the target,[5,9] such as ^{28}Si(p,n2p)^{26}Al and ^{56}Fe(α,n)^{59}Ni.

The paths traveled in certain crystalline dielectric phases (e.g., certain minerals like olivine and pyroxene) by individual cosmic-ray nuclei with $Z \geq 20$ near the end of their ranges contain enough radiation damage that they can be etched by chemicals and made visible as tracks.[1,2,11] The heavy cosmic-ray nuclei that produce tracks are usually classified as the VH (very heavy) group ($20 \leq Z \leq 28$, although mainly iron nuclei) and the VVH group ($Z \geq 30$), with the ratio of VH to VVH nuclei being $\approx$500-700.

The products from the interactions of cosmic-ray particles and their secondary particle have been measured in extraterrestrial matter with a wide distribution of sizes, ranging from small pieces of cosmic dust to meteorites (which typically have preatmospheric radii of ~5–50 cm) to lunar samples. Secondary particles made by the interaction of the primary GCR particles are usually important in producing cosmogenic nuclides: in the Moon there are about 7 neutrons produced per primary GCR particle.[5] In the Moon and most meteorites, nuclear reactions induced by secondary particles are more probable than those from the primary GCR particles.

FLUX VERSUS DEPTH PROFILES

The distributions of cosmogenic products in meteorites and lunar samples often have been studied. These measurements imply a build up of the fluxes of secondary particles from GCR interactions for some distance inside these objects, the amount of build up being dependent on the energies necessary to induce nuclear reactions.[5,6] Only for high-energy ($E \gtrsim 1$ GeV) particles or for large (radii greater than ~300 g cm^{-2}) bodies are there decreases in flux near the center of a meteorite.[6] The largest increases in the fluxes of cosmic-ray particles are for neutrons, with the amount of the increase tending to be inversely proportional to the neutron's energy. High-energy ($E_n \gtrsim 100$ MeV) neutrons have very little increase with depth near the surface of an extraterrestrial object, while thermal ($E_n \lesssim 1$ eV) neutrons tend to increase by large factors away from the surfaces of large (radii $\gtrsim 100$ g cm^{-2}) objects.[7,8] Most cosmogenic neutrons are made with energies of ~MeV, and it is difficult to slow them to thermal energies by scattering unless the object is big (radii greater than ~75 g cm^{-2}) and/or the hydrogen content of the object is high.[7,8] In such objects the flux of low-energy neutrons is very low near the surface because of neutron leakage into space.[7,8] Similar distributions of thermal neutrons will occur in large spacecraft, especially those with much hydrogen-containing material.

While the energetic GCR primary protons and α particles and their secondary particles are very penetrating in matter and have fluxes that vary slowly with the object's size or a sample's location, heavy nuclei ($Z \geq 20$) and solar energetic particles have flux-versus-depth profiles that vary considerably with depth. The heavy nuclei are rapidly stopped by ionization energy losses within ~ 10 cm for GCR energies and ~ 1 mm for the heavy nuclei from the Sun.[1,11] The relatively low-energy protons from solar-flare events also are also rapidly stopped in extraterrestrial matter, usually within less than 1 cm.[5,9]

The production rates of tracks and of several radionuclides made by a variety of nuclear reactions are shown as a function of depth in the Moon in Fig. 1. This figure illustrates the great spread in production profiles that exists for various cosmogenic products in matter exposed to cosmic-ray particles in space. VH nuclei in the cosmic rays only penetrate millimeters (for SCR) to centimeters (for GCR) in solid matter exposed in space before they are stopped. Similarly, protons and α particles emitted by solar flares are mainly stopped within a centimeter or so, although the fluxes of SCR particles vary considerably with time. The profile for ^{56}Co and for the top ~ 1 cm of ^{26}Al in Fig. 1 result from solar-proton reactions in the Moon. High-energy primary and secondary GCR particles, such as those that make ^{10}Be, show mainly a decrease with depth. The secondary particles, especially neutrons, made by the cascade induced in matter by the high-energy protons and α particles in the GCR can be important meters deep in solid bodies in space. Medium-energy ($E_n \sim 10$–50 MeV) neutrons contribute significantly to the production of ^{39}Ar and ^{26}Al, and their production profiles show increases to depths of $\sim$5–50 cm in the Moon. Neutron-capture-produced ^{60}Co has the greatest increase in a large object and has the deepest peak rate,[7,8] usually at depths of $\sim$50-100 cm.

TEMPORAL VARIATIONS IN COSMIC RAY FLUXES

Cosmogenic products have allowed us to determine the nature of the cosmic rays in the past. Several radionuclides with half-lives of the order of a million years have shown that the average fluxes of cosmic rays over the last 10^6 years are not very different than the contemporary fluxes. The main variations seen for GCR-produced nuclides, production rate changes of a factor of ≈ 2, have been due to the modulation of GCR particles by solar activity, mainly the 11-year solar cycle.[4] During the Maunder minimum, GCR production rates increased relative to those during a typical solar minimum.

Much larger fluctuations have been seen in the fluxes of energetic protons from the Sun over time periods up to a few million years.[1] The intensities of solar particles in individual solar flare events has only been studied for ≈ 30 years.[9,10] The similarity of long-term-averaged solar-proton fluxes with those observed now implies that huge solar-energetic-particle events, larger than those of 23 February 1956 and 4 August 1972, are rare.[1,10] These and several other large solar-particle events since 1956 had, for energies above 10 MeV, omnidirectional fluxes at their peaks of $\sim 10^6$ protons cm^{-2} s^{-1} and fluences integrated over the few days of the event of $\sim 10^{10}$ protons cm^{-2}.[9,10] The probabilities of events with integrated fluences much above $\sim 10^{10}$ protons cm^{-2} can not be predicted but appear to be fairly low.[10] As even events with peak fluxes

Fig. 1. Production rates for heavy (VH) nuclei tracks and various radionuclides as a function of depth in the Moon (density = 3.4 g cm^{-3}). The shaded region for tracks of VH nuclei reflects the uncertainties in the average fluxes of the low-energy VH nuclei in the SCR.[1] Deeper than ~0.1 cm, VH nuclei in the GCR dominate. The units for the production of ^{60}Co are atoms per minute per gram of cobalt;[7,8] the other radioactivities are in units of atoms per minute per kilogram of sample.[5] (The decay rates assume that the radionuclide's activity is in equilibrium with its production rate.) At depths of ≲1 cm, SCR production of nuclides usually dominates and appears as a steeply dropping curve (^{56}Co and ^{26}Al). Production of ^{10}Be mainly by high-energy GCR particles results in a profile with very little increase with depth, whereas production by low-energy secondary GCR neutrons results in big increases (^{39}Ar and ^{60}Co).

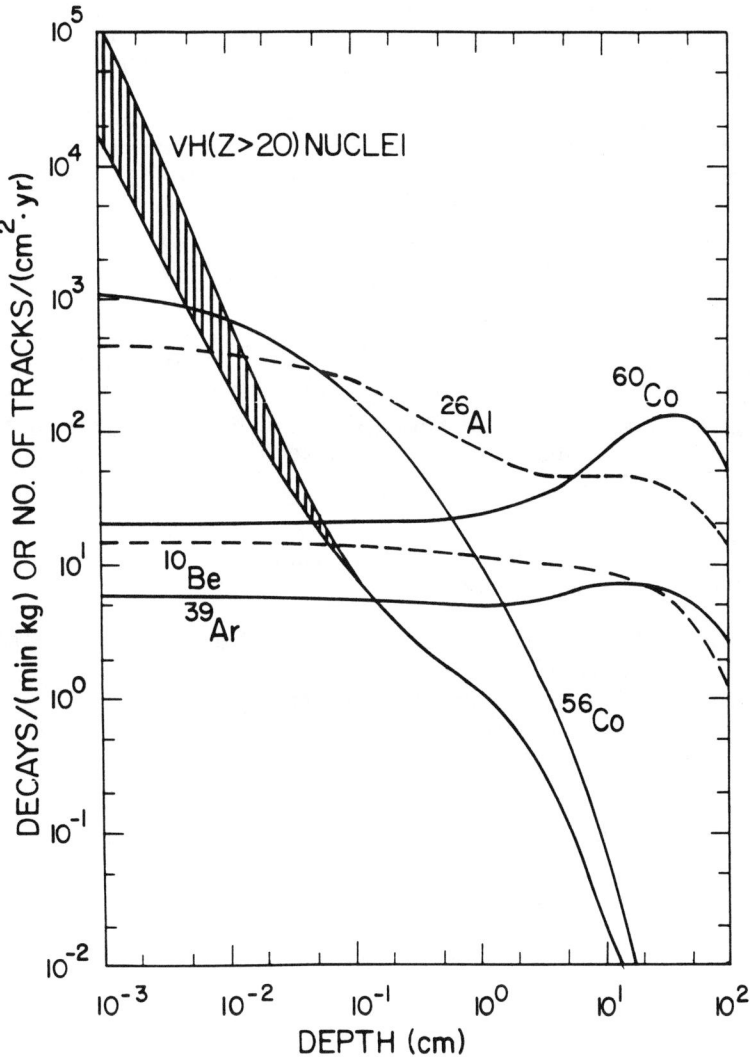

of $\sim 10^6$ protons cm^{-2} s^{-1} (which usually occur a few times per 11-year solar cycle) can cause significant radiation damage, spacecraft that will be in space away from the shelter of the Earth's magnetic fields for long periods of time will probably have a good chance of encountering such high particle fluxes.

SUMMARY

The concentration-versus-depth profiles of cosmogenic products measured in lunar samples and meteorites can be used to help to estimate the fluxes of heavy cosmic-ray nuclei, energetic primary and secondary GCR particles, thermal neutrons, and solar energetic particles in any matter in space as a function of location and the object's size. These measured concentrations of cosmogenic products and their distributions also can be used to test computer codes that model the interactions of cosmic-ray particles with spacecraft and instruments in space. The Sun controls the variations in the intensities of GCR particles, with the major fluctuation (by a factor of ≈ 2) being over the 11-year solar cycle. Energetic solar particles are very episodic, although they seldom occur during the few years around solar minimum. The probabilities for the occurrence of very large fluxes of solar particles (peak fluxes above 10 MeV of $>10^6$ protons cm^{-2} s^{-1} or event-integrated fluences $>10^{10}$ protons cm^{-2}) can not be well predicted, although the long-term record in lunar samples suggests that such huge flare-particle events are very rare.

ACKNOWLEDGMENTS

Many of the results presented here represent the results of productive collaborations with numerous colleagues, including J. R. Arnold, D. Lal, P. Englert, J. N. Goswami, G. F. Herzog, C. M. Hohenberg, M. Honda, R. E. Lingenfelter, K. Marti, R. E. McGuire, K. Nishiizumi, S. Regnier, L. Schultz, M. S. Spergel, J. I. Trombka, and R. M. Walker.

REFERENCES

1. R. C. Reedy, J. R. Arnold, and D. Lal, Annu. Rev. Nucl. Part. Sci. 33, 505 (1983); and Science 219, 127 (1983).
2. D. Lal, Space Sci. Rev. 14, 3 (1972).
3. P. Englert, R. C. Reedy, and J. R. Arnold, Nucl. Instrum. & Methods A262, 496 (1987).
4. R. C. Reedy, J. Geophys. Res. 92, E697 (1987).
5. R. C. Reedy and J. R. Arnold, J. Geophys. Res. 77, 537 (1972).
6. R. C. Reedy, J. Geophys. Res. 90, C722 (1985).
7. M. S. Spergel, R. C. Reedy, O. W. Lazareth, P. W. Levy, and L. A. Slatest, J. Geophys. Res. 91, D483 (1986).
8. R. E. Lingenfelter, E. H. Canfield, and W. N. Hess, J. Geophys. Res. 66, 2665 (1961).
9. R. C. Reedy, Proceedings of the 8th Lunar Science Conference (Pergamon Press, New York, 1977), p. 825.
10. J. N. Goswami, R. E. McGuire, R. C. Reedy, D. Lal, and R. Jha, J. Geophys. Res., in press.
11. R. L. Fleischer, P. B. Price, and R. M. Walker, Nuclear Tracks in Solids (University California Press, Berkeley, 1975).

BREMSSTRAHLUNG PRODUCTION BY ELECTRONS: CROSS SECTIONS AND ELECTRON-PHOTON TRANSPORT CALCULATIONS[*]

Stephen M. Seltzer
National Bureau of Standards, Gaithersburg, MD 20899 USA

ABSTRACT

This paper describes the rather accurate Monte Carlo model and cross sections that have been used to calculate effects from electrons and photons on measurement, electronic and biological systems in space. A number of applications are illustrated, with emphasis on effects from bremsstrahlung produced by electrons.

INTRODUCTION

Bremsstrahlung from interactions of energetic electrons with spacecraft material or nearby mass can constitute a radiation component much more penetrating than the electrons that produce it. Such a photon flux can cause significant untoward effects in radiation measurement systems, and can represent a serious threat to the reliability of radiation-sensitive electronic components and to the radiological safety of on-board personnel. To accurately calculate such effects, it is necessary not only to have accurate information on the cross section for the production of bremsstrahlung by electrons, but also to have reliable methods to calculate the penetration, diffusion and energy degradation of the electrons and the photons in bulk material.

Much of our work has been to develop the methods and the cross-section database for such calculations. This work has included the development of detailed Monte Carlo models for the solution of the coupled electron/photon transport problem. With these rather general Monte Carlo codes at hand, it has been possible to consider various problems of interest in space science, including also those in which bremsstrahlung may play only a minor role. The applications have included calculations of auroral bremsstrahlung produced in the Earth's atmosphere; calculations of the detector response to externally incident photons, to internally induced radioactivity, and to bremsstrahlung produced in the surrounding material, for the NaI gamma-ray spectrometer flown on Apollo 15, 16 and 17; calculations of the response to photons of high-purity Ge detectors; and space-shielding calculations of the absorbed dose from electrons and bremsstrahlung in tissue-equivalent and electronic-device-equivalent targets within a variety of assumed spacecraft geometries. In this paper we briefly describe the Monte Carlo model, outline our recent work in the preparation of improved bremsstrahlung cross sections, and highlight the work in some of these application areas by way of illustrative results.

---

[*]This work was supported in part by the Office of Health and Environmental Research of the US Department of Energy.

MONTE CARLO MODEL

Details of the Monte Carlo model, and fuller discussions of the underlying cross sections and distributions, can be found in Refs. 1-6. We include here only a brief description to indicate the level of accuracy and the comprehensiveness with which the transport processes are treated. The transport of photons is handled using conventional methods in which successive photon interactions are sampled individually in direct analogy to the physical processes. The energy and direction of the photon is appropriately changed in the case of a Compton interaction; the emission of characteristic x rays or Auger electrons are sampled in the case of inner-shell photoelectric absorption. The histories of the Compton, pair, photo-, and Auger electrons set in motion due to photon interactions are followed as described below.

The very large number of elastic and inelastic interactions between electrons and atoms are treated according to a "path-segment" model. The electron trajectories are divided into many segments in each of which numerous interactions occur. The net angular deflection from the combined effect of the elastic and inelastic collisions in each segment is sampled from the Goudsmit-Saunderson[7] multiple-scattering distribution, using as the underlying single-scattering cross section the Mott[8] cross section modified according to work of Molière[9] to take into account the effects of screening. The net energy loss due to ionization and excitation along each segment is sampled from the Landau[10] straggling distribution (as modified by Blunck and Leisegang[11]), which takes into account fluctuations in collision energy loss. The individual radiative energy losses along the segment are sampled from bremsstrahlung production cross sections. Also sampled along each segment is the production of energetic knock-on electrons (delta rays) and the production of characteristic x rays (or Auger electrons) from inner-shell electron-impact ionization events. Annihilation radiation is emitted when a positron slows down to rest.

The Monte Carlo model follows all generations of electrons and photons in the target with energies above chosen cut-off values. A weighting scheme allows for the sampling of bremsstrahlung photon histories in excess of the natural production rates so that statistical fluctuations in the bremsstrahlung results can be reduced without an increase in the number of time-consuming electron histories. The statistical accuracy of the photon results is also improved by scoring for each collision point in the photon history the probability that the photon crosses the boundaries of interest without further interaction.

BREMSSTRAHLUNG CROSS SECTIONS

Necessary for the calculation of electron-bremsstrahlung transport is, of course, the availability of accurate bremsstrahlung production cross sections. In earlier work we used a combination of Bethe-Heitler cross sections[3], obtained primarily from available analytical Born-approximation theory[12], and modified by rather poorly known empirical correction factors. More recently, we completed the preparation of a comprehensive set of cross sections (differential in emitted photon energy) for bremsstrahlung production by electrons on neutral atoms, taking explicit account of the contributions from interactions both with screened atomic nuclei and orbital electrons[13,14]. These cross sections go beyond the results from Bethe-Heitler theory in that they incorporate (a) the results from numerical phase-shift calculations for nuclear-field bremsstrahlung by Pratt et al.[15] at energies below 2 MeV; (b) the analytical high-energy nuclear-field bremsstrahlung theory of Davies et al.[16], including Coulomb corrections consistent with accurately-known high-frequency limits and including screening corrections evaluated with Hartree-Fock atomic form factors; and (c) the analytical electron-electron bremsstrahlung theory of Haug[17], combined with screening corrections derived from Hartree-Fock incoherent scattering factors, for bremsstrahlung in the field of the atomic electrons.

The differences between the new and the old bremsstrahlung cross sections are indicated in Table I, which gives the ratios of the new to the old cross-section values for electrons with kinetic energies T of 0.05, 0.5 and 5 MeV in Al and Pb, for emitted photon energies k from 0 to T. For this purpose, the "old" bremsstrahlung cross section is defined as

$$d\sigma/dk = c_R (1+1/Z) \{f_E\, d\sigma_{3BN}/dk + (d\sigma_{3BS}/dk - d\sigma_{3BNb}/dk)\} \quad , \quad (1)$$

where f_E is the Elwert[18] Coulomb correction factor. In Eq. (1), the subscripts follow the formula-designation scheme used in the review by Koch and Motz[12]: 3BN denotes the Born-approximation Bethe-Heitler result, with no energy approximation, for an unscreened nucleus; 3BS denotes the result in the high-energy approximation for a screened nucleus (using the analytical approximations for the Thomas-Fermi screening functions found in Ref. 19); and 3BNb is the high-energy approximation of 3BN or, alternatively, the result from 3BS in the limit of no screening. Scanning Eq. (1) from right to left, we note the following: the last two terms comprise a screening correction (valid in the high-energy approximation); multiplication of $d\sigma_{3BN}/dk$ by the Elwert factor f_E gives non-zero values for the cross section at the high-frequency limit (k=T) that are correct to first order in

Z/137; the multiplicative factor $(Z^2+Z)/Z^2$ approximately takes into account the additional bremsstrahlung produced in the field of the Z atomic electrons; and the empirical correction factor c_R (assumed independent of emitted photon energy) had been estimated to be 1.0 at all electron energies for Al, and 1.018, 1.182 and 1.0 at T=0.05, 0.5 and 5 MeV, respectively, for Pb.

From Table I we see that the deficiencies of the old cross section are more serious for high-Z materials. For low-Z atoms, the failure of the screening correction for soft photon emission does however become noticeable at low electron energies. This is somewhat mitigated by the fact that in a transport calculation these low- energy photons are rather quickly absorbed by the medium.

Table I. Ratio of new to old cross sections for the production of bremsstrahlung by electrons incident on neutral atoms of Al and Pb. Results are given for the cross section differential in emitted bremsstrahlung energy k and for electron kinetic energies T = 0.05, 0.5 and 5 MeV. The bases of the old and new cross sections are discussed in the text.

k/T \ T(MeV)	Al			Pb		
	0.05	0.5	5	0.05	0.5	5
0	0.415	0.744	1.086	0.297	0.653	0.963
0.1	0.812	0.944	1.009	0.625	0.903	0.957
0.2	0.883	0.961	1.001	0.728	0.974	0.963
0.3	0.909	0.966	0.995	0.794	1.030	0.971
0.4	0.924	0.971	0.993	0.845	1.084	0.980
0.5	0.934	0.972	0.995	0.889	1.143	0.996
0.6	0.941	0.973	0.999	0.930	1.206	1.022
0.7	0.949	0.973	0.998	0.972	1.275	1.062
0.8	0.956	0.978	0.995	1.015	1.353	1.125
0.9	0.962	0.985	0.984	1.059	1.447	1.208
1.0	0.962	0.993	0.922	1.103	1.558	1.331

AURORAL BREMSSTRAHLUNG

Early applications of the Monte Carlo codes included calculations of the energy deposition and of the electron and photon flux spectra due to the precipitation of electrons and the production of secondary bremsstrahlung in the Earth's atmosphere[19-23]. Such auroral phenomena are, of course, of strong interest in atmospheric physics, but the resultant bremsstrahlung flux may also represent a possible background component in x-ray astronomy.

In the auroral bremsstrahlung calculations, the atmosphere was represented by a semi-infinite air medium whose plane boundary forms the top of the atmosphere. The incident electron flux was assumed to be of uniformly broad spatial extent (i.e., wide-area precipitation), so that the only spatial variable in the problem is the atmospheric depth (in g/cm²). The initial electron flux, incident on the top of the atmosphere, was assumed to be isotropic over the downward hemisphere. Systematic calculations were done for monoenergetic incident electrons, so that these results could be later combined for any arbitrary incident electron spectra.

Examples of the results are shown in Figs. 1 and 2, for incident electron flux spectra of the form

$$F_0(T_0) = (\phi_0/\alpha) \exp(-T_0/\alpha) \quad , \qquad (2)$$

where α is the "e-folding" energy, and ϕ_0 is the incident <u>flux</u> differential only in solid angle (e.g., with units of cm^{-2} sec^{-1} sr^{-1}) which is assumed to be constant for pitch angles between 0 and $\pi/2$. It is worthwhile to point out that the results in Figs. 1 and 2 are normalized per unit total <u>current</u> j_0 of incident electrons crossing the plane boundary of the atmosphere (e.g., with units of cm^{-2} sec^{-1}) which in this case is equal to $\pi\phi_0$.

Figure 1 gives the resultant downward-directed bremsstrahlung flux spectra at a depth of 5 g/cm² from the top of the atmosphere, a depth much larger than the range of the incident electrons. The flux spectra differential in photon energy k are shown in Fig. 1a, and tend to vanish at photon energies below ~ 50 keV due to the onset of strong photoelectric absorption. Figure 1b gives the corresponding integral flux, i.e., the number of photons with energies greater than k. Further results, covering most all balloon altitudes, can be found in Ref. 21. Figure 2 gives results for the bremsstrahlung flux emerging upward through the top of the atmosphere. In this case the photons emerge from the production region in the medium with rather little attenuation, so that a turn-over in the spectrum occurs at much lower energies. Such results as these, multiplied by the appropriate incident electron current, can be used to gauge possible contributions to background effects in x-ray measurements made at balloon or satellite altitudes.

DETECTOR RESPONSE

In this section, some problems are described that involve the calculation of the response functions of NaI and Ge detectors which are used to measure photon spectra. The conversion of a measured pulse-height distribution into the true incident gamma-ray spectrum requires accurate and detailed information on the response of the detector to monoenergetic photons, usually over a rather broad range of energies. The limited number of suitable monoenergetic gamma-ray sources and the complex structure of the response function itself makes it extremely difficult to develop solely on an experimental basis the necessary body of data.

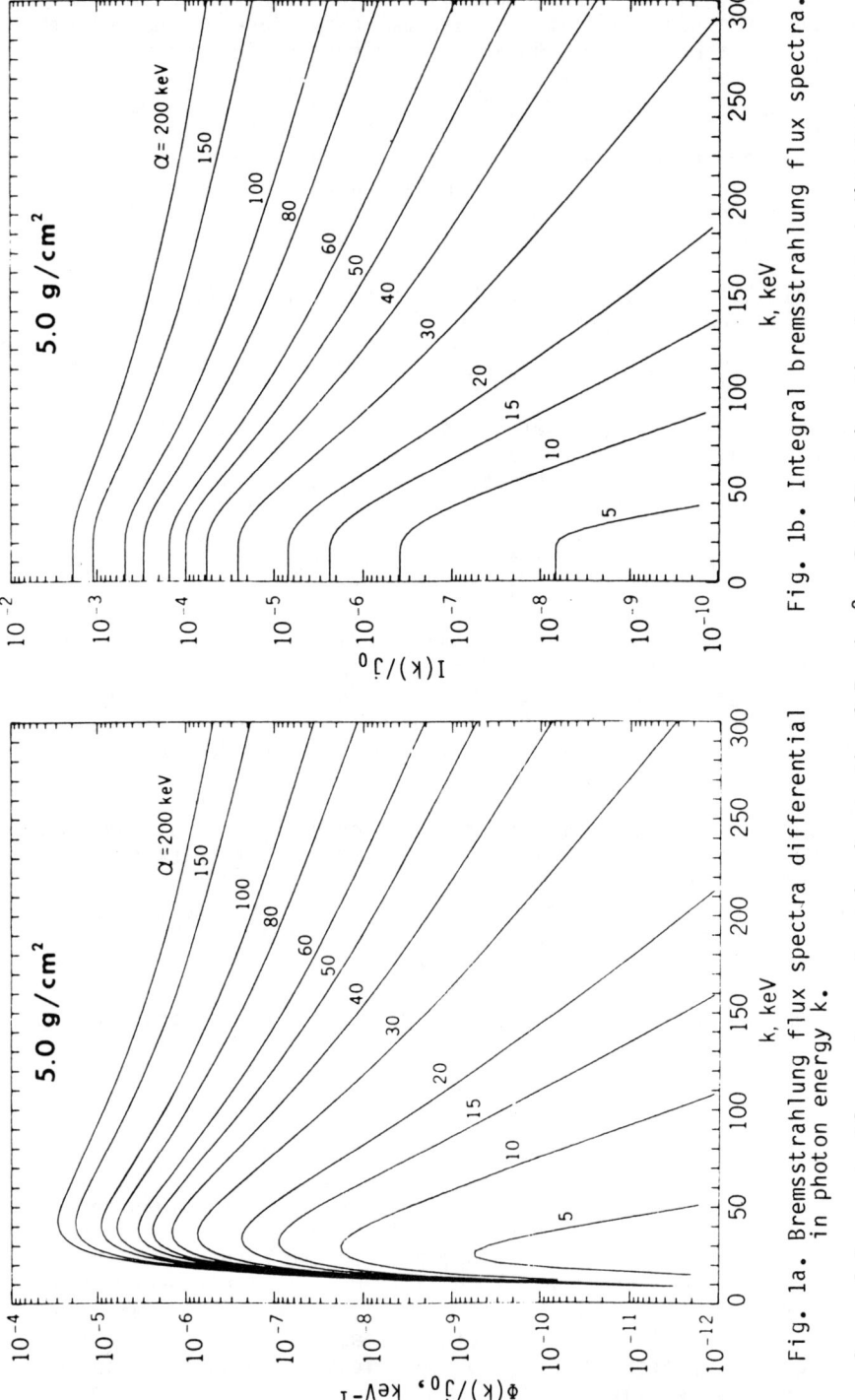

Fig. 1a. Bremsstrahlung flux spectra differential in photon energy k.

Fig. 1b. Integral bremsstrahlung flux spectra.

Auroral bremsstrahlung at an atmospheric depth of 5 g/cm². Results give the downward-directed bremsstrahlung flux spectra (i.e., integrated over all downward angles) per unit incident electron current, from electron beams incident on the top of the Earth's atmosphere with exponential energy spectra (with an e-folding energy α).

Fig. 2. Auroral bremsstrahlung at the top of the atmosphere. Results give the upward-directed bremsstrahlung flux spectra (i.e., integrated over all upward angles) per unit incident electron current, from electron beams incident with exponential energy spectra (with an e-folding energy α).

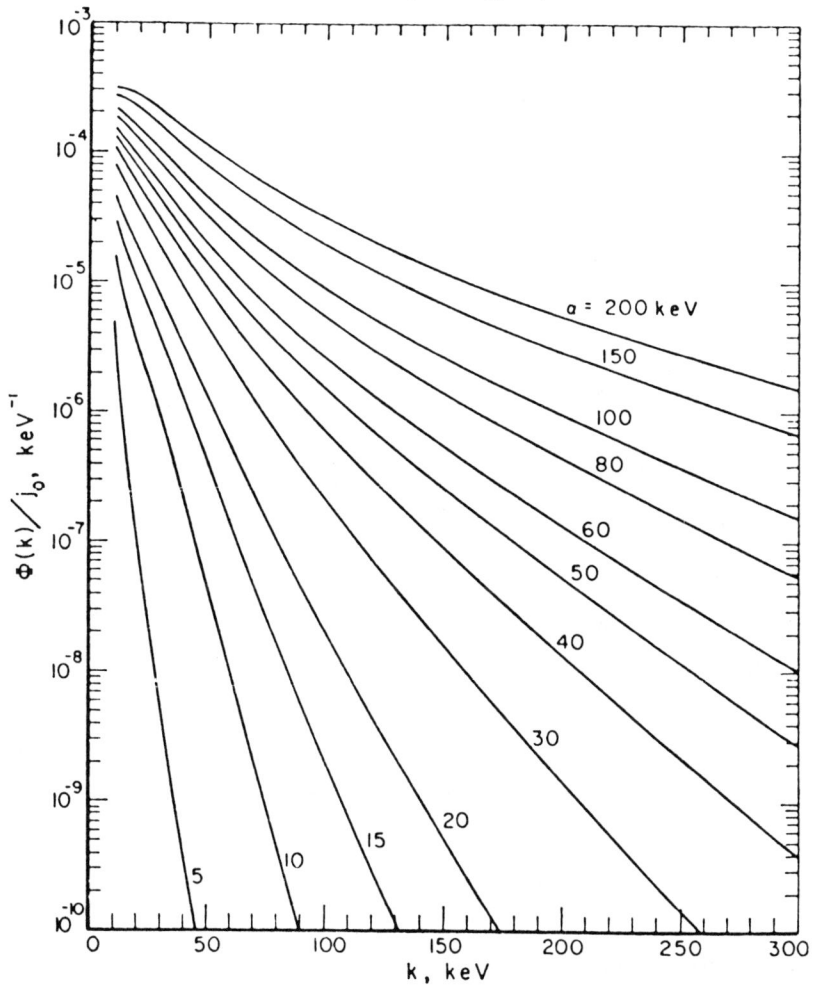

Fig. 2a. Bremsstrahlung flux spectra differential in photon energy k.

We can write the response function R(h,E) as the convolution of two distributions:

$$R(E,h) = \int_0^E D(E,\varepsilon)\, G(\varepsilon,h)\, d\varepsilon \quad , \tag{3}$$

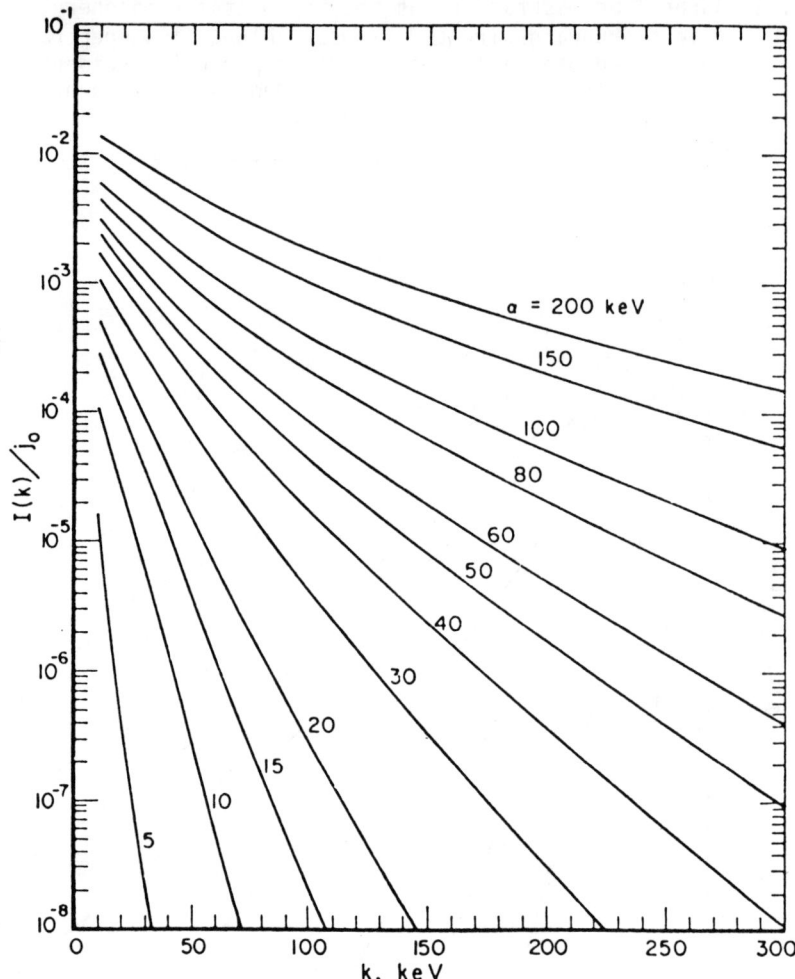

Fig. 2b. Integral bremsstrahlung flux spectra.

where $R(E,h)$ is the probability per unit pulse height that a photon incident with energy E will produce a pulse of height h, $D(E,\varepsilon)$ is the probability per unit deposited energy that the incident photon deposits energy ε in the detector, and $G(\varepsilon,h)$ is the probability per unit pulse height that the deposition of energy ε gives rise to a pulse of height h. The detector resolution function $G(\varepsilon,h)$, which depends on the efficiency and statistics of signal collection, is usually assumed to be a Gaussian whose full-width at half-maximum as a function of ε is best determined experimentally.

The energy-deposition spectrum $D(E,\epsilon)$ can be rather accurately determined from our Monte Carlo calculations. In general, this spectrum has the form of a line spectrum plus a continuum $C(E,\epsilon)$:

$$D(E,\epsilon) = C(E,\epsilon) + P_0\, \delta(\epsilon-E) + P_1\, \delta(\epsilon-E+mc^2)$$
$$+ P_2\, \delta(\epsilon-E+2mc^2) + P_x\, \delta(\epsilon-E+E_x) \ . \qquad (4)$$

In Eq. (4), δ is the Dirac delta function, P_0 is the probability that the photon energy E is completely absorbed (i.e., the area under the total-absorption peak or the photopeak efficiency), and P_x is the probability that the full energy is deposited except for the escape of a fluorescence x-ray of energy E_x produced in photo-electric absorption events (fluorescence escape peak). For incident photon energies above ~ 1 MeV where pair production can occur, P_1 and P_2 are the probabilities that the entire energy is deposited except for the amounts mc^2 and $2\,mc^2$ carried out of the detector by one or two unscattered annihilation quanta (single-annihilation and double-annihilation escape peaks, respectively). Depending on detector dimensions and photon energy, these components can depend not only on the escape of the photon scattered in the detector, but also on the escape of secondary electrons and their bremsstrahlung.

The detection efficiency η is the probability that the incident photon will have at least one interaction in the detector leading to the deposition of energy, and is thus equal to the integral of $D(E,\epsilon)$ over all ϵ (or, equivalently, the integral of $R(E,h)$ over all h). However, the evaluation of the detection efficiency can be accomplished independently — and much more simply — through the solution of the geometrical problem, without knowledge of the energy-deposition spectrum. This can be used to advantage because the detection efficiency is a good scaling parameter: dividing $D(E,\epsilon)$ by η reduces its dependence on detector dimensions and on the photon angle of incidence.

A program was undertaken in the early 1970's to characterize rather completely the response functions for a 3" (diam.) x 3" (height) cylindrical NaI detector[24]. The goal of this work was to provide the means to generate an extensive, reliable library of monoenergetic response functions for use in the pulse-height deconvolution algorithm developed by Trombka[25], which was used in the analysis of gamma-ray measurements made on the Apollo flights 15 and 16. These Apollo measurements had two purposes: those from lunar orbit measured the cosmic-ray-induced and natural emission of gamma-ray lines to provide information for the elemental analysis and geochemical mapping of the Moon's surface; and the measurements during trans-Earth flight were of the diffuse cosmic gamma-ray spectrum whose shape has significant astrophysical implications.

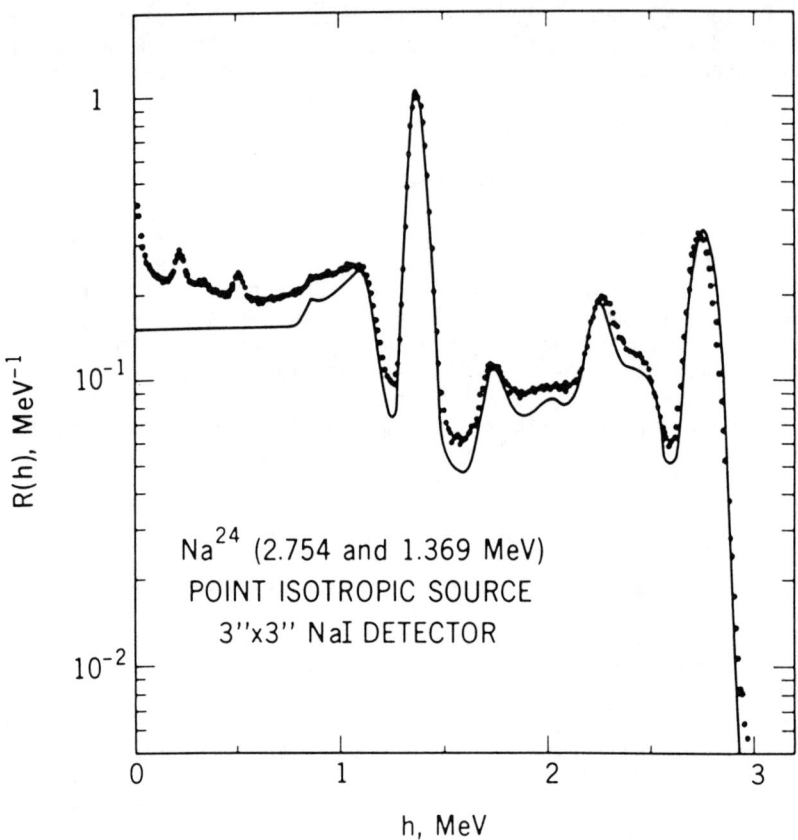

Fig. 3. Response function for a 3" × 3" NaI detector irradiated with gamma rays from a ^{24}Na point-isotropic source. The curves are from the Monte Carlo calculation, and the points are from the measurements of Heath[36].

The validity of our Monte Carlo calculations was tested through comparisons of calculated and experimental response functions for various sized detectors at photon energies up to 20 MeV. A typical comparison is given in Fig. 3. Systematic calculations of the energy-deposition spectrum were then done for the bare 3" × 3" NaI crystal (with no surrounding material) exposed to broad parallel beams of gamma rays incident along the cylinder axis with energies from 0.1 to 20 MeV. The calculated components of the energy-deposition spectrum were refined into a scaled, smooth dataset from which one could easily interpolate to any incident photon energy and, through the evaluation of Eq. (3), generate the complete response function.

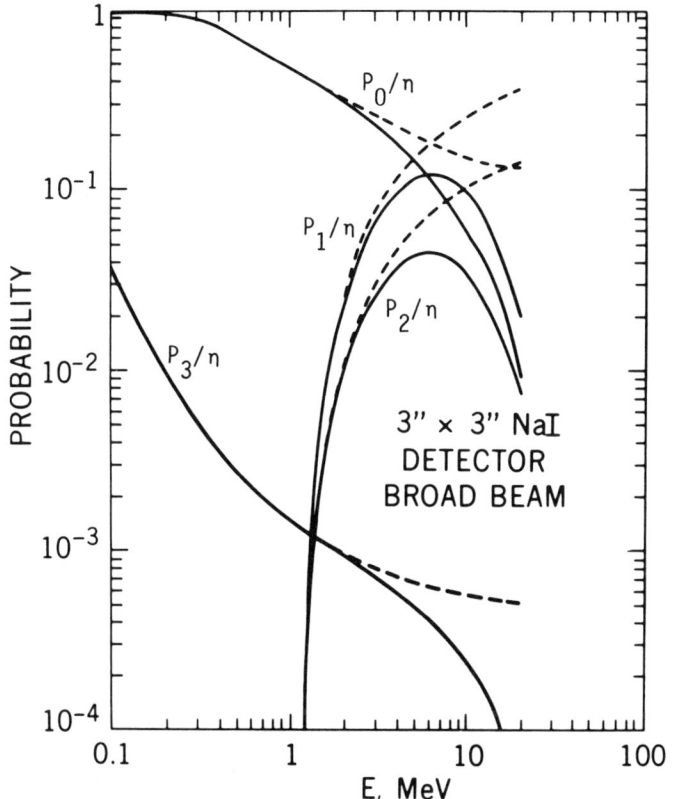

Fig. 4. Response of a NaI detector to broad, parallel, monoenergetic photon beams. Total absorption peak (P_0), single and double annihilation radiation escape peaks (P_1 and P_2) and iodine K-shell fluorescence escape peak (P_x), for a 3" × 3" NaI detector. All quantities are divided by the detector efficiency η. Dashed curves are calculated disregarding the escape of bremsstrahlung and secondary charged particles.

Figure 4 shows for the 3" × 3" detector the resultant peak probabilities divided by the detection efficiency. Corresponding results calculated without taking into account the escape of secondary electrons and bremsstrahlung are shown also, and indicate the rather significant effect above about 3 MeV of including electron and bremsstrahlung transport. Because the dimensions of the actual flight detector were somewhat different, the dataset was re-calculated for a 2.75" × 2.75" crystal. Figure 5 shows a family of complete response functions generated for this detector, assuming a resolution of 7.5% for the ^{137}Cs 662-keV gamma ray.

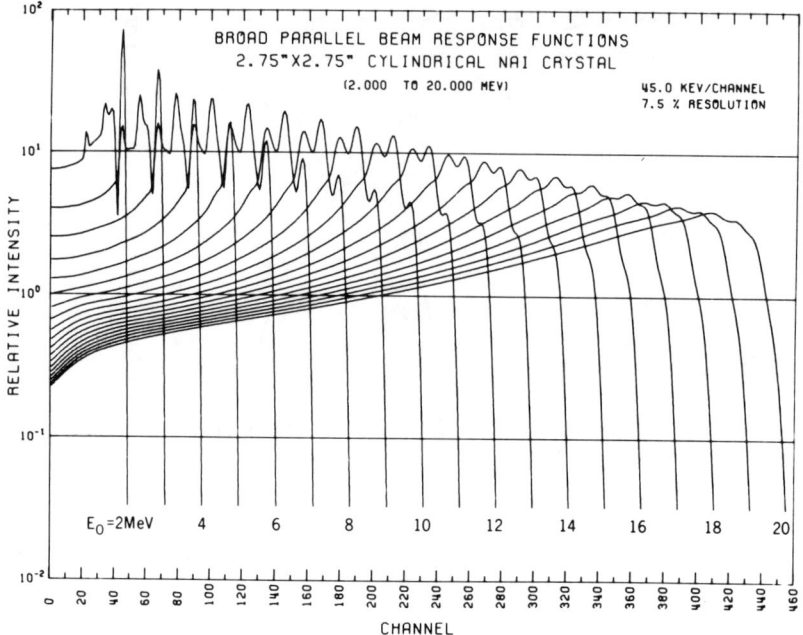

Fig. 5. Family of response functions for a 7 cm × 7 cm NaI detector, with an assumed resolution of 7.5% at 662 keV. Curves are normalized such that the area under each is equal to 45 times the detection efficiency. Results pertain to the case of a broad parallel beam incident on the flat end face of the detector.

In the analysis of the flight data, a number of background effects had to be subtracted before the measured data were converted to true incident photon spectra. This was especially important in the analysis of the diffuse cosmic gamma-ray measurements in which about 80% of the measured counts were from unwanted background sources. The various background components and, in some cases, the Monte Carlo calculations done to study them are discussed in Trombka et al.[26] These included: (a) direct charged-particle counts due to insufficient rejection by the anti-coincidence shield; (b) gamma rays from natural and induced radioactivity in the spacecraft and in the material surrounding the detector; (c) bremsstrahlung photons produced by high-energy electrons (presumably of Jovian origin) in the aluminum casing of the detector; and (d) the pulse-height distribution from the decay of radioactivity that was induced in he detector's central crystal due to bombardment by the protons, cosmic rays, and secondary neutrons present in the space environment.

The theoretical estimation of the background due to induced activity in the detector (see Dyer et al.[27-29]) consists of two parts: the calculation of the rate of isotope production and decay in the detector (i.e., the source function), and the calculation of the pulse-height distribution due to the decay of the internal radioactive products. The second part was done with electron-photon Monte Carlo methods that involved the sampling from the complete decay scheme of radioactive atoms assumed distributed uniformly over the detector volume (see Ref. 30). Some typical results for the response of the Apollo detector are shown in Fig. 6. The decay of ^{22}Na (primarily by β^+ to an excited state of ^{22}Ne which then de-excites by gamma-ray emission) gives rise to the pulse-height distribution shown in Fig. 6a. The periodic nature of the distribution is a result of the shift of the β^+ spectrum by combinations of the total absorption of the 1.275-MeV gamma ray and the 0.511-MeV photons produced in the annihilation of the β^+ particles. The gamma ray appears as a discrete line only due to total absorption events following the decay to ^{22}Ne by electron capture (about 9.5% of the decays). Figure 6b shows the pulse-height distribution for the decay of ^{124}I (77.3% by electron capture, 22.7% by 3 β^+ groups, and by 30 gamma rays in the de-excitation of ^{124}Te). The prominent peaks are due to combinations of the total absorption of the K-shell binding energy and of the three most-probable emitted gamma rays.

Fig. 6. Pulse-height distribution produced by radioactive products uniformly distributed within a 7 cm × 7 cm NaI detector, with a resolution of 7.5% at 662 keV. The results are normalized to one decay.

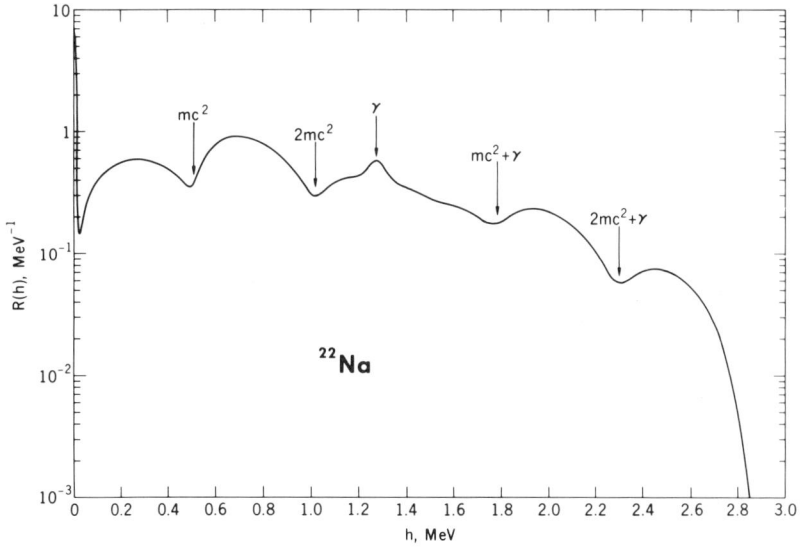

Fig. 6a. ^{22}Na. The energies of annihilation quanta (mc^2) and the 1.275-MeV gamma ray (γ), involved in the decay are indicated.

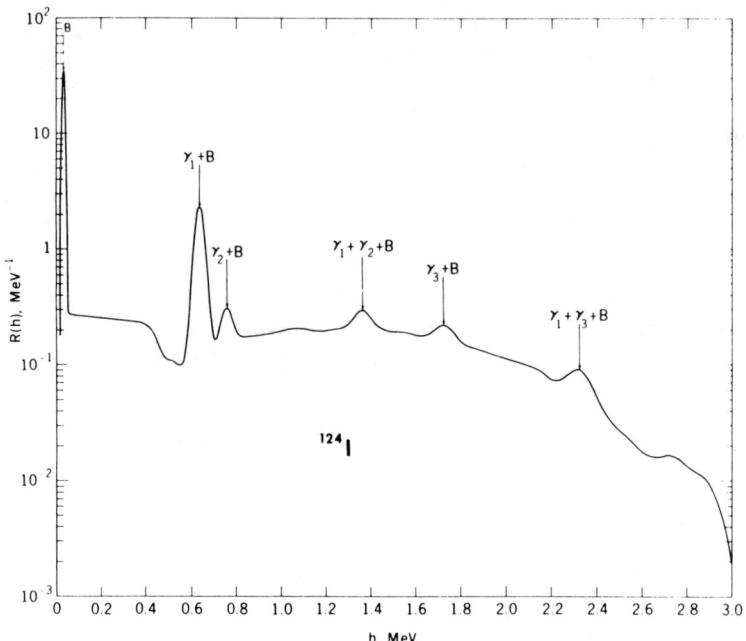

Fig. 6b. ^{124}I. Indicated are the peaks corresponding to the absorption of the K-shell binding energy (B: 0.032 MeV) and some high-intensity gamma rays (γ_1: 0.603 MeV, γ_2: 0.723 MeV, γ_3: 1.691 MeV).

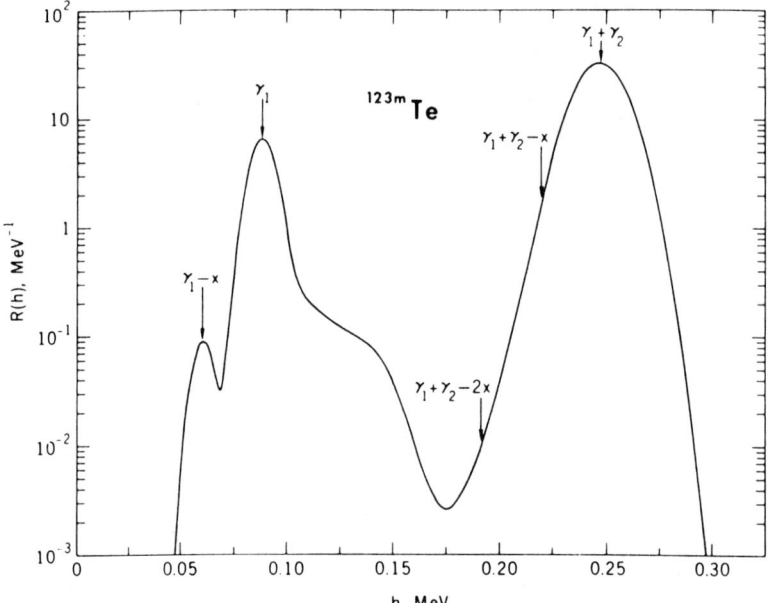

Fig. 6c. ^{123m}Te. The energy of the K x-ray (x: 0.028 MeV) and the transition energies (γ_1: 0.088 MeV, γ_2: 0.159 MeV) are indicated.

Figure 6c shows results for ^{123m}Te which decays by two successive transitions: an 88-keV transition preferentially by internal-conversion-electron emission and an 159-keV transition preferentially by gamma-ray emission. Notable features here are the Compton shoulder due to incomplete absorption of the 159-keV gamma ray, and the small satellite peak due to the escape of the K x-ray associated with the internal conversion of a K-shell electron.

A 5.5 cm × 5.5 cm high-purity Ge detector has been selected in the design of the Gamma-Ray Remote Sensing Spectrometer to be flown on NASA's Mars Observer Mission. Extending the techniques used on the Apollo flights, this experiment will measure from orbit the gamma- ray line emission induced in the Martian surface by incident cosmic rays, solar protons and secondary neutrons (as well as that due to natural radioactivity), for the compositional analysis and mapping of the Martian surface. Because the high resolution of the Ge detector will separate rather well the peaks associated with the line emissions, the analysis requires knowledge mainly of the delta-function components of the detector response (total-absorption and the single- and double-annihilation escape peaks). These have been calculated with our Monte Carlo code for broad beams of gamma rays incident with energies up to 20 MeV[31]. Calculations done as a function of incident angle confirm that although the absolute peak efficiencies vary by as much as about 50%, if they are divided by the detection efficiency then the resultant peak fractions are virtually independent of the incident angle. The relative peak probabilities for the Ge detector are shown in Fig. 7.

SPACE SHIELDING

In some regions of the space radiation environment, the radiation dose delivered to personnel or to electronic components within spacecraft by incident electrons and their secondary bremsstrahlung represents a significant hazard. The latter can pose a particularly troublesome threat because of the ability of bremsstrahlung photons to penetrate deep into the target. Electron/photon Monte Carlo calculations, necessary for accurate estimates of these radiation doses, are capable of including complete geometrical detail of a spacecraft and its contents[32]. However, such a complicated calculation is very expensive, particularly for the dose in a relatively small interior volume, and the results for a specific configuration are unlikely to be applicable to another case of interest. An alternative approach is to develop data in a somewhat simple, but more generally applicable, geometry in which the accuracy of the transport results can be maintained at modest effort.

Thus, in order to provide an efficient utility for dose estimates, Monte Carlo calculations were done of the one-dimensional distribution in depth of the absorbed dose, and of the forward- and backward-directed energy-degradation flux spectra, produced by isotropic incident fluxes of monoenergetic electrons (with energies up to 20 MeV) and their secondary bremsstrahlung in semi-infinite

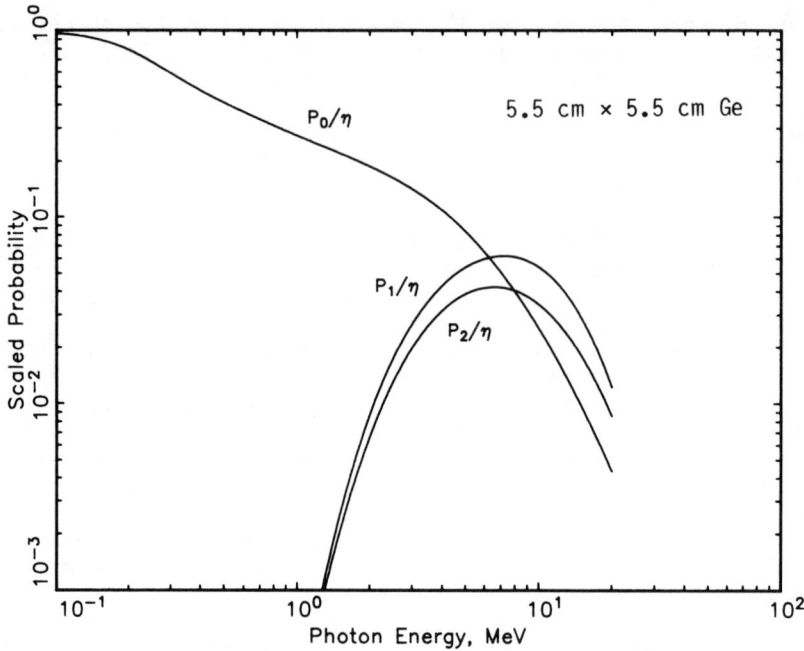

Fig. 7. Peak probabilities for total absorption, P_0, single-annihilation escape, P_1, and double-annihilation escape, P_2, for the 5.5 cm × 5.5 cm Ge detector irradiated by broad beams of gamma rays. The results have been scaled by the detection efficiency to remove dependence on the angle of incidence.

aluminum slabs. With these data it was possible to generate a database of dose kernels for use in an algorithm which quickly interpolates and integrates for any spectrum of incident electrons likely to be encountered, and whose output is given separately for electrons and bremsstrahlung in terms of dose to Al, H_2O, Si and SiO_2, at points in semi-infinite, as well as behind finite-thickness, Al slabs. Further details can be found in Ref. 33, which describes also the computer code, named SHIELDOSE, that uses the database.

Although the assumption that spacecraft are composed mostly of Al would seem adequate, at least in mission planning and design stages, there has been a fair amount of ambiguity as to how slab results may be applied to three-dimensional objects irradiated from all directions. It turns out that the assumption in space-shielding problems that the incident electron flux is isotropic, at least in a time-averaged sense, leads to very useful relations between the doses in slab targets and those in solid spheres and hollow spherical shells, which are more realistic configurations in this application. These connections were explored in Ref. 34.

Consider solid spheres and hollow spherical shells, where R is the outer radius R of the sphere or shell, and t is either the radial depth from the surface of the solid sphere or the radial thickness of the spherical shell. It is assumed that there exists a distribution that gives the energy deposited per unit pathlength in the target as if the penetrating radiation travels in a straight line. For protons which do travel nearly in straight lines, this distribution is simply the stopping power. For electrons and bremsstrahlung which do not have straight trajectories, we can nevertheless define such a distribution implicitly in terms of the one-dimensional slab depth-dose.

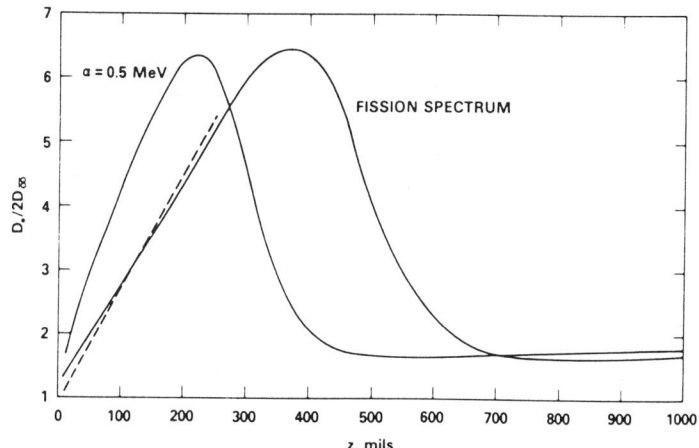

Fig. 8. Ratio of the electron dose $D_.$ at centers of aluminum spheres (radius z) to twice the dose D_∞ in a semi-infinite aluminum medium (at depth z). The results are plotted out to large z where the bremsstrahlung dominates. Solid curves are from the transformation of our slab results, given both for incident electrons with a fission spectrum and for electrons with an exponential spectrum characterized by an e-folding energy α = 0.5 MeV. Dashed curve is the fission-spectrum results of Jordan[35] who considered explicitly the two geometries.

The results are that one can estimate the dose as a function of t within either spheres or shells in terms of a simple integral over the slab dose. For the case t = R (i.e., the center of a solid sphere), the relationship reduces to one involving a derivative of the slab dose with respect to depth. The dose at centers of spheres (as a function of sphere radius R = z) is shown in Fig. 8 in terms of its ratio to twice the corresponding slab dose at depth z*.

___
*The factor of two comes about because we assume for the slab case that the dose point would be sandwiched between two slabs of thickness z, irradiated from both sides.

Results are shown for an exponential spectrum of incident electrons with an e-folding energy of $\alpha = 0.5$ MeV, and for a fission electron spectrum. For the fission spectrum, there is good agreement with the corresponding ratios obtained by Jordan[35], who calculated the electron dose (but not the bremsstrahlung tail) explicitly for both geometries using his adjoint Monte Carlo code. To pin down the accuracy of the approximate transformation of doses in slabs to spheres for the bremsstrahlung component, we did calculations for 2-MeV electrons incident on an Al sphere of radius $R = 20$ g/cm² using a general-geometry code incorporating our electron-photon Monte Carlo model. The results, as a function of depth in the sphere, are compared to those from the transformed SHIELDOSE dose kernel in Fig. 9. Agreement is reasonably good, to within about 20%, which is nearly within the statistical errors of the general-geometry Monte Carlo results. It should be pointed out here that the general-geometry calculation required approximately one hour on a Cyber 205 super computer in order to get sufficient scores in the small volumes near the center of the sphere, while the equivalent accuracy can be obtained from a slab calculation requiring less than about 4 minutes.

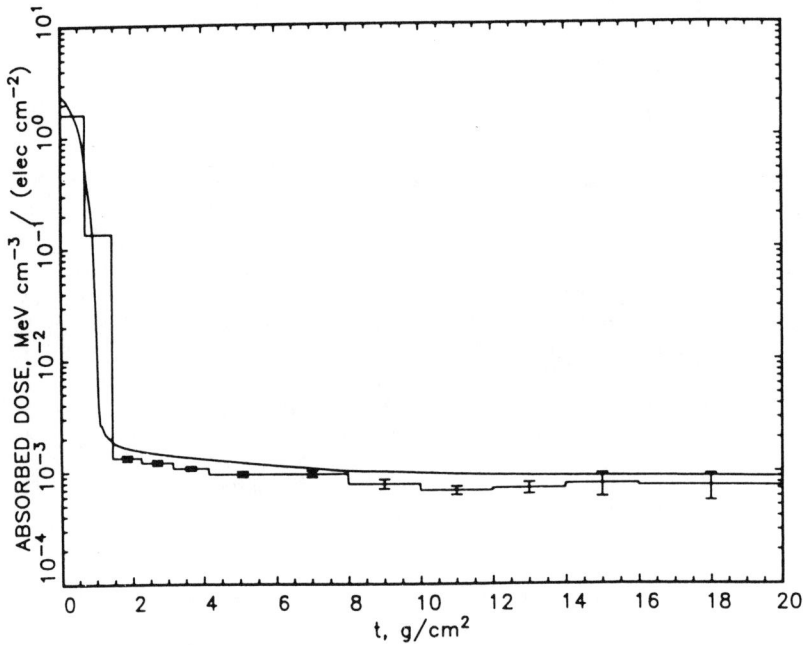

Fig. 9. The spatial distribution of absorbed dose from 2-MeV electrons incident on an Al sphere with a radius of 20 g/cm². The results of Monte Carlo calculations done for the sphere (histogram) are compared to the results obtained by converting the slab depth-dose distribution to apply to the sphere (curve).

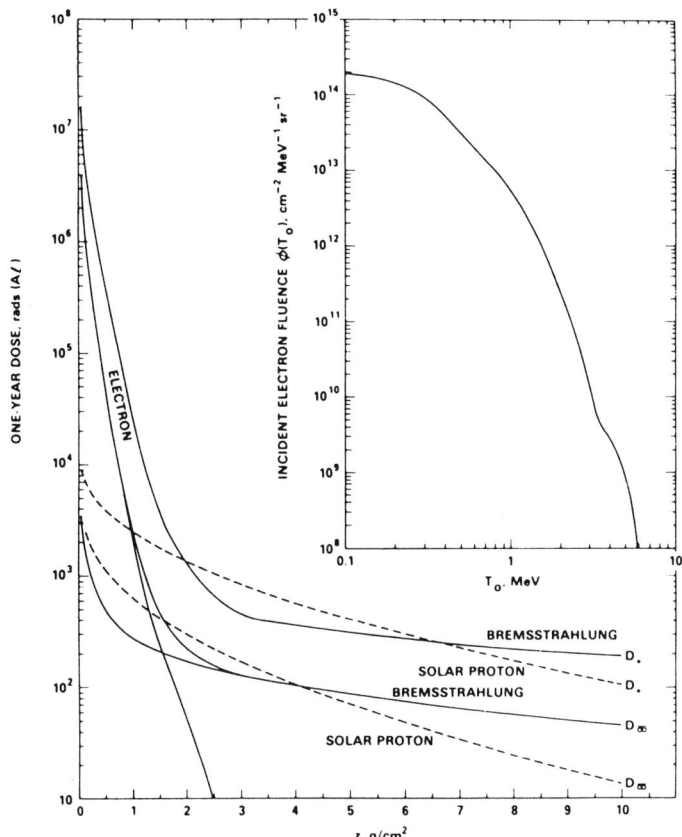

Fig. 10. Depth-dose distributions in aluminum targets for the radiation encountered during one year in a geosynchronous orbit. The orbit parameters are given in the text; the incident electron fluence is given in the insert. Results are given both for the dose D_∞ at depth z in a semi-infinite medium and for the dose $D_\bullet$ at the center of a sphere of radius z. The solar-proton dose, for one anomalously-large event, is given by the dashed curves.

Figure 10 gives an example of the application of these methods to dose estimates for a geosynchronous orbit with an altitude of 35790 km, an inclination of 0°, and a parking latitude of 160° W. For the same geosynchronous spectrum, Fig. 11 illustrates the relationship to the slab case of the dose at points in a solid sphere or at the inside surface of a hollow spherical shell, both with outer radius R = 25 g/cm².

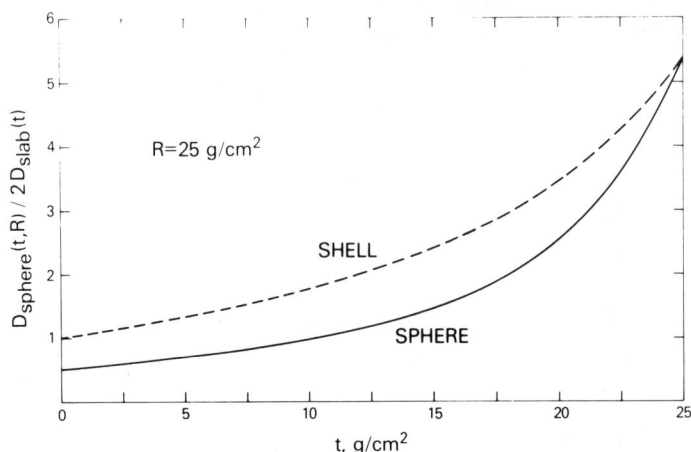

Fig. 11. Ratio of the dose from electrons and secondary bremsstrahlung in spheres to twice the corresponding dose in a semi-infinite slab. Results pertain to aluminum targets irradiated by an electron flux expected in a geosynchronous orbit, and are given at depths t in a solid sphere (solid curve) and at points on the inside surface of a hollow spherical shell of thickness t (dashed curve), both with an outer radius of 25 g/cm².

CONCLUSIONS

Rather accurate Monte Carlo codes and cross sections are available for calculating electron and photon transport in extended media, including space-borne structures and instrumentation. Such calculations can be of great value in determining the effects of this radiation on systems of interest. With the availability of these calculational methods, the major source of uncertainty in their results lies then in the specification of the incident primary radiation. Critical applications may therefore require better knowledge of the pertinent radiation environment than is available from past information.

REFERENCES

1. M.J. Berger, "Monte Carlo Calculation of the Penetration and Diffusion of Fast Charged Particles", in Methods in Computational Physics, Vol. 1, Eds. B. Alder, S. Fernbach and M. Rotenberg, Academic Press, New York (1963) p. 135.
2. M.J. Berger and S.M. Seltzer, NBS Reports 9836 and 9837 (1968).
3. M.J. Berger and S.M. Seltzer, Phys. Rev. C $\underline{2}$, 621 (1970).
4. M.J. Berger and S.M. Seltzer, NASA SP-169 $\overline{285}$ (1968).
5. S.M. Seltzer and M.J. Berger, NBS Publ. NBSIR 84-2931 (1984).

6. S.M. Seltzer, "An Overview of ETRAN Monte Carlo Methods of Coupled Electron/Photon Transport Calculations", Proceedings of the Course on Monte Carlo Transport of Electrons and Photons Below 50 MeV, International School of Radiation Damage and Protection, Ettore Majorana Centre for Scientific Culture, Erice, Italy, Sept. 24-Oct. 3, 1987, Plenum Press (in press).
7. S. Goudsmit and J.L. Saunderson, Phys. Rev. 57, 24 (1940).
8. N.F. Mott, Proc. Roy. Soc. (London) A124, 425 (1929); see also J.A. Doggett and L.V. Spencer, Phys. Rev. 103, 1597 (1956).
9. G. Moliere, Z. Naturforsch. 3a, 78 (1948).
10. L. Landau, J. Phys. (USSR) 8, 201 (1944).
11. O. Blunck and S. Leisegang, Z. Physik 128, 500 (1950).
12. See, e.g., H.W. Koch and J.W. Motz, Rev. Mod. Phys. 31, 920 (1959).
13. S.M. Seltzer and M.J. Berger, Nucl. Instr. and Meth. B12, 95 (1985).
14. S.M. Seltzer and M.J. Berger, Atomic Data and Nuclear Data Tables 35, 345 (1986).
15. R.H. Pratt, H.K. Tseng, C.M. Lee, L. Kissel, C. MacCallum and M. Riley, Atomic Data and Nuclear Data Tables 20, 175 (1977); errata in 26, 477 (1981).
16. H. Davies, H.A. Bethe and L.C. Maximon, Phys. Rev. 93, 795 (1954); and H. Olsen, Phys. Rev. 99, 1335 (1955).
17. E. Haug, Z. Naturforsch. 30a, 1099 (1975).
18. G. Elwert, Ann. Physik 34, 178 (1939).
19. M.J. Berger and S.M. Seltzer, J. Atmosph. Terr. Phys. 32, 1015 (1970).
20. M.J. Berger and S.M. Seltzer, J. Atmosph. Terr. Phys. 34, 85 (1972).
21. S.M. Seltzer, M.J. Berger and T.J. Rosenberg, NASA SP-3081 (1973).
22. M.J. Berger, S.M. Seltzer and K. Maeda, J. Atmosph. Terr. Phys. 36, 591 (1974).
23. S.M. Seltzer and M.J. Berger, J. Atmosph. Terr. Phys. 36, 1283 (1974).
24. M.J. Berger and S.M. Seltzer, Nucl. Instr. and Meth. 104, 317 (1972).
25. J.I. Trombka and R.L. Schmadebeck, Nucl. Instr. and Meth. 62, 253 (1968).
26. J.I. Trombka, C.S. Dyer, L.G. Evans, M.J. Bielefeld, S.M. Seltzer and A.E. Metzger, Astrophys. J. 212, 925 (1977).
27. C.S. Dyer, J.I.Trombka and S.M. Seltzer, NBS Spec. Publ. 425, 480 (1975).
28. C.S. Dyer, J.I. Trombka, R.l. Schmadebeck, E. Eller, M.J. Bielefeld, G.D. O'Kelly, J.S. Eldridge, K.J. Northcutt, A.E. Metzger, R.C. Reedy, E. Schonfeld, S.M. Seltzer, J.R. Arnold and L.E. Peterson, Space. Sci. Instr. 1, 279 (1975).
29. C.S. Dyer, J.I. Trombka, S.M. Seltzer and L.G. Evans, Nucl. Instr. and Meth. 173, 585 (1980).
30. S.M. Seltzer, Nucl. Instr. and Meth. 127, 293 (1975).
31. S.M. Seltzer, NBS Publ. NBSIR 87-3548 (1987).

32. See, e.g., G. Barnea, S.M. Seltzer and M.J. Berger, NBS Publ. NBSIR 86-3429 (1986); and G. Barnea, M.J. Berger and S.M. Seltzer, J. Spacecraft and Rockets $\underline{24}$, 158 (1987).
33. S.M. Seltzer, IEEE Trans. Nucl. Sci. NS-$\underline{26}$, 4896 (1979); and NBS Technical Note 1116 (1980).
34. S.M. Seltzer, IEEE Trans. Nucl. Sci. NS-$\underline{33}$, 1292 (1986).
35. T.M. Jordan, Exptl. and Math. Phys. Consultants Report EMP.L77.023 (1977).
36. R.L. Heath, USAEC Publ. IDO-16889-1 & 2 (1964).

INTERACTION OF MULTI-MEV GAMMA RAYS WITH MATTER

R.L. Coldwell and F.E Dunnam
Institute for Astrophysics and Planetary Exploration
University of Florida
Alachua, FL 32615

M.W. Katoot and P.S. Haskins
Space Astronomy Laborotory
University of Florida
Gainesville, FL 32609

ABSTRACT

The code BSIMUL is used to simulate the interaction of a 10.76-MeV gamma ray with a 5 cm x 6 cm germanium detector crystal. The Monte Carlo calculation includes the following interactions: the photoelectric effect, Compton scattering, pair-production, pair annihilation, and bremsstrahlung. In this work we investigate the effect of the simplifying assumption that the electrons and positrons do not travel significant distances in the material.[1] The simulation is compared to experiment by following the decay of the 12.54-MeV resonance produced by 992-keV protons on ^{27}Al in the reaction ^{27}Al$(p,\gamma)^{28}$Si.[2] In 75% of the reactions this resonance decays by a 10.76-MeV gamma ray whose interactions with the detector dominate the spectrum in the energy region between 8 and 11 MeV.

INTRODUCTION

The absorption of energy in a detector from interactions of radiation with matter can in principle be simulated by following in detail each and every interaction throughout the detector. By making reasonable assumptions one can obtain reliable results without using excessive amounts of computer time. The goal of this work is a fast computer code, BSIMUL,[1] that produces useful simulations. The possible interactions that radiation undergoes in the detector material are well understood and calculation of the relevant cross sections is straightforward using standard QED techniques.[1,3] Included in the present code is the differential cross section for each type of interaction. The total cross sections are evaluated analytically.

© 1989 American Institute of Physics

The code was tested by comparing the simulation to experimental data. A 10.76-MeV gamma ray is produced in the decay of the 12.54-MeV resonance to the 1.78-MeV 2^+ state in ^{28}Si.[2] This resonance is excited by a beam of 992-keV protons on ^{27}Al to give the ^{27}Al$(p,\gamma)^{28}$Si reaction.[2] The measurement was made at the University of Florida's 4-MeV Van de Graaff accelerator. The detector was a 5 cm diameter by 6 cm long n-type high purity germanium detector.

A 10.76 MeV gamma ray passing near a germanium nucleus is more likely to produce an electron-positron pair than to have any other interaction. Of the original energy, 1.02 MeV is used to produce the pair, leaving 4.5 MeV of kinetic energy for the electron and positron, respectively. In this work we investigate the effect of the simplifying assumption that the electrons and positrons do not travel significant distances in the crystal.

RESULTS

The spectrum consists of a full energy peak (FEP) at the input energy of 10.76 MeV, a single escape peak (SEP) at 10.25 MeV and a double escape peak (DEP) at 9.74 MeV. Variation of the cutoff energy (below which one assumes all remaining energy to be absorbed by the crystal) changes the shape of the simulated spectrum as well as the ratios of the peak areas. A prominent spectral feature on which this work focuses is the discontinuous nature of the background above and below the second escape peak. In the simulation it is assumed that 9.74 MeV is always absorbed by the crystal when a pair is produced. Of the three absorption processes, only Compton scattering is capable of depositing less than 9.74 MeV in the crystal after an interaction and it is far less likely than pair production. If one assumes that 100% of this energy is transferred to the crystal by collection of the electrons (no bremsstrahlung), the code BSIMUL yields the result shown in Fig. 1. Note that, in the simulation, a larger fraction of the energy is included in the peaks and less is included in the background than in the experimental data.

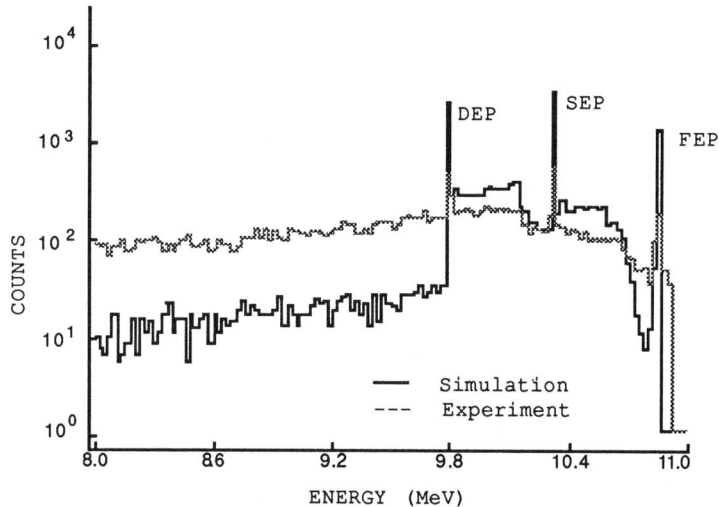

Figure 1. INTENSITY VS. ENERGY This simulation does not include any bremsstrahlung.

For the simulation shown in Fig. 2, the assumption is made that neither the electron nor the positron move significant distances from where they are created, but that they emit secondary gamma rays by means of bremsstrahlung. (The positron is not likely to be annihilated at these large energies.) Those gamma rays do travel away from the creation site. This is accomplished by selecting an energy, E_γ, for the emitted photon from the theoretical energy distribution of bremsstrahlung by an electron of energy E_e near a nucleus. The energy, E_γ, is then subtracted from E_e and the process is repeated until a cutoff energy is reached below which the electron energy is assumed to be absorbed by the crystal. The result for a cutoff energy of 400 keV is compared with the experimental spectrum in Fig. 2.

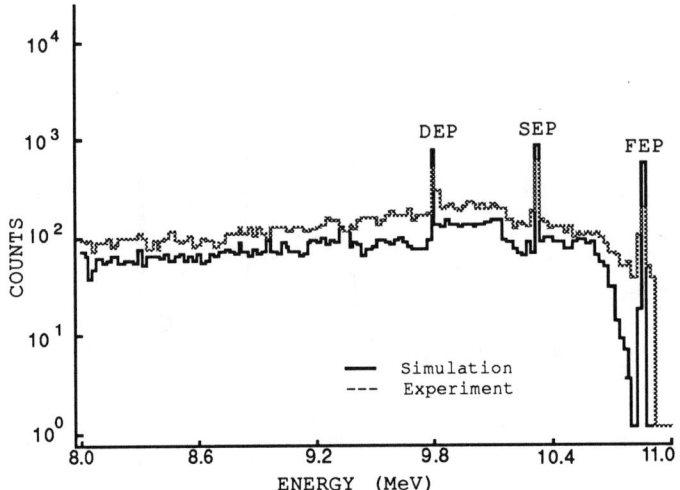

Figure 2. INTENSITY VS ENERGY In this simulation, electrons and positrons with kinetic energies above 400 keV produce photons by bremsstrahlung. Note that on a log scale a multiplicative constant merely moves one curve up or down relative to the other.

The simulation of gamma rays resulting from the bremsstrahlung of electrons and positrons with kinetic energies above 400 keV has brought the continuous spectrum between 9.8 MeV and 10.4 MeV as well as that below 9.3 MeV into substantial agreement with the experimental results. The most obvious disagreements are the peak heights and the depths of the dips preceding the Compton edges on the low energy sides of the photopeak and the two escape peaks. These results are somewhat sensitive to the energy below which the electrons are assumed to undergo no bremsstrahlung; this can be seen in Fig. 3 where the cutoff kinetic energy for bremsstrahlung has been reduced to 150 keV.

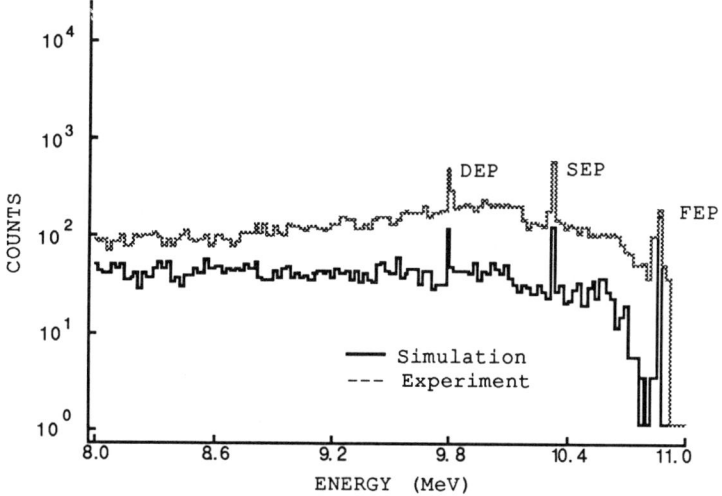

Figure 3. INTENSITY VS ENERGY In this simulation electrons and positrons with kinetic energies above 150 keV produce photons by bremsstrahlung.

The ratio of the sum of the intensities in all three peaks to the total counts in the region from 8 - 11 MeV as well as the ratio of the full energy peak to the single escape peak and the double escape peak to the single escape peak are summarized in Table I. While the peak parameters listed appear to unambiguously support the lowest cutoff energy used, the background is best reproduced by the 400-keV cutoff shown in Fig. 2. In tracing the general deposition of energy the extra energy found in the peaks is not of great significance while the general shape is quite important. A detector simulation, however, would do better using the low cutoff since the peak intensities are then more important than the general shape of the background. It should be mentioned that because of the large number of low energy photons the code takes about four times longer to process the same number of incoming photons with a cutoff of 150 keV than with no bremsstrahlung.

Table I Variation of peak ratios with kinetic energy below which electrons and positrons are assumed to transfer their energy directly to the crystal without radiating.

Brem Cut,	$\sum$ peaks/$I_{(8-11\ MeV)}$	FEP/SEP,	DEP/SEP
Experiment	0.069 ± 0.004	0.36 ± 0.06	0.78 ± 0.11
No Brem	0.41 ± 0.02	0.46 ± 0.04	0.76 ± 0.06
600 keV	0.24 ± 0.01	0.68 ± 0.09	0.94 ± 0.12
400 keV	0.19 ± 0.01	0.68 ± 0.08	0.84 ± 0.09
250 keV	0.13 ± 0.01	0.89 ± 0.12	0.92 ± 0.16
200 keV	0.095 ± 0.010	1.11 ± 0.23	1.08 ± 0.31
150 keV	0.076 ± 0.008	0.69 ± 0.24	1.71 ± 0.44

While the ratio of the peaks to the background and the general shape of the background improves with the addition of bremsstrahlung, the ratio of the photopeak to the first escape peak appreciably worsens. These peaks contain only a small fraction of the total intensity but are of great importance to detector modelling. This effect may be the first instance where the rather drastic approximation that the electrons do not move makes itself known. The photopeak at 10.76 MeV results largely from those pair-production events near the center of the crystal in which both 511-keV gamma rays produced are reabsorbed along with all bremsstrahlung. If the positrons were to move as they radiate (as they do in nature) the 511-keV gamma rays would be more likely to escape. This would enhance the escape peaks and lower the photopeak by making it less likely that all the energy would be absorbed.

CONCLUSION

The results reported here indicate that photons travelling through matter can be reasonably simulated by considering them to interact only by photo-absorption, Compton scattering, and pair production followed by bremsstrahlung without taking into account electron/positron transport. The kinetic energy of the electron and positron created in pair production is reduced by producing photons until it is below a critical value where it is absorbed into the crystal. The positron then decays into two 511-keV gamma rays. All bremsstrahlung and 511-keV gamma rays were assumed to originate at athe point where the pair was created.

This research was sponsored by the Defense Advanced Research Projects Agency through Grant # N00014-87-G-1259, monitored by the Office of Naval Research.

1. R.L. Coldwell, and M.W. Katoot, Bull Am. Phys. Soc. <u>32</u>, 1045 (1987).
2. M. A. Meyer, I. Ventner, and D. Reitmann, Nucl. Phys. <u>A250</u>, 235 (1975).
3. Bjorken and Drell, <u>Relativistic Quantum Mechanics</u>, McGraw-Hill Book Company, New York, New York (1985)

THREE-DIMENSIONAL MONTE-CARLO SIMULATION OF GAMMA-RAY SCATTERING AND PRODUCTION IN THE ATMOSPHERE

Daniel J. Morris
Space Science Center, University of New Hampshire, Durham, N.H. 03824

ABSTRACT

Monte Carlo codes have been developed to simulate gamma-ray scattering and production in the atmosphere. The scattering code simulates interactions of low-energy gamma rays (20 to several hundred keV) from an astronomical point source in the atmosphere; a modified code also simulates scattering in a spacecraft. Four incident spectra, typical of gamma-ray bursts, solar flares, and the Crab pulsar, and 511 keV line radiation have been studied. These simulations are consistent with observations of solar flare radiation scattered from the atmosphere. The production code simulates the interactions of cosmic rays which produce high-energy (above 10 MeV) photons and electrons. It has been used to calculate gamma-ray and electron albedo intensities at Palestine, Texas and at the equator; the results agree with observations in most respects. With minor modifications this code can be used to calculate intensities of other high-energy particles. Both codes are fully three-dimensional, incorporating a curved atmosphere; the production code also incorporates the variation with both zenith and azimuth of the incident cosmic-ray intensity due to geomagnetic effects. These effects are clearly reflected in the calculated albedo by intensity contrasts between the horizon and nadir, and between the east and west horizons.

INTRODUCTION

An important part of the high-energy radiation background encountered in near-earth orbit is the earth albedo, radiation scattered from the atmosphere or produced by cosmic-ray interactions in the atmosphere. Two sets of calculations which produced detailed descriptions of the gamma-ray albedo are discussed here. While these calculations, motivated by their importance to gamma-ray astronomy, have been described elsewhere[1,2] it is appropriate to give a short presentation here, to bring this work to the attention of a wider community of researchers concerned with the space radiation environment. The presentation will consist of brief descriptions of the physical processes modeled and methods used, followed by a representative sample of the results. Details of the calculation are discussed only if they represent a significant innovation over similar calculations done in the past.

The first set of calculations deals with the scattering in the atmosphere of low-energy gamma rays (20 to several hundred keV) from an astronomical point source. With each of four incident spectra the ratio of the scattered intensity to the incident flux was calculated as a function of energy, direction, and geographical location relative to the source nadir. Calculations were also done with incident 511 keV line radiation. The results are consistent with satellite observations of scattered solar flare radiation. A modified code was also developed to simulate scattering from a spacecraft. The spacecraft is approximated by a semi-infinite slab of aluminum, with the scattered intensity calculated as a function of the source direction, scatter direction, energy, and slab thickness.

The second set of calculations deals with the production of high-energy gamma rays (above 10 MeV) in cosmic-ray interactions. For two locations,

© 1989 American Institute of Physics

Palestine, Texas (site of the National Scientific Balloon Facility) and the equator, the gamma-ray intensity was calculated as a function of zenith and azimuth directions, energy, and atmospheric depth. The results are in reasonable agreement with satellite measurements of the albedo and balloon observations of the gamma-ray intensity in the upper atmosphere. These calculations also produced estimates of the high-energy albedo electron flux. With minor modifications the code could be used to study the production of other energetic particles as well.

Because of the complexity of the processes involved, particularly in the gamma-ray production calculations, Monte Carlo simulation methods were used rather than attempting an analytical calculation. A simulation begins with a photon or cosmic-ray entering the atmosphere. This particle is then propagated through the atmosphere to a point of interaction. The results of the interaction are determined and the emerging particles are in turn propagated through the atmosphere to secondary interactions; this may continue through several generations of interactions until all particles have escaped from the atmosphere, penetrated below a specified altitude, or fallen below a low energy limit. The quantities describing a cascade, including characteristics of the incident particle, propagation distances, and characteristics of particles emerging from each interaction, are randomly chosen from probability distributions for each quantity. A large number of such cascade simulations are performed until sufficient statistics are accumulated to determine the gamma-ray intensity or other quantity of interest.

While there have been other attempts to model these processes[3,4] these are the first which are fully three-dimensional. In both calculations a curved model of the atmosphere was used. In the gamma-ray production calculations the effect of the earth's magnetic field on the trajectories of incident cosmic rays was also included in the model. This geomagnetic effect is approximated by placing a lower limit on the magnetic rigidity of particles entering the atmosphere. Magnetic rigidity is the ratio of a particle's momentum to its charge, and determines the particle's trajectory in a magnetic field. The cutoff rigidity depends on both geographic location and the cosmic-ray arrival direction. The gamma-ray production calculations also incorporate a new model of high-energy nucleon-nucleus interactions giving more accurate results for large-angle scattering events which contribute disproportionately to the albedo. No results of these calculations have been normalized to fit observations.

ATMOSPHERIC SCATTERING CALCULATIONS

The coordinate system for the atmospheric scattering calculations is shown in figure 1. A uniform flux of gamma rays is incident on the atmosphere from the direction of the source. Positions are given in terms of altitude and colatitude relative to the source direction; all longitudes are equivalent. Directions are given in terms of a local zenith angle, ζ, and azimuth, ψ, with a photon direct from the source having azimuth 0°. Energies of incident particles were chosen from one of four spectra, with a low-energy threshold of 20 keV. One spectrum, $dN/dE=(1/E)\exp(-E/500\text{keV})$, is typical of gamma-ray bursts. The other three are power laws, $dN/dE=E^{-\alpha}$. One, with index $\alpha=2.1$, simulates the flux from the Crab pulsar. The other two, with indices 3 and 4 are typical of solar flares. Calculations were also done with an incident flux of 511 keV annihilation photons.

Four types of photon interaction are included in the simulations: Compton scattering, photoelectric absorption, coherent scattering, and pair production. The total cross section decreases monotonically with energy from 20 keV to tens of MeV. Between 100 keV and 3 MeV Compton scattering dominates the cross section. Near

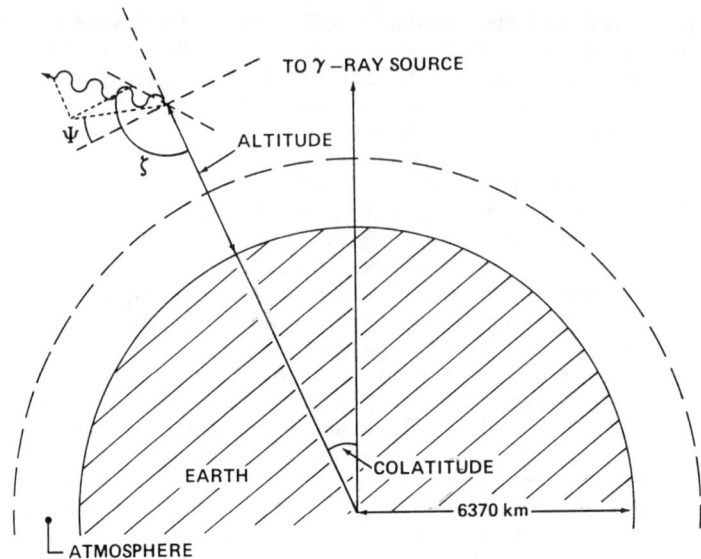

Fig. 1. Coordinate system used in atmospheric scattering calculations. Locations are given by colatitude and altitude; directions by a locally defined zenith, ζ, and azimuth, ψ.

and below 50 keV photoelectric absorption is also important. Coherent scattering is significant at low energies, but its effect is similar to Compton scattering. Pair production is significant only at energies of several MeV or more. Electrons produced in these interactions are ignored since the bremsstrahlung photons they may produce can be shown to be insignificant for spectra considered here. The annihilation of pair-production positrons is included, since it may produce a significant flux in a narrow 511 keV line.

    Photon propagation distances are chosen on the basis of the total cross section from an exponential distribution in the integrated density along the photon's trajectory. Particle propagation is the most time consuming part of these calculations. The computing time was minimized by using an approximation to the model atmospheric density along each particle trajectory which can be integrated analytically; this approximation is within 1% of the model density along a straight trajectory of 40 km through the curved atmosphere.

    Some examples of the results obtained with the atmospheric scattering simulation are shown in figure 2. The results are given in terms of the flux observed by a planar detector shielded from behind, similar to the detectors of the Burst and Transient Source Experiment (BATSE)[5] to be flown aboard the Gamma-Ray Observatory (GRO). The quantity plotted is the ratio of the observed flux, both albedo and direct, to the flux which would be observed by a detector pointed directly at the source in the absence of scattering. The detector is at 400 km altitude and the given colatitude with orientation indicated by the zenith and azimuth. In all these results there is a peak in the flux ratio at 50-100 keV; at lower energies the albedo flux is reduced by photoelectric absorption, at higher energies by decreases in the Compton cross-section and the fraction of the energy retained by a scattered photon.

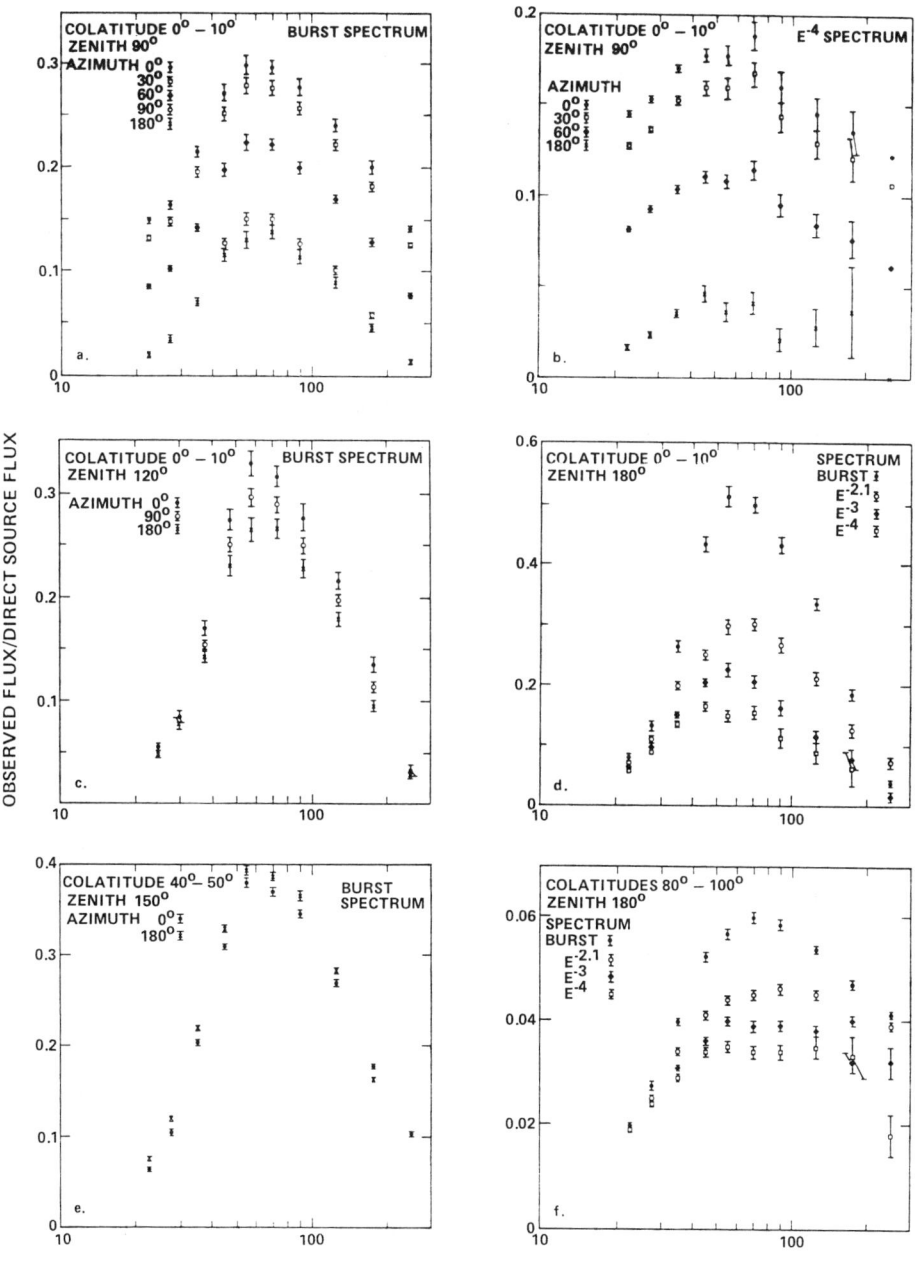

Fig. 2. Examples of atmospheric scattering calculations with continuous spectra. The quantity plotted is the ratio of the flux observed by an ideal BATSE detector at 400 km altitude and the indicated colatitude, with the indicated orientation, to the flux such a detector pointed at the source would see in the absence of scattering.

The albedo is more important for harder spectra, as can be seen in figures 2d and 2f, or by comparing figures 2a and 2b. The flux ratio is greatest for the situation illustrated in figure 2d with the source near the zenith and the detector pointed at the nadir; with the burst spectrum the albedo is as much as half the direct flux.

A combination of geometrical effects is seen in the dependence of the flux ratio on colatitude and detector orientation. The direct flux is limited in the results presented in figure 2, since the source is never more than a few degrees inside the detector's field of view. Still, the azimuthal variation seen in figures 2a and 2b is nearly all due to variation in the direct flux received; the smaller azimuthal variation in albedo alone is seen in figures 2c and 2e. The observed albedo increases with zenith as the fraction of the field of view filled by the atmosphere increases. Many albedo photons undergo a single Compton scattering so their directions reflect the Compton scattering angular distribution, which is peaked in the forward and backward directions at low energies. This is apparent in figure 2e, where the flux below 150 keV is smaller for azimuth 0°, with the detector oriented perpendicular to the source direction, than for azimuth 180°, with the detector oriented opposite to the source. The Compton angular distribution also contributes to the decrease in observed flux at zenith 180° as the colatitude increases (compare figs. 2d and 2f) though it is probably less important than the greater atmospheric pathlength for scattered photons at large colatitudes.

Some of these geometric effects are also seen in calculations with incident 511 keV line flux, including the increase in observed albedo with zenith and decrease with colatitude. The characteristics of Compton scattering are seen more clearly with an

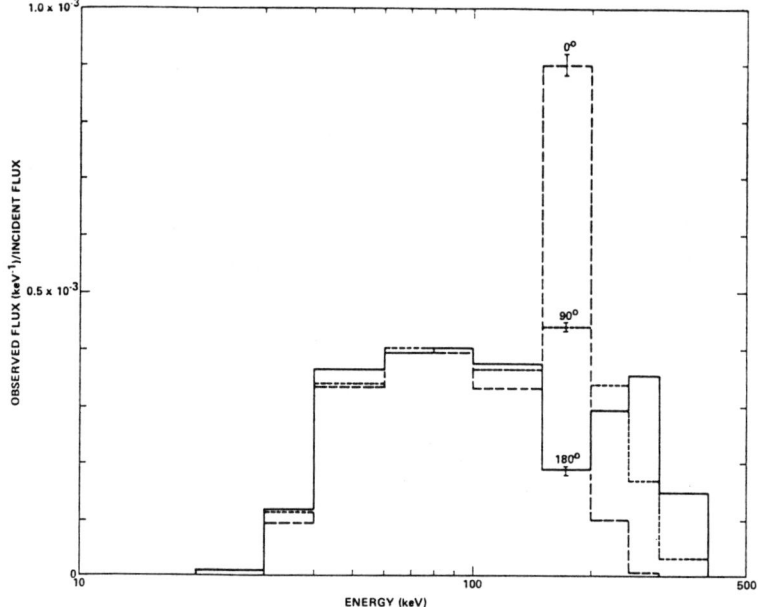

Fig. 3. Examples of atmospheric scattering calculations with 511 keV line flux. The quantity plotted is the ratio of observed intensity to incident flux for an ideal BATSE detector at 400 km altitude and 40°-50° colatitude. The detector is oriented with a zenith angle of 90°; the azimuth is indicated for each curve.

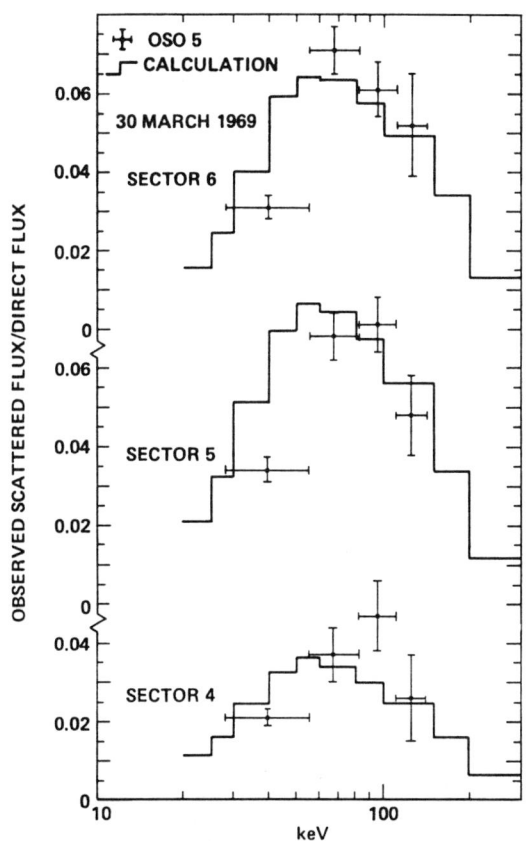

Fig. 4. Comparison of the observations of scattered radiation from the solar flare of 30 March 1969 by the OSO 5 hard X-ray experiment with results of these calculations.

incident line flux. This is evident in Figure 3, showing results for colatitudes 40°-50°, detector zenith 90°, and three different azimuths. Each spectrum shows a narrow single-scattering peak together with a wider multiple scattering peak at lower energy. The width of the single scattering peak reflects the width of the detector's field-of-view.

In figure 4 these calculations are compared with observations of scattered radiation from the solar flare of 30 March 1969 by the OSO 5 hard X-ray experiment[6]. The OSO spacecraft rotated about an axis perpendicular to the Earth-Sun line with a period of 1.9 s, with observations integrated over 40°-wide sectors; atmospheric scattering was observed primarily in sectors 3 through 6, over 100° from the solar direction. The calculated albedo has been averaged over angle to simulate the angular response and rotation of the experiment. The spectrum of the 30 March flare was exceptionally hard, with power-law index 2.3; the observations are compared to the calculations with index 2.1. Observations above 150 keV, which were affected by strong but uncertain 511 keV line radiation, are not shown. Similar agreement is found with observations during flares of 24 November 1969 and 12 March 1969.

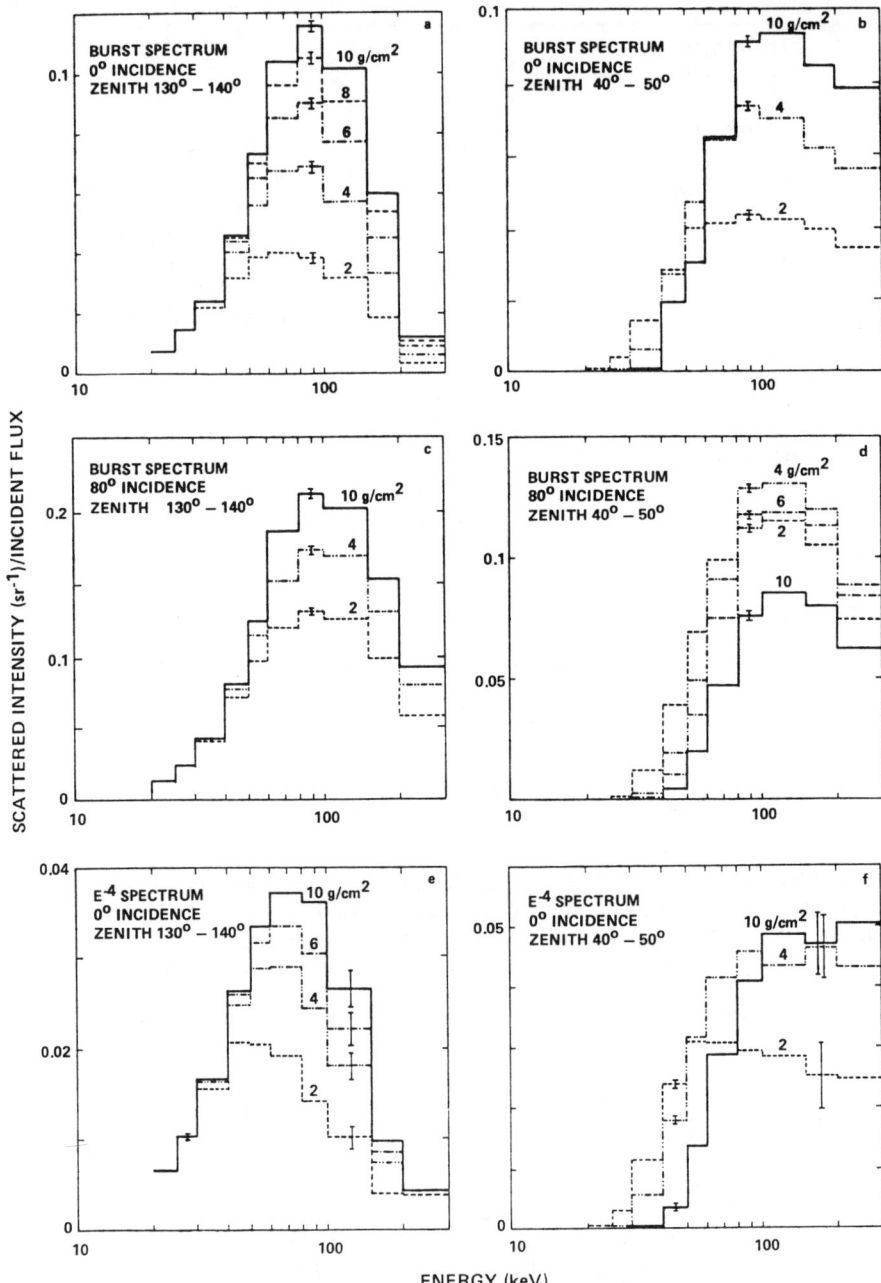

Fig. 5. Examples of backscattered (zenith 130°-140°) and forward scattered (zenith 40°-50°) radiation from a semi-infinite aluminum slab. Each curve shows the ratio of scattered intensity to incident flux for a slab of specified thickness, averaged over azimuth.

SPACECRAFT SCATTERING CALCULATIONS

A modified scattering code was used to investigate scattering from a spacecraft. The geometry was very simple, with the spacecraft approximated by a semi-infinite aluminum slab; the slab thickness was varied from 2 g/cm^2 to 10 g/cm^2. Calculations were done for three incidence angles: 0°, 40°, and 80°. The results are ratios of the scattered intensity to the incident flux as a function of zenith angle, azimuth relative to the incident flux, and energy. Both forward scattering through the slab, and backscattering were considered.

Figure 5 presents examples for backscattered (zenith 130°-140°) and forward scattered (zenith 40°-50°) radiation, averaged over azimuth. In the backscattered spectra there is a clear peak which is more prominent and occurs at higher energy for thicker slabs. In most cases there is a peak in the forward scattered spectra as well, but it is less prominent. The backscattered intensity increases with slab thickness, though the increase is leveling off by 10 g/cm^2, particularly at large incidence angles (figure 5c). While the forward scattered intensity is smaller than the backscattered at lower energies, it can greatly exceed the backscattered intensity at high energies, particularly for small incidence angles (compare figures 5a and 5b). At 80° incidence, with the effective slab thickness for incident photons increased by a factor of 5.76, forward scattering is suppressed at thicknesses over 4 g/cm^2 (figure 5d). Spacecraft scattering is less important for softer spectra (compare figures 5a and 5e or 5b and 5f); this is most evident for backscattering.

The angular variation in spacecraft-scattered radiation is quite complex, depending on the incidence angle, slab thickness, and spectrum, but a few general statements can be made. The angular variation is generally negligible at energies below the peak in the scattered-to-incident ratio. For backscattered radiation the zenith dependence is greatest for thin slabs. The intensity of forward scattered radiation increases away from the incidence direction in thin slabs but decreases in thick slabs. The azimuthal variation increases with incidence angle, and is somewhat more significant for a soft spectrum or a thin slab.

Above 100 keV spacecraft backscattering for slabs of thickness 6 g/cm^2 or more is comparable to atmospheric backscattering. At lower energies spacecraft backscattering is suppressed by the larger photoelectric absorption cross section of aluminum.

ATMOSPHERIC PRODUCTION CALCULATIONS

The gamma-ray production simulation is much more complex than the scattering simulation. Each cascade begins with the selection of an energy and incidence direction for a proton or nucleus entering the atmosphere. When the particle's rigidity falls below the cutoff for the given direction, another is chosen. For Palestine, Texas cutoff rigidities were obtained from tables of Shea and Smart[7]; at the equator they were calculated using an approximation of Smart and Shea[8]. The incident cosmic-ray is propagated to a point where it interacts with an atmospheric nucleus. The interaction can be quite complex; the resultant particles relevant to gamma-ray production are pions, and nucleons with kinetic energy above 300 MeV, the pion production threshold. The nucleons are propagated to secondary interactions. Neutral pions decay immediately to two photons. Charged pions are propagated until they decay to a muon; muons are in turn propagated until they decay to electrons. Electrons and photons initiate electromagnetic cascades in which more electrons and photons are produced by pair production, bremsstrahlung, and

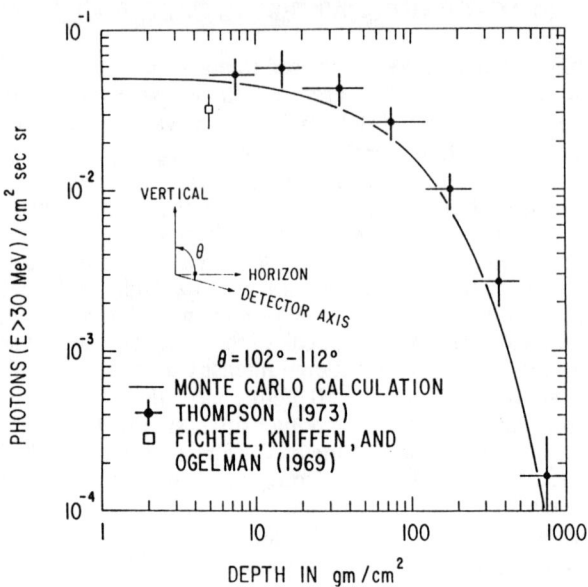

Fig. 6. Calculated horizontal gamma-ray intensity above 30 MeV at Palestine, Texas, with corresponding data from balloon experiments.

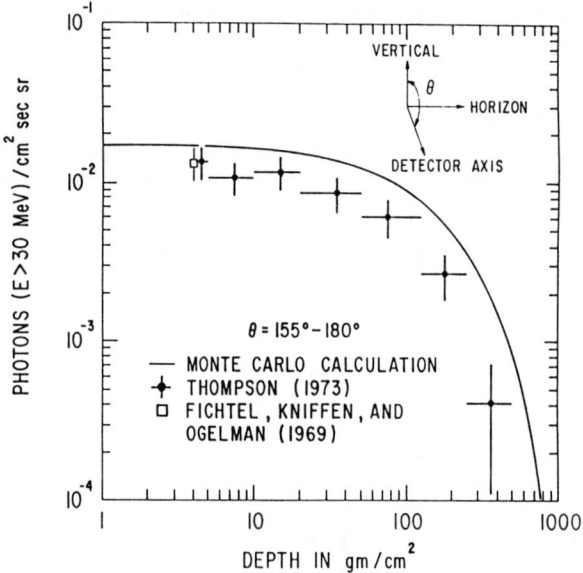

Fig. 7. Calculated upward gamma-ray intensity above 30 MeV at Palestine, Texas, with corresponding data from balloon experiments.

Compton scattering. Particle propagation is complicated by several factors which are not relevant to the scattering calculations. First, the energy loss of charged particles during propagation must be included. For electrons the energy loss rate is a strong function of energy. In addition, electrons are subject to multiple scattering. Because of these two effects the step size in electron propagation must be limited. For particles which decay in flight the proper time elapsed during propagation must be calculated; it is possible to do this analytically. There is a final complication involving albedo electrons: only the highest energy electrons can escape the earth's magnetosphere. Most electrons follow a complicated trajectory in the magnetosphere, reentering the atmosphere at a different location, but one having a similar cutoff rigidity. This situation is approximated by changing the zenith angle for the trajectory of each albedo electron to its supplement, sending the electron back into the atmosphere.

Some results for Palestine are presented in figures 6-8 and compared with balloon observations[9,10,11]. The gamma-ray intensity in given energy and zenith-angle intervals is plotted as a function of atmospheric depth. Figures 6 and 7 present results for energies above 30 MeV; figure 6 shows the intensity just below the horizon, figure 7 that near the nadir. Lower energy results, at 10-25 MeV, are shown in figure 8 for a wide angular range near, but excluding, the nadir. In all these cases the intensity varies little with atmospheric depth at very high altitudes, so the intensity near 10 g/cm^2 is essentially the same as the albedo. Above 30 MeV the agreement between calculation and observation is quite satisfactory at high altitudes for all zenith angles. Comparison of figures 6 and 7 shows the intensity is much greater near the horizon than the nadir. The strong horizontal flux is produced by cosmic rays entering the atmosphere on near horizontal trajectories. The upward flux results from relatively rare large-angle scattering events, or cascades of several

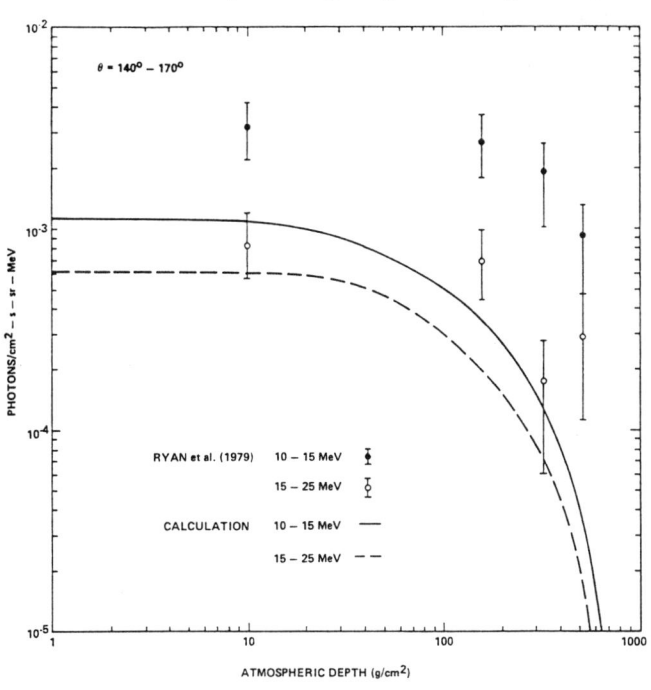

Fig. 8. Calculated upward gamma-ray intensity at Palestine, Texas for energies 10-25 MeV, with corresponding data of Ryan et al. (ref. 11).

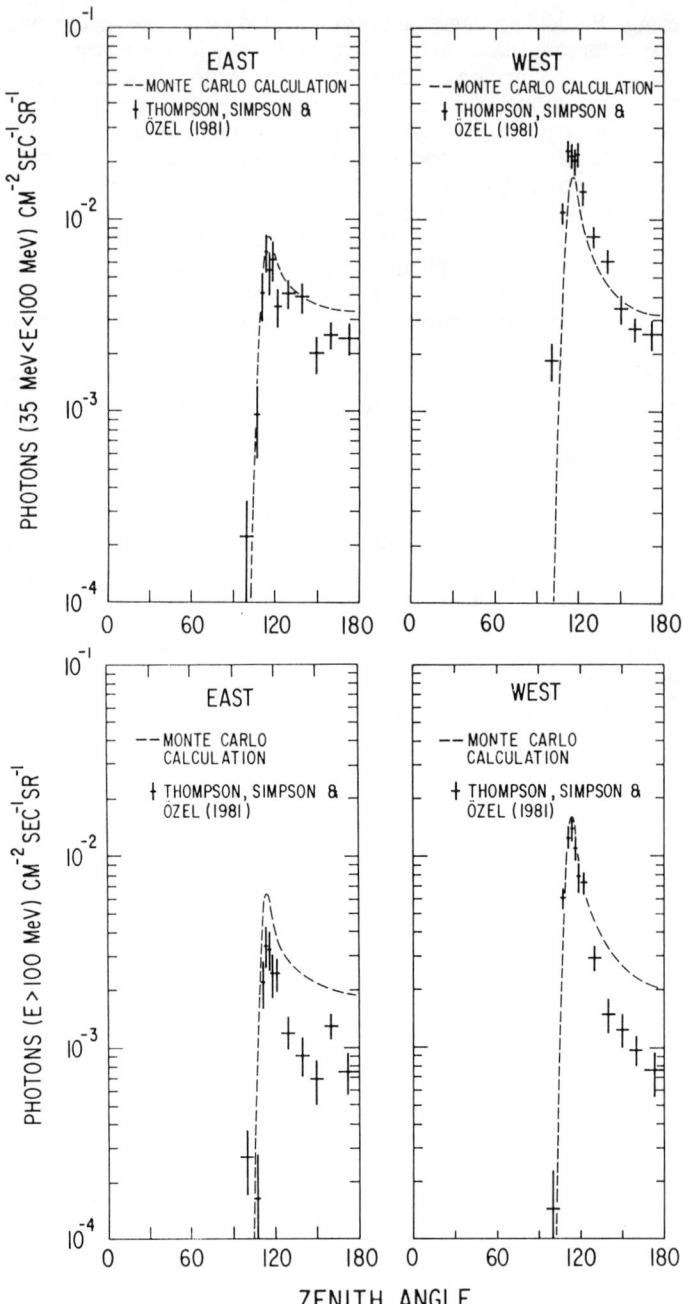

Fig. 9. Comparison of the calculated equatorial gamma-ray albedo with SAS 2 satellite data. The calculated intensity is averaged over zenith to simulate the limited angular resolution of the experiment.

generations with moderate scattering in each interaction. Thus the uncertainty in the calculations is greatest for the upward intensity. The calculated albedo is also consistent with the high-altitude observations at 15-25 MeV in figure 8, but the calculated upward intensity at 10-15 MeV is smaller than that observed by a factor near 3. The observed 10-15 MeV intensity may include a substantial background produced by neutron interactions within the telescope. A fraction of the observed intensity may also be nuclear de-excitation gamma rays which are important below 10 MeV, but are not included in these simulations.

The calculated albedo intensity at the equator is compared with observations by the SAS 2 satellite[12] in figure 9; the calculated albedo has been averaged over zenith to simulate the angular resolution of the experiment. Again, there are strong contrasts between horizon and nadir. Strong contrasts are also seen between the east and west horizons, due to an east-west contrast in the geomagnetic cutoff. Overall the calculated intensity is consistent with the observations at 35-100 MeV, but is somewhat too high for energies above 100 MeV. For both energy ranges the horizon-nadir and east-west contrasts are smaller in the calculations than the observations. The geomagnetic cutoff is relatively uncertain for the equator, being

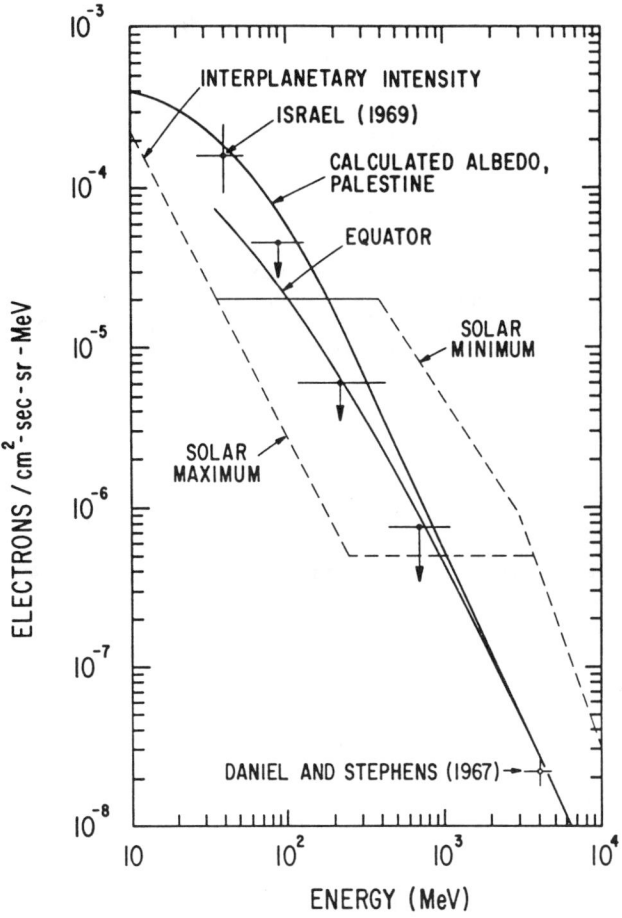

Fig. 10. Calculated albedo electron spectra, at Palestine and the equator. Data points show vertically incident reentrant electron intensities observed at Palestine (Israel, 1969) and Hyderabad, India (Daniel and Stephens, 1967). Also shown for comparison is an approximation to the interplanetary electron intensity.

derived from a simple approximation rather than detailed calculations and averaged over the orbit; this may detrimentally affect the calculated east-west contrast.

In the course of these calculations estimates of the electron albedo were also obtained for both locations. These estimates are compared in figure 10 with observations of Israel at Palestine[13] and Daniel and Stephens at Hyderabad, India[14], and with approximations to the interplanetary electron intensity at solar minimum and solar maximum. The calculated albedo agrees with the observations at the two energies for which there were positive measurements.

CONCLUSION

These calculations show that characteristics of atmospheric albedo radiation can be calculated including the detailed angular distribution. These details are needed for the correct interpretation of astronomical gamma-ray observations, and may well be of interest to other researchers concerned with the near earth radiation environment. Certain parts of the gamma-ray production simulation can be identified for which better models or data are particularly desirable. One is the production of particles at large angles in high-energy nucleon-nucleus interactions. Others are details of the geomagnetic cutoff rigidity and intensity of reentrant albedo electrons. With minor modification the production code could be used to determine atmospheric intensities of other particles including nucleons, pions, and muons. Another useful, but much more involved, extension of this work would be a production model continuing to lower photon energies at which a number of processes ignored here become important.

ACKNOWLEDGEMENTS

The scattering calculations were done at NASA Marshall Space Flight Center under an NAS-NRC Research Associateship. The production calculations were done at NASA Goddard Space Flight Center under NASA Grant 21-002-316 as part of a doctoral thesis in the Department of Physics and Astronomy of the University of Maryland, College Park.

REFERENCES

1. D. J. Morris, in High Energy Transients in Astrophysics, edited by S. E. Woosley (American Institute of Physics, New York, 1984), pp. 665-668.
2. D. J. Morris, J. Geophys. Res., 89, 10685 (1984).
3. H. Horstmann, L. Bassani, and E. Horstmann-Moretti, Astrophys. Space Sci., 52, 265 (1977).
4. D. J. Thompson, J. Geophys. Res., 79, 1309 (1974).
5. G. J. Fishman, C. A. Meegan, T. A. Parnell, R. B. Wilson, and W. Paciesas, in High Energy Transients in Astrophysics, edited by S. E. Woosley (American Institute of Physics, New York, 1984), pp. 651-664.
6. S. H. Langer, V. Petrosian, and K. J. Frost, Astrophys. J. 235, 1047 (1980).
7. M. A. Shea and D. F. Smart, Tables of Asymptotic Directions, Cutoff Rigidities, and Reentrant Albedo Calculations for Palestine, Dallas, and Midland, Texas, (Rep. AD/A-005 408, Air Force Cambridge Res. Lab., Hanscom AFB, Bedford, Mass., 1975).
8. D. F. Smart and M. A. Shea, in Proceedings of the 14th International Cosmic Ray Conference (Munich, 1974), 4, 1309.
9. D. J. Thompson, Ph.D. thesis, University of Maryland, College Park, 1973.

10. C. E. Fichtel, D. A. Kniffen, and H. B. Ögelmann, Astrophys. J., 158, 193 (1969).
11. J. M. Ryan, M. C. Jennings, M. D. Radwin, A. D. Zych, and R. S. White, J. Geophys. Res., 84, 5279 (1979).
12. D. J. Thompson, G. A. Simpson, and M. E. Özel, J. Geophys. Res., 86, 1265 (1981).
13. M. H. Israel, J. Geophys. Res., 74, 4701 (1969).
14. R. R. Daniel and S. A. Stephens, Proc. Indian Acad. Sci. Sect. A, 65A, 319 (1967).

HIGH-ENERGY RADIATION ENVIRONMENT DURING MANNED SPACE FLIGHTS

R. Silberberg, C. H. Tsao, J. H. Adams, Jr.,
E.O. Hulburt Center for Space Research
Naval Research Laboratory, Washington, DC 20375

J. R. Letaw
Severn Communications Corp., Severna Park, MD 21146

ABSTRACT

Radiation doses from cosmic rays on long-duration space missions, at solar minimum, with 4 g/cm^2 Al shielding and 5 g/cm^2 tissue self-shielding exceed by a factor of about 60 the allowable dose of 0.5 rem/year to the general public and by a factor of about 6 the dose of 5 rem/year allowable to radiation workers. The annual dose of 30 rem nearly equals the allowable dose to a few volunteer astronauts which is 40 rem/year. Rare solar flare particle events would deliver doses of about 200 to 300 rem, in a period of about one day (like the 1956 and 1972 flares) for the above conditions of shielding. These doses are lethal to about 20 percent and 40 percent of the recipients respectively.

We shall present here detailed flux and dose calculations with shielding, using the fluxes and energy spectra of all the nuclei from hydrogen to iron, nuclear spallation cross sections, and our NRL propagation code for cosmic ray nuclei, and the HETC and MCNP codes developed at Oak Ridge for protons, mesons and neutrons. Radiation quality factors, appropriate to the rate of ionization of various cosmic-ray and secondary nuclei and of nuclear recoils, will be included in the dose calculations.

Our calculations show that both the cosmic-ray and solar particle doses can be brought down to acceptable levels. For thin shielding, the contribution of heavy nuclei to the dose dominates since the rate of ionization is proportional to Z^2, and for heavy ions, the quality factor Q=20. For thick shielding, the dose from neutron-generated nuclear recoils dominates.

© 1989 American Institute of Physics

INTRODUCTION

Radiation hazards in space are due to three sources; (1) The Galactic cosmic rays that permeate the whole Galaxy, (2) solar flare particles within the heliosphere, that extends beyond the orbit of the planet Pluto, and (3) the trapped radiation in the magnetospheres of the earth and other planets with magnetic fields, e.g. Jupiter and Saturn.

The papers presented in the session on Radiation Environment discussed in detail these sources of radiation, their elemental and isotopic components and energy spectra. Here we shall only present a compact comparison and contrast of these sources of radiation. The cosmic rays have the hardest energy spectrum with many high energy particles. Also cosmic rays have the largest fraction of highly ionizing heavy nuclei, so that the dose equivalent of cosmic rays is mainly from heavy nuclei, with atomic numbers $6 \leq Z \leq 26$. The shielded dose equivalent is about 50 rem/year near times of sunspot minimum. The solar flare particles are most variable in intensity, and hardest to predict in terms of time. They occur mainly during four or five years, around the time of the sunspot maximum of the 11-year solar cycle. They are sporadic, of about two-day duration. The unshielded dose of the largest flares of a solar cycle is about 1000 rem. The energy spectrum is generally steep; hence the heavy nuclei are readily absorbed. However, the flux of secondary neutrons from mainly proton interactions is intense and highly penetrating. The trapped radiation consists mainly of protons. The energy spectrum is relatively steep. At altitudes near 2000 nautical miles, the dose is high, about 10^2 to 10^3 rem/day.

The doses discussed in the previous paragraph have to be compared with the permissible doses which are 5 rem/year for radiation workers and 40 rem/year for a few volunteer astronauts, with a lifetime limit of 200 rem. The biological effects of higher doses are described in several papers in these conference proceedings.

Radiation transport calculations are essential for evaluating the dose, since the nuclear reaction cross sections are sufficiently large for transforming the elemental composition of the particles, and generating secondary particles, including highly penetrating neutrons and highly ionizing nuclear recoils. Radiation transport calculations require accurate knowledge of the intensity, composition and energy spectra of the incident nuclei, of their interaction cross sections, and of the production rates of secondary nuclei, protons and neutrons as well as the energy spectra for the latter. Furthermore, the rate of energy deposition of all these radiation components as a function of the energy of the particle enters into the radiation transport calculations. Environmental conditions like the phase of the solar cycle and the geomagnetic cutoffs at given geographic coordinates of the spacecraft enter into the radiation transport calculations. We shall discuss these calculations in greater detail in the section on radiation transport calculations below.

RADIATION TRANSPORT CALCULATIONS

Our radiation transport calculations are based on extensive collections and reanalysis of experimental data on cosmic rays and solar flare particles. They are also based on extensive collections of total and partial cross section measurements, and semiempirical formulations. A great deal of electromagnetic and nuclear theory developed over the past 50 years is part of the transport model.

The data on cosmic ray environment were developed into the CREME model by Adams et al.[1], also presented in the present Proceedings. The model gives the fluxes and energy spectra of all elements near times of sunspot minimum and maximum, i.e. at cosmic ray maximum and minimum. Solar flare environments are also given in the report, based on the giant flare of August, 1972. Also a composite worst case intensity, based on a combination of the above flare with that of February 1956, with many high energy protons near energies of 10^3 MeV is part of the model. The latter worst case flare has been explored by Adams and Gelman[2].

(a) Spallation Cross Sections.

Our partial inelastic cross section formalism uses Rudstam's[3] equation as a starting point. We (Silberberg and Tsao[4]) constructed a semiempirical equation applicable for calculating the partial cross sections of target nuclides in the range of mass numbers $9 \leq A_t \leq 209$ and products with $6 \leq A \leq 200$.

$$\sigma = \sigma_0 \, f(A) f(E) \exp(-P\Delta A) \exp(-R|Z-SA + TA^2|^\nu) \Omega \eta \xi \quad (1)$$

The factors and parameters of this equation and their numerical values are defined by Silberberg and Tsao[4] and Silberberg et al.[5] Here σ_0 is a cross section normalization factor. $f(A)$ and $f(E)$ are introduced for the fission of heavy nuclei and for the production of light nuclear fragments from heavy nuclei. The effects of the exponential factors of Rudstam's equation are shown in Fig. 1, for the spallation of iron nuclei into isotopes of V and Ar. Here P describes the reduction of cross sections with increasing ΔA, the difference of target and product mass, A_t - A. The second exponential factor describes the gaussian-like distribution of the product isotopes. The width of these distribution curves is described by the parameter R. The parameter S describes the location of the peaks of these distribution curves for small values of A. The parameter T describes the shift of the distribution curves toward greater neutron excess as the atomic number of the product increases. The last three factors of Eq. 1 are related to nuclear structure, and the nuclear evaporation process.

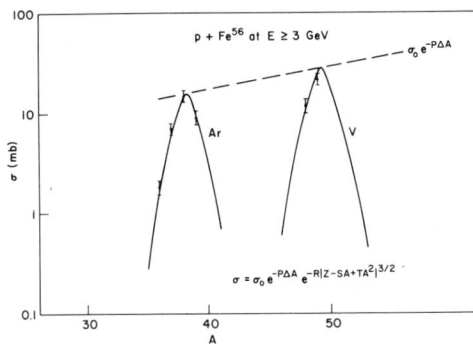

Figure 1. Illustration of the terms of Rudstam's spallation equation. The example shows the experimental and calculated partial cross sections of iron into isotopes of Ar and V, at energies $\geq$ 3 GeV.

For the spallation of nuclei with $Z_t \leq 20$ we had to introduce parameters that differed appreciably from those of Rudstam[3]. Likewise for the peripheral reactions with $\Delta A \leq 5$, and for nuclear fission reactions, we had to modify Rudstam's basic equation developed for spallation reactions.

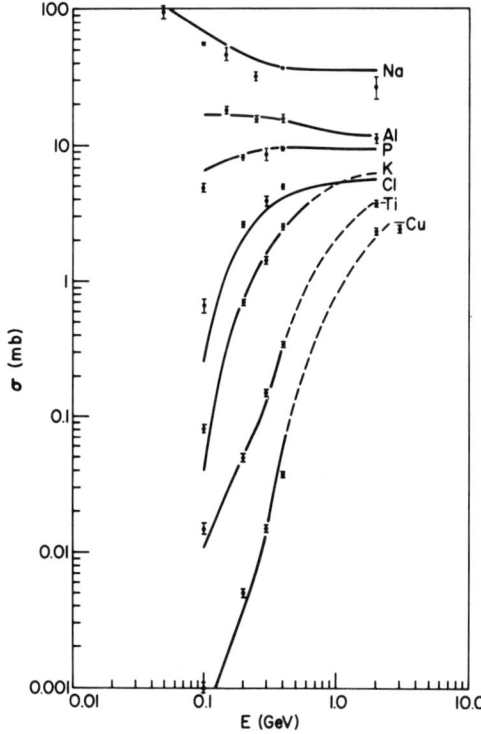

Figure 2. Comparison of calculated and experimental yields of ^{22}Na for E $\geq$ 100 MeV and targets $11 \leq Z_t \leq 29$.

Spallation reactions are highly energy dependent as shown in Fig. 2. This figure shows the energy dependence of the production of ^{22}Na from various nuclei, that range from Na to Cu. The experimental values, as well as the semi-empirical curves based on our above equation are shown. The energy dependence of the curves is derived from the energy dependence of the parameter P.

This year we completed the analysis of published cross section measurements, with special emphasis on data that have standard deviations near 10%, for spallation of nuclei with $Z_t \geq 21$, at high energies, $E \geq 3$ GeV. A preliminary version was published at the 20th International Cosmic Ray Conference by Silberberg et al.[6] A term μA^3 is added to the second exponential factor of the cross section equation. The new parameters for $Z_t > 21$ are tabulated below in Table 1. The parameters will be extended soon (1) down to energies of 0.1 GeV/nucleon; (2) to peripheral reactions; and (3) to targets $3 \leq Z_t \leq 20$.

Table 1

New parameters for $Z_t > 21$

$P = 1.97 A_t^{-0.9}$ for $E > 3000$ MeV or E_0

$S = 0.482 - 0.7(A_t - \bar{A})/Z_t$

$T = 0.00028$

$\nu = \begin{cases} 1.3 & Z* < -1 \\ 1.5 & \text{for } -1 < Z* < 1 \\ 1.75 & Z* > 1 \end{cases}$

$Z* = Z - SA + TA^2 + \mu A^3$

$\mu = 3 \times 10^{-7}$

$E_0 = 20.3 A_t^{1.169}$ $E_0 < 4000$ MeV

$f_1(E) = 1$ for $E > E_0$. (see eq. 4 of Silberberg and Tsao[4].)

Figure 3 shows the ratio of calculated to experimental cross sections of the spallation of Cu, as a function of ΔA. The experimental data are from Cumming et al.[7] We note that the calculated cross sections have a standard deviation of ~10%. In general, the standard deviations of the new calculations are 2 or 3 times less than in Silberberg and Tsao[4]. The total inelastic cross sections of the transport equation are calculated from the equations of Letaw et al.[8]

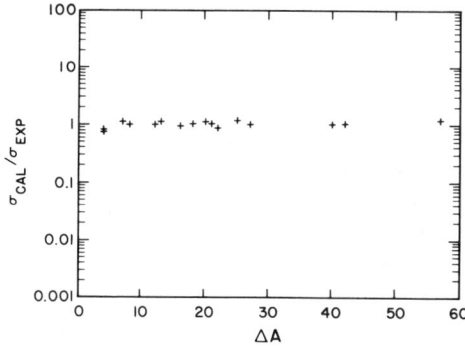

Figure 3. The ratio of calculated to experimental cross sections of Cu, as a function of $A_t - A$, for $E > 3$ GeV, using the new parameters.

For nucleus-nucleus reactions we use the partial cross sections of Silberberg and Tsao[9]. These essentially use a scaling factor relative to proton-nucleus cross sections, with enhanced production of light product nuclei with $A < 0.5 A_t$, and enhanced single-nucleon stripping due to giant dipole resonance. The latter was observed by Lindstrom et al[10], and a procedure for calculating this reaction cross section was developed by Townsend et al.[11] There are many recent measurements of nucleus-nucleus reactions using heavy ion beams. We have collected these into a computerized file and plan to develop new cross section equations after completing the work on proton-nucleus reactions. The total inelastic nucleus-nucleus cross sections are calculated using the procedures of Karol[12], supplemented by corrections for skin depth.

(b) Transport Models

The radiation transport of cosmic-ray nuclei and nucleons through shielding materials and in astronaut tissue is modeled using three separate computer codes: UPROP, HETC and MCNP. Many years have been invested into the development of each of these. The codes are executed on the DEC VAX 11/785s, and the CRAY XMP computer at the Naval Research Laboratory.

UPROP ("Universal Propagation Code") describes the transport of cosmic ray primary nuclei with atomic numbers $Z < 29$, and their nuclear secondaries. (Nuclei with $Z > 28$ are neglected, their abundance is 0.003 of that of iron). Ionization losses are computed from 1 MeV/nucleon to 100 GeV/nucleon, using the model of Ahlen[13]. Nuclear transformations are evaluated using the procedures for calculating cross sections as outlined in the subsection above. The production of hydrogen and helium nuclei is given by Letaw[14]. (An alternative procedure for calculating the latter with highly accurate applications to radio-therapeutic beams has been developed by Townsend and Wilson[15]. Production of nuclear fragments and other secondaries is recomputed in target materials at intervals of 0.1 g/cm^2 (about 0.5 mm aluminum). Some of the procedures of these transport computations have been summarized by Letaw et al.[16], however, the procedures outlined in that paper are directly applicable only to high-energy (> 1 GeV/nucleon) cosmic rays.

HETC ("Monte Carlo High-Energy Nucleon Meson Transport Code") is a well known procedure for calculating the transport of moderately high energy protons and neutrons (20 MeV to 3000 MeV) through shielding materials. It does not transport heavy ions, but supplements UPROP by following in detail the production from proton interactions, and transports the target secondaries such as protons, neutrons, alpha particles, and pions. It also tabulates the energy and composition of recoil nuclei emanating from the target, and of neutrons below the 20 MeV transport cut off. HETC was originally developed by Oak Ridge National Laboratory. We have used the Los Alamos (LANL) version of the code of Prael[17].

MCNP ("Monte Carlo Neutron and Photon Transport Code") describes the transport of relatively low energy neutrons (< 20 MeV) through materials. MCNP also computes the energy deposited (dose in rads) by secondary neutrons. MCNP was also developed at Oak Ridge. The LANL version of the code is well integrated with both HETC and ENDF nuclear data base of Brookhaven National Laboratory.

(c) Shielding and Radiation Dose

The above radiation transport models and codes allow the flux of charged particles to be computed at any position in the target (Planetary soil and atmosphere, spacecraft and astronaut). In addition, they provide either the energy deposition rate (LET spectrum) or the actual energy deposited by each particle species. One of our shielding models simulates the protection of blood-forming organs (bone marrow, using layered plane slab geometry. The layers are: a variable thickness spacecraft (aluminum) shielding, 4.5 cm water approximating body self-shielding, and a 1 cm water absorption region. The density of energy deposited in the absorption region is the dose (in units of 100 ergs/g or rads).

The biological effect of radiation depends both on the density of energy deposited in the tissue and the quality factor Q that represents the relative biological effectiveness (RBE) of the radiation. The dose equivalent (rem) equals Q x dose (rads). The quality factor depends on the rate at which a particle deposits energy in the tissue. A future, more refined definition that relates it more closely to RBE, may also include the distribution of energy deposited in a cell, the rate at which the dose is received, and the type of biological damage considered.

The quality factors used in our calculations are based on those proposed by the International Commission on Radiological Protection (ICRP) and International Commission of Radiation Units and Measurements (ICRU). The quality factor of cosmic ray iron nuclei is about 20; that of neutrons (that generate highly ionizing nuclear recoils) was defined earlier as 10, but based on the recent ICRU Report 40[18], we have adopted 20 for neutrons.

RESULTS

In this section we present our calculated results of LET distributions and doses for a wide variety of environments: (1) outside the earth's magnetosphere, with minimal EVA-type shielding, (2) in various earth orbits, below the trapped radiation belts, (3) in a sub-surface Lunar habitat, and (4) in a shielded spacecraft on an extended mission, e.g. to Mars. Our publications on items 1, 2, 3, 4 respectively are: Silberberg et al.[19], Silberberg et al.[20], Silberberg et al.[21], and Letaw et al.[22].

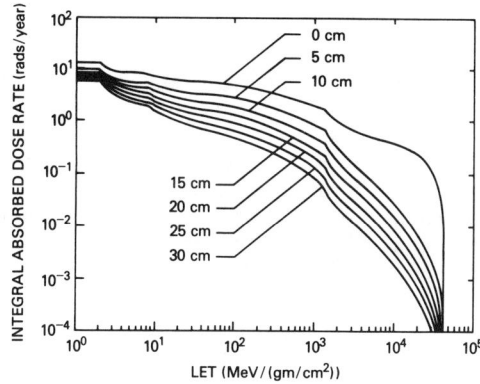

Figure 4. The integral LET distributions without shielding of the annual absorbed dose of cosmic rays at solar minimum outside the magnetosphere, at centers of spheres of water, for radii of about 0 to 30 cm.

Figure 4 shows the unshielded absorbed dose and the integral LET distributions of the absorbed dose of cosmic rays at solar minimum, outside the earth's magnetosphere at centers of spheres of water (representative of biological tissue self-shielding) with radii of 0, 5, 10, 15, 20, 25 and 30 cm for a 1-year exposure. The calculations are done with the UPROP code. We note the greater attenuation of iron nuclei (at LET $> 10^3$) than of protons (LET ~ 2).

Figure 5 shows the annual dose equivalent due respectively to cosmic ray elements H, He, Li to B, C to O, F to S, Cl to Mn and Fe, at various depths in a water phantom of 30 cm diameter, at solar minimum outside the earth's magnetosphere.

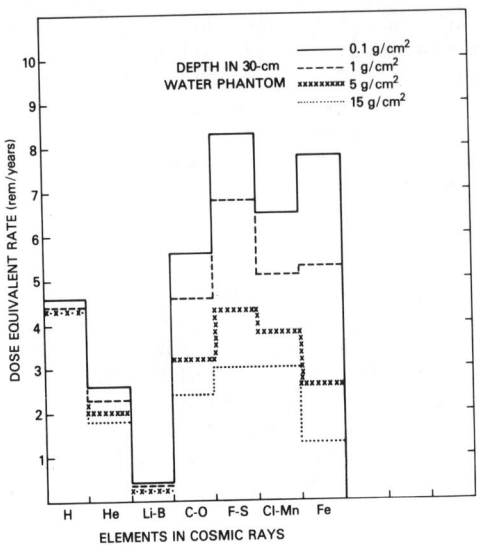

Figure 5. The annual dose equivalent due to various sets of cosmic ray elements, from H to Fe, at various depths of a water phantom of 30 cm diameter, at solar minimum, outside the earth's magnetosphere.

Figure 6 shows the annual dose equivalent and its integral LET-distribution at the surface of water phantom of 15 cm radius, at solar minimum. The four curves are respectively (a) for outside the magnetosphere, (b) in a polar 90° orbit, (c) in an orbit of 50° inclination and (d) of 30° inclination. These calculations have been integrated over the latitude dependent geomagnetic cutoffs. We note that the latter case does not have the high-LET <u>slow</u> (E < 1 GeV/nucleon) iron nuclei.

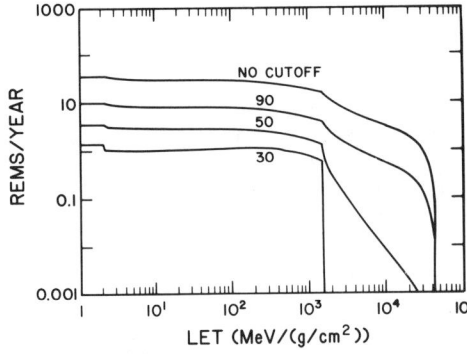

Figure 6. The integral LET distributions of the annual dose equivalent at solar minimum, at the surface of a water phantom of 30 cm diameter, with no geomagnetic cutoff, i.e. outside the magnetosphere and below the magnetosphere in orbits of 90°, 50° and 30° inclination.

Figure 7 shows a comparison of the annual dose equivalent due to cosmic-ray nuclei and neutrons, as a function of depth in lunar material, down to 500 g/cm². Also the annual absorbed dose due to cosmic rays is shown. We note the rapid attenuation of the dose due to cosmic rays, and the buildup of that due to neutrons. The dose rate due to cosmic rays is calculated with UPROP, and that due to neutrons from the measured neutron depth profile measured by Woolum and Burnett[23]. It can be seen that for a shielding > 400 g/cm^2, the annual dose is brought down to 5 rem/year, the level permissible to radiation workers. Alternatively, if one spends 40 hours a week on the surface, the remainder has to be spent under even deeper shielding, to maintain an average of 5 rem/year.

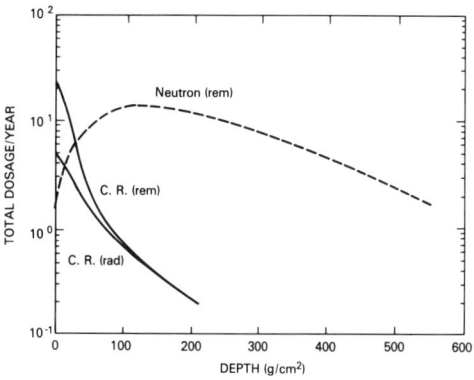

Figure 7. A comparison of the annual dose equivalent due to cosmic ray nuclei and secondary neutrons, as a function of depth in lunar soil, at solar minimum. Also the absorbed dose rate due to cosmic ray nuclei is shown.

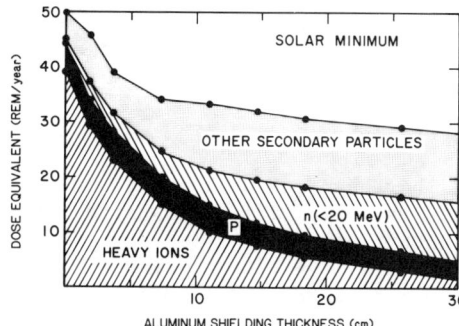

Figure 8. Components of the dose equivalent due to cosmic rays at solar minimum, as a function of aluminum shielding thickness. Doses from heavy ions, primary protons, (p), secondary neutrons (E < 20 MeV), and other secondary particles (including recoil nuclei and high energy neutrons) are shown.

Expected radiation doses to astronauts on extended spaceflight outside the magnetosphere (e.g. to Mars) are shown in Fig. 8 as a function of spacecraft shielding. This figure has been calculated using the combined codes UPROP, HETC and MCNP, at solar minimum. We note that without aluminum shielding, the annual dose equivalent is 50 rem/year; it is reduced to less than 35 with 8 cm of Al, which is less than that allowable to a few volunteer astronauts[24]. At solar maximum, the dose rate is about half of the above, but attenuation by shielding is weaker due to less low energy particles. However, the risk of large solar flares is much higher at solar maximum. We note that the character of the radiation changes with shielding; after 10 cm of Al, secondary particles (with large quality factors) dominate over primary protons and heavy nuclei.

A different set of quality factors has been proposed[18], but not yet adopted. The dose equivalent with these factors is higher.

The respective annual dose equivalent values with the standard and the proposed qualifications, at solar minimum, in the center of a water sphere with a diameter of 5 cm, are 45 and 61 rem/year.

The radiation dose during the August, 1972 solar flare would have been about 960 rem with no spacecraft shielding, using the quality factors adopted here. This falls to 40 rem with 9 cm aluminum shielding. The higher dose is lethal, while the shielded dose would have resulted in no short-term health problems for astronauts in general.

The production of secondary particles, such as neutrons, can be reduced by using hydrogen-rich shielding material, e.g. water (or methane). Hydrogen rich materials are also more effective per unit weight at fragmenting and slowing down heavy ions, and for moderating neutrons. Shielding considerations of this type are essential for reducing weight requirements in future spacecraft.

REFERENCES

1. J. H. Adams, Jr., R. Silberberg and C. H. Tsao, Cosmic Ray Effects on Microelectronics, Part I: The Near-Earth Particle Environment, NRL Memo Report 4506 (1981).

2. J. H. Adams, Jr. and A. Gelman, IEEE Trans. Nucl. Sci. NS-31, 1212 (1984).

3. G. Rudstam, Z. Naturforschung 21A, 1027 (1966).

4. R. Silberberg and C. H. Tsao, Ap. J. Suppl. 25, 315 (1973).

5. R. Silberberg, C. H. Tsao and J. R. Letaw, Ap. J. Suppl. 58, 873 (1985).

6. R. Silberberg, C. H. Tsao and J. R. Letaw, 20th Internat. Cosmic Ray Conf. (Moscow), 2, 133 (1987).

7. J. B. Cumming, R. W. Stoenner, and P. E. Haustein, Phys. Rev. C14, 1554 (1976).

8. J. R. Letaw, R. Silberberg, and C. H. Tsao, Ap. J. Suppl. 51. 271 (1983).

9. R. Silberberg and C. H. Tsao, 15th Internat. Cosmic Ray Conf. (Plovdiv) 2, 89 (1977).

10. P. J. Lindstrom, D. R. Greiner, H. H. Heckman, B. Cork, and F. S. Bieser, Lawrence Berkeley Lab. Report, LBL-3650 (1975).

11. L. W. Townsend, J. W. Wilson, F. A. Cucinotta, and J. W. Norbury, Phys. Rev. C34, 1491 (1986).

12. P. J. Karol, Phys. Rev. C11, 1203 (1975).

13. S. P. Ahlen, Rev. Mod. Phys. 52, 121 (1980).

14. J. R. Letaw, Phys. Rev. C28, 2178 (1983).

15. L. W. Townsend and J. W. Wilson, to be publ. in Health Physics (1988).

16. J. R. Letaw, R. Silberberg and C. H. Tsao, Ap. J. Suppl. 56, 369 (1984).

17. R. E. Prael, Users Guide to HETC, Los Alamos National Lab. Draft Report (1985).

18. ICRU Report 40, the Quality Factor in Radiation Protection, Bethesda, MD (1986).

19. R. Silberberg, C. H. Tsao, J. H. Adams, Jr., and J. R. Letaw, Rad. Res. 98, 209 (1984).

20. R. Silberberg, C. H. Tsao, J. H. Adams, Jr., and J. R. Letaw, Advances in Space Research 4, 143 (1984).

21. R. Silberberg, C. H. Tsao, J. H. Adams, Jr., and J. R. Letaw, p. 663 in Lunar Bases and Space Activities of the 21st Century, ed. W. W. Mendell, publ. by Lunar and Planetary Institute, Houston (1985).

22. J. R. Letaw, R. Silberberg and C. H. Tsao, Nature 330, 709 (1987)

23. D. S. Woolum and D. S. Burnett, Earth Planet. Sci. Letters 21, 153 (1974).

24. W.H. Langham, ed., Radiation Protection Guides and Constraints for Space-Mission and Vehicle-Design Studies Involving Nuclear Systems, Natl. Acad. Sci., Washington, DC 1970.

HIGH-ENERGY OUTER RADIATION BELT DYNAMIC MODELING

Y. T. Chiu, R. W. Nightingale and M. A. Rinaldi
Lockheed Palo Alto Research Laboratory
3251 Hanover St., Palo Alto, Ca. 94304

ABSTRACT

Specification of the average high-energy radiation belt environment in terms of phenomenological montages of satellite measurements has been available for some time. However, for many reasons both scientific and applicational (including concerns for a better understanding of the high-energy radiation background in space), it is desirable to model the dynamic response of the high-energy radiation belts to sources, to losses, and to geomagnetic activity. Indeed, in the outer electron belt, this is the only mode of modeling that can handle the large intensity fluctuations. Anticipating the dynamic modeling objective of the upcoming Combined Release and Radiation Effects Satellite (CRRES) program, we have undertaken to initiate a study of the various essential elements in constructing a dynamic radiation belt model based on interpretation of satellite data according to simultaneous radial and pitch-angle diffusion theory. In order to prepare for the dynamic radiation belt modeling based on a large data set spanning a relatively large segment of L-values, such as required for CRRES, it is important to study a number of test cases with data of similar characteristics but more restricted in space-time coverage. In this way, models of increasing comprehensiveness can be built up from the experience of elucidating the dynamics of more restrictive data sets. The principal objectives of this paper are to discuss issues concerning dynamic modeling in general and to summarize in particular the good results of an initial attempt at constructing the dynamics of the outer electron radiation belt based on a moderately active data period from Lockheed's SC-3 instrument flown on board the SCATHA (P78-2) spacecraft. Further, we shall discuss the issues brought out and lessons learned in this test case.

FORMULATION

Some two decades of satellite observations have shown that the electron radiation flux (including the satellite-damaging electrons of >500 keV energy) in the outer belt fluctuates by orders of magnitude in response to geomagnetic activity. Some typical fluxes and spectra measured by the SCATHA SC-3 instrument[1,2,3] are shown in Figure 1. The practical need to characterize and predict these large radiation fluctuations for satellite survivability considerations requires an understanding of their dynamics beyond phenomenological correlations; thus, the need for dynamic modeling of the outer electron belts was initiated.

In its most comprehensive form dynamic modeling involves a theoretical accounting of the flux responses in terms of fundamental processes of electron transport and loss in the magnetosphere. Since we are dealing with energetic electrons, a Fokker-Planck description of the transport processes naturally leads to a diffusion picture in terms of several independent variables characterizing

© 1989 American Institute of Physics

the population.

Diffusive transport of energetic particles in the magnetosphere is an excellent theoretical approach to outer-belt dynamic modeling, but the latter involves much more. The constraints of satellite data and conditions of the magnetosphere play roles in dynamic modeling which are as important as the diffusion theory itself. In the outer belt the magnetic field is transient. Plasma and field conditions encountered by a drifting electron vary drastically and parts of the energetic particle population may even cross outside of the magnetosphere. Thus, dynamic modeling involves an accounting of fields and magnetospheric boundary configurations as much as diffusive particle transport. All of these elements are integral parts of a dynamic model of the outer electron belt.

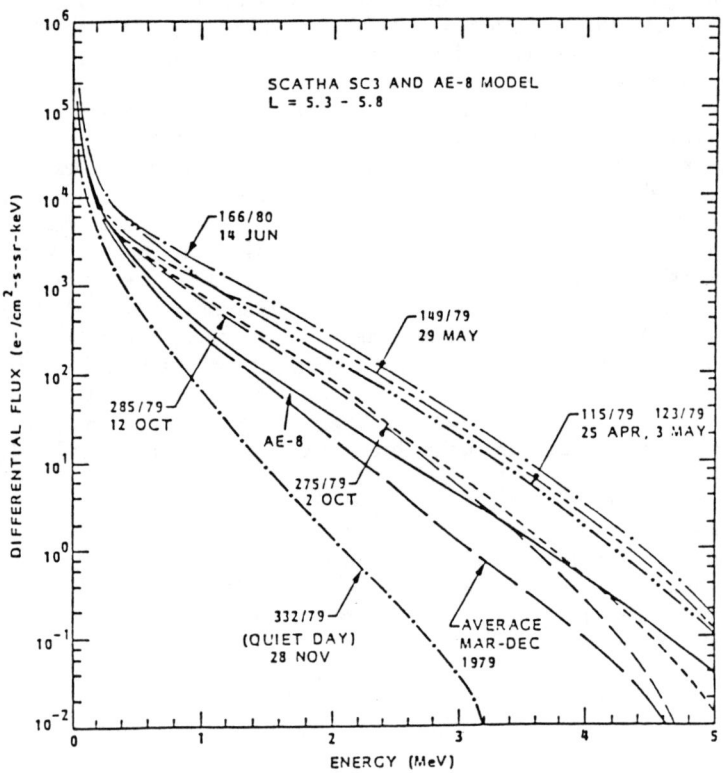

Figure 1. Sample spectra showing the variation of outer-belt electrons obtained in 1979 and 1980 by the Lockheed SC3 instrument. Phenomenological models based on 6-month averages of the data can be in error by an order of magnitude or more.

To illustrate the complexity that a dynamic model must deal with, Figure 2 shows a survey of the high-energy electron characteristics encountered by SCATHA in an active period on day 115 of 1979 in which a moderate magnetic storm occurred, with a sudden commencement just prior to the start of the day and a maximum Dst of -148 γ. Just before 18:00 h LT there was a brief enhancement of electron fluxes that may have been an adiabatic compression event. "Butterfly" pitch-angle distributions, indicating loss of 90° pitch-angle particles at the magnetospause, were seen at all energies for almost 2 hours, after which the lower-energy electron fluxes recovered to their previous levels. Shortly after

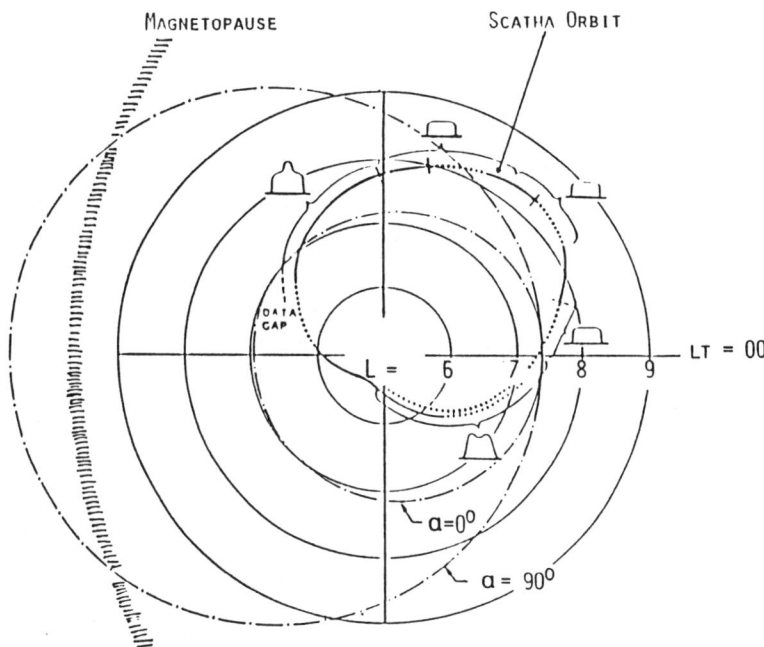

Figure 2. A montage of electron flux characteristics encountered by SCATHA on day 115 in 1979. The inset figures are schematics of pitch-angle distribution with 90° pitch angle placed at the center.

midnight, there began a succession of substorms, with deep flux dropouts over the portions of the orbit shown as dotted lines in Figure 2. The dropouts exhibit the expected correlation with the geomagnetic activity indices, particularly AE. These were followed by rises in the flux levels that can be interpreted as a fresh arrival of electrons on the observed flux tubes by either injection or compressional changes of the magnetic field. Such active periods are usually recognized by decreases in the total flux and the occasional appearance of nearly isotropic pitch-angle distributions as shown on the Figure. In the morning hours, typical "hat" distributions, indicating enhanced fluxes of 90° pitch-angle particles, are encountered.

In view of the complexity of the above data, it is sensible to separate the elements of outer-belt dynamic modeling into modules which can be individually formulated. Thus, our approach to dynamic modeling is to initially treat each of the following elements as if they are not related, but bearing in mind all the time that they are to be integrated into a whole. These elements are:

(i) Transport, injection and loss of trapped energetic electrons in terms of diffusion in phase space;

(ii) Description of the space-time variations of the geomagnetic field;

(iii) Physics of transport parameters.

The principal topics of this paper deal only with issues arising out of the diffusion physics [item (i)]. The diffusion physics, which is the fundamental basis of a dynamic model, appears initially to be mostly straightforward, although mathematically cumbersome. However, it turns out that the boundary conditions corresponding to satellite data can become a major issue to be resolved irrespective of the chosen diffusion formulation. Since the boundary data issue is central to all schemes of dynamic modeling, we discuss it at the outset.

On the least ambitious level, dynamic modeling can be viewed as a means of projecting radiation belt data obtained along a satellite track by physical means to specify the radiation environment throughout some reasonable regions not on the satellite track. As such, the selected diffusion treatment is the projection mechanism and the in-track satellite data is the raw material to be processed: the "boundary" values for the diffusion physics. However, satellite measurements, obtained on a space-time trajectory, are not standard boundary conditions stated at fixed time or space coordinates to define either initial-value problems or boundary-value problems of diffusion theory. The situation for dynamic modeling is different from these simple boundary-value or initial-value cases. Figure 3 illustrates the satellite data span and its relation to the appropriate solution volume in (x, L, t) space, where x is the cosine of the pitch angle and L is the McIlwain parameter. The fundamental nature of nonsynchronous satellite data spans is that they lie on a data surface (ABCD on the figure) defined by the satellite trajectory $L(t)$. Further, the time span defines the (L, x) boundary surfaces of the solution volume labelled by the Roman letters A to H on the figure. In such a source-free volume, particles are transported into and out of the surfaces ABGH and CDEF by radial diffusion; in other words, these

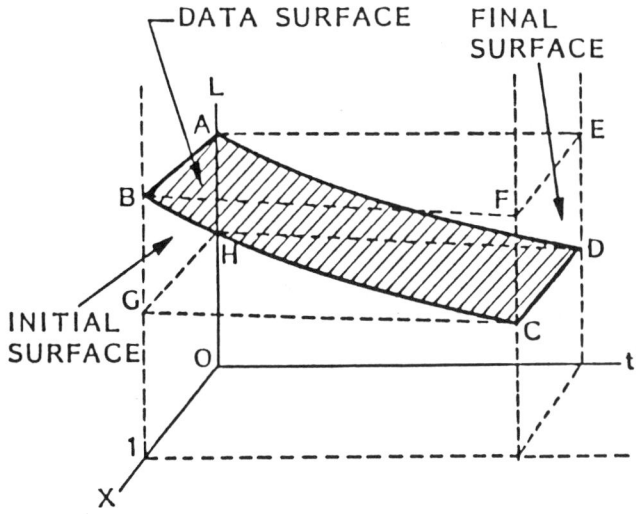

Figure 3. Schematic illustration of the boundary volume in (x, L, t) space in which the general bi-modal diffusion solution is developed. Satellite data to be matched to the solution span lie on the data surface.

boundary surfaces are also source surfaces. Particles are lost through the surface BGCF through pitch-angle diffusion into the loss cone, approximately signified by $x = 1$ here. If the data time span is less than or comparable to the injection and loss times in the volume, as is the case in most satellite observations such as SCATHA, the appropriate solution to be developed is the general source-free solution in a finite space-time interval, and not an initial-value solution in the semi-infinite time domain corresponding to response to a transient (i.e., delta function) source. In the finite space-time domain, the entire spectrum of time constants (rising as well as falling) must be included in the general complete-set solution expansion because unknown sources (and sinks) outside of the volume are diffusing particles into (and out of) the finite volume causing the phase-space density to increase (decrease) with time scales that can be comparable to the data time span. In contrast, in the semi-infinite time domain only the falling time response function is appropriate. To be sure, this Green's function can be applied to obtain the finite space-time solution by applying Green's theorem onto the boundary surfaces[4]. This crucial integration imparts the source time scales onto the solution. The important point is that this integration convolves the rising and falling time constants of the surface sources with the natural falling time constants of diffusion to result in a spectrum of both rising and falling time constants for the spatial eigenfunctions. As a result of this time convolution, the final solution for finite volume need not look anything like the Green's function that would be developed for the initial-value problem over semi-infinite time. To be sure, all excited states eventually decay by diffusion to equilibrium, but with the satellite data span, over a finite space-time interval, we do not know whether it has caught the rising or falling phase of a modal disturbance of unknown time scale. For the SCATHA orbit, space and time are relatively distinct since the range of L covered by the satellite is small. However, for highly elliptical orbits (such as for CRRES) space and time on an orbit are not linearly related. We shall see that under such circumstances the satellite data set acts as a highly nonlinear boundary for the dynamic model.

The consequence of the esoteric points discussed above is that it would be wrong to force satellite data obtained over a finite space-time volume to fit an impulsive transient response solution such as the Green's function. For purposes of satellite data analysis in a finite space-time interval, the appropriate solution to diffusion theory is the solution corresponding to a general boundary-value problem[4], and not corresponding to a global response to an impulsive source. Since a general solution can be obtained by eigenfunction expansion with unrestricted separation constants, which will be constrained by data, this convenient property forms the basic principle of data charactrization in dynamic modeling. It will be shown later, based on the success of our data analysis, that this principle can be used to characterize the state of the entire radiation belt in a piece-wise-continuous fashion.

Given the above general property of radiation belt dynamic modeling, let us proceed to consider our test case[4] for which we choose to specify the diffusion physics in a static magnetosphere in (x, L, t) space as illustrated in Figure 3. [The choice of spaces to characterize radiation belt diffusion is not arbitrary. This complex issue is treated elsewhere[4].] Our choice of diffusion description is that of simultaneous multi-mode diffusion theory. The multimode diffusion equation can be simplified[4,5,6] under a number of reasonable assumptions. One of these is that the L-dependence of the radial diffusion coefficient D_{LL} can be parameterized as $D_{LL} = \xi L^\mu$ where ξ and μ are parameters to be determined by the on-trajectory satellite data. The diffusion equation for the electron distribution function f can be written

$$\frac{\partial f}{\partial t} = D_{xx}\left[\frac{\partial^2 f}{\partial x^2} + \frac{1}{x}\frac{\partial f}{\partial x}\right] + \xi L^{-2+\frac{3}{2}\nu}\frac{\partial}{\partial L}\left[L^{\mu+2-\frac{3}{2}\nu}\frac{\partial f}{\partial L}\right] \qquad (1)$$

where D_{xx} (the pitch-angle diffusion coefficient), ξ, μ and ν are considered to be constant parameters.

This simplified equation is to be solved in a small finite domain $L_l \leq L \leq L_u$ expected to be in the outer belt and in the pitch-angle domain $0 \leq x \leq x_c$, where x_c is cosine of the loss-cone angle. The effect of pitch-angle diffusion is to define pitch-angle behavior of f inside and outside of the loss-cone boundary $x = x_c(L)$. The dependence of x_c on L is yet again another factor which mixes the x and L operators in (1); however, for the range in our test case in outer zone region, where $(L_u - L_l) \leq 2$, x_c is very nearly constant and nearly unity.

We assume that x_c is constant in the data span so that separable solutions can be obtained for (1) with the present test case. This is another important point when we deal with data from a large domain of L (such as that from CRRES) which involves large variations of the loss cone both from the large variations in L and in distances to the magnetic mirror points. For the SCATHA test case the loss-cone dependence on L is entirely negligible.

Under the approximations discussed above, a general solution to (1) in finite space-time can be developed[4] in terms of complete-set eigenfunction expansions. In general, there are three modal expansions: (1) single-mode pitch-angle diffusion, (2) single-mode radial diffusion and (3) simultaneous bi-modal diffusion. Using the device of the Kronecker delta δ_{n0}, the three modal expansions can be written in terms of a single eigenfunction expansion

$$f = \sum_{n=0}^{\infty} \left[\delta_{n0} + (1-\delta_{n0}) \frac{\sqrt{2}}{x_c} \cdot \frac{J_0(k_n x/x_c)}{J_1(k_n)} \right] \left(c_n(1-\delta_{n0}) e^{t/\tau_n} + h(L,t) \right) \quad (2a)$$

$$h(L,t) = e^{t/t_n} [a_n Y_n(L) + b_n Z_n(L)] \quad (2b)$$

where k_n is the nth zero of the Bessel function J_0. The parameters and new functions are defined as follows:

$$\epsilon \equiv 3(1 - \frac{\nu}{2})/(\mu - 2)$$

$$\sigma \equiv -\frac{\mu}{2} - \frac{1}{2} + \frac{3}{4}\nu \quad (3)$$

$$\lambda \equiv \frac{\mu}{2} - 1$$

$$Y_n(L) = L^\sigma \cdot [J_{1+\epsilon}(|\beta_n|/L^\lambda)]_> \cdot [I_{1+\epsilon}(|\beta_n|/L^\lambda)]_< \quad (4)$$

$$Z_n(L) = L^\sigma \cdot [N_{1+\epsilon}(|\beta_n|/L^\lambda)]_> \cdot [K_{1+\epsilon}(|\beta_n|/L^\lambda)]_< \quad (5)$$

J is the regular Bessel function, I and K are the modified Bessel functions, and N is the Neumann function. In the above, the separation constants are t_n and β_n. The usual notation $[g]_>$ is defined in this case by

$$[g]_> = g \quad \text{if} \quad t_n > 0$$
$$\quad (6)$$
$$[g]_> = 1 \quad \text{if} \quad t_n < 0$$

and analogously for $[g]_<$. The separation constants and parameters of the solution are related by the algebraic relations

$$t_n^{-1} = (1 - \delta_{n0})\tau_n^{-1} - \xi\beta_n^2 \lambda^2 \qquad (7)$$

$$\tau_n^{-1} \equiv -D_{xx} k_n^2/x_c^2 \qquad (8)$$

which will be referred to as separation equations. It is seen that the single pitch-angle mode time constants are negative as expected, since single-mode pitch-angle diffusion always leads to decay. For bi-modal and radial diffusion, however, the time constants t_n can be either positive or negative over a finite time interval. This is due to the fact that the phase-space density in the finite L domain can rise or fall in time in response to particle transport into and out of the adjacent volumes, as discussed above. With the sign of each t_n not predetermined, the separation constant β_n^2 can be positive or negative. This is the origin of the (J, I) and (N, K) options of (4) and (5). The solution (2) is not the most general solution, for a separate sum of pure radial modes can be added. We have not done so for two reasons: (1) we expect that episodes of radial diffusion must accompany pitch-angle diffusion because of the association of magnetospheric fluctuations of all scales with each other and (2) this separate sum creates more parameters. As it presently stands, only one pure radial mode is included, i. e. for $n = 0$.

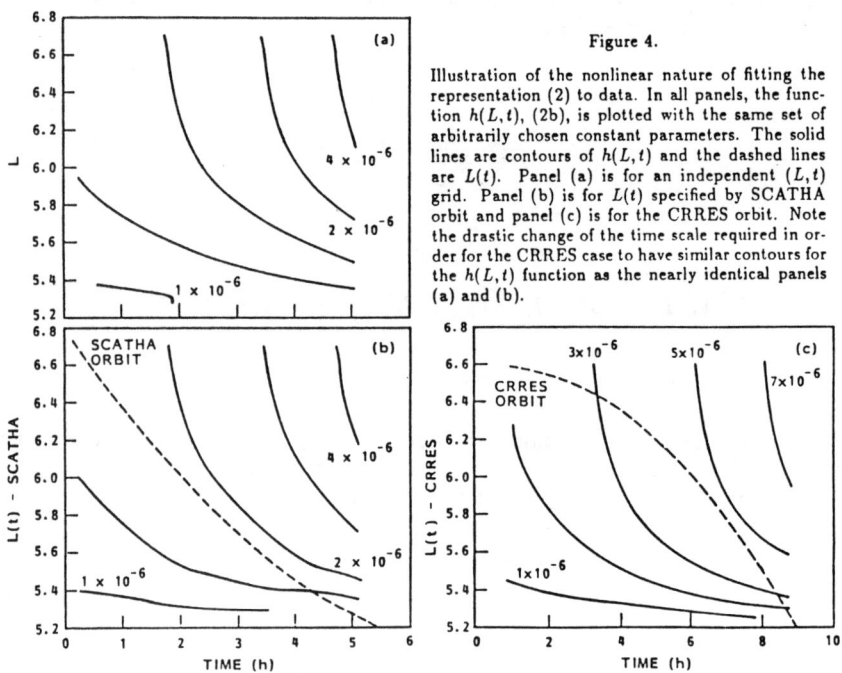

Figure 4.

Illustration of the nonlinear nature of fitting the representation (2) to data. In all panels, the function $h(L, t)$, (2b), is plotted with the same set of arbitrarily chosen constant parameters. The solid lines are contours of $h(L,t)$ and the dashed lines are $L(t)$. Panel (a) is for an independent (L, t) grid. Panel (b) is for $L(t)$ specified by SCATHA orbit and panel (c) is for the CRRES orbit. Note the drastic change of the time scale required in order for the CRRES case to have similar contours for the $h(L,t)$ function as the nearly identical panels (a) and (b).

Having now specified a general solution of the chosen diffusion theory, we can quantitively illustrate how the satellite trajectory $L(t)$ impacts the characterization of radiation fluxes away from the trajectory. The idea is of course central to dynamic modeling and is very simple: determine the parameters of the solution with on-trajectory satellite data then the solution yields fluxes at all other (x, L, t) points. But as pointed out above, the problem is that the satellite trajectory mixes L and t, i.e., $L(t)$, which complicates the fit. For near-synchronous satellites such as SCATHA for which only a small range of L is covered in an orbit, the effect of space-time mixing is quite minimal. This allows us to complete the test case shown in the next section. For the more comprehensive situation of highly-elliptic orbits such as that for CRRES, which attempts to specify the radiation belts over a wide range of L, the trajectory $L(t)$ is a non-linear function of t, requiring a non-linear fit for the determination of the diffusion parameters. This significant complication is illustrated in Figure 4 which shows plots of the function $h(L, t)$ in (2b) for three situations: (a) L and t independent, (b) SCATHA orbit on which L is nearly independent of t and (c) CRRES-like orbit on which L is strongly dependent on t. The strong distortions on panel (c) for $h(L(t), t)$ relative to the same function $h(L, t)$ on panel (a) signifies that dynamic modeling in conjunction with a satellite data base involves much more than obtaining solutions to diffusion theory.

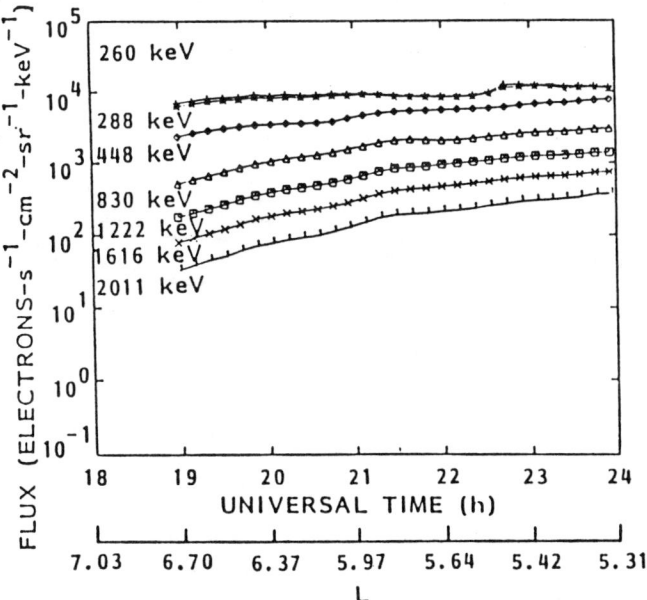

Figure 5. Ten-minute averaged electron fluxes integrated over all pitch angles for 7 energy channels measured by the Lockheed SC3 instrument on board SCATHA during day 165, 1980. This is the data set to be analyzed in this paper.

TEST CASE VERIFICATION

Notwithstanding the difficulties of applying radiation belt diffusion theory to dynamic modeling in the comprehensive situation of the expected CRRES data base, it is of interest to test whether multi-mode diffusion theory adequately describes the simpler on-trajectory SCATHA data base. This section summarizes the results of such an investigation[4].

The Lockheed SCATHA SC3 data base[1,2,3] consists of ~ 950 days of digitized data on magnetic tape from launch in 1979 to the present in several intervals. The pitch angle for each accumulation period is determined using data from the on-board vector magnetometer (SC11). From the data tapes a condensed data base of 10-min averaged, pitch-angle binned electron fluxes for the 24 energy channels over the full energy range of 47 to 4970 keV, has been generated for on-line storage on a VAX 11/780 computer. For each 10-min time interval the data are grouped by pitch-angle values into 3° bins from 0° to 90°, making 30 pitch-angle bins for each energy channel in the interval.

The SCATHA SC3 temporal data set for this verification was from the latter portion of day 165, June 13, 1980, a moderately active period with $Kp = +4$ and $Dst = -50$. Two days earlier a geomagnetic storm had occurred, but the magnetosphere had since relaxed enough that the observed electron fluxes on day 165 were smoothly varying with no pitch-angle signatures of particle loss in the magnetopause. Figure 5 shows the 10-min averaged data integrated over all pitch angles for 7 channels of electron flux. During this period the spacecraft

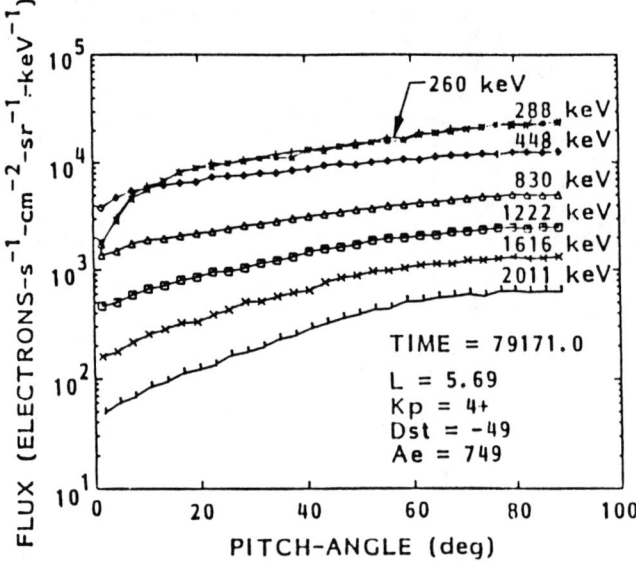

Figure 6. Typical pitch-angle distributions for the 7 energy channels of the data set referred to in Figure 5.

was drifting inward from $L = 7.2$ to $L = 5.3$. The L-shell range included in the analysis has been restricted to $L \leq 6.48$ to eliminate the large variability of the electron fluxes beyond the geosynchronous region[7] and the low count rates due to the approximately exponential fall off with increasing L-shell[1]. In addition we have selected for the formal analysis only those data channels above 256 keV so as to eliminate channels which are most likely to be affected by convective forces resulting from consequences of the major storm preceding the data period. Data with electron energies above 2208 keV have low statistics and are also excluded from the data set. Figure 6 illustrates typical pitch-angle distributions during this time for the selected energies. Although the period was moderately active, the magnetic field was smoothly varying, with a field direction that remained perpendicular to the spacecraft spin vector. This allowed SC3 to look into the loss cone for the whole period, providing data over the full pitch-angle range. In addition, the pitch-angle distributions were smoothly varying and without signatures that indicate loss at the magnetopause.

The application of the diffusion theory summarized above to the SC3 electron data for day 165 has been discussed in detail in an earlier paper[4]. The verification that the theory is a good description of the on-trajectory SCATHA data is indicated in Figure 7, in which the entire data set is shown in Panel(a) as a three-dimensional plot in terms of pitch angle and $L(t)$. Panel(b) shows the reconstructed data set after it has been decomposed into a sum of pitch-angle eigenfunctions truncated at the fourth order. Comparison of Panels (a) and (b) shows that the pitch-angle eigenfunction series is a very efficient representation of the data set. In Panel (c) we also show the reconstruction of the data set after decomposition in (x, L, t) space, using the property that L and t are approximately decoupled for the SCATHA orbit as illustrated by Panels (a) and (b) of Figure 4. In order to show the worst case scenario we show in Panel (c) of Figure 7 the reconstruction after the first iteration of the decomposition; we have not attempted to obtain the best solution at this time. As we can see by a comparison of the panels in Figure 7, we conclude that multi-mode diffusion theory in a fairly simplified form is a good representation of the electron outer belt fluxes. This implies that multi-mode diffusion theory can be used as a fundamental basis of radiation belt dynamic modeling.

Verification that multi-mode diffusion theory is a good basis for radiation belt dynamic modeling involves more than just data representation and the implied power of flux mapping throughout the entire belt, although the applicational needs are focused onto this area. From the science view point, an important part of dynamic modeling is the determination of transport parameters which are related to the microscopic physics of wave-particle interactions in the magnetospheric plasma. The above data representation exercise yields physical diffusion parameters with maximum error spread factors of .2-5 that can be compared to other determinations. The radial diffusion coefficient D_{LL} determined in our exercise and shown on Figure 8 as the shaded trapezoid region compares well with other determinations. Indeed, our determination is able to obtain the energy dependence from 0.8 MeV to 2.0 MeV. Our pitch-angle diffusion coefficients are also in good agreement with other studies[4]. These give confidence that our dynamic modeling exercise is not in conflict with independent expectations.

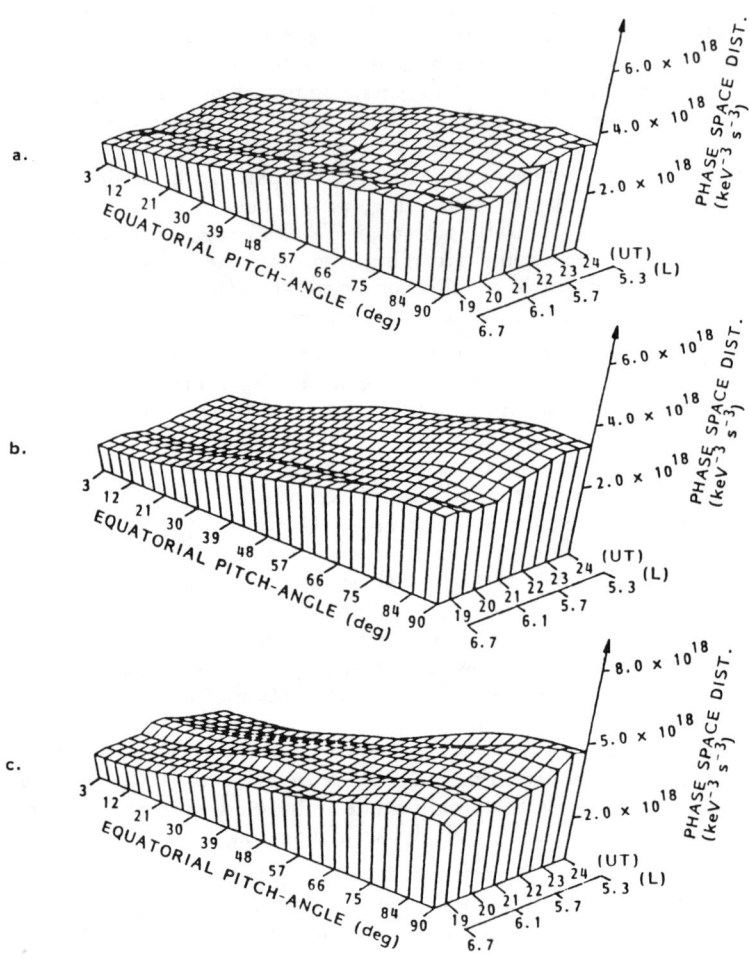

Figure 7. Three-dimensional presentation of the phase-space density of the data set and its reconstitutions for the 830 keV channel in pitch angle and space-time. Panel (a) shows the data set. Panel (b) is the reconstitution after fitting the pitch-angle eigenfunction decomposition. Panel (c) is the reconstitution with the bi-modal diffusion analysis but without the benefits of an iterative fitting procedure.

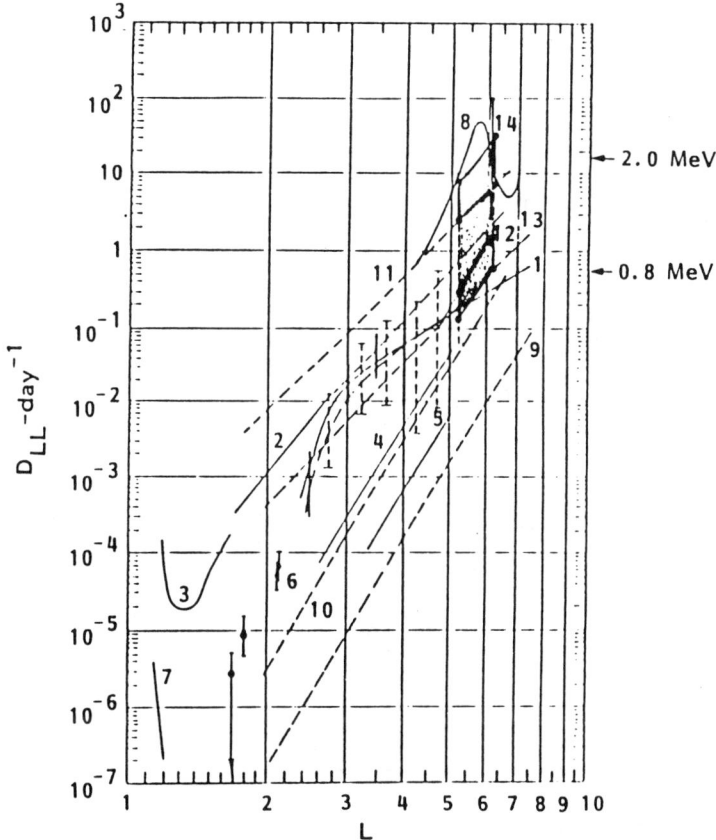

Figure 8. Comparison of the radial diffusion coefficient determined for the energy range 0.8-2.0 MeV (mottled area showing decreasing shading with increasing energy) with a previous compilation[7] which yields curve 1. Curves 2-8 are experimental[8-14]. Curves 9-14 are theoretical or semitheoretical[15-19].

FURTHER ISSUES IN DYNAMIC MODELING

The test case exercise[4] summarized above may have resolved an important issue in radiation belt dynamic modeling, namely that multi-mode diffusion theory is a viable basis for dynamic modeling. However, there are several other issues that must be resolved before CRRES-like applications can be made. Here we shall discuss what we preceive to be the two most important issues.

(A) Flux mapping: The representational use of dynamic modeling must clear another hurdle before it can be made operational. The issue arises because of the multi-dimensional nature of phase space. Instead of basing the discussion on phase space mapping in the abstract, we can illustrate the issue with our familiar (L,t)-space. Armed with a diffusion theory representation tested on-trajectory, as in the exercise above, we can obtain fluxes at any given off-trajectory point by "mapping" the on-trajectory flux utilizing the representation. Without a general representation such as (2), the mapping is not unique if it is not based on a solution of the assumed diffusion physics. Figure 9 illustrates this mapping based on the assumed diffusion theory of the SCATHA test case above. The fluxes at Point 2, outside of the SCATHA trajectory, are uniquely specified by the diffusion parameters based on the fitted fluxes (dashed curves) on all points along the SCATHA trajectory between Points 1 and 3. In this regard, uniqueness is somewhat relative because it depends on the errors involved in specifying the on-trajectory diffusion parameters. A case of extreme constrast to Figure 9 is to imagine that no solution of diffusion theory was available as the basis for dynamic modeling. In such a case, Point 2 can still be mapped in (L,t)

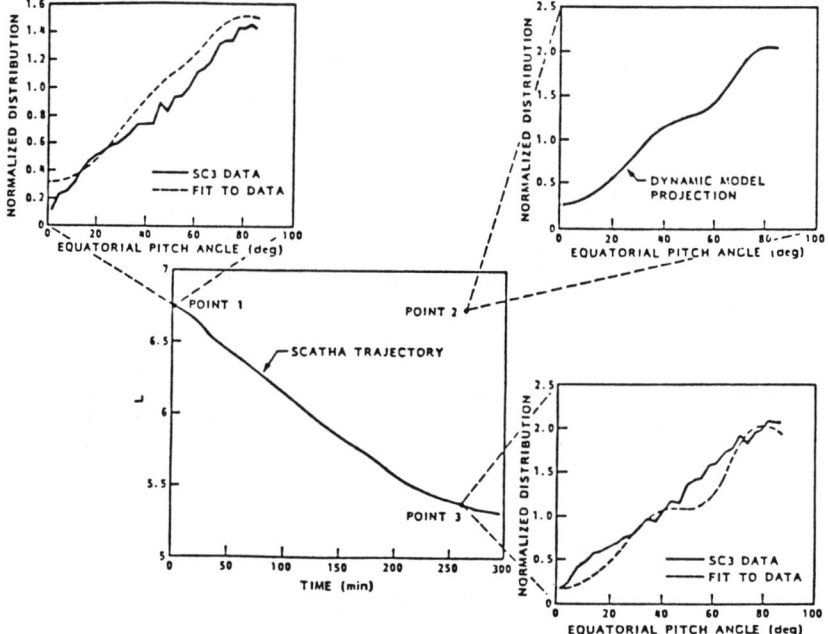

Figure 9. Illustration of flux mapping using a nominal dynamic model: our test case verified on the SCATHA trajectory. To the extent that the on-trajectories fits are good, the mapping by the diffusion representation (2) is "unique".

from Point 1 or Point 3 along t and L axes, respectively, by some phenomenological construct. Indeed, a large multitude of such mappings can be made to Point 2 from data on the trajectory. Since these individual mappings need not be consistent with each other, the uniqueness (within observational errors) of the predicted flux at Point 2 will be lost without the diffusion representation as a consistent basis for the mapping. Having a diffusion representation, however, does not free us from the fetters of non-uniqueness. Indeed, the errors incurred in finding the on-trajectory diffusion parameters operate in the same way as the inconsistent multiple mappings. Similar to these, the errors tend to vitiate the validity of a given representation for dynamical modeling. There are two necessary elements in ensuring "uniqueness" in data mapping. The first is to minimize errors in diffusion parameters for a given test diffusion representation. The second is to determine the most "unique" representation among several theoretical specifications of the diffusion physics. The first element is much easier to carry out than the second since the only way to insure that a diffusion representation is "unique" is to show that it works in practice. The theory of transport coefficients in the magnetosphere is presently not advanced enough to allow us to choose the "unique" physics on an a priori basis.

(B) Invariant Representations: An area of dynamic modeling not touched upon above is the validity of the implicit assumption in the test case that fluxes on the same L-shell are the same at the same time and energy. We know from Figure 2 that on the time scale of a day this is not true as effects of magnetic and electric shell-splitting, local injections and losses introduce a local time (azimuthal) dependence upon the data. This crucial aspect of radiation data is under current investigation on a case by case basis[20]; however, there is currently little work done on how these effects can be incorporated into a dynamic response model. A possible solution to this important problem may be to return to the most fundamental representation of radiation belt diffusion physics, i.e., that of diffusion in adiabatic invariant space rather than in its various reductions to configuration space[4], such as in the above. The abstraction required to treat data at such a level is somewhat of a drawback; however, a test case attempt is needed to identify the detailed issues involved.

CONCLUSIONS

Various elements required in radiation belt dynamic modeling have been discussed. It is seen that dynamic modeling differs fundamentally from radiation belt data analysis in that a theoretical representation of the on-trajectory data set is required to predict fluxes outside of the satellite orbit. The faithfulness of the projection depends on whether the representation is a good description of the physical diffusive processes seen in the on-trajectory data. We have summarized the results of a test case study[4] on such a tractable representation which yields results consistent with basically all other measurements in the moderately active outer electron radiation belt. The resolution of a number of important issues brought out by the examination of the test case awaits further work before operational dynamic models can be constructed.

ACKNOWLEDGMENT

This work is supported by the Air Force Geophysics Laboratory under contract F19628-85-C-0073. It is a pleasure to thank Drs. S. Gussenhoven, M. Schulz, and G. Davidson for discussions on various aspects of this work. The SCATHA SC-3 payload and initial data analysis were sponsored by the Office of Naval Research under contract N00014-76-C-0444. The P78-2 SCATHA spacecraft is sponsored and operated by the USAF Space Test Program.

REFERENCES

1. Reagan, J.B., R.W. Nightingale, E.E. Gaines, W.L. Imhof, and E.G. Stassinopoulos, J. Spacecraft Rockets, 18, 83, 1981.
2. Reagan, J.B., R.E. Meyerott, E.E. Gaines, R.W. Nightingale, P.C. Filbert, and W.L. Imhof, IEEE Trans. Elect. Insul., EI-18, 354, 1983.
3. Davidson, G.T., P.C. Filbert, R.W. Nightingale, W.L. Imhof, J.B. Reagan and E.C. Whipple, J. Geophys. Res., 93, Jan. 1988.
4. Chiu, Y.T., R.W. Nightingale and M.A. Rinaldi, J. Geophys. Res., 93, accepted, 1988.
5. Schulz, M., and L.J. Lanzerotti, in Physics and Chemistry in Space 7: Particle Diffusion in the Radiation Belts, Springer, New York, 1974.
6. Walt, M., in Particles and Fields in the Magnetosphere, edited by B.M. McCormac, D. Reidel Hingham, Mass., 1970.
7. West, H.I., Jr., R.M. Buck, and G.T. Davidson, J. Geophys. Res., 86, 2111, 1981.
8. Tomassian, A.D., T.A. Farley, and A.L. Vampola, J. Geophys. Res., 77, 3441, 1972.
9. Farley, T.A., J. Geophys. Res., 74, 1969.
10. Newkirk, L.L., and M. Walt, J. Geophys. Res., 73, 1013, 1968a.
11. Fälthammar, C.G., in Particles and Fields in the Magnetosphere, edited by B.M. McCormac, p. 387, D. Reidel, Hingham, Mass., 1970.
12. Fälthammar, C.G., Walt, M., J. Geophys. Res., 74, 4184, 1969.
13. Newkirk, L.L., and M. Walt, J. Geophys. Res., 73, 7231, 1968b.
14. Kavanaugh, L.D., Jr., J. Geophys. Res., 73, 2959, 1968.
15. Nakada, M.P., and G.D. Mead, J. Geophys. Res., 70, 4777, 1965.
16. Tverskoy, B.A., Geomagn. Aeron. Engl. Transl., 5, 617, 1965.
17. Birmingham, T.J., J. Geophys. Res., 77, 2169, 1969.
18. Cornwall, J.M., Radio Sci., 3, 740, 1968.
19. Holzworth, R.H., and F.S. Mozer, J. Geophys. Res., 84, 2559, 1979.
20. Sibeck, D.G., R.W. McEntire, A.T.Y. Lui and R.E. Lopez, J. Geophys. Res., 92, 13485, 1987.

III. Data Bases

NUCLEAR CROSS SECTIONS FOR ESTIMATING SECONDARY RADIATIONS PRODUCED IN SPACECRAFT

L. W. Townsend and J. W. Wilson
NASA Langley Research Center, Hampton, VA. 23665-5225

ABSTRACT

Methods for estimating nuclear absorption (reaction) and breakup (fragmentation) cross sections for space radiation transport calculations are presented, and comparisons with experimental data are made. For the fragmentation problem, the importance of electromagnetic dissociation is discussed and calculational methods described.

INTRODUCTION

In the approaching era of a permanently manned Space Station, proposed Lunar bases, manned missions to Mars, and routine operations of high-altitude aircraft into the upper atmosphere and beyond, there is a crucial need to develop methods for estimating the complex radiation fields inside spacecraft and high-altitude aircraft. Because of atomic and nuclear interactions within the vehicle's structure and interior components, the composition of the interior radiation fields can differ appreciably from that of the external environment. To adequately describe these interior fields, accurate values for nuclear absorption (reaction) and fragmentation (breakup) cross sections are needed. Unfortunately, the experimental data bases for these cross sections are sparse. Because of the large number of possible incident particle/target material combinations, the complexity of the reaction products, and the broad range of energies involved, it is unlikely that the required data base will ever be obtained from experiments alone. Hence, theoretical and semiempirical methods, validated and guided by experimental measurements, must be developed to provide the needed cross sections.

Methods for estimating nuclear absorption (reaction) cross sections include fundamental theories based upon quantum mechanics,[1] and simple parameterizations based upon classical collision models.[2-5] The main advantages of the quantum mechanical models are their accuracy and generality. Excellent agreement between theory and experiment is attainable for any projectile/target combination over a wide range of energies. Their principal disadvantages are the non-trivial nature of the calculations and the extensive computer storage requirements if they are stored in tabular form. As an alternative to the more detailed quantum mechanical models, fully energy-dependent parameterizations of proton-nucleus[3] and nucleus-nucleus[4,5] absorption cross sections have been published. These parameterizations typically agree to within 10 percent of the more detailed quantum mechanical calculations. Comparisons between these calculational methods and recent experimental measurements,[5]

© 1989 American Institute of Physics

which confirm the predicted energy dependence of the total absorption cross sections, will be presented.

Unlike absorption processes, the physics underlying nuclear fragmentation is not yet well understood. Much of the sparse experimental data base is relevant to narrowly focussed studies of exotic phenomena, and not suitable for determining secondary particle production cross sections. Questions concerning the validity of the factorization approximation and the energy dependence, if any, of the secondary particle production cross sections remain to be resolved. In addition, the physical picture is complicated by competing processes such as fission and electromagnetic dissociation, which can become dominant breakup mechanisms for some nuclei and/or at high energies. Fundamental theories based upon quantum mechanics are too complex and expensive to use for repetitive calculations. Semiempirical formulae have been developed[6] with typical inaccuracies of ~ 30 percent. Recently, a semiempirical abrasion-ablation fragmentation model was developed[7] which uses only a single adjustable parameter. Its predicted cross sections agree with experimental data to the extent that the experiments agree among themselves. Further improvements in these calculational methods, however, await additional experimental data.

NUCLEAR ABSORPTION CROSS SECTIONS

We have developed a quantum mechanical optical model formalism[1] based upon a microscopic optical potential approximation to the exact nucleus-nucleus multiple scattering series.[8] The methods are applicable to any projectile nucleus colliding with any target nucleus at energies greater than 25 MeV/nucleon. The methods use no arbitrarily adjusted parameters. Typical accuracies, when predictions are compared with experimental measurements, are within 3 percent for energies greater than 80 MeV/nucleon and within 10 percent for energies as low as 30 MeV/nucleon.

From eikonal scattering theory, the absorption cross sections are given by

$$\sigma_{abs} = \int d^2\vec{b} \; \{1 - \exp[-2 \, \text{Im} \, \chi(\vec{b})]\} \tag{1}$$

where the complex phase function, χ, written in terms of the nucleus-nucleus optical potential, $V(\vec{b},z)$, is

$$\chi(\vec{b}) = -[mA_pA_T/k(A_p + A_T)] \int_{-\infty}^{\infty} V(\vec{b},z) \, dz \tag{2}$$

with

$$V(\vec{b},z) = A_pA_T \int d^3\vec{\xi}_T \rho_T(\vec{\xi}_T) \int d^3\vec{y} \; \rho_p(\vec{b}+\vec{z} + \vec{\xi}_T + \vec{y}) \; \tilde{t}(e,\vec{y})[1-\tilde{C}(\vec{y})] \tag{3}$$

Symbols in Eqs. (1)-(3) are
m - nucleon mass
$\vec{b}$ - impact parameter
A_i (i = P,T) - projectile (P) and target (T) mass numbers
k - projectile momentum
$\tilde{t}$ -constituent-averaged, energy-dependent two-nucleon transition amplitude
ρ_i(i = P,T) - nuclear densities of colliding nuclei

The potential given by Eq. (3) was derived from an optical model potential multiple scattering series and does not inherently depend on the eikonal approximation, although we are using it in that context. In Eq. (3), $\tilde{t}$ is the constituent-averaged energy-dependent two-body transition amplitude

$$\tilde{t}(e,\vec{y}) = -(\frac{e}{m})^{1/2} \sigma(e)[\alpha(e) + i][2\pi B(e)]^{-3/2} \exp\left[\frac{-y^2}{2B(e)}\right] \quad (4)$$

and the correlation function is taken to be

$$\tilde{C}(\vec{y}) = 0.25 \exp\left(\frac{-k_F^2 y^2}{10}\right) \quad (5)$$

For the analyses of this work, the Fermi momentum is assumed to be that of infinite matter, $k_F = 1.36$ fm^{-1}.

The correct nuclear density distributions ρ_i(i = P,T) to be used in Eq. (3) are the nuclear ground state, single-particle number densities for the collision pair. Since these are not experimentally known, the number densities are obtained from their experimental charge density distributions by assuming that

$$\rho_c(\vec{r}) = \int \rho_p(\vec{r}') \rho_A(\vec{r} + \vec{r}') d^3r' \quad (6)$$

where ρ_c is the nuclear charge distrubution, ρ_p is the proton charge distribution, and ρ_A is the desired nuclear single-particle density. All density distributions in Eq. (6) are normalized to unity. The proton charge distribution is taken to be the usual Gaussian form

$$\rho_p(\vec{r}) = \left(\frac{3}{2\pi r_p^2}\right)^{3/2} \exp\left(\frac{-3r^2}{2r_p^2}\right) \quad (7)$$

where $r_p = 0.87$ fm is the proton root-mean-square charge radius.

When the projectile is a nucleon, Eq. (6) yields a delta function for ρ_A:

$$\rho_A(\vec{r} + \vec{r}') = \delta(\vec{r} + \vec{r}') \tag{8}$$

since ρ_c and ρ_p are identical.

For nuclei lighter than neon (A < 20), the nuclear charge distribution is the harmonic well (HW) form given by

$$\rho_c(\vec{r}) = \rho_0 \left[1 + \gamma \left(\frac{r}{a}\right)^2\right] \exp\left(\frac{-r^2}{a^2}\right) \tag{9}$$

where ρ_0 is the normalization constant, r is the radial coordinate, and a and γ are charge parameters. Values for a and γ used herein are listed in table 1 of Ref. 1. Substituting Eq. (7) and (9) into Eq. (6) yields

$$\rho_A(\vec{r}) = \frac{\rho_0 a^3}{8s^3} \left(1 + \frac{3\gamma}{2} - \frac{3\gamma a^2}{8s^2} + \frac{\gamma a^2 r^2}{16 s^4}\right) \exp\left(\frac{-r^2}{4s^2}\right) \tag{10}$$

where

$$s^2 = \frac{a^2}{4} - \frac{r_p^2}{6} \tag{11}$$

For neon and heavier nuclei (A > 20), the nuclear charge distribution is taken to be the Woods-Saxon (WS) form given by

$$\rho_c(\vec{r}) = \frac{\rho_0}{1 + \exp[(r - R)/c]} \tag{12}$$

where R is the radius at half density, and the surface diffuseness c is related to the nuclear skin thickness t through

$$c = \frac{t}{4.4} \tag{13}$$

Values for R and t used herein are listed in table 1 of Ref. 1. Inserting Eq. (7) and (12) into Eq. (6) yields, after some simplification, a number density ρ_A that is of the WS form (see Eq. 12) with the same R, but different overall normalization factor ρ_0 and surface thickness. The latter is given by

$$t_A = \frac{8.8 r_p}{3^{1/2}} \left[\ln\left(\frac{3\beta - 1}{3 - \beta}\right)\right]^{-1} \tag{14}$$

where

$$\beta = \exp\left(\frac{4.4 r_p}{t_c 3^{1/2}}\right) \tag{15}$$

with t_c denoting the charge skin thickness obtained by using Eq. (13) and the charge distribution surface diffuseness values listed in Ref. 1.

The nucleon-nucleon cross sections, $\sigma(e)$, used in the energy-dependent two-body transition amplitude [Eq. (4)] are obtained by performing a spline interpolation of values taken from various compilations.[1] Since scattering at these energies is mainly diffractive, the nucleon-nucleon slope parameters, $B(e)$, are those appropriate to purely diffractive scattering. From Ref. 1 these are

$$B(e) = 10 + 0.5 \ln \left(\frac{s'}{s_0}\right) \tag{16}$$

where s' is the square of the nucleon-nucleon center-of-mass energy and $s_0 = 1(\text{GeV } c^{-1})^{-2}$. Values for the parameter $\alpha(e)$ are not required since only the imaginary part of Eq. (4) is utilized in Eq. (1).

Using the formalism described in the previous paragraphs, absorption cross sections for nucleons, deuterons, and selected heavy ions colliding with various target nuclei have been calculated. Extensive tables of results are given in Ref. 1 where detailed comparisons with available experimental data are made. Representative results from that work are displayed in Figures 1 and 2.

As an alternative to these somewhat detailed cross section calculations, simple parameterizations based upon classical collision models have been developed. The most commonly used ones are of the Bradt-Peters form[2]

$$\sigma_{abs} = \pi r_0^2 (A_p^{1/3} + A_T^{1/3} - \delta)^2 \tag{17}$$

where r_0 and δ are energy-independent parameters adjusted to fit available experimental data. These parameterizations are reasonably accurate (within 5 percent) at energies above ≈ 1 GeV/Nucleon where the cross sections are nearly asymptotic. For lower energies, substantial differences exist due to the cross section energy dependence. Recently, fully energy-dependent parameterizations of proton-nucleus[3] and nucleus-nucleus[4] absorption cross sections have been developed for use in space radiation transport studies. These energy-dependent parameterizations typically agree to within 10 percent of the more detailed optical model potential results obtained with Eq. (1). From Ref. 4, the energy dependent nucleus-nucleus cross section parameterization is

$$\sigma_{abs}(A_p, A_T, E) = \pi r_0^2 \beta(E) [A_p^{1/3} + A_T^{1/3} - \delta(A_p, A_T, E)]^2 \tag{18}$$

where
$$\beta(E) = 1 + 5 E^{-1} \tag{19}$$

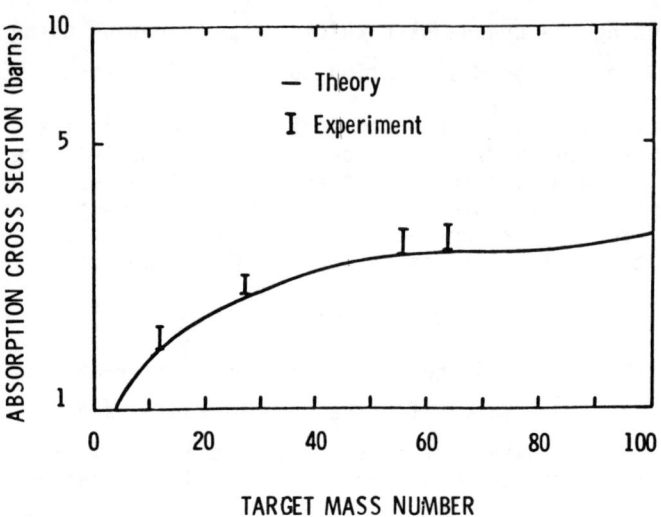

Figure 1. - Absorption cross sections for neon projectiles at 30 MeV/nucleon (from Ref. 1)

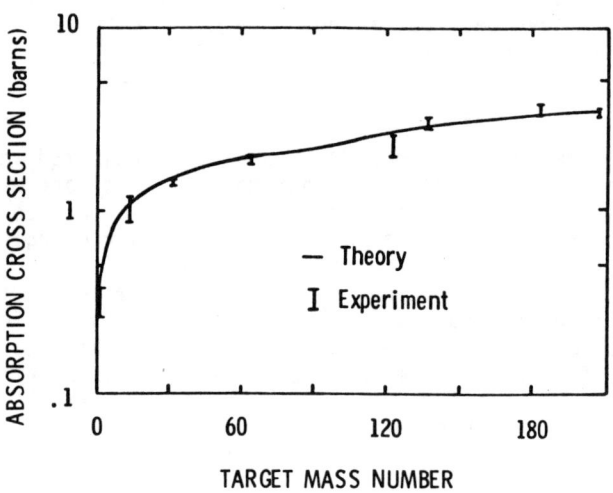

Figure 2. - Absorption cross sections for oxygen projectiles at 2.1 GeV/nucleon (from Ref. 1)

and

$$\delta(A_p, A_T, E) = 0.2 + A_p^{-1} + A_T^{-1} - 0.292\, e^{-E/792} \cos(0.229 E^{0.453}) \quad (20)$$

with $r_0 = 1.26$ fm and E in units of MeV/nucleon. Note that Eq. (18) reduces in form to Eq. (17) for high energies (in the limit as $E \to \infty$). Typical results obtained from Eqs. (17) and (18), for $^{12}C - ^{12}C$ collisions, are displayed in Figure 3. Also displayed for comparison are predictions from the optical model [Eq. (1)] as well as representative experimental measurements.[5,9-11] Note that both energy dependent calculations reproduce the experimentally observed energy dependence for the cross sections including the observed minimum[5] near 250-300 MeV/nucleon. Further validation of this general cross section dependence upon incident projectile energy would be useful for collision pairs other than carbon-carbon; for example, iron-carbon and iron-lead collision pairs would be of interest for space radiation protection purposes.

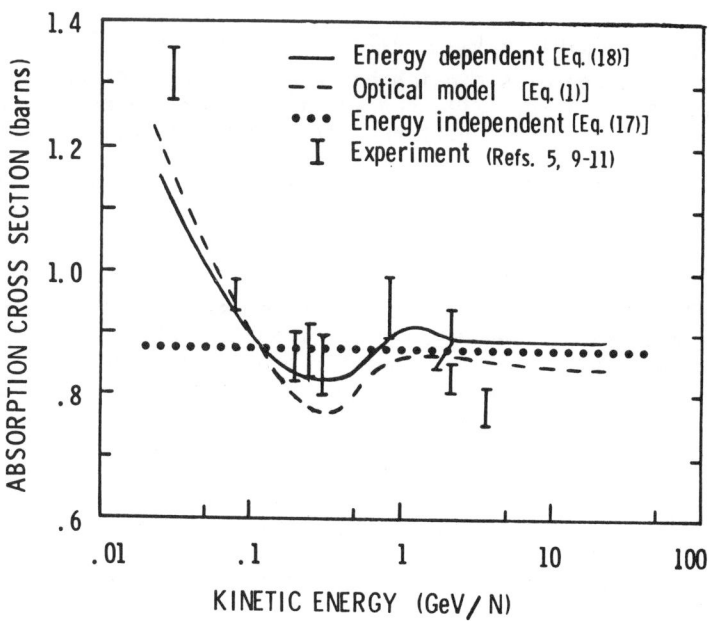

Figure 3. - Absorption cross sections for carbon-carbon collisions

NUCLEAR FRAGMENTATION CROSS SECTIONS

Methods of estimating nuclear fragmentation cross sections for secondary particle production generally fall into one of two categories: semiempirical parameterizations;[6] or abrasion-ablation formulations based upon quantum-mechanical[12,13] or geometric collision models.[14] Recently, an alternative, semiempirical abrasion-ablation model was developed.[7] In this section, the quantum-mechanical[13] and semiempirical[7] abrasion-ablation fragmentation models will be briefly described. The semiempirical parameterization due to Silberberg, et al[6] will not be described here since it is being discussed elsewhere in this conference (R. Silberberg, paper no. 17).

In previous work, the quantum-mechanical optical model reaction theory was extended to describe nuclear fragmentation within the context of an abrasion-ablation formalism[12,13] which includes contributions from frictional-spectator-interactions (FSI). The latter results when some of the abraded nucleons are scattered into rather than away from the prefragment, thereby depositing additional excitation energy in the prefragment nucleus. In the theory, the nuclear fragmentation cross section for production of a particular secondary particle species f is given by

$$\sigma_{nuc}(Z_f, A_f) = \sum_n \sum_p \alpha_{fn}(p) \, \sigma_{abr}(Z_n, A_n, p) \quad (21)$$

where $\sigma_{abr}(Z_n, A_n, p)$ is the cross section for producing by abrasion a prefragment of species n which has undergone p frictional-spectator interactions. In Eq. (21), the summation over n accounts for the contributions to f from different prefragment species n, and the summation over p accounts for the distribution of excitation energies resulting from the FSI'S. Depending upon its excitation energy, the excited prefragment may decay by emitting one or more nucleons, composites, or gamma rays. This is the ablation step. The probability, $\alpha_{fn}(p)$, for formation of a particular final fragment of type f as a result of the deexcitation of a prefragment of type n which has undergone p frictional-spectator interactions, is obtained using the EVA-3 computer code as modified by Morrissey[15]. Extensive calculational details are given in Refs. 12 and 13.

Although completely general in applicability, and reasonably accurate for predicting heavy-fragment production cross sections, the optical model methods are of questionable accuracy for light-fragment production because of the restrictive nature of the assumption that the collision is peripheral. In addition, these calculations are far too complex to perform repeatedly within a transport code and appear to be better suited for generating data tables for access as needed. Recent comparisons[13] between the energy-dependent quantum-mechanical abrasion model and an energy-independent geometric one, suggest that the fragmentation cross sections above 200 MeV/nucleon are essentially independent of the incident ion's kinetic energy. Motivated by these considerations, a much simpler energy-independent fragmentation model was developed for use in heavy ion transport codes.[7] In this model, the classical, geometric abrasion-ablation model of Bowman[14] is modified to include FSI contributions through the use of semiempirical higher-order corrections to the abraded prefragment excitation energies. In this method, the nuclear fragmentation cross sections are given by

$$\sigma_{nuc}(Z_f, A_f) = F_1 \exp\left[-R|Z_f - SA_f + TA_f^2|^{3/2}\right] \sigma(\Delta A) \quad (22)$$

where according to Rudstam[16] $R = 11.8 \, A_F^{-0.45}$, $S = 0.486$, $T = 3.8 \times 10^{-4}$, and F_1 is a normalizing factor such that

$$\sum_{Z_f} \sigma_{nuc}(Z_f, A_f) = \sigma(\Delta A) \quad (23)$$

which ensures charge and mass conservation. The Rudstam formula for $\sigma(\Delta A)$ is not used because his ΔA dependence is too simple and breaks down for heavy targets. Instead, the cross section for removal of ΔA nucleons is estimated using

$$\sigma(\Delta A) = \pi b_2^2 - \pi b_1^2 \quad (24)$$

where b_2 is the impact parameter at which Δ_{abr} nucleons are abraded by the collision and Δ_{abl} nucleons are ablated, in the subsequent prefragment deexcitation, such that

$$\Delta_{abr}(b_2) + \Delta_{abl}(b_2) = \Delta A - 1/2 \quad (25)$$

and similarly for b_1

$$\Delta_{abr}(b_1) + \Delta_{abl}(b_1) = \Delta A + 1/2 \quad (26)$$

The number of abraded nucleons is estimated from the geometric overlap volume and the mean free path in nuclear matter, λ, as

$$\Delta_{abr} = FA_p[1 - 0.5 \exp(-C_p/\lambda) - 0.5 \exp(-C_t/\lambda)] \quad (27)$$

where F is the fraction of the volume in the geometric overlap between projectile and target nuclei, and C_p and C_t are the maximum chord lengths of the intersecting surface in the projectile and target, respectively. Expressions for F, which differ depending on the relative sizes of the colliding nuclei and on the nature of the collision (peripheral versus central), are given in Ref. 7. The number of ablated nucleons, Δ_{abl}, is computed from

$$\Delta_{abl} = (E_s + E_{FSI})/10 \text{ MeV} \qquad (28)$$

which assumes that a nucleon is ablated (evaporated) for every 10 MeV of excitation energy. In Eq. (28), the E_s represents excitation energy associated with the surface energy contribution from abrasion, and E_{FSI} represents the contributions resulting from frictional-spectator interactions. The only arbitrarily adjusted parameter in this model is a second-order correction to the expression for the surface energy term. Further details of the calculational details are presented in Ref. 7 along with extensive comparisons with experimental measurements. A computer code, NUCFRAG, based upon this semiempirical, abrasion-ablation fragmentation model has been developed,[19] and is available for use either with mainframes or with desktop personal computers.

Because the dissociation of projectile and target nuclei by their interacting Coulomb fields may be important for some heavier nuclei at high energies, the electromagnetic dissociation contributions, σ_{em}, must be added to the nuclear fragmentation cross section, σ_{nuc}, to yield the total fragmentation cross section

$$\sigma_{frag} = \sigma_{nuc} + \sigma_{em} \qquad (29)$$

Preliminary studies suggest that cross sections for multiple-nucleon removal by Coulomb dissociation are approximately one to two orders of magnitude smaller than single-nucleon removal cross sections for the same projectile-target collision pairs. Therefore, we limit ourselves for the present analyses to single-nucleon removal cross section contributions for σ_{em} in Eq. (29). The appropriate reactions are those satisfying

$$\gamma + {}^Z_A \begin{cases} {}^{(Z-1)}(A-1) + p \\ {}^Z(A-1) + n \end{cases}$$

where γ symbolizes the virtual quanta exchanged in the electromagnetic interaction. Details of the electromagnetic dissociation cross section calculations will be described in the next section.

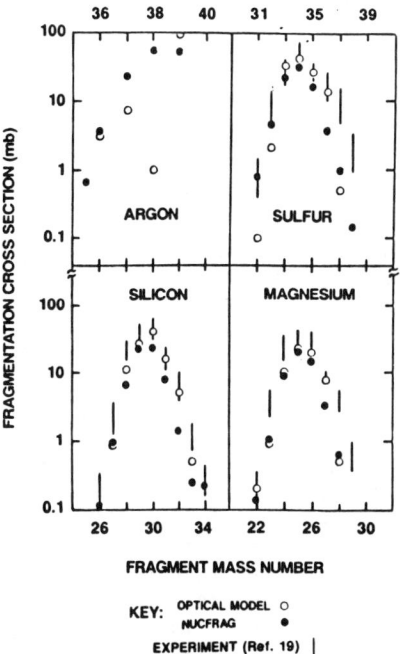

Figure 4. - Representative argon - carbon fragmentation cross sections

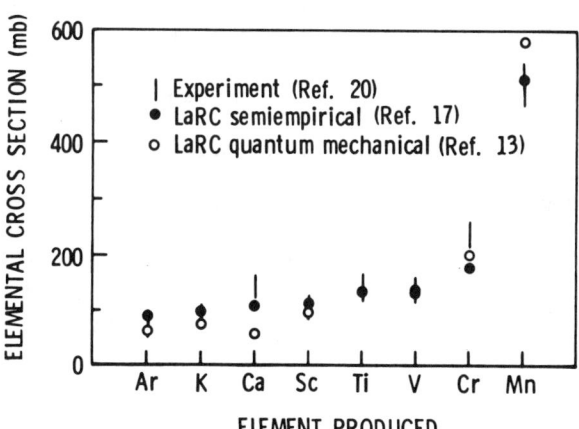

Figure 5. - Elemental production cross sections for 1.88 GeV/nucleon iron fragmenting in lead

Figures 4 and 5 display sample predictions utilizing the quantum-mechanical, optical potential, abrasion-ablation-FSI model,[13] and the semiempirical abrasion-ablation model (NUCFRAG).[17] Also displayed in these figures are experimental results obtained at the Lawrence Berkeley Laboratory.[19,20] The calculated cross sections, including the electromagnetic dissociation one-nucleon removal contribution, were obtained using Eq. (29). As is clearly evident, the cross section predictions obtained using either calculational model compare favorably to the experimental data, although significant uncertainties in the data are evident from the error bars. Further improvements to these calculational methods require additional experimental data. Two of the major issues remaining to be resolved are the energy dependence of the fragmentation cross sections and the validity of the factorization hypothesis. To unambiguously answer questions concerning the energy dependence, if any, of the fragmentation cross sections, careful measurements of complete sets of secondary fragment production cross sections for several different projectile-target collision pairs, as a function of incident ion kinetic energy, are required. Such a set of measurements might involve measuring all the secondary fragments produced by iron or argon, at incident energies, for example, of 100, 300, 600, 1000, and 2000 MeV/nucleon, fragmenting in carbon targets. By extending these measurements for the projectiles to other targets, such as aluminum and lead, the dependence of the cross sections on the target nucleus and incident energy could be established and the validity of the factorization hypothesis tested. (The factorization hypothesis suggests that the fragment production cross sections can be factored into a product of two terms, one depending only on the target nucleus, and the other depending only on the projectile and fragment.) In any case, the current sparse data base limits calculational uncertainties to approximately 30-50%, which are not yet accurate enough for precise transport studies. Finally, we note that very little is known for these reactions about particle multiplicites. Most experimental measurements to date are inclusive and therefore do not measure multiplicities, and there are no simple and accurate calculational methods for predicting them.

ELECTROMAGNETIC DISSOCIATION CROSS SECTIONS

Methods for estimating electromagnetic dissociation cross sections using the Weiszacher-Williams method of virtual quanta have been developed.[17,20,21] In this theory, the cross section is obtained from

$$\sigma_{em} = \int \sigma(E) \, N(E) \, dE \qquad (30)$$

where N(E) is the virtual photon number spectrum, $\sigma(E)$ is the photonuclear cross section for neutron or proton knockout, and E is the virtual photon energy. The virtual photon number spectrum is given by[21]

$$N(E) = \frac{1}{E}\frac{2}{\pi} Z_t^2 \alpha \frac{1}{\beta^2} \{x K_0(x) K_1(x) - \frac{1}{2} x^2 \beta^2 [K_1^2(x) - K_0^2(x)]\} \quad (31)$$

where N(E) is the number of virtual photons per unit energy E, Z_t is the number of protons in the target nucleus, β is the velocity of the target in units of c, and α is the electromagnetic fine structure constant. The $K_0(x)$ and $K_1(x)$ are modified Bessel functions of the second kind. The parameter x in Eq. (31) is defined as

$$x = \frac{E b_{min}}{\gamma \beta (\hbar c)} \quad (32)$$

where γ is the usual relativistic factor and b_{min} is the minimum impact parameter. The dependence of N(E) on b_{min} in Eq. (31) comes from the fact that there is a finite maximum momentum transfer in the collision process. The photonuclear cross sections $\sigma(E)$ are usually obtained from experiment or a parameterization due to Westfall.[20] The electromagnetic dissociation contributions to the results displayed in Figs. 4 and 5 are 258 mb $(^{56}Fe \rightarrow ^{55}Mn + p)$ and 1.2 mb $(^{40}Ar \rightarrow ^{39}Ar + n)$. To further illustrate the potential importance of σ_{em} to the total fragmentation cross section, Fig. 6(a) displays the one proton removal cross section contribution to σ_{em} for several different collision pairs. Note the monotonic increase in cross section with increasing incident ion kinetic energy. In Fig. 6(b), the percent contribution of these σ_{em} to the total (nuclear + electromagnetic) fragmentation cross section for one proton removal is plotted as a function of incident energy. For low mass collision pairs (C+C), the contribution is small (< 5%) at all energies. For C+Pb, there is a rapid increase in the electromagnetic dissociation contribution with increasing incident ion energy. For heavier-mass collision pairs (Fe+Pb), most of the single proton removal cross section is electromagnetic in origin. Clearly, further studies of this important breakup mechanism are needed, including experimental measurements of multiple-nucleon removal processes.

CONCLUDING REMARKS

Theoretical methods for calculating nuclear absorption (reaction) cross sections from fundamental quantum mechanics are well-developed and yield accurate estimates when compared to experimental data. In addition, reasonably accurate energy-dependent parameterizations have been developed for use in obtaining

rapid cross section estimates. The existing absorption cross
section database is marginally adequate. Additional cross sections
in the energy range 0.1-1.0 GeV/nucleon are needed, for projectiles
other than carbon, to confirm the general energy dependence of the
cross sections. For fragmentation cross sections, the experimental
data base is sparse, but improving. Improved theoretical
formulations based either upon fundamental physical principles or
upon semiempirical parameterizations are needed for precise
transport studies. Significant theoretical improvements, however,
will require significantly enlarged data bases--ones which include
systematic determinations of the energy dependence, if any, of the
secondary fragment production cross sections, and which include
actual multiplicities of the secondary products. Finally, the
importance of electromagnetic dissociation processes, especially for
multiple nucleon emissions, requires further theoretical and
experimental investigation.

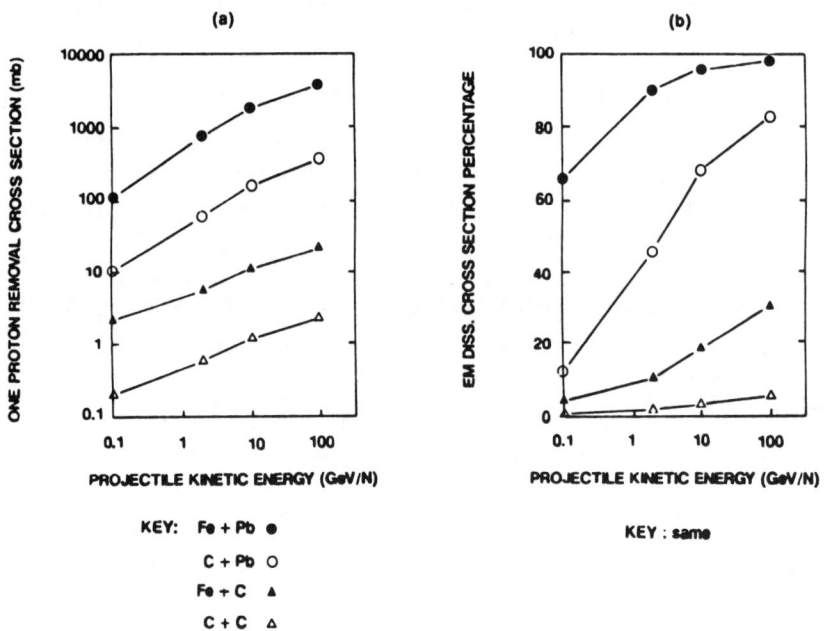

Figure 6. - Electromagnetic dissociation contributions to one proton removal

REFERENCES

[1] L. W. Townsend and J. W. Wilson, National Aeronautics and Space Administration Reference Publication No. RP-1134, 1985.

[2] H. L. Bradt and B. Peters, Phys. Rev. 77, 54 (1950).

[3] J. R. Letaw, R. Silberberg, and C. H. Tsao, Ap. J. Supp. 51, 271 (1983).

[4] L. W. Townsend and J. W. Wilson, Radiat. Res. 106, 283 (1986).

[5] S. Kox et al., Phys. Rev. C 35, 1678 (1987).

[6] R. Silberberg, C. H. Tsao, and M. M. Shapiro, in Spallation Nuclear Reactions and Their Applications, eds. B.S.P. Shen and M. Merker (Reidel, Boston, 1976) pp. 49-82.

[7] J. W. Wilson, L. W. Townsend, and F. F. Badavi, Nucl. Inst. and Meth. B18, 225 (1987).

[8] J. W. Wilson and L. W. Townsend, Can. J. Phys. 59, 1569 (1981).

[9] J. Jaros et al., Phys. Rev. C 18, 2273 (1978).

[10] V. D. Aksinenko et al., Nucl. Phys. A 348, 518 (1980).

[11] H. H. Heckman et al., Phys. Rev. C 17, 1735 (1978).

[12] L. W. Townsend, J. W. Wilson, and J. W. Norbury, Can. J. Phys. 63, 135 (1985).

[13] L. W. Townsend, J. W. Wilson, F. A. Cucinotta, and J. W. Norbury, Phys. Rev. C 34, 1491 (1986).

[14] J. D. Bowman, W. J. Swiatecki, and C. F. Tsang, Lawrence Berkeley Laboratory Report No. LBL-2908, 1973.

[15] D. J. Morrissey et al., Phys. Rev. Lett. 43 1139 (1979).

[16] G. Rudstam. Z. Naturfors. 21a, 1027 (1966).

[17] F. F. Badavi, L. W. Townsend, J. W. Wilson, and J. W. Norbury, Computer Physics Communications, 1987 (in press).

[18] F. A. Cucinotta, J. W. Norbury, and L. W. Townsend, Bull. Am. Phys. Soc. 31, 1765 (1986).

[19] Y. P. Viyogi et al., Phys. Rev. Lett. 42, 33 (1979).

[20] G. D. Westfall et al., Phys. Rev. C 19, 1309 (1979).

[21] J. W. Norbury and L. W. Townsend, National Aeronautics and Space Administration Technical Paper No. TP-2527, 1986.

NUCLEON INTERACTION DATA BASES FOR BACKGROUND ESTIMATES

John W. Wilson and Lawrence W. Townsend
NASA Langley Research Center, Hampton, VA 23665

ABSTRACT

Nucleon interaction data bases available in the open literature are examined for potential use in a recently developed nucleon transport code. Particular attention is given to secondary particle penetration and the multiple charged ion products. A brief description of the transport algorithm is given.

INTRODUCTION

It is well established that 30 to 50 percent of the energy absorbed in human exposure to high energy protons (> 400 MeV) is due to secondary radiations produced in nuclear reactions. There are several transport codes for making exposure estimates and two primary data bases which are readily found in the open literature.[1-4] The first such data base is an intranuclear cascade Monte Carlo derived data set (sometimes the intranuclear cascade code resides within the transport code itself), and the second is an empirical fit to experimental or theoretical cross sections. A review of these data bases is given in the Proceedings of the Second International School of Radiation Damage and Protection.[3] An interesting question concerns the relative differences in results attributable to these data bases. For the present consideration, we use the analytic cross sections of Ranft[5] and Bertini[6] associated with two well-known transport codes: namely, FLUKA and HETC, respectively. Although the Bertini data set to which Alsmiller and Barish made their curve fit is quite old, we compared their data base to results of the most recent version of Bertini's code (MECC7) and found agreement to within statistical fluctuations. It was shown by Wright et al. that many meaningful results are obtained by solving the proton transport equation in the straightahead approximation.[2] That one may further neglect the neutron coupling in estimating the proton transport solution was further demonstrated by Wilson and Lamkin.[7,8] Such a simplified treatment should be adequate for the purpose of evaluating possible differences between the data sets.

THEORETICAL DETAIL

The Boltzmann equation in the straightahead approximation is given as

$$[\frac{\partial}{\partial x} - \frac{\partial}{\partial E} S(E) + \sigma] \phi(x,E) = \int_E^\infty f(E,E')\phi(x,E')dE' \quad (1)$$

where $\phi(x,E)$ is the proton flux, $S(E)$ is the charged particle stopping power, σ is the macroscopic interaction cross section, and

$f(E,E')$ is the secondary particle production spectrum. Making the following definitions

$$r = \int_0^E dE'/S(E') \qquad (2)$$

$$\psi(x,r) = S(E)\phi(x,E) \qquad (3)$$

$$\tilde{f}(r,r') = S(E)f(E,E') \qquad (4)$$

allows Eq. (1) to be rewritten as

$$\psi(x,r) = e^{-\sigma x}\psi(0,r+x) \qquad (5)$$

$$+ \int_0^x dz\, e^{-\sigma z} \int_{r+z}^\infty dr'\, \tilde{f}(r+z,r')\psi(x-z,r')$$

where the boundary condition is

$$\psi(0,r) = S(E)\phi(0,E) \qquad (6)$$

A numerical algorithm for equation (5) is found by noting that

$$\psi(x+h,r) = e^{-\sigma h}\psi(x,r+h) + \int_0^h dz\, e^{-\sigma z} \int_r^\infty dr'\, \tilde{f}(r+z,r'+z)\psi(x+h-z, r'+z) \qquad (7)$$

which can be simplified by using

$$\psi(x+h-z,r) \approx e^{-\sigma(h-z)}\psi(x,r+h-z) + O(h) \qquad (8)$$

yielding

$$\psi(x+h,r) \approx e^{-\sigma h}\psi(x,r+h) + e^{-\sigma h} \int_0^h dz \int_r^\infty dr'\, \tilde{f}(r+z,r'+z)\psi(x,r'+h) \qquad (9)$$

to terms $O(h^2)$. Equation (9) is accurate for distances such that $\sigma h \ll 1$ and may be used to relate the spectrum at some point x to the spectrum at x+h. Therefore, one may begin at the boundary (x = 0) and propagate the solution to any arbitrary interior point using equation (9).

In the event that the boundary has a discrete spectrum such as

$$\phi(0,E) = \delta(E-E_o) \tag{10}$$

then

$$\psi(0,r) = \delta(r - r_o) \tag{11}$$

When discrete spectra are present at the boundary, the solution contains both singular and continuous components which we label ψ_s and ψ_c, respectively. The corresponding singular term in the solution is then

$$\psi_s(x,r) = e^{-\sigma x}\delta(r+x-r_o) \tag{12}$$

while the continuous term satisfies

$$\psi_c(x+h,r) = e^{-\sigma h}\psi_c(x,r+h)$$
$$+ \int_0^h dz\, e^{-\sigma z}\int_r^\infty dr'\, \bar{f}(r+z,r'+z)[\psi_s(x+h-z,r'+z)+\psi_c(x+h-z,r'+z)] \tag{13}$$

The first term under the integral may be evaluated using equation (12) to obtain

$$\psi_c(x+h,r) = e^{-\sigma h}\psi_c(x,r+h) + e^{-\sigma(x+h)}\int_0^h dz\, \bar{f}(r+z,r_o-x-h)$$
$$+ \int_0^h dz\, e^{-\sigma z}\int_r^\infty dr'\, \bar{f}(r+z,r'+z)\psi_c(x+h-z,r'+z) \tag{14}$$

The solution over small values of $(h-z)$ may be approximated as

$$\psi_c(x+h-z,r) \simeq e^{-\sigma(h-z)}\psi_c(x,r+h-z) + O(h-z) \tag{15}$$

for which

$$\psi_c(x+h,r) \simeq e^{-\sigma h}[\psi_c(x,r+h)+e^{-\sigma x}\bar{F}(r,h,r_o-x-h)] \tag{16}$$
$$+ e^{-\sigma h}\int_r^\infty dr'\, \bar{F}(r,h,r')\psi_c(x,r'+h)$$

where

$$\bar{F}(r,h,r') = \int_0^h dz\, \bar{f}(r+z,r') \tag{17}$$

$$= F_c[\varepsilon(r+h),E'] - F_c(E,E')$$

$\varepsilon(r)$ is the energy for range r, and

$$F_c(E,E') = \int_0^E f(E,E')dE' \qquad (18)$$

which is the cumulative secondary particle spectrum. Note that equation (16) only requires one numerical integration per step in x. To test the algorithm, we consider two test problems for which analytic solutions can be obtained.

DISCRETE SPECTRUM

Nucleon-nucleon scattering can be well approximated[9] by

$$f(E,E') = ce^{-\alpha(E'-E)} \qquad (19)$$

where α is the spectral parameter and $c/\alpha = \sigma$. This spectrum is related to the quasi-elastic spectrum of nucleon-nucleus reactions. Similar to this spectrum is

$$f(r,r') = ce^{-\alpha(r'-r)} \qquad (20)$$

As a model problem, the spectrum of equation (20) is realistic and can be solved using perturbation theory. The first term is the uncollided beam term

$$\psi_0(x,r) = e^{-\sigma x}\delta(r_0-r-x) \qquad (21)$$

The first generation term is

$$\psi_1(x,r) = xe^{-\sigma x}ce^{-\alpha(r_0-r-x)} \qquad (22)$$

and the higher order terms are

$$\psi_n(x,r) = \frac{1}{n!}x^n e^{-\sigma x}\frac{c^n}{(n-1)!}(r_0-r-x)^{n-1}e^{-\alpha(r_0-r-x)} \qquad (23)$$

This problem is solved numerically using equation (16) and is compared with the analytic solution in table 1. The incident beam is for 500 MeV protons on a water shield with

$$\sigma = 0.01 \text{cm}^2/\text{g} \text{ and } \alpha \approx 0.0123 \text{cm}^2/\text{g}.$$

As seen from the table, solutions with discrete spectra are limited in accuracy to 5 percent, independent of the depth of penetration. This error arises from the energy interpolation formula as the spectrum is highly discontinuous. Special interpolation methods could be developed to greatly reduce this error.

Table 1 - Ratio of Numerical Solution to Analytic Solution of Equation (13) for 500 MeV Protons on a "Water" Shield.

	Ratio for x, g/cm^2				
E, MeV	10	20	40	60	80
0.1	1.000	1.000	1.004	.998	1.023
19.5	1.000	.999	1.004	.997	1.024
120.9	1.000	.999	1.002	1.008	1.046
333.3	1.003	.994	1.037	.963	-
454.1	1.031	-	-	-	-

CONTINUOUS SPECTRUM

For this test, a spectrum similar to a solar proton event is taken as

$$\psi(0,r) = e^{-\beta r} \tag{24}$$

where β is the solar spectral parameter. The leading term in the perturbation theory is

$$\psi_0(x,r) = e^{-\sigma x} e^{-\beta(r+x)} \tag{25}$$

with successive collision terms given by

$$\psi_n(x,r) = \frac{1}{n} x \frac{c}{(\alpha+\beta)} \psi_{n-1}(x,r) \tag{26}$$

This problem is solved numerically and compared with the analytic result in table 2. It is seen from the table that the agreement for the two solutions in this case is generally within 1 percent. Clearly, high quality numerical solutions are available for continuous spectra at the boundary. The algorithm developed herein provides adequate solutions to proton beam problems with discrete spectral components and highly accurate solutions for typical space applications involving continuous spectra. The computation times for each of these test problems were less than several minutes on a Cyber mainframe, a very favorable comparison to Monte Carlo or previous methods based on the perturbation theory.[7,8] This algorithm should prove very useful in future applications.

Table 2 - Ratio of Numerical Solution to Analytic Solution of Equation (13) for Continuous "Space" Proton Spectral Input on a "Water" Shield

E, MeV	Ratio for x, g/cm^2				
	10	20	40	60	80
0.1	.994	.995	.998	.999	.999
11.2	.994	.996	.999	1.001	1.000
36.3	.997	.998	1.002	1.003	1.005
118.1	1.001	1.003	1.005	1.006	1.004
383.9	.997	1.000	.991	1.000	.996

SECONDARY PROTON SPECTRA

Both the Ranft[5] (equations 17 to 19 of Ranft) and Bertini spectra show a characteristic low-energy proton evaporation spectrum of similar magnitude and shape (see figure 1). We note that Ranft represents the low-energy spectrum although FLUKA does not transport particles below 50 MeV. There is a slight hardening of each spectrum above the evaporation peak for each of the two bases, but the main difference between the two data sets is the lack of a quasi-elastic region for Ranft's formulas. This fact, as we shall see, strongly affects the

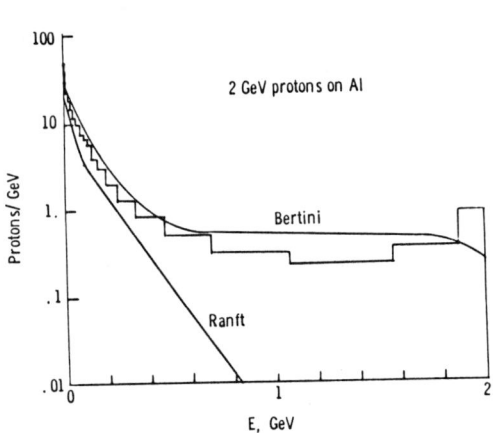

Fig.1   Comparison of the secondary proton spectra produced by 2 GeV proton interaction with aluminum according to Ranft and Bertini.

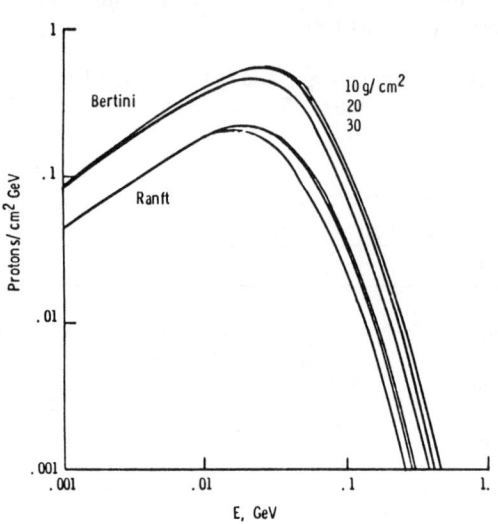

Fig. 2 Secondary proton spectra accumulated to three different depths in aluminum according to the Bertini and Ranft production spectra.

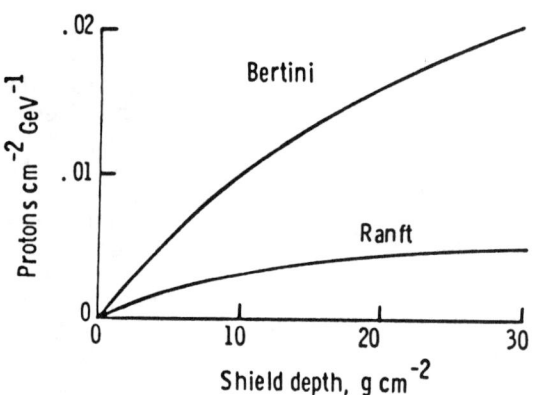

Fig. 3 Buildup of 200 MeV secondary protons in aluminum shield according to the Bertini cross sections and Ranft cross sections.

calculated penetrating characteristics of the proton transport problem.

The transition of 1 GeV protons through aluminum was studied for the two data sets. The general shapes of the transported secondary proton distributions for the two data sets were surprisingly similar (as shown in figure 2) at each of the three depths of 10, 20, and 30 g cm^{-2}. Although the low-energy proton spectrum quickly establishes equilibrium with the passing beam, the number of intermediate energy protons generated by the Bertini spectrum as represented by the Alsmiller and Barish formula increases rapidly even at depths of 30 g cm^{-2}, whereas the Ranft' solution approaches equilibrium more quickly. This is seen in figure 3 where the 200 MeV secondary proton fluence is shown.

The continued buildup of intermediate energy

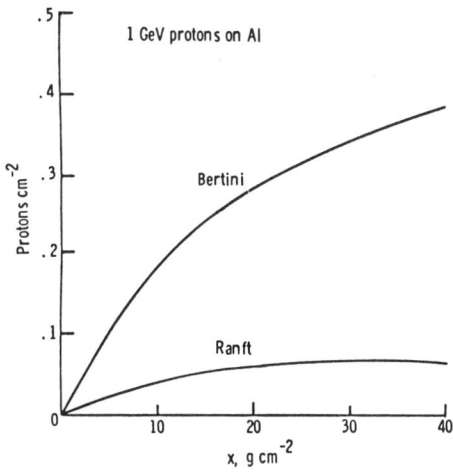

Fig. 4 Number of transmitted secondary protons as a function of shield thickness for the two different cross section sets.

protons associated with the Bertini cross sections results in large differences in the total proton fluence as seen in figure 4. The total fluence, according to the Ranft data base, peaks at ~ 33 g cm^{-2} while the Bertini data base predicts transport results which are much larger than for Ranft's data and continue to increase even for depths beyond 40 g cm^{-2}.

SECONDARY HEAVY ION PRODUCTS

There is a similar concern over the treatment of the high linear energy transfer (LET) nuclear reaction products which are not

Table 3 - Comparison of Fragmentation Cross Sections (mbn) and Fragment Energy Transfer Cross Sections (MeV - mbn) of Bertini with experiments (Greiner and Lindstrom et al.).

A_F	$\sigma_{Bertini}$	$\sigma_{Greiner}$	$\overline{E}\sigma_{Bertini}$	$\overline{E}\sigma_{Greiner}$
16	4.69	.02	5.04	.0006
15	103.4	61.5	60.6	56.9
14	40.0	35.4	48.8	51.7
13	18.5	22.8	37.6	48.3
12	32.2	34.1	85.8	68.2
11	8.2	26.4	37.9	99.1
10	11.0	12.7	52.8	62.0
9	1.2	5.2	6.5	25.7
8	.56	1.23	2.5	7.1
7	1.06	27.9	6.11	153.4
6	5.46	13.9	31.4	73.4
Total	226.3	241.2	375.1	645.8

represented in the Ranft data base.[5] An accurate description of these products is quite important since they make a substantial contribution to biological response.[10] The fragmentation cross sections and energy transfer cross sections for 1 GeV protons on oxygen nuclei as calculated from the Bertini code MECC7 are compared to the experiments of the Heckman group at Lawrence Berkeley Laboratory[11,12] in table 3.

One may first note that the fragment distribution predicted by MECC7 is quite different from those observed by Greiner and Lindstrom et al. although the total cross section for each is nearly the same. The underestimate of beryllium fragment products by the Bertini code is well known;[13] we consider here its biological implication. Since the light fragment recoil energy is substantially greater than that for the heavy fragments, an even greater difference results when comparing the energy transfer cross sections especially for those (lighter) lithium through boron fragments.

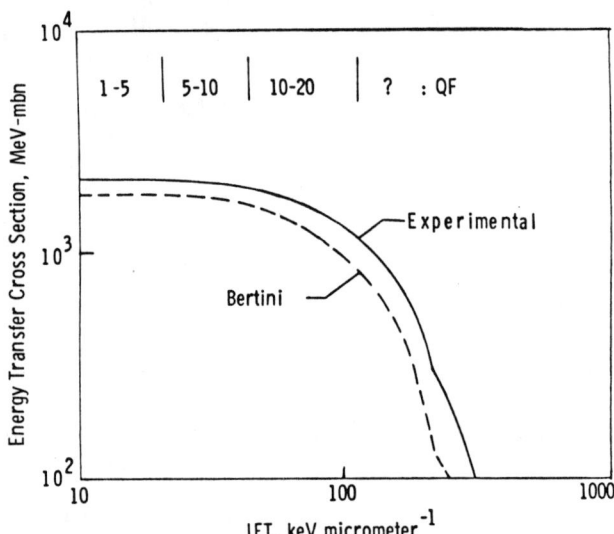

Fig. 5 Energy transfer cross section for 1 GeV protons on ^{16}O resolved into LET components in water according to the two data sets in table 3.

The total energy transfer to ion fragment products measured experimentally is nearly twice that obtained from the Bertini model. The energy transfer cross section is resolved into its LET components in water in figure 5 for the two data sets in table 3 (the method is that described elsewhere).[14] In calculating the results, the Bertini cross sections for α production was used for each of the two curves. The range of presently accepted Quality Factors (QF) is shown at the top scale of the figure 5. It is seen in figure 5 that the Bertini cross sections greatly underestimate the high LET components. Such differences will play an important role especially as Quality Factors in the highest LET region are revised upward as has been recently proposed.[15]

It is clear from the present results that some care must be exercised in using available transport codes with regard to the nuclear data set incorporated. Clearly, efforts to generate evaluated data sets are important.[16] Until they are available, the cross sections associated with Bertini-like data are probably the most reliable. However, one should bear in mind that high LET components are in general not well treated by the Bertini code.

REFERENCES

1. Alsmiller, R. G., Nucl. Sci and Eng. 27, 158 (1967).
2. Wright, H. A.; Anderson, V. E.; Tapner, J. E.; Neufeld, J.; and Snyder, W. S.: Health Physics, 16, 13 (1969).
3. Nelson, W. R.; Jenkins, T. M., Computer Techniques in Radiation Transport and Dosimetry. New York and London: Plenum Press; 1980.
4. Mohring, H. J.; Ranft, J., Nucl. Inst. Meth. 201, 323 (1982).
5. Ranft, J., The FLUKA and KASPRO Hadronic Cascade Codes. In: Computer Techniques in Radiation Transport and Dosimetry, ed. by W. R. Nelson and T. M. Jenkins, New York: Plenum Press; 1980: p. 339-371.
6. Alsmiller, R. G.; Barish, J., Analytic Representation of Nucleon- and Pion-emission Spectra from Nucleon-Nucleus Collisions in the Energy Range 750-2000 MeV. Oak Ridge National Laboratory; 1968. Oak Ridge, TN, ORNL TM-2277.
7. Wilson, J, W.; Lamkin, S. L., Nucl. Sci. and Eng. 57, 292 (1975).
8. Lamkin, S. L.: A Theory of High-Energy Nucleon Transport in One Dimension, Thesis, Old Dominion University, Norfolk, VA (1974).
9. D. V. Bugg, D. C. Salter, G. H. Stafford, R. F. George, K. F. Riley, and R. J. Topper, Phys. Rev., 146, 980 (1966).
10. Alsmiller, R. G.; Armstrong, T. W.; and Coleman, W. A. The Absorbed Dose and Dose Equivalant from Neutrons in the Energy Range 60 to 3000 MeV and Protons in the Energy range 400 to 3000 MeV. Oak Ridge National Laboratory; 1970. Oak Ridge, TN. ORNL-TM-2924.
11. Greiner, D. E.; Lindstrom, P. J.; Heckman, H. H.; Cork, B.; and Bieser, F. S. Phys Rev Lett, 35, 152 (1974).
12. Lindstrom, P. J.; Greiner, D. E.; Heckman, H. H.; Cork, B.; and Bieser, F. S. Isotope Production Cross Sections from the Fragmentation of ^{16}O and ^{12}C at Relativistic Energies. National Aeornautics and Space Administration; 1975. Washington, D.C. NASA CR-142589.
13. Bertini, H. W.: Spallation Reactions: Calculations. In: Spallation Nuclear Reactions and Their Applications, ed. by B. S. P. Shen and M. Merker, D. Reidel Publishing, Dordrecht-Holland/Boston USA; 1976: pp. 27-48.
14. Wilson, J. W.; Buck, W. W.; Townsend, L. W., Health Phys. 50, 666 (1986).

15. International Commission on Radiation Units and Measurements. The Quality Factor in Radiation Protection. Bethesda, MD. ICU Report No. 40, 1986.
16. Pearlstein, S., Summary of the Meeting of the Medium Energy Nuclear Data Working Group, Brookhaven National Laboratory, 1986. Long Island, NY. BNL-NCS-38404.

REFERENCE NUCLEAR DATA FOR SPACE APPLICATIONS

Sol Pearlstein
National Nuclear Data Center, Brookhaven National Laboratory,
Upton, NY 11973

ABSTRACT

The National Nuclear Data Center (NNDC) at Brookhaven National Laboratory is active in the development of high energy data bases for space applications. Validated data and methods of interaction analysis are needed to explain and predict radiation patterns. The NNDC uses methods consisting of nuclear systematics and nuclear model codes to provide neutron and proton induced reaction data from 1 MeV to 1 GeV. The data can be placed in convenient form for use by radiation transport codes. In addition to cross-sections, nuclear structure and radioactive decay data are also stored in data bases. Data are distributed by the NNDC in a variety of ways including on-line access through computer networks or telephone lines.

Nuclear reaction, nuclear structure, and radioactive decay data are needed to explain the origin and composition of background radiation in space. Since it is not possible to measure everything of importance, measurements must be supplemented by validated data and methods of interaction analysis to explain and predict radiation patterns. The National Nuclear Data Center (NNDC) at Brookhaven National Laboratory is active in the development of high energy data bases for radiation transport applications.

A nuclear data file is useful in two ways. Cross-section data are stored and accessed as needed instead of being calculated on demand by lengthy deterministic or Monte Carlo nuclear model calculations. This can not only save time, but also yield unique cross-section values compared to Monte Carlo cross-section calculation methods. The file of reference cross-sections can also be used for sensitivity studies where the effect on calculated results due to discrete changes in input data are observed. This aids cost-benefit analysis since only those data that significantly effect applications need to be seriously reviewed and improved where necessary.

Some features of a nuclear data file are illustrated for high energy protons incident on an Fe-56 target. The nonelastic cross-section for protons on Fe-56 is shown in Figure 1 together with the cross-section for producing secondary neutrons. Empirical fits to the nonelastic cross-section[1] on targets having atomic mass A are similar, varying as $A^{2/3}$, i.e., proportional to the nuclear radius squared. The neutron production cross-section increases with energy as the energy thresholds for releasing neutrons are exceeded. Although the sum of the $\sigma(p,n)$, $\sigma(p,pn)$, $\sigma(p,2n)$, $\sigma(p,3n)$, etc., cross-sections cannot exceed the nonelastic cross-section, the inclusive cross-section for producing neutrons,

© 1989 American Institute of Physics

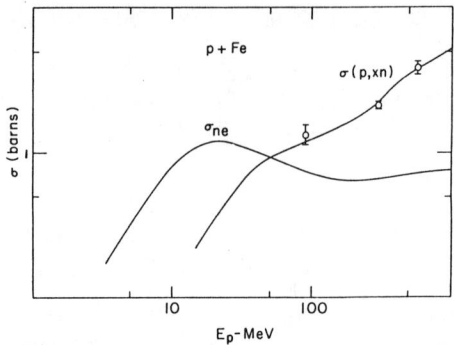

Fig. 1. The iron proton induced nonelastic and (p,xn) cross-sections. The nonelastic cross-section is empirically determined by Letaw et al[1]. The (p,xn) curve is calculated using an extended version of the ALICE code. The experimental (p,xn) cross-sections are inferred from 90 MeV neutron and 160 MeV proton data[4] and 590 MeV proton data[5].

σ(p,xn), can, since it contains neutron multiplication factors and is equal to the sum of the σ(p,n), σ(p,pn), 2σ(p,2n), 3σ(p,3n), etc., quantities. The solid curve for the σ(p,xn) cross-section was generated using an extended version[2] of the ALICE code[3].

The properties of exclusive reactions such as the σ(p,n) cross-section are also important since they determine the production of specific isotopes. A comparison of extended ALICE calculations with experiment[6] for the Fe-56 (p,n) cross-section is shown in Figure 2. Cross-sections for Co isotopes produced by proton bombardment of Fe-56 were calculated. For isotopes with decreasing numbers of neutrons, the peak cross-section decreases while the cross-section energy threshold increases, as expected. In comparing the calculated isobaric cross-sections for atomic mass A=54, the cross-section energy threshold is approximately the same but the peak cross-section for atomic number Z=25 is highest in the range Z=24 to 27. The calculated cross-sections are in general agreement with measurements.

The intranuclear cascade and evaporation nuclear model is generally used to calculate the spallation products of high energy proton reactions. Measured high energy spallation neutrons from the reaction of 590 MeV protons with Fe were compared with intranuclear cascade calculations[5]. The high energy neutrons in the forward and backward directions are underestimated by calculations.

The experimental data[5] can be accurately fit at each angle by simple 8 parameter formulas[7] as shown in Figure 3. The usefulness of the systematics is determined by how well the parametrization can describe measured data at other proton beam energies that were not used in the fitting. Figures 4 and 5 show the results of using systematics determined from 590 MeV proton data to compare with

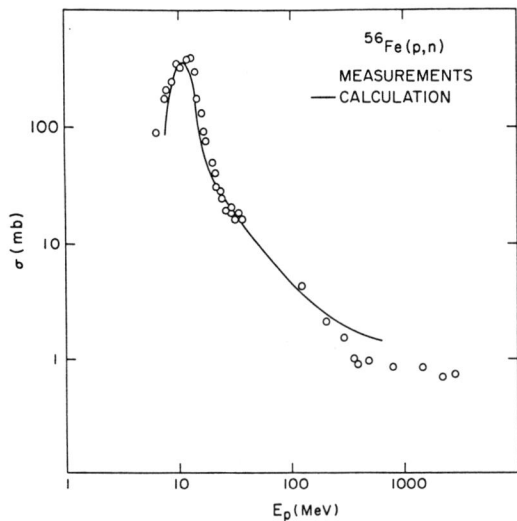

Fig. 2. The iron (p,n) cross-section as a function of incident proton energy. The solid curve is the result of calculations using an extended version of the ALICE code.

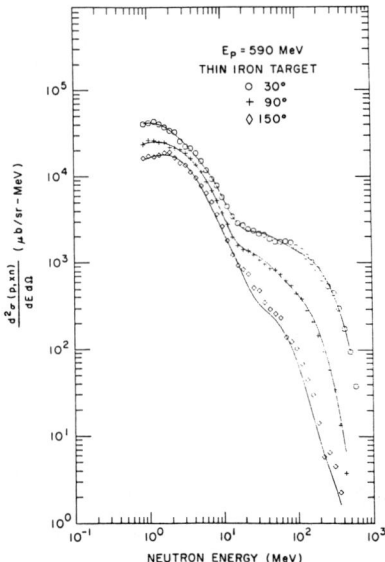

Fig. 3. Double differential neutron emission cross-sections from a thin iron target bombarded by 590 MeV protons. The solid curves are empirical fits to the experimental data.

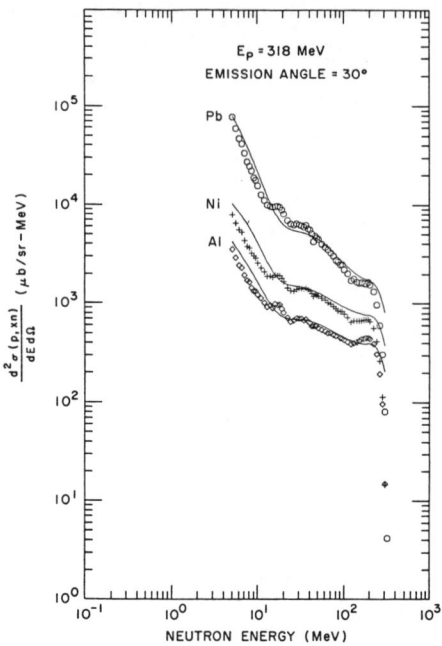

Fig 4. Double differential neutron emission cross-sections at 30° from thin lead, nickel and aluminum targets bombarded by 318 MeV protons. The solid curves are derived from nuclear systematics.

measurements at 318 and 800 MeV, respectively. The energy and angular distributions of high energy neutrons calculated from systematics are significantly closer to experiment than those calculated by the intranuclear cascade nuclear models. The empirically determined parameters may be used to define the cross-section continuously in energy and space as required for radiation transport calculations. The double differential (p,xn) cross-section for 300 MeV protons on Fe-56 is shown in Figure 6 at several neutron emission angles.

The interaction of the neutron fields generated by proton induced spallation reactions with materials in space require the knowledge of high energy neutron cross-sections. The neutron total cross-section has been measured to several GeV for several atomic mass targets. The measured data are stored in NNDC files. A simple parametrization[2] of the neutron cross-section can describe the data over the atomic mass range equal to or greater than A=6 and neutron energies above 1 MeV as shown in Figure 7.

Producing a nuclear data file can include several steps. A search of the bibliography to experimental data is useful to determine what quantitative information is available. A centralized tabulation of experimental data from different

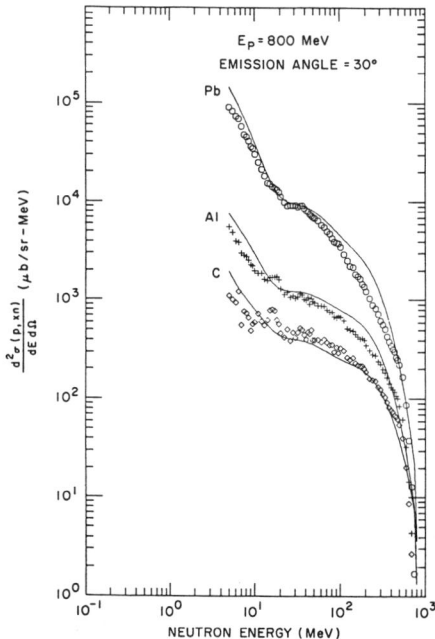

Fig. 5. Double differential neutron emission cross-sections at 30° from thin lead, aluminum and carbon targets bombarded by 800 MeV protons. The solid curves are derived from nuclear systematics.

measurers, and especially if computerized, can facilitate an intercomparison of measurements and the calculation of additional parameters of interest, e.g., mean energies, integrals, etc. Evaluators must recommend values to use in applications even if experimental results differ, with gaps in measurements filled in by nuclear model calculations or systematics derived from data patterns. The evaluated data should be formatted for convenient use in transport calculations. The data file must be validated for use in applications through calculations of benchmark experiments. Documentation describing the physics content and the computer formatting of the data file ensures correct use of the data and provides reference material when reporting results.

Three indexes to the bibliography are maintained by the NNDC: 1) Neutron Data (CINDA) - bibliographic references to measurements, calculations, reviews and evaluations of microscopic neutron data; 2) Charged Particle Data (CPBIB) - bibliographic references for integral charged-particle reaction data for energies up to 100 MeV; and 3) Nuclear Structure and Radioactive Decay Data (NSR) - reference-oriented bibliography of the literature pertaining to nuclear structure, nuclear reactions and radioactive decay data.

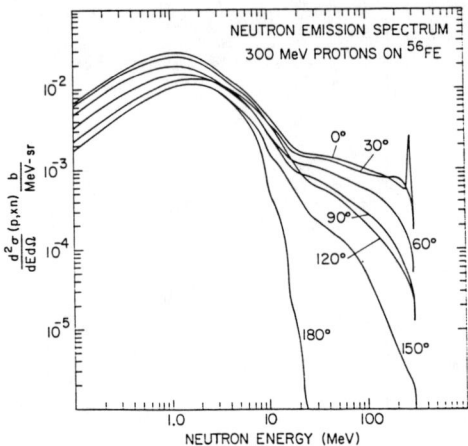

Fig. 6. Evaluated data file contents for the double differential (p,xn) cross-section at several neutron emission angles for 300 MeV protons on Fe-56.

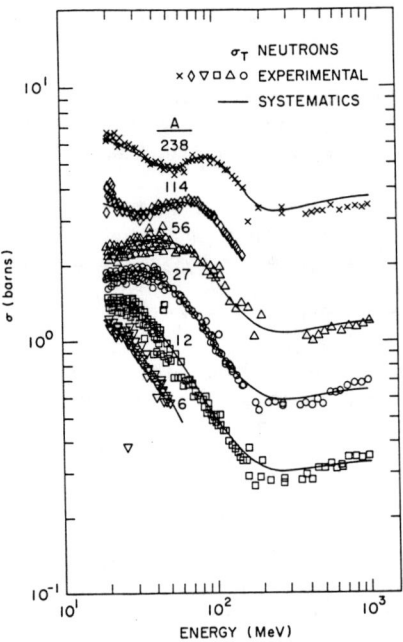

Fig. 7. The neutron induced total cross-section for targets of atomic mass 6, 12, 27, 56, 114, and 238. Experimental data are taken from the NNDC data files. The solid curves are the result of nuclear systematics[2].

The CINDA and NSR data bases can be accessed on-line as can other NNDC data bases containing experimental and evaluated data. An atlas of experimental neutron data up to 200 MeV with curves has recently been completed. The complete NNDC services and publications will not be described here, but information can be obtained by contacting the National Nuclear Data Center, Building 197D, Brookhaven National Laboratory, Upton, NY 11973 or by calling 516-282-2902, (FTS) 666-2901.

REFERENCES

1. J. R. Letaw, C. H. Tsao, and R. Silberberg, Astrophys. J. Suppl. Ser. 51, 271 (1983).
2. S. Pearlstein, private communication.
3. M. Blann et al., Phys. Rev. C28, 1648 (1983).
4. W. E. Crandall and G. P. Millburn, J. App. Phys. 29, 698 (1958).
5. D. Filges et al., Validation of the Intra-Nuclear Cascade Evaporation Model for Particle Production, Jul-1960, November 1984.
6. R. Silberberg and C. H. Tsao, Cross-Sections of Proton-Nucleus Interactions at High Energies, NRL Report 7593, December 1973.
7. S. Pearlstein, Nucl. Sci. Eng., 95, 116 (1987).

ENVIRONET: AN INTERACTIVE SPACE-ENVIRONMENT INFORMATION RESOURCE

A. L. Vampola
The Aerospace Corporation, P.O. Box 92957, Los Angeles, CA 90009

William N. Hall
Air Force Geophysics Laboratory, Bedford, MA 01731

Michael Lauriente
NASA Goddard Space Flight Center, Greenbelt, MD 20771

ABSTRACT

EnviroNET is an interactive menu-driven system set up as an information resource for experimenters, program managers, and design and test engineers who are involved in space missions. Its basic use is as a fundamental single-source of data for the environment encountered by Shuttle and Space Station payloads, but it also has wider applicability in that it includes information on environments encountered by other satellites in both low altitude and high altitude (including geosynchronous) orbits. It incorporates both a text-retrieval mode and an interactive modeling code mode. The system is maintained on the ENVNET MicroVAX computer at NASA/Goddard. It's services are available at no cost to any user who has access to a terminal and a dial-up port. It is a tail-node on SPAN and so it is accessible either directly or through BITNET, ARPANET, and GTE/TELENET via NPSS.

INTRODUCTION

The extensive use of space for platforms for communications, surveillance such as weather and earth resources, science research, military objectives, and manned activities is continuing to increase. With this increase comes an equivalent increase in the number of personnel who have to have knowledge about or access to information about the space environment and the local environment encountered on space platforms such as Shuttle or Space Station. Initially, many of these individuals do not have the appropriate technical background to be familiar with sources for the space environment data they require. There is also a need for a focal point of such information so that groups working on the same mission at different institutions have a common data base for use in their respective portions of the mission. Additionally, the common source should be easily modified and maintained with the most recent data available. EnviroNET has been created to perform this role.

DESCRIPTION

EnviroNET is an information resource for experimenters, design and test engineers, and program managers who are involved with space missions. Its basic use is as a fundamental single repository of information about the environmental areas of concern encountered by Shuttle and Space Station payloads, but it also has wider applicability for information on the somewhat hostile space environments encountered by satellites in both low altitude and high (including geosynchronous) orbits. It is

© 1989 American Institute of Physics

maintained by NASA through cooperative efforts of industry, other government agencies, academia, and of the NASA community.

EnviroNET incorporates a combination of expository text and numerical tables amounting to about one million characters (bytes) plus FORTRAN programs that model the neutral atmosphere, ionosphere, geomagnetic field and the energetic electron and proton environments. This text is under continuous review, correction, and augmentation by ten subpanels of technical experts -- one for each of the main topics dealt with. The aim is to keep it as accurate and current as possible. The EnviroNET files are stored on a MicroVAX II computer at Goddard Space Flight Center and may be accessed on a 24 hour dial-up basis at 300/1200 baud with ordinary telephone connections and at 9600 baud for users on the Space Physics Analysis Network (SPAN). The SPAN network includes several hundred computers in the U.S. and in other countries.

EnviroNET is ideally suited for the science users who find it desirable and feasible to perform an increasing amount of their work by computer networking with their colleagues from their "remote" home laboratories and computers. This is an expansion of the concept started with the Atmosphere Explorer and Dynamics Explorer programs wherein remote scientists were connected over dedicated phone lines to a central "remote" computer site containing their data and computer programs. With the advent of SPAN, the remote Dynamics Explorer scientists could communicate with one another directly and offload calculations and data analysis to their home systems, thereby improving productivity with simultaneous analysis on remote, distributed computer systems. Following this example, we are creating a facility to permit the user to conduct teleanalysis, i.e., perform analysis on the Space Shuttle/Space Station environment data and use the space environment models on computers at remote institutions. This effort will include the NASA centers, other government laboratories, industry, and universities.

The academic community is also involved because it provides important opportunities for testing and evaluating new ideas, techniques and concepts before they have reached the state of maturity considered by contractors and project managers as being suitable for implementation. This testbed program provides a valuable way of training the graduate students who represent the future scientists and engineers of the nation, and who need to be at the leading edge of our developing technology to ensure our economic survival.

The various facilities in EnviroNET are accessed by a menu-driven system which includes a number of options: Retrieval and reading or downloading of text; summaries and/or plots of environmental parameters; on-line computations of magnetic field parameters, particle fluxes, atmospheric constituents, etc. For more detailed studies, software can be downloaded to the user's computer for use at his/her facility.

When the system is accessed for information, the Table of Contents is displayed and the user is instructed to select a topic. When the user has finished his/her activities related to the selected topic, the user is returned to the Table of Contents for additional topic selections. The menu-driven system includes the following options:
-- retrieval and reading or downloading of text;
-- downloading of high resolution graphics summaries and/or environmental parameters;
-- on-line computations of magnetic field parameters;
-- on-line computations of particle fluxes, atmospheric constituents, etc.

Data flow in the EnviroNET system is shown in Figure 1. Text, data, and environmental models reside in a number of files on the ENVNET computer. A number of modeling groups, including the Natural Environments group with which we are associated, are responsible for the text, data bases, models, and interactive computation programs. These modules are installed and maintained on the EnviroNET system by NASA personnel who work directly on the ENVNET computer.

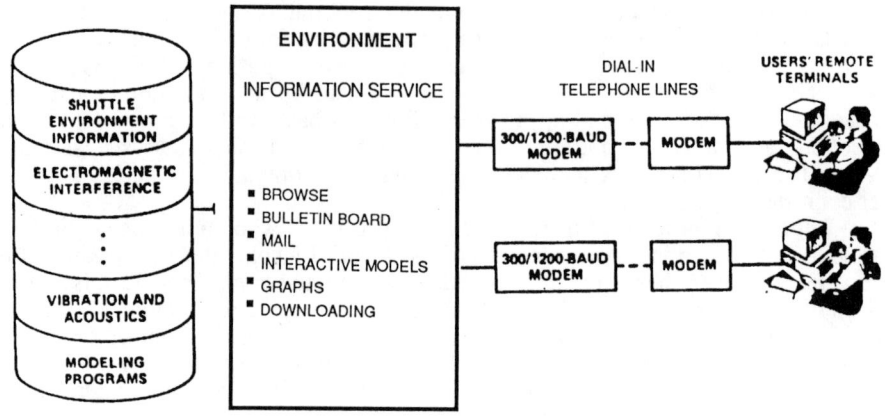

Figure 1. Data Flow in EnviroNET

USER ACCESS

User access to EnviroNET is shown in Figure 2. EnviroNET is a tail node on SPAN. Thus anyone who has access to SPAN either directly or through BITNET or ARPANET can access EnviroNET simply (e.g., using the SET HOST feature). Those who do not have access to SPAN directly can get access through the local

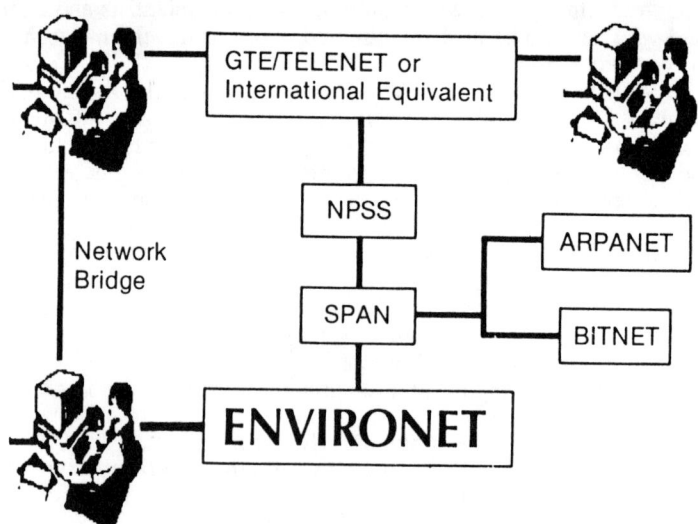

Figure 2. User access to EnviroNET

GTE/TELENET system. To do this, they must obtain the GTE/TELENET local phone access number and enter through NPSS. Details for this access are available from NASA/Goddard. No charge is made for accessing and using EnviroNET, but users should avoid overloading or otherwise abusing the system. Rather than reading through many pages of text on-line, the text should be down-loaded to the user's local system and accessed there.

BROWSE

For an introduction to the system (or for very short inquiries), the BROWSE system is available. The BROWSE system is menu-driven and permits fast, easy access to specific information in the system. It permits a number of activities:
-- display of main topic headings
-- display of index of key-words and topics with chapter and page numbers
-- direct access to any specified page, plus forward and backward paging through text
-- search on indexed key-words or phrases
-- search of any character string in text.

For more detailed studies, text and software files can be downloaded to the user's computer for use at one's facility.

DOWNLOADING TEXT AND FIGURES

Text may be downloaded by a number of options: KERMIT; direct copy to the screen with capture software at the user's end; or, by using the DEC file transfer protocols available on SPAN. Note that if it is downloaded by copying it to the screen, it is up to you to capture it on your local computer as it is displayed. The DEC file transfer method is much faster and can be done in batch mode, provided your terminal has access to SPAN. The chapters which may be downloaded as text are the following:

> Thermal and Humidity
> Vibration and Acoustics
> Electromagnetic Interference
> Loads and Low Frequency Dynamics
> Microbial and Toxic Contaminants
> Molecular Contamination
> Natural Environment
> Orbiter Motion
> Particulate Environment
> Surface Interactions
> Definitions and Acronyms

The technical content of the information is constantly improved to keep it current. After flight data has been extracted, analyzed and verified by other scientists, the information is then entered into EnviroNET. The inclusion of models makes EnviroNET an interactive system instead of just an archive of information. Panels are contributing new information on a continuing basis. They are also trying to work with principal investigators on extraction of flight data from experiments and participating in technical meetings and workshops.

Data graphs and figures can be downloaded as bitmaps from EnviroNET for viewing on a user's terminal if a color board and a color monitor are available. The KERMIT protocol is used. First, the graphics software is downloaded; then the figure is selected and may be downloaded either using KERMIT or by a direct copy to the user's terminal using the DEC file transfer protocols.

INTERACTIVE SOFTWARE AND MODELS

The current interactive computation software includes a magnetic field tracing routine, several energetic particle models, MSIS-86, and the International Reference Ionosphere. The models are accessed by entering the Function Calculation System selected from the main menu. When this system is selected, a new menu is displayed from which one can select the MSIS-86 Neutral Thermosphere Model, the International Reference Ionosphere, the Magnetic Field Model, or Energetic Particles Models. A brief description of each of these follows.

The MSIS-86 Neutral Thermosphere Model is the 1986 COSPAR International Reference Atmosphere and is based on in-situ composition and temperature measurements and ground-based radar measurements covering a complete solar cycle. The inputs required, which are prompted for, are: day, altitude, latitude, longitude, local time, $F_{10.7}$ flux (both 3-month and previous day averages), and the magnetic index A_p. The model, which is valid over the altitude range of 85 km to 1000 km, produces the following outputs: number densities of H, N, He, N_2, O_2, and Ar in cm^{-3}, total mass density in gm/cm^3, and exospheric temperature and the temperature at the selected altitude, both in °K.

The International Reference Ionosphere Model (IRI-86) provides the ionospheric density and temperature, electron density profiles, electron and ion temperatures, ion composition (O^+, H^+, He^+, O^{2+}, and NO^+) and a 12-month running mean sunspot number. Again, temperatures are in °K and compositions are in cm^{-3}. The model prompts for geographic latitude, longitude, altitude, month, local time, and solar activity (quiet, moderate, or active).

The magnetic field model used is a much-modified version of a code originally written by G. Mead. For the internal field, the model permits the user to select a dipole field or any of the standard internal field coefficient sets: the Definitive Geomagnetic Reference Fields (DGRF) for 1965, 1970, 1975, and 1980 and the International Geomagnetic Reference Field 1985. The user may opt against using an external contribution to the field or may select a number of options: Mead-Fairfield Quiet, M-F Disturbed, M-F Super Quiet, M-F Super Disturbed, Olson-Pfitzer No Tilt, or O-P Tilted. Calculations may be performed either at a point or along a field trace. The program prompts for the type of trace (up, down, north, or south), type of field model(s), the epoch, and the latitude, longitude, and altitude for the start of the trace or for the point. The output is the latitude, longitude, altitude and total field at the point or at various points along the trace. Three orthogonal components of the field (outward, south, and east) are also returned at each point. If a trace is requested, the equatorial value of B and McIlwain's parameter L are also provided if the equatorial region is crossed during the trace.

The energetic electron models that are currently (November 1987) in EnviroNET are the AE6 electron model for the region $1.4 \leq L \leq 2.2$ and AE7-Hi for $2.2 \leq L \leq 8.25$. AE7-Hi consists of a number of two-component exponentials defined at the equator for a number of L intervals. They are terminated at 7.5 MeV. The model calculation uses logarithmic interpolation in E and L and a $Sin^2\lambda$ interpolation along the field line. The proton model used is AP6 for the intervals $1.2 \leq L \leq 6.0$ and $0.1 \leq E \leq 170$ MeV. Tabular interpolation at the equator and along a field line are similar to those used in the electron calculation. Both unidirectional differential and integral flux is returned for the electrons. Only omnidirectional integral fluxes are returned for the protons. Values returned by these subroutines are within a factor of 2 to 3 of the values which AP8 and AE8 would predict. This accuracy is within the confidence limits of AP8 and AE8, and so can be used without reservation until the more comprehensive models are available.

FUTURE PLANS

Future plans include adding the following to EnviroNET: downloading of all codes and models; incorporation of the AE8 electron and AP8 proton models; orbital integrations of fluxes; addition of the ORB and ORP codes from NSSDC. The orbital integrations will have limited orbital position and energy resolution in order to avoid having users overload the system by attempting to do detailed calculations. The intent will be that a user will be able to determine whether the energetic particle environment might be a problem or not. If it might be, the user then can download the appropriate codes and models and do more detailed calculations at his/her own facility.

More distant plans include the addition of dose calculations as a function of shielding and position in orbit and calculation of cosmic ray fluxes as a function of mass, energy, and position in orbit.

ACKNOWLEDGEMENTS

We wish to thank D. Bilitra, J. Green, A. Hedin, N. Thomson, and J. Vette for the various modeling codes used in EnviroNET. Funding was provided by NASA Headquarters, the AFGL Space Systems Environmental Interactions Technology Office, and by the U.S. Air Force Systems Command's Space Division under Contract No. F04701-86-C-0087.

SEL MONITORING OF THE EARTH'S ENERGETIC PARTICLE RADIATION ENVIRONMENT

Herbert H. Sauer
NOAA Space Environment Laboratory, Boulder, Colo. 80303

ABSTRACT

The Space Environment Laboratory (SEL) of the National Oceanic and Atmospheric Administration (NOAA) maintains instruments on board the GOES series of geostationary satellites, and aboard the NOAA/TIROS series of low-altitude, polar-orbiting satellites, which provide monitoring of the energetic particle radiation environment as well as monitoring the geostationary magnetic field and the solar x-ray flux. The data are used by the SEL Space Environment Services Center (SESC) to help provide real-time monitoring and forecasting of the state of the near earth environment and its disturbances, and to maintain a source of reliable information to research and operational activities of a variety of users.

The data, data sources, and products relevant to the characterization of the near- earth radiation environment and its response to solar cosmic ray events are briefly described, as are the laboratory's archives and uses of these data.

INTRODUCTION

The Space Environment Laboratory (SEL) of the National Oceanic and Atmospheric Administration (NOAA) provides monitoring and forecasting of solar and geophysical activity including the state of the near-earth space environment, and conducts research on solar terrestrial relationships in order to provide a better understanding of that environment and to improve the quality and utility of its services. These services are provided by the Space Environment Services Center (SESC) within SEL, which is charged to monitor and report "space weather", and to issue forecasts and warnings of geoeffective solar activity in support of national and public needs, in cooperation with a corresponding activity within the Defense Department. These services derive from an information base obtained from a global network of both ground- and satellite-based observatories providing primarily real-time observations of solar x-ray, visible and radio emissions, ground and satellite geomagnetic field observations, satellite observations of magnetospheric energetic particles and observations of ionospheric parameters. Through real-time analysis of these data, the SESC provides for rapid detection of environmental disturbances, on-line monitoring of geophysically significant parameters, and the generation and distribution of disturbance indices, alerts and forecasts of geophysical disturbances to those for whom knowledge of the earth's space environment is important.

The magnetospheric energetic particle data are obtained from instruments maintained by SEL aboard two series of satellites: the SMS/GOES series of geosynchronous meteorological satellites, and the NOAA/TIROS series of low-altitude, polar-orbiting, sun-synchronous satellites.

THE GOES ENERGETIC PARTICLE SENSORS

The GOES Energetic Particle Sensors customarily consist of three detector assemblies: the EPS Telescope, the EPS Dome Detector, and the High Energy Proton and Alpha Detector (HEPAD). Of the currently operational GOES-6 and -7 satellites, the GOES-7 EPS does not include the HEPAD, however the follow-on members of the GOES series planned

© 1989 American Institute of Physics

for launch beginning in 1990 will include the HEPAD. Together these detector assemblies measure the flux of protons and alpha particles at geostationary orbit from energies of about 1 MeV per nucleon to relativistic energies, as well as measuring the integral flux of trapped electrons above 2 MeV.

The components of the energetic particle populations observed by the GOES EPS at geosynchronous orbit principally include those due to solar cosmic ray events superimposed on a galactic cosmic ray background extending to highly relativistic energies, but at low intensity levels. The lowest proton energy channel (P1: $0.6 < E < 4.2$ MeV) also responds to trapped particles of the outer magnetospheric radiation belts, as does the electron channel. The higher energy proton and alpha particle channels primarily respond to galactic and solar cosmic rays, since the geostationary geomagnetic cutoff energy[1] seldom exceeds several MeV[2]. Table I summarizes some of the characteristics of the GOES energetic particle sensors.

The GOES EPS provides the principal source of real-time energetic particle data to the SEL-SESC, in that the satellites remain at an essentially fixed position in the sky and can therefore provide continuous data to the laboratory. The data are processed in real-time, and stored together with other real-time geophysical data streams in two temporary, disk-based data archives containing one-minute averages of the last week's data, and five-minute averages of the last month's data, respectively. These data then form the basis for real-time data displays. Through subsequent analysis the SESC determines the state of the near-earth space environment, establishes alerts and forecasts of distubances to that environment, and disseminates the resulting characterizations and products to its customers.

Prior to the temporary archive data being supplanted by more recent data, the GOES EPS one-minute averaged energetic particle data are transferred to magnetic tape, as are the corresponding one-minute averaged geostationary magnetic field and whole-sun x-ray emission data measured by the GOES Space Environment Monitor. These data tapes are archived by the National Geophysical Data Center (NGDC), Boulder, Colorado, where they are made publicly available for operational and scientific use.

In order to simplify public access to the GOES geostationary data, an additional data archive procedure has recently been established[3] for the user for whom five-minute resolution data are adequate. Five-minute averaged GOES observations of solar x-rays integrated over the solar disk, the vector geomagnetic field, and the energetic particle data of Table I, have been made available on standard personal computer format diskettes, through the NGDC. One diskette can contain an entire month of five-minute energetic particle data, or one month of five minute magnetometer and solar x-ray data. These data are made available with associated computer codes which provide simple and immediate data access and routine data plotting capability to the user. Figure 1 illustrates a specimen output of the PC-Archive routine, for which three day's data beginning November 7, 1987, were requested to be plotted for the four proton channels P2, P3, P4, and P5. Note that channels plotted are identified on the figure, and that the successive traces are scaled by factors of 100 so that they may be clearly separated for legibility.

TABLE I

GOES EPS DETECTOR CHARACTERISTICS

Channel Name	Average Energy (MeV)	Accumulation Time (s)	Eff. Geometric Factor (cm² –s–ster–MeV)⁻¹
Telescope:			
P1	2.4	3.06	0.202
P2	6.5	12.24	0.252
P3	11.6	12.24	0.325
A1	6.9	12.24	0.342
A2	16.1	12.24	0.638
A3	41.2	12.24	2.22
*E1	>2 MeV	3.06	0.034 (cm² –s–ster)⁻¹
Dome:			
P4	29.5	12.24	6.09
P5	60.5	12.24	20.
P6	168.	6.12	136.
P7	427.	6.12	891.
A4	120.	12.24	25.2
A5	210.	6.12	36.0
A6	435.	6.12	176.
HEPAD (Not on GOES-7):			
P8	390.	6.12	75.
P9	465.	6.12	46.
P10	660.	12.24	310.
P11	1000.	12.24	1390.
A7	825.	6.12	306.
A8	1000.	12.245	5000.

* Note: The channel names designate P(roton), A(lpha) and E(lectron) channels respectively. Note also that the geometric factor of the integral electron channel has appropriately different units.

THE NOAA/TIROS SPACE ENVIRONMENT MONITOR (SEM)

The NOAA/TIROS consist of a series of low-altitude (850 km nominal), polar-orbiting (98.8° inclination), sun-synchronous satellites, whose ascending nodes are variously in the morning or evening local time sector. The SEL maintains Space Environment Monitor (SEM) instruments aboard these satellites which consist of two detector assemblies; the Total Energy Detector (TED), which was designed to separately monitor the energy fluxes of electrons and protons in the energy range of 300 eV to 20 keV, which constitute the principal particle energetic partical flux into the upper atmosphere; and secondly, the Medium Energy Proton and Electron Detector (MEPED) which measures the electron fluxes from greater than 30 keV to greater than 300 keV, and proton fluxes from 30 keV to greater than 80 MeV. We will address only the energetic particle data from the MEPED as appropriate to radiation hazard environment considerations. The MEPED itself contains two detector assemblies, the Proton and Electron Telescope detectors and the Dome detectors which are comparable to those on the GOES satellite discussed previously. The Telescope sensors consist of four units, a pair of which view the local zenith (precipitating particles), and a pair that "look" at approximattely 90^0 to the zenith (locally mirroring, or trapped particles). Each pair consists of a proton telescope and electron detector. The Dome, or omnidirectional detectors, are independently mounted under spherical shell moderators to provide sensitivity in three integral proton energy channels. A summary of SEM energetic particle data channels is given in Table II.

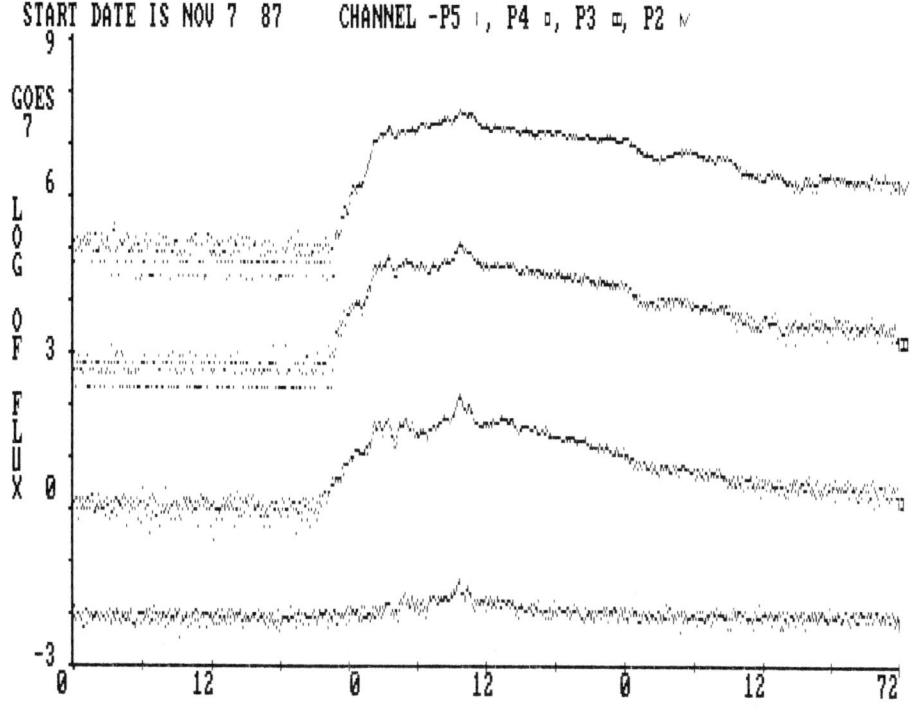

Fig. 1 Sample PC Archive Output Plot

The equality of the secondary energy response of channels P6 and P7 results from a design decision to have the out-of-aperture response of the three detectors equal, in order that secondary response correction may be accomplished through subtraction of the P8 output from the response of P6 and P7.

The NOAA/TIROS data do not constitute a primary energetic particle data source to the SESC because they are not available in real time. Since the NOAA/TIROS are in near-polar orbit, data must be accumulated for readout once per orbit of approximately 105 minutes. However, they provide valuable substantiation during the occurrence of solar cosmic ray events, and a direct assessment of energetic proton and electron precipitation into the high latitude regions of the earth's atmosphere.

TABLE II

NOAA/TIROS MEPED DETECTOR CHARACTERISTICS

Channel Name	Energy	Accumulation time(s)	Geometric factor $(cm^2 - s\text{-ster})^{-1}$
Telescope:			
P1	30–80 keV	1	0.0094
P2	80–250 keV	1	0.0094
P3	250–800 keV	1	0.0094
P4	800–2500 keV	1	0.0094
P5	>2.5 MeV	1	0.0094
E1	>30 keV	1	0.0095
E2	>100 keV	1	0.0095
E3	>300 keV	1	0.0095
Dome:			
P6	16–80 MeV	2	1.57
	80–215 MeV		5.4
P7	36–80 MeV	2	1.57
	80–215 MeV		5.4
P8	80–215 MeV	2	5.4

SOLAR COSMIC RAY EVENT MONITORING AND PREDICTION

The energetic particle data from the GOES geosynchronous and NOAA/TIROS polar satellite detectors together provide for reliable monitoring of solar cosmic ray (SCR) events. While the NOAA/TIROS instruments do not include alpha particle detectors, the geomagnetic cutoff energy (the minimum required energy for particle access through the geomagnetic field) at geosynchronous orbit is seldom above a few MeV, and energetic solar cosmic rays are permitted access to the geomagnetic polar regions at very low energies. Further, Liouville's Theorem[1] assures that if the particle energies are above local geomagetic cutoff and the particle fluxes outside the magnetosphere are isotropic, flux measurements at different locations within the magnetosphere will yield the same result. Interplanetary fluxes are usually found to be isotropic shortly after event onset. Therefore, alpha particle measurements above several MeV at geostationary orbit are representative of those that would have been obtained over the polar caps of the earth. Further confidence in that conclusion may be obtained through comparison of the respective proton observations: equality of the observed proton fluxes at geostationary and over the poles implies the equivalence of the corresponding alpha particle fluxes. While SCR events occur infrequently, these events produce significant terrestrial consequences. The association of SCR event occurence with solar activity levels in general, and with solar flares and solar coronal mass ejections specifically, allows the prediction of SCR events to be made at some level of confidence.

The occurrence of SCR events is well-correlated with the general solar activity as indicated by the sunspot number, with an occurrence frequency ranging from about one per year at solar minimum to one or more per month around solar maximum. Event duration ranges from several hours to several days, with the principal terrestrial effect of their ocurrence being felt in the high-latitude regions of the earth.

The main consequences of the enhanced energetic particle populations present are: first, the increased ionization levels of the high-latitude ionosphere, which strongly influences radio wave propagation and significantly effects upper atmosphere density, chemistry, and dynamics; and second, increased radiation hazard to both human activity at high altitude and to sensitive satellite instrumentation—sometimes to substantial levels. For example, the SCR event of August, 1972, produced estimated peak space-suit-shielded dose rates of the order of 500 REM per hour, and numerous satellite instrument upsets have been clearly assiociable with enhanced energetic particle fluxes.

Among its services, the SESC produces several types of proton event predictions based upon its real-time monitoring of the sun and interplanetary space and user needs. Weekly predictions are made of the general level of proton event probability for the following 27-day period, and 3-day solar flare and proton event probabilities are prepared and disseminated daily. Proton event duration predictions and event alerts are provided when solar events suspected of accelerating energetic particles are observed, and upon satellite proton event detection. Regularly updated predictions of event duration and time profiles are produced throughout the course of the event.

CONCLUSION

The Space Environment Laboratory, through its Space Environment Services Center, provides real-time monitoring of the near-earth space environment and forecasting of disturbances to that environment of consequence to man's activities. The data base from which these services derive also provides a long-term archive of solar and geophysical data which, together with the real-time products and services, are made readily available to a community of both operational and scientific users.

REFERENCES

1. B. Rossi and S. Olbert, Introduction to the Physics of Space, McGraw-Hill, N. Y. (1987).
2. L. J. Lanzeroti, Phys. Rev. Ltrs., 21,929,(1968).
3. H. H. Sauer, NOAA Tech. Mem. ERL-SEL-74,1987.

IV. Instrument Background and Dosimetry

GAMMA RADIATION BACKGROUND MEASUREMENTS FROM SPACELAB 2

William S. Paciesas and John C. Gregory
University of Alabama in Huntsville, AL 35899

Gerald J. Fishman
Space Science Laboratory, NASA/Marshall Space Flight Center
Huntsville, AL 35812

ABSTRACT

A Nuclear Radiation Monitor incorporating a NaI(Tℓ) scintillation detector was flown as part of the verification flight instrumentation on the Spacelab 2 mission, July 29 - August 6, 1985. γ-ray spectra were measured with better than 20 s resolution throughout most of the mission in the energy range 0.1 to 30 MeV. Knowledge of the decay characteristics and the geomagnetic dependence of the counting rates enable measurement of the various components of the Spacelab γ-ray background: prompt secondary radiation, Earth albedo, and delayed induced radioactivity. We summarize herein the present status of the data analysis and present relevant examples of typical background behavior.

INTRODUCTION

The usefulness of the Space Shuttle as a platform for high energy astrophysics experiments is highly dependent upon the spectrum and intensity of the ambient radiation background. The Nuclear Radiation Monitor (NRM) was developed to characterize the γ-ray, proton, electron and neutron environment inside the Shuttle payload bay during orbital operations. To this end, the NRM was included in the Verification Flight Instrumentation on the Spacelab 2 mission, where it operated continuously during the orbital portion of the flight. The data are currently being analyzed to distinguish among the various background components: primary radiation incident on the spacecraft (generally well-known from previous measurements) and secondary radiation from the spacecraft or the Earth's atmosphere. The secondary spacecraft radiation spectrum is difficult to calculate or predict because its production depends in a complex way on the composition and distribution of materials. Although this implies that the NRM measurements hold in the strictest sense only for Spacelab 2, other Shuttle missions may be sufficently similar to make the data relevant to other experimenters, either in instrument design or data analysis. Data from the NRM is available to other experimenters for this purpose.

EXPERIMENT DESCRIPTION

The NRM consists of a 12.7 cm diameter, 12.7 cm thick, NaI(Tℓ) central detector shielded on the front and side by a 1 cm thick plastic scintillator. The configuration resulted in a nearly omnidirectional response with an effective energy threshold of ~80 keV and an energy resolution of 9.0% at 662 keV. The NRM was installed in the payload bay of the Space Shuttle Challenger and operated during the Spacelab 2 mission (July 29 to August 6, 1985). The orbit for this mission was inclined at 49.5° and its altitude varied between 290 and 327 km. The NRM was mounted on a pedestal on pallet no. 3, approximately 1 m above and 1 m inside the payload bay sill, at least 2 meters away from any massive objects. The NRM thus had a representative "view" of the inside of the payload bay.

The data from the NRM consist of 510-channel spectra (dispersions of 1.5, 15 and 150 keV/channel) with 20.16 s time resolution and 16-channel rates with 5.25 ms time resolution. Separate plastic coincidence/anticoincidence spectra are accumulated simultaneously for each dispersion. The instrument and data system are described in more detail elsewhere[1].

DATA ANALYSIS

The data have been organized into two databases for efficient analysis. One of these combines the spectral data with livetime corrections and retains the inherent time resolution of 20.16 s. The other contains the 16-channel gamma ray and charged particle counting rates integrated over 0.504 s intervals. Figure 1 is an example of the available rate histories. Two different γ-ray energy ranges are shown together with the integral rate of events in the central detector in coincidence with the plastic scintillator (the latter are predominantly charged particle events). The characteristic sinusoidal modulation of the background results from the orbital position dependence of the cosmic ray secondary production in the atmosphere and the spacecraft. Significant increases over the basic orbital modulation are produced by at least two processes. In one case, passages through the South Atlantic Anomaly (SAA) region produce intense events which have a hard spectrum, smooth time structure, and durations of about 15 minutes. The other case is due to precipitating energetic electrons which produce bremsstrahlung x-rays by interacting with the atmosphere or the spacecraft. These events are seen only at high latitudes and have relatively soft spectra, as they are not prominent in the higher energy γ-ray channels. The electron events often feature rapid, complex temporal variation on timescales as short as 1 s, much faster than the proton events (an example of this is shown in Figure 3 of Ref. 1). Based on other studies of precipitating electrons at high latitudes (cf. Ref. 2), we conclude that the such rapid x-ray temporal structures reflect the time-variability of the electron flux rather than the spatial (orbital) motion of the detector.

Figure 2 shows a typical uncorrected spectrum at the time of an electron event compared with a spectrum taken at a background minimum. The softness of the bremsstrahlung spectrum is evident, the e-folding energy of the excess being ~100 keV. In contrast, the γ-ray spectra of SAA passages typically differ from those of low background regions in intensity but not in shape. The spectral

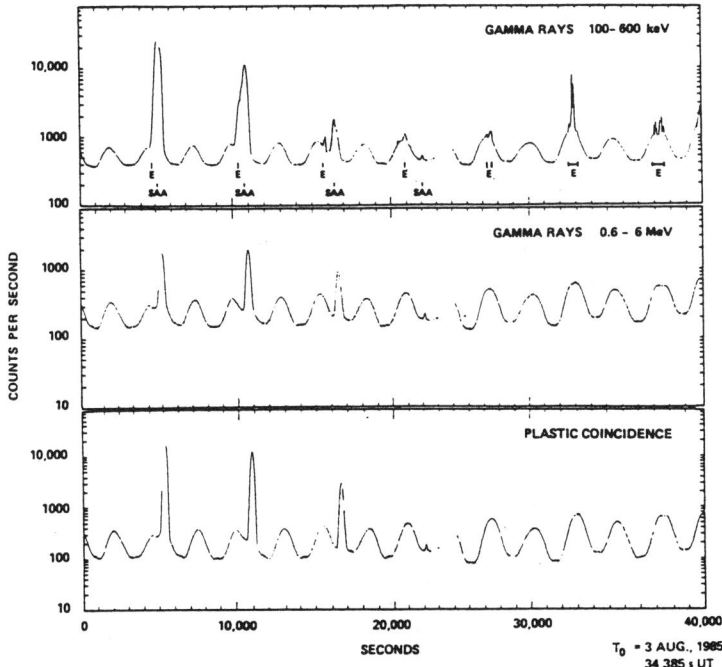

Figure 1. Background data from the NRM during the Spacelab 2 mission. Shown are the detected count rates in two γ-ray energy regions and the integral counting rate from the NaI in coincidence with the plastic (due mostly to charged particles). Several regions of increased background due to trapped electrons (E) and protons (SAA) are indicated, superimposed on the underlying modulation due to geomagnetic latitude variations. The data span ~7.5 orbits with 20.16 s time resolution.

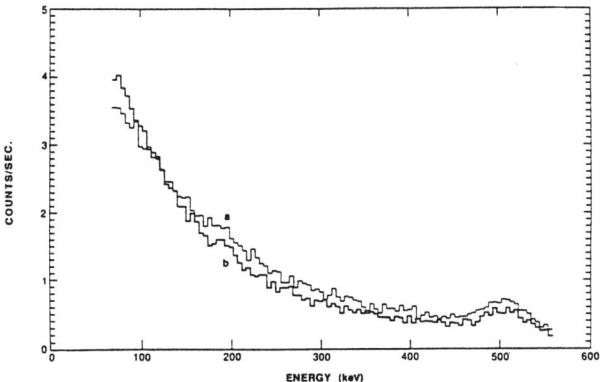

Figure 2. Spectrum of low energy γ-rays during an electron enhancement compared with a normal background spectrum. The energy scale is ~1.5 keV/channel.

distinction of proton enhancements shows up primarily in the highest-energy charged particle spectra.

The contribution of earth albedo to the background effects is illustrated best by examining data taken during an episode on July 31 in which the orbiter was rolled continuously for approximately one hour. Figure 3 shows the γ-ray rates in three energy intervals during this episode. Despite some interfering SAA passages, the roll rate of about one revolution per 300 s is readily apparent. The lowest energy channel shows the smallest effect and, more importantly, it is 180° out of phase with the higher channels, due to the fact that the sky-background spectrum is steeper than that of the earth albedo. Spectra taken at these two extremes are shown in Figure 4. The crossover point (where the earth and the sky have the same brightness) occurs at ~120 keV.

Cosmic rays, trapped protons, and energetic neutrons can also contribute to the γ-ray background by producing radioactivity in the detector and surrounding materials via spallation reactions. We attempted to measure such effects in the NRM by observing the spectral lines produced by activation of the central detector crystal. Though too weak to be seen in flight, the lines show up in post-flight spectra taken under sufficiently low background conditions. Similar measurements on other heavy spacecraft have been interpreted[3,4] as evidence of a substantial flux of energetic neutrons, since the detected activity was too large to be explained by cosmic ray and trapped protons alone (the latter contributions are assumed to be much better-known than the neutron flux).

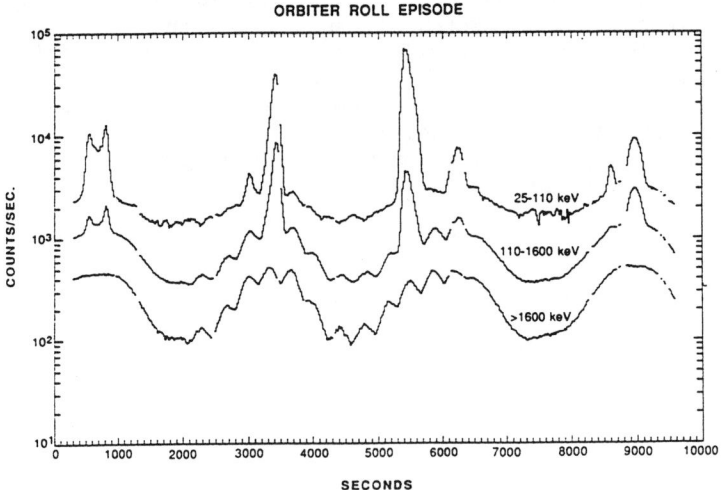

Figure 3. NRM γ-ray counting rates spanning an episode of orbiter roll. The roll period of ~300 s is superimposed on the orbital variations. The phase reversal due to the difference in spectral hardness between sky background and earth albedo is evident in the lowest energy band near the minimum of the orbital variation. The time resolution of the data is 20.16 s.

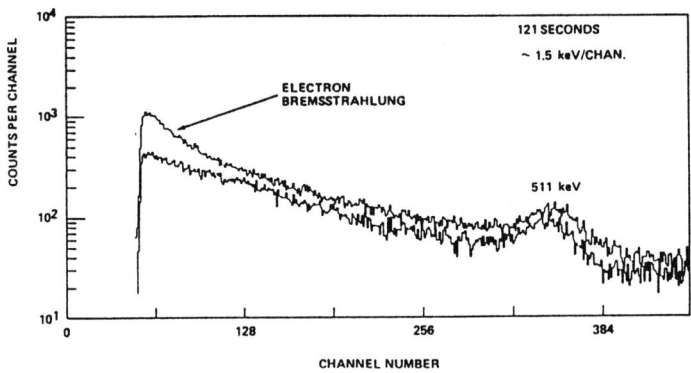

Figure 4. A comparison of background spectra taken during a) earth-pointing and b) sky-pointing intervals. The relative hardness of the earth albedo is evident.

The NRM was received at MSFC approximately three weeks after the end of the mission, at which time certain shorter-lived iodine spallation products such as ^{124}I ($t_{\frac{1}{2}}$ = 4.2 d) had decayed below detectability. With the NRM in a low-background counting facility at MSFC, we observed a γ-ray line which was consistent in energy and half-life with the most prominent line expected from ^{126}I decay within the detector (699 keV and 13 d, respectively; because the decay mode is electron capture within the detector, the apparent line energy is the sum of the ^{126}I γ-ray energy of 667 keV and the daughter tellurium K-shell binding energy of 32 keV). The line can be seen in figure 5, which shows spectra taken at various intervals after the end of mission. It is evident from the figure that the line at ~699 keV is decreasing in intensity relative to the other background lines. Extrapolation of these data back to the end of the mission implies an induced ^{126}I activity of ~0.5 decays/s/kg-NaI, with an uncertainty of ~50% due predominantly to systematic errors.

Similar measurements performed on Apollo 17[3] and the Apollo-Soyuz Test Project (ASTP)[4] are compared with the present result in Table I, from which it appears that evidence for neutron activation on Spacelab 2 is less strong than on the Apollo-17 and ASTP missions. More direct measurements of the energetic neutron fluence on Spacelab 2 by Parnell et al.[5] support this conclusion; their results imply an average energetic neutron flux of ~0.8 cm^{-2}-s, compared with values of 2 and 5 cm^{-2}-s derived for ASTP and Apollo 17, respectively[3,4]. Part or all of this discrepancy may be due to

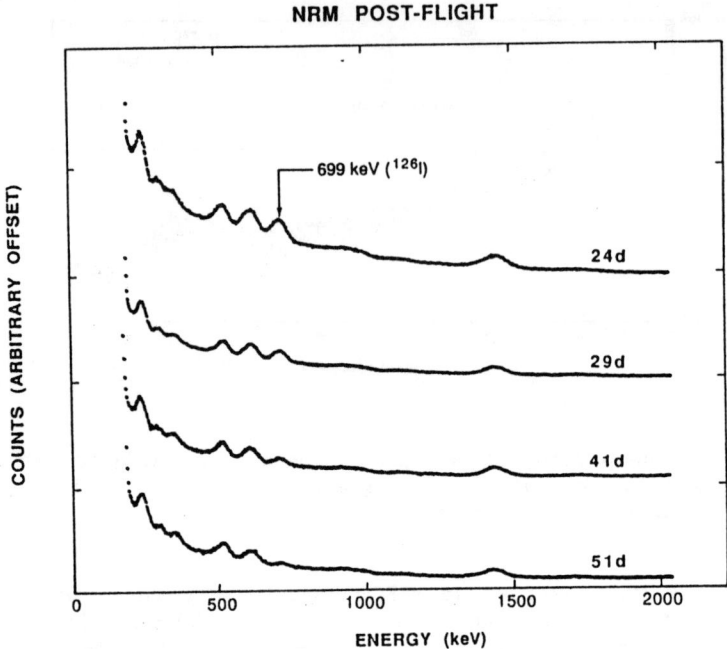

Figure 5. Post-mission γ-ray spectra accumulated at various intervals with the NRM inside a low-background shield. The line at 699 keV is identified with ^{126}I ($t_{\frac{1}{2}}$ = 13 d). The latter is produced in orbit by interactions of energetic neutrons and protons with ^{127}I in the central detector.

Table I. Measurements of ^{126}I Activity

Mission	^{126}I activity (decays/s/kg-NaI)
Apollo 17	5.0 ± 2.5
ASTP	~2.5
Spacelab 2	~0.5

different assumptions regarding the spectrum of the neutrons, which is poorly known. Nevertheless, the data are consistent with the conclusion that the energetic neutron flux on Spacelab 2 was a factor of 2 to 4 smaller than on ASTP and a factor of 5 to 10 smaller than on Apollo 17.

SUMMARY

We have seen spectral and temporal variations in the NRM γ-ray and charged particle event rates which may be used to distinguish among the major components of the radiation background on Spacelab 2. Further analysis of these variations will enable us to determine the contribution due to secondary radiation in the spacecraft, a result of importance not only for high energy astrophysics applications but also for other classes of experiments and materials such as photographic films and biological samples.

ACKNOWLEDGMENTS

We are grateful to the following people for the successful operation of the NRM: F. Berry, Jr., W. Selig, R. Austin, W. Hammon, S. Dothard, L. Wallace, R. Asquith, F. Emens, M. Teal, J. Gattis, R. Eakes, and C. Messer. We also thank T. Parnell for useful discussions regarding neutron fluence measurements.

REFERENCES

1. Fishman, G. J., Paciesas, W. S., and Gregory, J. C., *Adv. Space Res.* 7, (5)231 (1987).
2. Imhof, W. L. *et al.*, *J. Geophys. Res.* 91, 3077 (1986).
3. Dyer, C. S. *et al.*, *Space Sci. Instr.* 1, 279 (1975).
4. Trombka, J. I. *et al.*, ASTP Preliminary Science Report, NASA TM X-58173, p. 7-1 (1976).
5. Parnell, T. A. *et al.*, *Adv. Space Res.* 6, 125 (1986).

RADIOACTIVITY OBSERVED IN SCINTILLATION COUNTERS DURING THE HEAO-1 MISSION

D. E. Gruber, G. V. Jung[1] and J. L. Matteson
Center for Astrophysics and Space Sciences
University of California, San Diego, La Jolla, CA, 92093

ABSTRACT

The UCSD/MIT Hard X-Ray and Low-Energy Gamma-Ray Experiment (A4) was carried on the HEAO-1 satellite to perform a survey of the sky at energies between 10 keV and 10 MeV. Observation of radiation from cosmic sources in this energy range has been difficult because of high internal background counting rates from radioactivity induced in the material of the detectors. The activation is caused by ambient charged particles and neutrons, and leads to complex, time-variable energy-loss spectra. We report here on the results of an extensive analysis of the radioactivity induced in the NaI Medium Energy Detectors of this experiment during the 500-day HEAO-1 mission. These 7.5 cm dia by 2.5 cm thick crystals operated from 40 keV to 2 MeV. Various radioactive nuclides were identified through their observed spectra and decay times. The incident charged particle flux was monitored with three NE 102 plastic scintillators. Model activities at a variety of expected half-lives were calculated from this charged particle input. The amplitude of each of these activation terms was treated as a free parameter in standard multiple regression fits at each energy. Since the dominant behavior of each term is exponential with time, the problem is ill-conditioned. Judicious selection of data from the entire mission nevertheless permitted a reliable separation of activities. The resulting energy spectra and nuclide identifications are discussed.

INTRODUCTION

In the observation of cosmic x- and gamma-ray sources at energies between 10 keV and 10 MeV the limiting factor has been the high internal background counting rate from radioactive species induced in the detectors by ambient protons and neutrons at balloon and satellite altitudes. A wide variety of species are produced, each with its own decay modes and half life, thus modeling of the internal background is a formidable task. Although such modeling has been employed for certain measurements, such as the Apollo-17 measurement[1] of the spectrum of the diffuse component, the observation of individual sources is more usually accomplished with chopping techniques using a shutter or by comparing the source region with nearby blank sky. Since the internal background is variable on a variety of time scales, it is necessary to chop on other time scales. Generally this is done by selecting data intervals on which only slow variability is expected, then chopping as rapidly as practical.

The A4 experiment on HEAO-1 was designed at a time when detector activation and the resultant variability were not yet fully appreciated. Enough was understood so that that the dwell time on an individual source in the Medium Energy Detectors was kept short (three minutes) compared to the 45 minute background variability expected from the orbital motion in the geomagnetic field. Moreover, the instrument carried

---

[1] Present Address, Naval research Laboratory, Washington, D. C., 20375

special detectors for monitoring charged particles, and an elaborate telemetry scheme returned very complete information on all detector and active shield counters on time scales of 41 seconds or less. Thus it was possible to make a rather complete analysis of the radiation environment, its variability, and the responses of the various parts of the detector system. This study formed a key role in the formation of a sky map from the A4 detectors: knowledge of the variability permitted the formulation of a scheme of data selection and accumulation which was relatively insensitive to the background changes. Furthermore, the size of residual variability in the sky map could be estimated. We report here on various aspects of this analysis. Much of this work has been reported earlier[2].

INSTRUMENTATION

The UCSD/MIT Hard X-Ray and Low-Energy Gamma Ray Experiment (Fig. 1) was a massive collection of inorganic scintillator counters with NaI detectors surrounded by anticoincidence CsI shielding in all directions except for the entrance apertures. With thickness below 10 cm in very few directions, this shielding was very effective against background charged particles and photons, but was quite permeable to neutrons, with an optical depth for scattering of the order of unity. Anticoincidence for charged particles was extended to 4π by covering the aperture with a thin sheet of NE 102 scintillator. The two low energy detectors, with thicknesses of only 3 mm, were optimized for operation below an energy of 100 keV. The four medium energy detectors (MED) had thicknesses of 2.5 cm, resulting in photon detection probabilities near unity up to several hundred keV. The single high energy detector had a thickness of 7.5 cm. The ambient charged particle flux was monitored with three 10 mm diameter spherical NE 102 scintillators surrounded by absorbers of different thicknesses to permit crude energy resolution. A complex telemetry scheme returned counting rate averages on several time scales, none longer than 41 seconds. Arrival times and energies of individual photon detections were

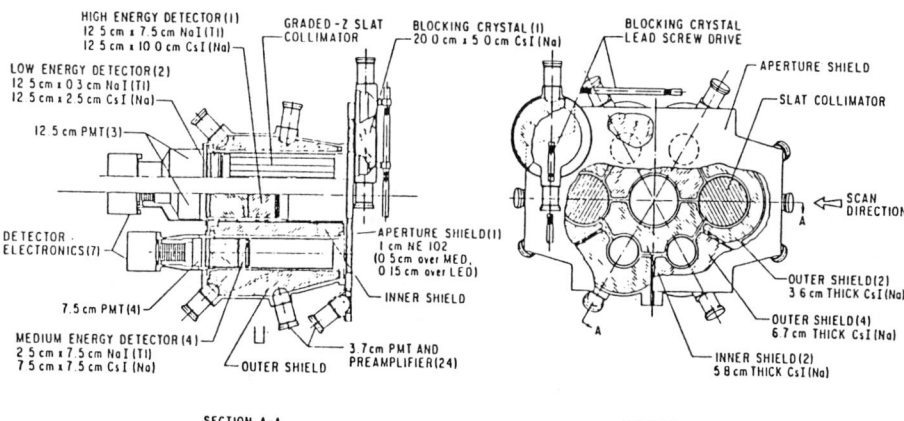

Figure 1: The A4 instrument on HEAO-1 consisted of seven NaI/CsI phoswich scintillation detectors surrounded by massive CsI shielding. Additional organic scintillators provided for monitoring of charged particles. The instrument was designed for observing cosmic sources of x- and gamma-radiation at energies between 13 keV and 10 MeV.

also returned.

The HEAO-1 satellite was launched into a highly circular orbit of inclination 23° and altitude 400 km. At this altitude drag from the tenuous exosphere was sufficient to cause reentry after only 580 days. Total mass of HEAO-1 was three tons.

ORIGIN AND VARIABILITY OF BACKGROUND

HEAO-1 and its experiments were subjected to two dominant sources of ionizing radiation: cosmic rays and geomagnetically trapped particles. The cosmic ray flux was modulated on two time scales as the satellite moved through the geomagnetic field: twice per orbit the satellite crosses the geomagnetic equator where the shielding effect of the geomagnetic cutoff is greatest; and once per day the orbital plane lies closest to the geomagnetic equator. Geomagnetically trapped particles were encountered only at one point in the orbit, the so-called south atlantic anomaly (SAA), where the inner belts dip unusually low. This region was traversed six to eight times per day. Changes of a few tens of kilometers in the satellite orbit caused large changes in the particle fluences encountered because of strong gradients of the population of trapped particles. This is immediately evident in Figure 2, where periodic changes are seen to modulate an overall strong decline which results from orbital decay. The wobbles about the general decline were resolvable into two periodicities at 28 and 43 days. These are probably identifiable with cyclic changes of two elements of the orbit: the argument of the perigee at 28 days and the right ascension of the ascending node at 48 days. The 28-day motion moves the satellite apogee in and out of the SAA region; more intense fluxes are encountered at apogee. A modulation with this period of 11.5% was measured. The 48-day orbital precession causes a variation of the local time of maximum daily SAA encounter with synodic period 43 days. Since the satellite spin axis was kept fixed near the sun, the satellite orientation with respect to the local vertical at SAA also varied with this period. Proton fluxes on the inner edge of the belt are strongly anisotropic with apex to the local west. The particle monitors are not completely isotropic in response, but are partially shielded by the massive instrument and spacecraft. Therefore a modulation of the observed proton flux may be expected. In fact, a highly significant 43-day modulation of amplitude 5.6% was measured, with maximum at times when the SAA was encountered near the dawn terminator.

After allowance for these two periodic terms, the monitored daily particle fluences showed no further signs of periodic structure, and evidently the considerable remaining variability was random. Sizeable random variability of particle populations has been reported by Stassinopoulos[3].

In Figure 2 normalized counting rates for two MED detectors are also shown. These rates are measured for a strong internal background line of I^{123}, which results from spallation interactions of protons with the I^{127} of the detector. This activity decays with a half-life of 13 hours, making it effectively an instantaneous monitor of proton dosage at the detector on the time scales of interest here. Moreover, the decay is predominantly by internal capture, and the energy of 193 keV is a sum peak of a 159 keV gamma from the Te^{123} daughter and the Te x-rays, thus the gammas leaking from the CsI shields, where I^{123} is also produced, are separately counted. The wobbles on the counting rate curves of the two detectors are obviously slower than for the particle monitor. In fact, Fourier analysis shows that the two same frequencies are present, but

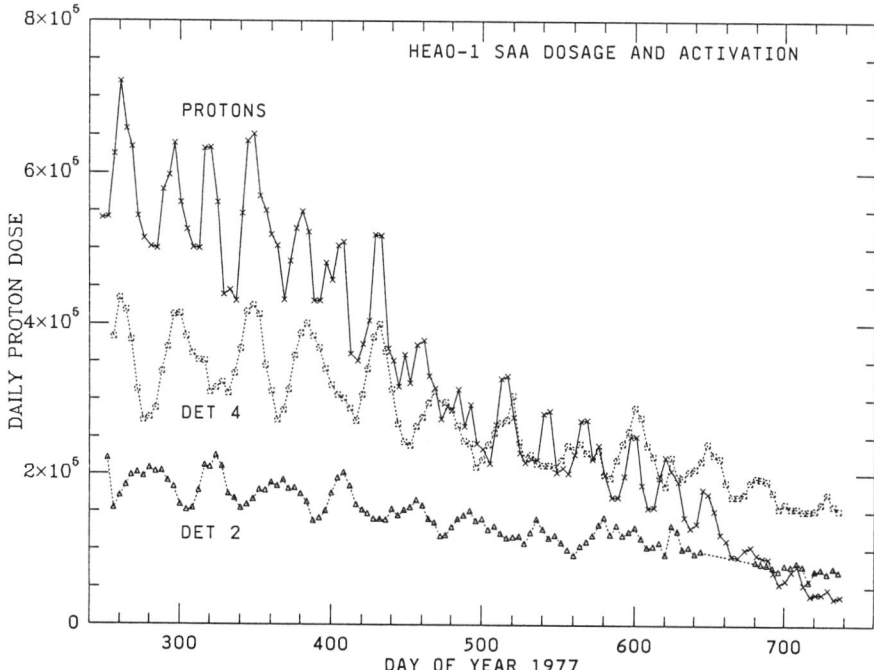

Figure 2: Daily dosage of trapped protons in the SAA region and counting rates of two detectors in a strong internal background line at 191 keV, arbitrarily normalized. The overall decline is due to orbital decay. The wobbles result from 28-day cyclic changes in the eccentricity, and a 43-day precession of the orbital plane.

the relative strengths are reversed: in both detectors the 28-day modulation has an amplitude near 6% and the 43-day modulation has amplitude 15%.

It is also obvious that the wobbles in the two MED curves of Figure 2 are anti-correlated. Analysis shows that the Detector 2 maxima occur when the SAA passages are encountered at the sunset terminator, and the Detector 4 maxima at the sunrise terminator. The explanation is simple, involving only the orientation of the spacecraft and anisotropies in the particle fluxes and detector shielding. Each MED has a direction of minimum shielding (plan view, Figure 1) corresponding to a particle threshold of roughly 80 MeV. Because the spacecraft spin axis is fixed near the sun, this direction of minimum shielding is in the solar direction for detector 4 and antisolar for detector 2. At the sunrise terminator the westward peak of particle flux is pointed at the thin spot of detector 4's shielding, and likewise for detector 2 at the sunset terminator.

In Figure 2 it is seen that the MED count rates drop only by a factor of two in the course of the mission, whereas the particle flux drops by an order of magnitude. An extra, relatively constant activation component is therefore present, which is attributable to cosmic rays. It is clear that at the beginning of the mission activation from trapped particles and cosmic rays occurs in roughly equal amounts.

Variability of internal background was, of course, not confined to longer time scales, but was quite pronounced in the course of a day. The SAA region was traversed in six to eight successive satellite passes out of the daily fifteen. Cosmic ray fluxes also varied with time as different geomagnetic latitudes were traversed, but the dynamic range of this variability was probably not more than a factor of two. The net daily variability of MED internal background is indicated in Figure 3. The daily variability of internal background is shown in more detail in Figure 4, which shows the response of the HED detector and one shield segment to trapped particle doses and to cosmic rays, whose flux is parametrized by the McIlwain L-parameter.

SEPARATION OF ACTIVITIES

Given the nearly complete data on particle input from the particle monitors, it was decided to attempt a straightforward modeling of the detector response, at least for the species which were recognized as major contributors to the background. For

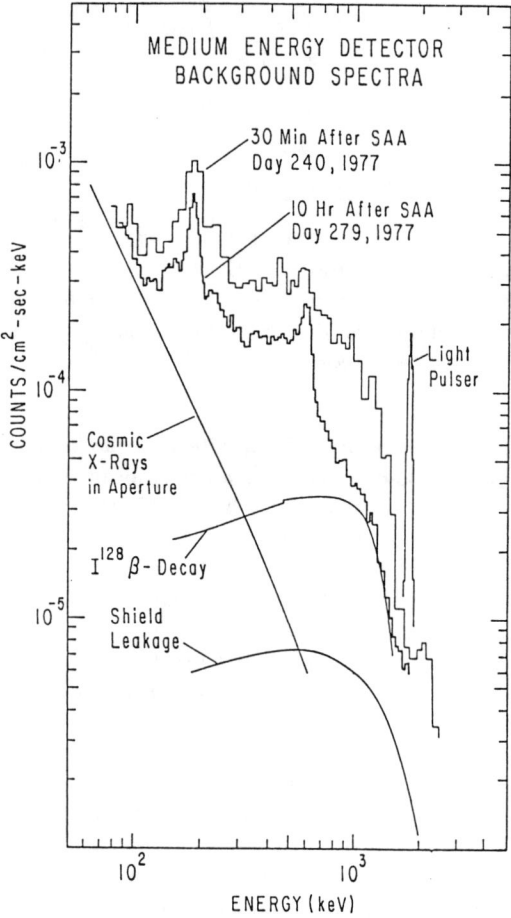

Figure 3: MED background spectra at two different times of day, showing activation from resulting from exposure to SAA trapped protons and its decay. Except below 100 keV the detector background is dominated by this internal radioactivity.

selected half-lives, then, calculations were made of model functions representing the instantaneous activity at half-lives τ_i resulting from particle dosage $D(t)$

$$A_i(t) = \int_{t_0}^{t'} D(t')exp(-(t-t')/\tau_i)dt'.$$

A distinction must be preserved between SAA and cosmic ray dosages, because of differing spectra and average energies. For convenience, the McIlwain L-parameter was used to represent the cosmic-ray dosage; the low counting rate of the 1 cm^2 particle monitors introduced unnecessary statistical noise. Only four half-lives were employed: 25 min (I^{128}), 2.1 hr (Xe^{123}), 13.3 hr (I^{123}), and 4.2 days (I^{124}). A prompt response ($\tau_i < 25$ min) function was also added for the cosmic ray (L) input. A prompt response term was not needed for SAA dosage because the intrument was turned off at these times. The behavior of these functions during a typical day of operation is shown in Figure 5. Only the prompt and 25 min terms were retained for the cosmic ray input; longer half-life terms tended to average out the 45 min variations of L. Computer limitations necessitated the elimination of one of the remaining seven terms (including constant). This was achieved by joining the 0.4 hr functions for SAA and cosmic-ray input, using

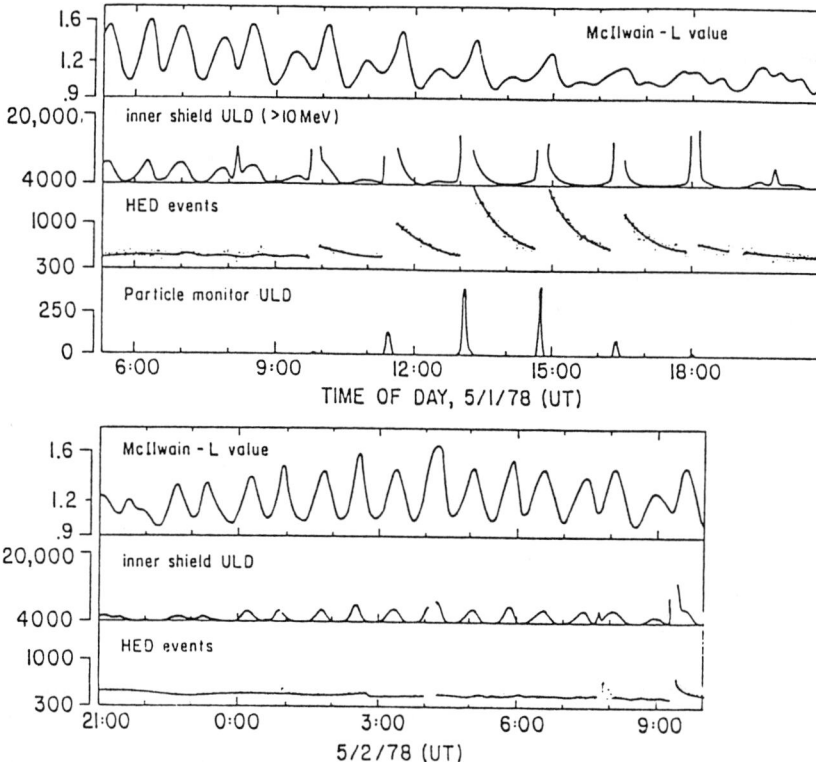

Figure 4: A typical daily history of fluxes of activating particles and the responses in selected instrument registers. Fluxes of cosmic rays and equilibrium neutrons are proportional to McIlwain L. The inner shield ULD is sensitive to the incident particles, while the HED events counter shows a delayed response dominated by the decay of I^{128}.

a normalization parameter, determined to be of the order of unity, which represents the relative efficiency of activation by the two particle populations. Although selected because of certain dominant activities, this set of half lives rather evenly sampled the range of time scales amenable to analysis; the half lives of 0.4, 2.1, 13 and 100 hrs stand in the ratios 5:6:7.

The analysis then was reduced to a very standard problem of multiple linear regression. For selected data intervals one accumulated the usual weighted sums of observed counting rates, $R(t_i, E)$, their products with the activity models, $A_j(t_i) \cdot R(t_i, E)$, and the cross-products $A_j(t_i) \cdot A_k(t_i)$. The matrix equations

$$((\sum A_j \cdot A_k)) \cdot (\sum S_k(E)) = (\sum A_j \cdot R(E))$$

are then solved for $S_k(E)$, the activation spectra at each half life, plus time-average spectrum $S_0(E)$. The detector background spectrum is then

$$B(t, E) = S_0(E) + \sum_{i=1}^{5} S_i(t, E)$$

As a practical matter, the various sums were accumulated with the spectral data during the preparation of the sky map for each detector. Great circle scans divided into ninety sky bins were accumulated on each four-day interval of the mission. For

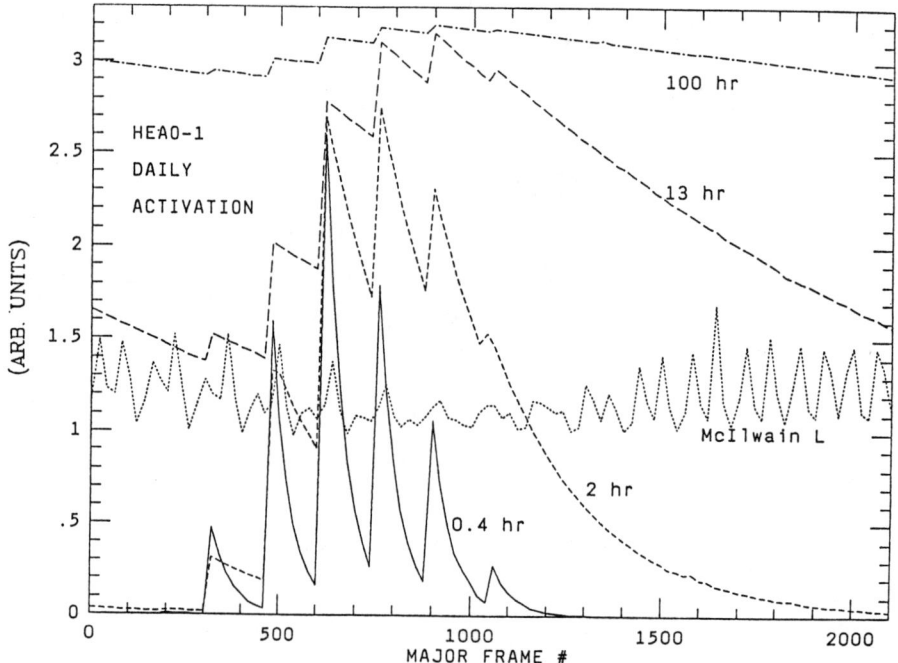

Figure 5: Model activity functions at several half-lives for SAA dosages received in a typical day of operation. The curves represent populations of radioactive species induced in the detectors by trapped particles in the SAA region. The "McIlwain L" curve represents prompt response to cosmic ray dosages. The longer half-life terms fail to decay fully in a day.

each 4-day sky strip a single background spectral file was generated by adding all of the blank sky data, using as selection criterion a distance of 20° or more from the galactic plane, Cen A, NGC 4151 and 3C273. Since the problem is linear, 4-day spectral files could be added in any combination before solution.

Solution, however, was not immediate. Two difficulties were present: statistical precision on shorter data stretches was rather limited for solution into six separate spectra; and the five activity functions were far from independent basis functions. Fortunately, data from the entire mission was sufficient to overpower the first difficulty. Overcoming the second difficulty required some careful manipulations. As a glance at Figure 5 will show, the 13 hour and 100 hour functions are very similar when allowance is made for arbitrary offsets and multiplicative factors S_k. The 13 hour and 2 hour functions are also similarly coupled. The degree of similarity can be quantified in a cross-correlation factor obtained from the error matrix following solution. The procedure employed was to find a subset of the data for which the cross-correlation factor between the 13 hour and 100 hour terms was as small as possible. A reduction from -0.98 to -0.70 was eventually obtained, which was taken as an indicator of a sufficiently stable and reliable solution. The best-fit solution was then used to subtract the 100 hour dependence from the data. Thus the most pathological term in the problem was isolated and removed. The procedure was then re-applied using the remaining terms to find the 2 hour activity spectrum and subtract it. The four terms still included were now sufficiently independent so that a good direct solution was obtained. The fitting was performed unconstrained, and a good indicator of the correctness of the fits was given by the absence of large, unphysical, negative excursions of the spectra, and also by their approach to zero at high energies. A final but very important verification comes from the expected spectral shapes. For example, the 25 min spectrum showed a very convincing representation of the I^{128} beta spectrum with end point at 2.12 MeV.

By this method reliable activation spectra, Figure 6 for example, were obtained. The non-independence of the basis functions adds an important benefit: each function collects, in the fit, all the activity at a wide range of related half-lives. If 13 hour and 100 hour activity can be distinguished only with difficulty, then 13 hour and, say, 25 hour activity must appear almost identical. Thus the solution extends to a full range of variability between a few minutes and many days, provided that the initial assumptions of SAA and cosmic-ray activation are correct. It is evident, though, from Figure 2 that the particle monitors gave an incorrect measure, by up to 30%, of the SAA dosage at the detectors, at least for time scales near one month. This difficulty could be overcome, however, by selection of 4-day spectral files only at times of peak activation.

The activation spectra thus obtained were examined for spectral features, usually line features but occasionally beta continua, and identifications of radioactive nuclides attempted. The results are in Table I. Many observed lines are broader than the instrumental resolution, indicating blending. Most of the identifications are already familiar from earlier work[4,5], but some new identifications are proposed. Many of the lines result from electron capture and therefore include the K-level energy of the daughter. A few lines are attributed to inelastic scattering of neutrons.

This work was supported by NASA grant NAG-8-499.

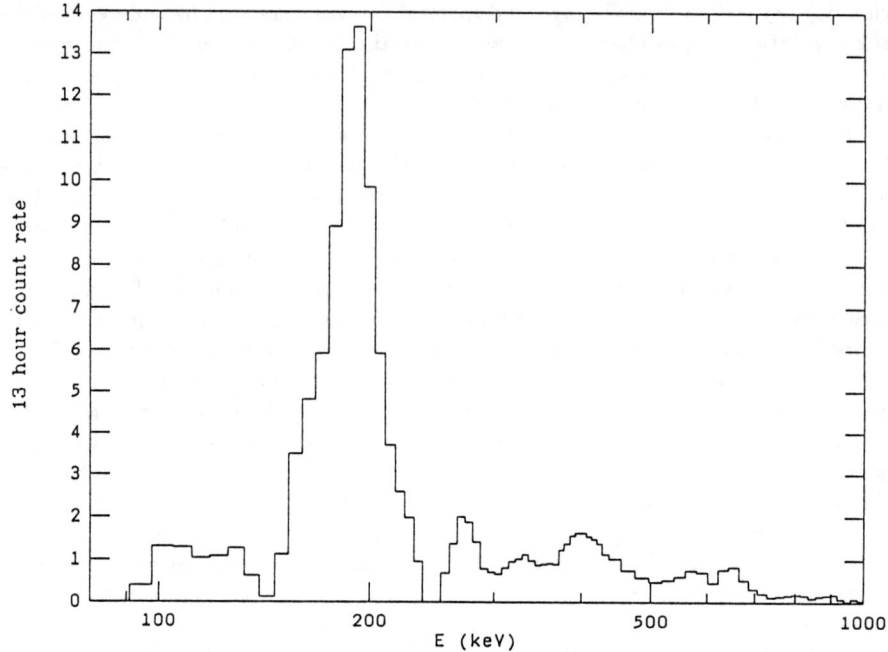

Figure 6: Activation spectrum at 13 hour nominal half life. Line and continuum deexcitation spectra are present from nuclides with half lives between roughly 5 hours and 30 hours.

REFERENCES

1. Trombka, J. I., Dyer, C. S., Evans, L. G., Bielefeld, M. G., Seltzer, S. M., and Metzger, A. E., Ap. J., 212, 925 (1977).

2. Jung, G. V., unpublished dissertation, UCSD (1986).

3. Stassinopoulos, E. G., these Proceedings (1988).

4. Dyer, C. S., and Morphill, G. E., Astrophys. Space Sci., 14, 243, (1971).

5. Fishman, G. J., unpublished NASA report NASA CR-150237 (1977)

Table I Background Lines and Identifications

Observed				Candidate			
Energy (keV)	FWHM (keV)	Flux (arb)	Time	Energy (keV)	Species	Half Life	Reference
63		21104	long	63	I^{125}	60d	5
71	8	1550	short	75	Sb^{122m}	4.2m	
88		1452	long	89	Te^{123m}	104d	4
				90	Sb^{120}	5.8d	
91	13	1311	short	88	Ag^{109m}	40s	
93	10	561	13h	85	Xe^{125}	17h	
				90	Xe^{122}	20h	
125		208	2h	127	Cs^{134m}	3h	
125	8	281	13h	127	Cs^{134m}	3h	
				125	Cs^{127}	6d	4
142	22	2424	short	141	Xe^{125m}	1m	
150	11	4000	long	145	Te^{125m}	58d	5
				159	Te^{123m}	104d	4
				160	Sn^{117m}	14d	4
151	8	232	2h	148	Xe^{123}	2h	4
179	9	736	short	172	Xe^{127m}	1m	2
191	11	1385	2h	191	Sb^{117}	2.8h	4
191	10	3472	13h	192	I^{123}	13h	4
				187	Xe^{125}	17h	4
191	11	11123	long	196	Xe^{129m}	8d	4
206	8	1004	short	203	I^{127*}	prompt	
219	15	955	13h	217	Xe^{125}	17h	4
244	11	958	2h	245	I^{121}	2h	5
244	16	5398	long	247	Te^{123m}	117d	5
				230	Sb^{120}	6d	4
				233	Xe^{127}	34d	4
246	22	2500	short	253	Xe^{125m}	1m	2
282	18	1006	13h	273	Xe^{125}	17d	4
291	17	1003	short	285	Cs^{128}	3m	4
296	12	1890	long	293	Te^{121m}	154d	5
328	11	155	13h				
353	50	3192	short	356	Cs^{133*}	prompt	
358	18	206	13h	350	Xe^{122}	20h	2
389	24	3263	long	386	I^{126}	13d	4
				375	Xe^{127}	36d	4
				405	Xe^{127}	36d	4
				392	Sn^{113}	120d	4
414	35	1381	13h	406	Cs^{127}	6h	4
				436	Cs^{127}	6h	4

Table I (cont.)

OBSERVED				CANDIDATE			
Energy (keV)	FWHM (KeV)	Flux (arb)	Time	Energy (keV)	Species	Half Life	Reference
433	26	1282	short	417	I^{127*}	prompt	
				439	Na^{23*}	prompt	
469	45	71	25m	440	I^{128}	20m	4
				470	Cs^{128}	3m	
489	22	2151	short	472	Na^{24m}	.02s	2
				470	Cs^{128}	3m	4
				511	e^+	many	
498	25	330	13h	511	e^+	many	
506	48	7887	long	511	e^+	many	
				507	Te^{121}	17d	4
581	52	1320	4d	603	I^{124}	4d	4
				566	Sb^{122}	2.8d	4
589	22	2810	long	605	Te^{121}	17d	5
				575	Te^{121}	17d	5
597	51	737	13h				
651	21	1440	4d	668	Cs^{132}	6.2d	4
663	29	9669	long	668	Cs^{132}	6d	4
				667	I^{126}	13d	4
				697	I^{126}	13d	4
693	21	328	13h	675	Te^{119}	16h	
817	50	140	long				
976	20	73	long				
1283	99	1532	long	1275	Na^{22}	2.6y	
1365	45	363	long	1450	I^{126}	13d	5

BACKGROUND OBSERVATIONS on the SMM HIGH ENERGY MONITOR at ENERGIES > 10 MeV

D. J. Forrest
Physics Dept., University of New Hampshire, Durham, NH 03824

ABSTRACT

The background rate in any gamma ray detector on a spacecraft in near earth orbit is strongly influenced by the primary cosmic ray flux at the spacecraft's position. Although the direct counting of the primary cosmic rays can be rejected by anti-coincident shields, secondary production can not. Secondary production of gamma rays and neutrons in the instrument, the spacecraft and the earth's atmospheric are recorded as background. We have used a 30 day data base of 65.5 second records to show that some of the background rates observed on the Gamma Ray Spectrometer can be ordered to a precision on the order of 1%. This ordering is done with only two parameters, namely the cosmic ray vertical cutoff rigidity and the instrument's pointing angle with respect to the earth's center. This result sets limits on any instrumental instability and also on any temporal or spacial changes in the background radiation field.

INTRODUCTION

The field of gamma ray astronomy can be characterized as a struggle to detect low signal fluxes in the presence of high backgrounds. As a consequence gamma ray astronomers spend considerable time and effort on background studies. Such studies have two objectives. First, an understanding of the background rates in existing instruments can hopefully be used in the design of new instruments to reduce or remove these backgrounds. Second, background studies of existing instruments can be used to predict these backgrounds in data sets with known or suspected signal fluxes and hence separate out these signals within the total observed rates.

The presence of background within a source data set can have two consequences. High background rates always result in a statistical uncertainty, even if the average background rate is known exactly. To be detected, a signal must always be significantly larger than this statistical "noise". However a second and more uncertain consequence of high backgrounds is the possibility of non-statistical variations of the background rates within the source observation interval. Hence not only must gamma ray astronomers minimize the background rates, but they must also be prepared to demonstrate by observational tests that the inevitable remaining background is well behaved and predictable.

In this paper we specifically study the behavior and predictability of the background in a orbiting gamma ray experiment. Our long term objective is to make the maximum use of the very long SMM Gamma Ray Spectrometer (GRS) data base for non-solar gamma ray sources. As an example of our needs we note the GRS has obtained over 10^6 seconds of observations for every source near the ecliptic plane since its launch in 1980. Given this observing time, our statistical precision in energy bands with average background rates of 1 and 10 counts/sec is 0.1% and 0.03% respectively.

© 1989 American Institute of Physics

The use of the GRS for the detection and study of non-solar narrow gamma ray lines has already been demonstrated for the galactic disk emission of ^{26}Al [1] and positron annihilation[2]. In a similar way the GRS data was used by Letaw[3] to quantify more than 20 gamma ray lines from the earth's atmosphere, a case where the background for one type of observation became the signal for another. The desire to extend these types of studies beyond gamma ray line observations to continuum emissions places strong demands on our understanding of the GRS backgrounds.

INSTRUMENT AND DATA BASE

The Gamma Ray Spectrometer[4] is one of seven instruments on the Solar Maximum Mission which was designed to study the emission from solar flares. The GRS consists of three separate detector systems covering the energy range 10 keV to over 140 MeV. One of these detector systems is the HIGH ENERGY MONITOR (HEM). The HEM is sensitive to gamma rays from 10 to 200 MeV and neutrons from 25 to 500 MeV. The detector is made up of two elements, a NaI front element and a CsI back detector. It has a maximum sensitive area for gamma rays incident from the forward (i.e. solar) direction of 300 cm^2 at 20 MeV and 55 cm^2 for neutrons at 200 MeV[5]. The output from the HEM allows us to separate the counting rates of the front NaI from the back CsI detector. Calculations[5] show that the gamma ray sensitivity of the NaI detector has a strong angular dependence with maxima in the forward or solar direction while the poorly shielded CsI has only a weak angular dependence. Because of the large sensitive area both detectors have a relatively high background rate and hence good statistical precision even for short observing times.

The background rates from the HEM were used for this study mainly because the nuclear activation produced in the S/C and instrument by passages through the South American Anomaly region of trapped particles does not extend above 10 MeV. The difficulty caused by this transient background source, which is very important at energies <10 MeV, has been discussed by Share et al.[1,6] and Dunphy et al.[7].

The data base of HEM rates used in this study consisted of 65.5 second records of all the data in the 30 day interval in April 1980. Each record contained the accumulated counts in several HEM energy bands and the spacecraft position as determined from the ephemeris data produced by the S/C on-board computer. The ordering parameters we have used are the Cosmic Ray Vertical Cut-off Rigidity and the GRS-Earth viewing or Aspect angle. The rigidity for each record was determined from the S/C position with a look-up table[8]. The Aspect angle is defined as the angle between the vectors from the earth's center to the S/C and from the earth's center to the sun. Hence the aspect angle is ~ 0° near orbital "noon" and near 180° near orbital "midnight".

After processing all the data in this 30 day interval, we obtained 20,532 records of useful data when the instrument was in it's normal data mode. Data was lost when the instrument was in low power while in the SAA region, when in a in-flight-calibration data mode and during intervals with bad or noisy telemetry. The distribution of the useful data with respect to Aspect Angle and Rigidity is shown in Table I. It is clear that not all aspect angle and rigidity intervals are sampled equally. Specifically the band of missing data with aspect angles from 121° to 130° is due to the two in-flight-calibration intervals which take place each orbit, one just after orbital sunset and one just before orbital sunrise.

Figures 3 and 4 show, in a similar way, the rigidity dependence for the 1865 records with $-50° < \text{ASP} < -30°$. The data in these two figures suggests that the rigidity dependence is similar among the four background rates. We test this by fitting this data, after binning them up in 1 GV intervals, to a function of the form:

$$\text{Rate} = \alpha + \frac{\beta}{\text{Rig}}.$$

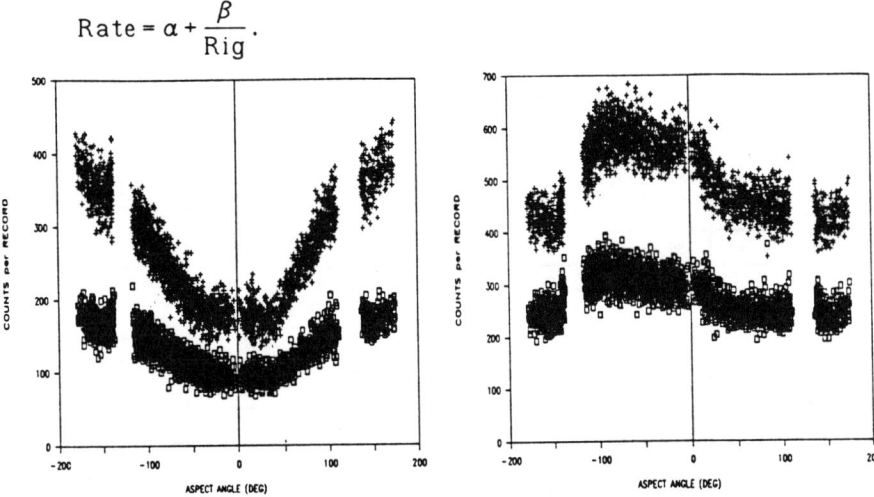

Fig. 1. NaI record rates plotted against aspect angle. (+) NaI(10-25 MeV) and (open sq) NaI(25-100 MeV).

Fig. 2. CsI record rates plotted against aspect angle. (+) CsI(10-25 MeV) and (open sq) CsI(25-100 MeV).

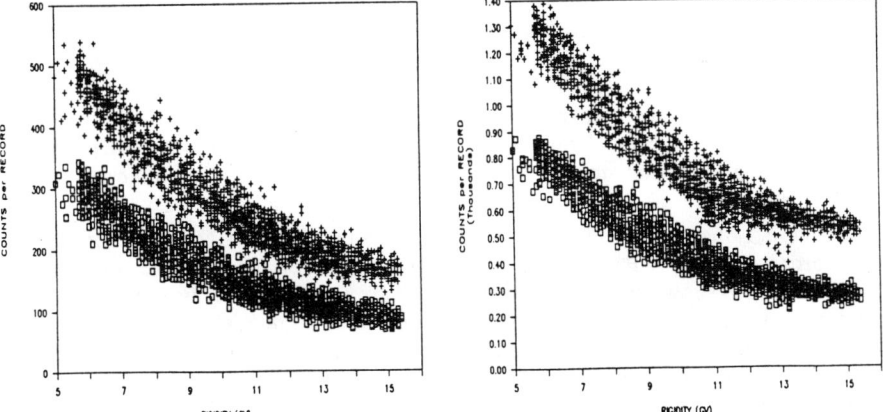

Fig. 3. NaI record rates plotted against rigidity. (+) NaI(10-25 MeV) and (open sq) NaI(25-100 MeV).

Fig. 4. CsI record rates plotted against rigidity. (+) CsI(10-25 MeV) and (open sq) CsI(25-100 MeV).

Table I Distribution of the number of 65.5 sec records with Normal Data Mode

Rig(GV) ------ Asp(deg)	4.1 - 5.0	5.1 - 6.0	6.1 - 7.0	7.1 - 8.0	8.1 - 9.0	9.1 - 10.0	10.1 - 11.0	11.1 - 12.0	12.1 - 13.0	13.1 - 14.0	14.1 - 15.0	15.1 - 16.0
1-10	0	0	0	0	1	8	16	46	68	90	79	20
11-20	0	1	3	11	19	22	32	105	194	208	145	25
21-30	6	22	26	30	51	57	113	241	368	275	119	19
31-40	10	42	129	122	162	179	244	325	290	216	135	19
41-50	2	105	181	158	156	165	273	297	310	213	98	25
51-60	9	38	132	118	109	138	224	278	310	244	119	22
61-70	8	53	71	105	121	119	213	268	316	244	152	12
71-80	9	39	85	92	92	126	207	300	319	288	150	27
81-90	10	48	77	78	81	105	192	281	349	297	162	35
91-100	8	42	78	97	98	103	177	290	354	296	178	39
101-110	13	45	79	68	76	101	190	273	318	302	189	37
111-120	4	28	33	52	25	30	46	88	105	94	50	7
121-130	0	0	0	0	0	0	0	0	1	0	0	0
131-140	24	31	35	46	46	44	73	106	122	98	19	1
141-150	0	17	26	36	53	66	86	157	172	196	149	19
151-160	0	9	38	39	46	57	98	146	146	172	152	52
161-170	0	0	0	1	7	17	35	149	159	139	83	13
171-180	0	0	0	0	0	0	1	30	77	65	17	0

ANALYSIS

In this section we present a study of the ordering and reproducibility of the counting rates of a subset of the data shown in Table I. Specifically we have studied the aspect angle dependence of all records within fixed rigidity intervals and the rigidity dependence of all records within fixed aspect angle intervals.

Figure 1 shows the aspect angle dependence of the front NaI detector rates for all records with 13.0 < RIG < 14.0 GV in the two energy bands 10-25 MeV and >25 MeV. Figure 2 shows the same data for the back CsI detector. In these two figures the aspect angle has been given negative values when the S/C is on the western or orbital sunrise side of the earth and positive values when on the eastern or orbital sunset side. We see in Figure 1 that the relatively well shielded NaI detector rates have a minimum when pointed away from the earth (i.e. asp = 0°) and are reasonably symmetric around 0° and 180°. The poorly shielded CsI rates, on the other hand, show much less aspect angle dependence and are asymmetric. The asymmetry of the CsI rates must be produced by cosmic ray interactions in the massive spacecraft material. This material is not distributed symmetrically with respect to the GRS instrument. Figures 1 and 2 each contain 3250 records.

The results are shown in Figure 5, where the observed and fitted rates have been normalized to the the average rate, <C>. The similarity of the fitted parameters shows that the background rate >10 MeV is well behaved. However it is also clear that the equation does not accurately describe the data within the precision of the measurements. The reason for this may be either that the form of the equation is incorrect, or that the rigidity values associated with the S/C position is not correct for the SMM orbit.

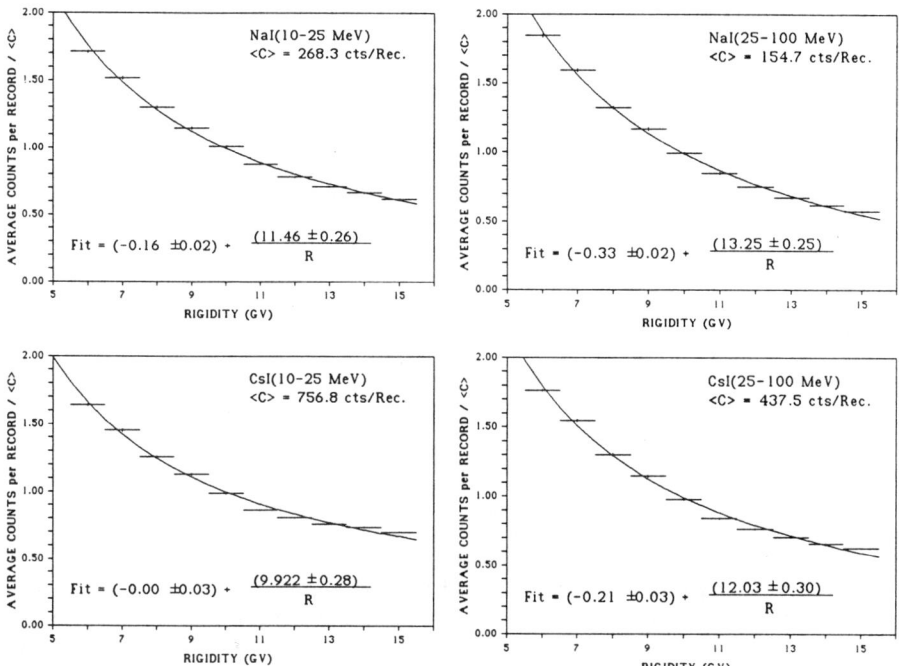

Fig. 5. The four panels show the four rates binned up in 1 GV intervals. The error bars reflect the observed 1 sigma errors of the mean rate within each interval.

In order to demonstrate the precision with which these rates can be defined, we show in Figure 6 the frequency distribution of the 153 record rates contained within the rigidity band 11.0 to 12.0 GV and the aspect band -40o to -31o. In the four panels of Figure 6, the step histogram curve is the observed distribution while the two smooth curves represent a.) the expected Gaussian distribution from a "true" constant rate with a value <C> and b.) the Gaussian distribution fitted to the observed data. The expected fractional standard deviation, σ_f, of the record population ranges from 4% to 9%, while the observed σ_f range from 9% to 13%. In each case the excess broadening is entirely accounted for by the finite rigidity and aspect bands, which introduce a broadening of ~ 7 - 9% that must be added in quadrature with the statistical σ_f. We conclude from this test that the combined variations produced by either instrument instabilities or by real changes in the gamma ray background over this 30 day interval must be less than or comparable to the σ_f of our experimentally determined mean rates. Among the four rates considered here, the mean rates range from 122 ± 2 to 623 ± 5 counts per record and have a σ_f range of 0.8% to 1.6%.

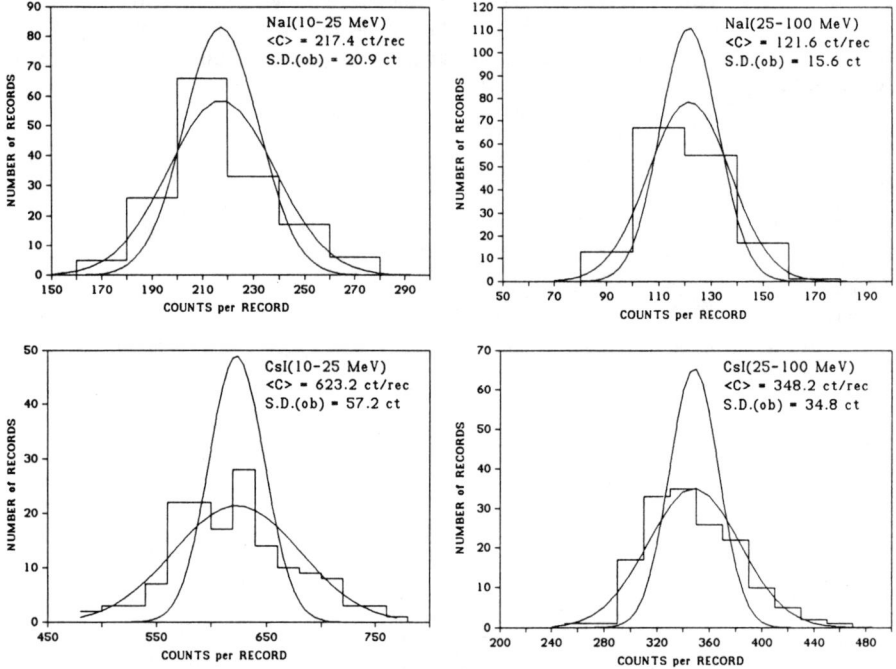

Fig. 6. The four panels show the observed distribution of 153 records in each of the four energy bands. The narrow smooth curve is the expected distribution from a fixed rate = <C>. The broader smooth curve is the distribution obtained form the observations. The measured standard deviation (S.D.(ob)) is given in each panel.

RESULTS

We have used a 30 day data base of 65.5 second HEM records to show that at least some of the background rates observed on GRS can be ordered to a precision comparable to our statistical uncertainty of ~1%. This ordering is done with only two parameters, namely the cosmic ray vertical cutoff rigidity and the instrument's pointing angle with respect to the earth's center. This result shows that a serious effort to construct a mathematical model which can accurately describe and predict at least some of the GRS rates is warranted. The effort required to construct such a model, which will be valid over a 8 year interval, will be formidable. The data presented here shows that both the aspect angle and rigidity dependence is complex and can be different from one output to another even when they cover the same energy band. It is also clear that there are real long-term variations which are not noticeable in a 30 day interval. Some of these long-term variations have been presented by Kurfess et al.[9].

ACKNOWLEDGMENTS

I would like to acknowledge discussion on the background problem on SMM with my colleagues W. T. Vestrand at UNH, G. Share at NRL and E Rieger at MPI. This work was supported by NASA Grant NAG 5720 at UNH.

REFERENCES

1. G. H. Share, R. L. Kinzer, J. D. Kurfess, D. J. Forrest, E. L. Chupp and E. Rieger, Ap. J., 292, L61-L65 (1985).
2. G. H. Share, R. L. Kinzer, J. D. Kurfess, D. C. Messina, W. R. Purcell, E. L. Chupp, D. J. Forrest and C. Reppin, Ap. J. , in press (1988).
3. J. R. Letaw, G. H. Share, R. L. Kinzer, R. Silberberg, E. L. Chupp, D. J. Forrest and E. Rieger, Adv. Space Res., 6, 133-137 (1986).
4. D. J. Forrest, E. L. Chupp, J. M. Ryan, M. L. Cherry, I. U. Gleske, C. Reppin, K. Pinaku, E. Rieger, G. Kanback, R. L. Kinzer, G. Share, W. N. Johnson and J. D. Kurfess, Solar Phys., 65, 15, (1980).
5. J. F. Cooper, C. Reppin, D. J. Forrest, E. L. Chupp, G. H. Share and R. L. Kinzer, 19th ICRC, 5, 474 (1985).
6. G. H. Share, R. L. Kinzer, M. S. Strickman, J. R. Letaw, E. L. Chupp, D. J. Forrest and E. Rieger, These Conf. Proc., (1988).
7. P. P. Dunphy, D. J. Forrest, E. L. Chupp and G. H. Share, These Conf. Proc., (1988).
8. J. E. Humble, D. F. Smart and M. A. Shea, 16th ICRC, 5, 303 (1979).
9. J. D. Kurfess, G. H. Share, R. L. Kinzer, E. L. Chupp, D. J. Forrest and C. Reppin, These Conf. Proc., (1988).

LONG-TERM VARIATIONS IN THE GAMMA-RAY BACKGROUND ON SMM

J. D. Kurfess, G. H. Share, R. L. Kinzer,
W. N. Johnson, J. H. Adams, Jr.
E.O. Hulburt Center for Space Research
Naval Research Laboratory, Washington, D. C. 20375

E. L. Chupp, D. J. Forrest
Univ. of New Hampshire, Durham, NH 03824

C. Reppin
Max-Planck-Institute for Physics and Astrophysics, Garching, FRG

ABSTRACT

Long-term temporal variations in the various components of the background radiation detected by the gamma-ray spectrometer on the Solar Maximum Mission are presented. The SMM gamma-ray spectrometer was launched in February, 1980 and continues to operate normally. The extended period of mission operations has provided a large data base in which it is possible to investigate a variety of environmental and instrumental background effects. In particular, several effects associated with orbital precession are introduced and discussed.

INTRODUCTION

The background associated with gamma-ray detectors in low-earth orbit has been investigated in considerable detail (for a review see reference 1). The primary sources of background are due to the gamma-ray background environment (cosmic and atmospheric) and background components produced by cosmic ray or geomagnetically trapped particles in the detector or other spacecraft components.

The data base obtained with the SMM satellite extends over many years and has shed new light on some long-term variations in the background rates which have not been previously observed or understood. These longer-term variations may be important in properly understanding certain observations obtained on past missions or in the planning of future instruments.

SMM GAMMA RAY EXPERIMENT

The SMM satellite was launched on 14 February 1980 into a 570-km circular, 28.5° inclination orbit. The orbit has decayed to about a 485-km altitude at the present time (November, 1987). The satellite carries seven instruments to study solar activity and has been continuously pointed toward the Sun except during short periods of offset pointings for observations of Cyg X-1 and SN1987A.

© 1989 American Institute of Physics

The SMM gamma ray experiment has been described in detail by Forrest et al.[2] The main detector consists of an array of seven 3"-dia. by 3"-thick NaI detectors. Pulse-height spectra in the 0.3-9 MeV energy range are accumulated every 16.38 seconds using a 476-channel quadratic PHA. The NaI detectors are shielded toward the rear by a 3"-thick CsI crystal and on the sides by a 1"-thick CsI annular shield. A plastic anticoincidence detector covers the front aperture. This aperture is also blocked by a ~0.9 g/cm^2 thick graded passive absorber to prevent pulse pile-up at low energies during intense flares. The field-of-view of the instrument is about 120 degrees FWHM.

BACKGROUND EFFECTS

Several of the effects which contribute to variations in the background environment are listed below. The time scales indicated in this table are appropriate for satellites in a 28.5° inclination, 500-km altitude orbit.

Effect	Variable Parameter	Time Scale
A. Cosmic Ray Irradiation	Rigidity, Latitude	Orbital (94 Min)
B. SAA Transits	Geographical Position	Orbital (94 min) 6-7 Transits Daily
C. Detector Orientation	Earth Angle	Instrument Dependent
D. Orbital Precession	System Orientation in SAA	53.5 days wrt stars 46.7 days wrt Sun
E. Orbital Decay	Altitude	Years
F. Orbital Eccentricity	Altitude	32.8 Days
G. Solar Modulation	Solar Magnetic Field	11 Years
H. Latitude/Precession Correlation	Geomagnetic Latitude, Detector Orientation	1/2 Precession Period

The first two items on this list are those most familiar as producing background variations in orbiting gamma-ray detectors. The first is the changing background due to the varying geomagnetic latitude and the associated change in the cosmic ray fluence with vertical cutoff rigidity. This results in two background maxima and two background minima during each orbit. The second is the increased background associated with the satellites' daily passes through the South Atlantic Anomaly (SAA) on each of typically 6-7 orbits/day. These phenomena as they relate to SMM are discussed, in part, by Share et al.[3] and will not be discussed further here.

Because the major factors which govern the background rates in the SMM detector are the time-dependent cosmic ray flux, the daily passes through the SAA, and to a lesser extent the aspect of the instrument with respect to the Earth's atmosphere, the pulse height spectra were grouped as a function of these parameters. Specifically, data were binned in a 10x10x10 array which included 10 vertical cutoff rigidity bins, 10 approximately 1-hour intervals covering about 10 hours each day which were free of SAA passes, and 10 equal bins of Earth Angle defined by the angle between the viewing axis of the detector and the center of the Earth. Using these data, it was then convenient to study effects associated with these several background parameters.

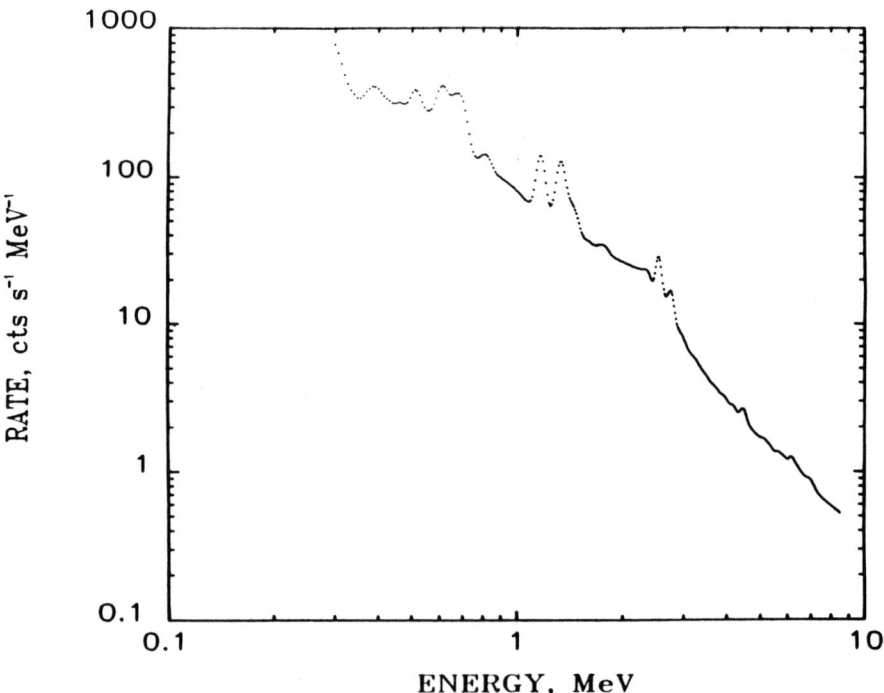

Figure 1. A gamma-ray spectrum obtained by the SMM spectrometer acquired during 3 x 10^{+6} seconds of orbits without SAA passes.

Figure 1 shows a long term background spectrum obtained with the SMM detector. This spectrum shows many features associated with different components of the background. Several lines associated with radioactive spallation components are seen in the 0.3-1.0 MeV region. The energy resolution of the NaI spectrometers is not adequate to separate out many of the lines, so that very often, e.g. in the 0.6- 0.7 MeV region, the two broad features seen are, in reality, due to a mixture of several spallation products from iodine and cesium.[3] The individual features can often be separated by studying the different temporal behavior which results from the different half-lives of the lines in question. For example, the intensity of lines associated with very long half-lives (years) build up slowly with time and are independent of the short-term variations in the irradiation exposures due to the SAA, while the shorter-lived features will reflect the time dependence of the exposures. Other features include calibration lines at 1.17 MeV and 1.33 MeV and a sum peak at 2.5 MeV associated with an onboard ^{60}Co calibration source. Also, prominent lines in the atmospheric albedo spectrum are seen in the 4.4-7.1 MeV region. In all, about twenty atmospheric lines have been identified and have been discussed by Letaw et al.[4]

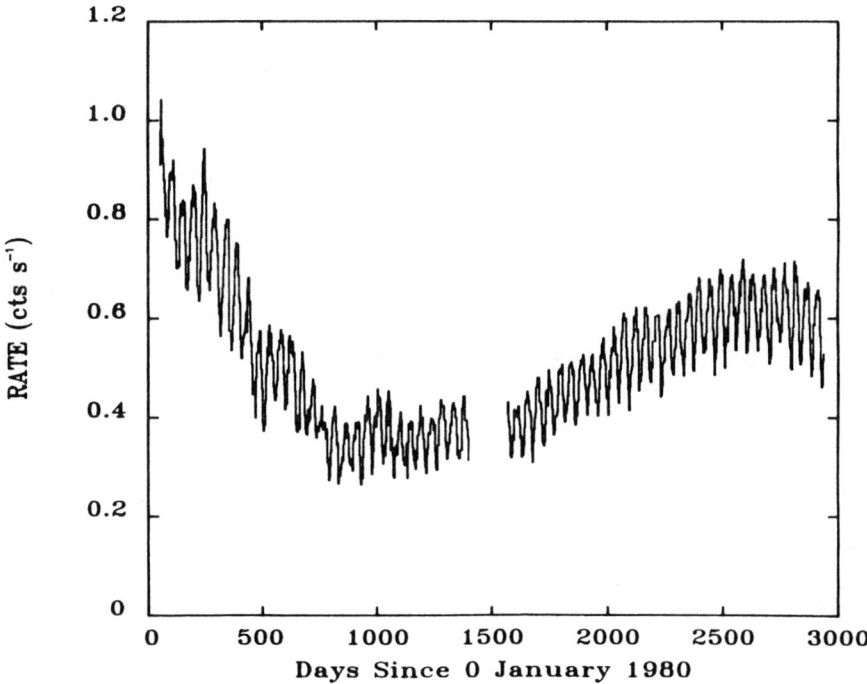

Figure 2. Time history of the ^{24}Na line at 2.75 MeV from 3-day summed spectra. The interval of no data near day 1500 represents a period when the SMM tape recorders were turned off prior to the SMM Repair Mission in April, 1984.

The background feature at 2.75 MeV, associated with the 15-hr decay of ^{24}Na produced by (n,alpha) reactions on aluminum in the spacecraft, illustrates several of the longer-term variations of background lines. Figure 2 shows the intensity time history, obtained from fits to 3-day summed spectra, for this line during the nearly eight-year period of the mission. Several aspects of the figure should be noted. The first is the decrease in the intensity of this line during the first 2-3 years of the mission. This is a result of the decrease in the orbital altitude from the initial altitude of 570-km after launch to an altitude of about 510-km after 1000 days. The SAA irradiation to energetic protons is reduced by about a factor of two for each 50-km reduction in altitude in this altitude range. As the period of solar minimum was entered, the upper atmospheric density decreased significantly, and further reduction in the altitude has been minimal. Associated with the reduced atmospheric density is an increasing intensity of trapped protons in the SAA at the spacecraft altitude, resulting in the gradual increase in the intensity from day 1500 until solar minimum which occured near day 2600. Because ^{24}Na has a 15-hr half life, the intensity of this line is dominated by spallation production which occured in the latest SAA passes. This is clearly evident by the altitude dependence of the line.

Another striking feature of the intensity of the ^{24}Na emission is the periodic nature of the emission with a period of about 46-48 days. The amplitude of the variation is also surprisingly large, and is seen to be roughly proportional to the intensity in the line, i.e. confirming the association with spallation production in the SAA.

The 47-day periodicity results from the anisotropic distribution of trapped protons as the satellite passes through the SAA coupled with the solar-oriented nature of the SMM satellite. As discussed by Watts et al.[5] large anisotropies occur in the distribution of trapped particles at the lower altitudes in the SAA. The trapped protons at the SMM altitude have a pitch angle distribution which is strongly peaked near 90 degrees and with arrival directions weighted heavily from the West. This results from the rather large gyroradii of trapped protons (about 100 km at 300 MeV) and the fact that particles arriving from the East have trajectories with guiding centers at altitudes below the SMM altitude. Hence these particles are preferentially lost by scattering in the atmosphere. West/East flux ratios of 5-10 are expected[5]. Thus, the background environment induced in a detector is very dependent on the orientation of the detector relative to local shielding as the instrument passes through the SAA.

For SMM, the satellite is always pointed towards the Sun, so that the orientation of the instrument as the satellite passes through the SAA varies as the orbital plane of the satellite precesses. The orbital precession results in a westward motion of the ascending node, Ω, of the orbit as given by Wertz[6]:

$$d\Omega/dt = -2.064 \times 10^{14} * A^{-3.5} * \cos(I)$$

where A is the radius (in km) of the orbit, I is the inclination, and $d\Omega/dt$ is given is degrees/day. For SMM the orbital precession period declined (as the orbital altitude decayed) from about 55 days at the beginning of the mission to about 53 days after several years. This represents a 6.7 degree/day motion of the orbital plane in inertial space. However, since SMM is always pointed toward the Sun, the change in the orbital plane relative to the direction of the Sun is 0.985 degrees/day higher. This results in a precession period relative to the Sun which declined from 48 days to 46 days over the period of the SMM observations. On this period, the orientation of the SMM as it passes through the SAA undergoes a slow rotation relative to the SAA. This phenomenon can be understood by reference to Figure 3. This results in a varying exposure to the radiation environment and the associated periodic behavior seen in Figure 2.

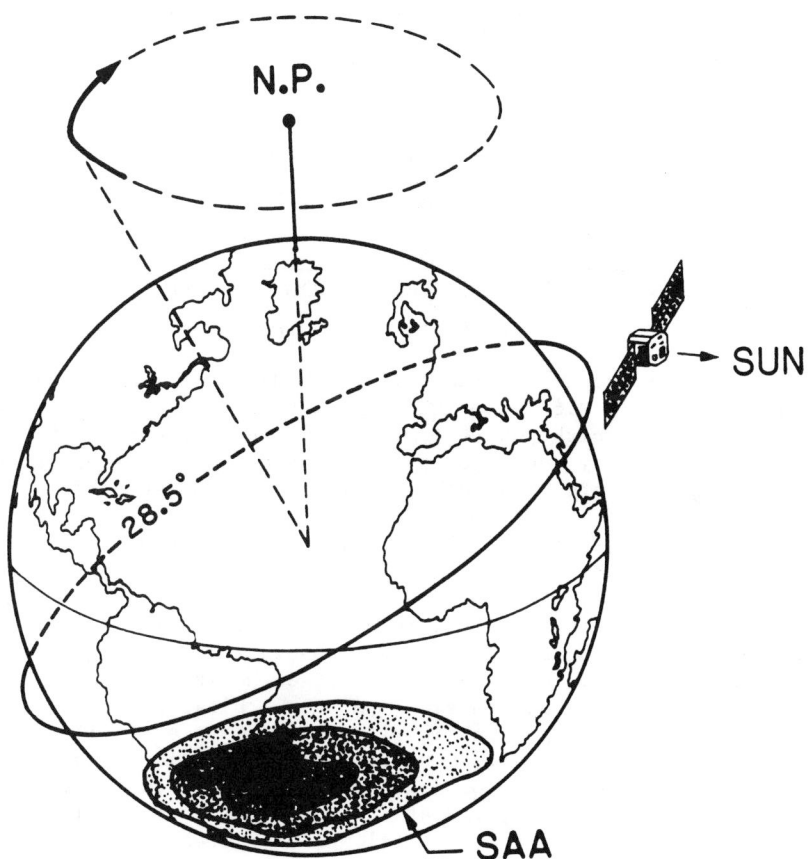

Figure 3. Illustration indicating the changing locations of the SAA exposures relative to SMM orientation. The SMM spacecraft, oriented toward the Sun, passes through the SAA with a different relative orientation as the orbit precesses. The satellite passes through the SAA with the instruments pointed eastward when the SAA is located near the sunrise terminator (as shown in this figure), and passes through the SAA facing westward when, about 24 days later, the SAA passes are located near the sunset terminator.

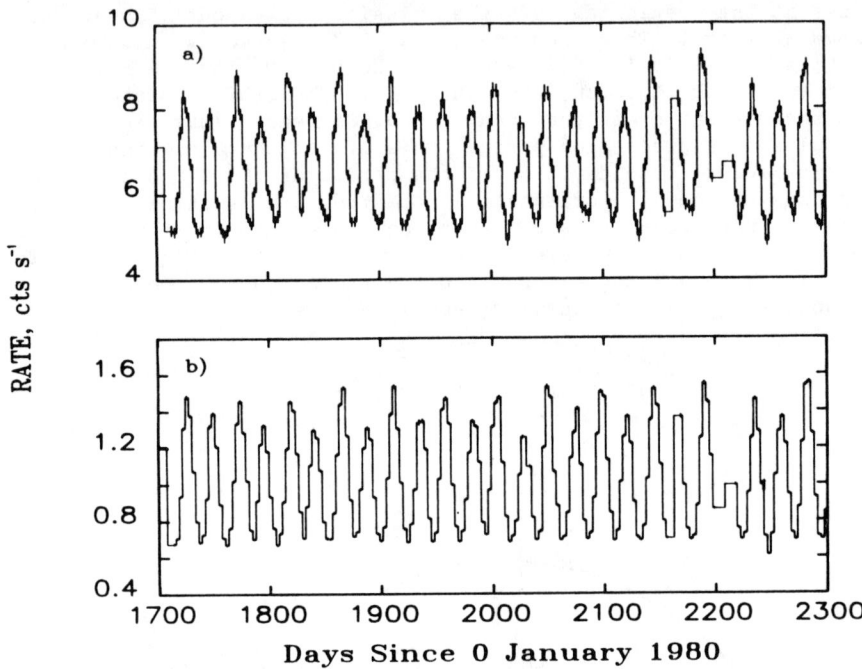

Figure 4. Variation of the a) 0.511 MeV positron annihilation line and b) the 5-8.5 MeV integral rate, obtained during viewing periods near the night/day terminator. The intensity exhibits a 24-day periodicity due to a correlation with the geographic latitude of the satellite (cosmic ray produced background).

A related phenomenon is observed due to a correlation between geographic latitude and Earth Angle, which is related to orbital precession. Recall that all SMM data are binned in Earth Angle bins. This results, for example, in all daytime (solar) data being acquired with Earth Angles (absolute magnitude) which are between 70 and 180 degrees (since the horizon is depressed by about 20° at 500 km altitude). As the orbital plane precesses, (see Figure 3) the geographic latitude of the satellite as it crosses the noontime meridian undergoes a periodic behavior. For example starting with the orbital plane as indicated in Figure 3, the noontime location occurs at the northernmost orbital latitude (+28.5 degrees). As the orbital plane precesses, the noontime position moves through the equator, to southern latitudes, and then re-traces back through the equator to the northern latitudes again. This results in a movement through two cycles of geomagnetic cutoff rigidity. Thus, background phenomena associated with cosmic rays undergo two minima and two maxima during this period; i.e. they will exhibit a temporal dependence with a period of one half the precession period. This can be seen in Figure 4, which shows the temporal behavior of the background line at 0.511 MeV, and 5-8.5 MeV gamma ray rates, the intensities of which are dominated by atmospheric albedo during the selected periods and which reflect the instantaneous cosmic ray flux. Both of these features show an ~24-day periodicity. Note that this period is seen in part due to a selection effect; that is, selection of data based on solar pointing, hence Earth angle, results in a correlation with the geomagnetic latitude of the satellite.

Finally, we note that a long-term variability due to orbital eccentricity, included in Table 1, was not significant for SMM which has a nearly circular orbit. Satellites in an eccentric, low-earth orbit will be exposed to a changing SAA irradiation as the rotation of the line of apsides results in SAA transits at varying altitudes. For a 28.5 degree, 400 x 500 km orbit, this period will be ~31 days with minimum irradiation occurring when perigee moves through the latitudes of the SAA.

DISCUSSION

We have presented evidence for several long-term temporal variations in the background observed by the gamma-ray detector on the SMM. Some of these variations have been studied in considerable detail previously. The variations associated with the precession of the orbital plane represent, to our knowledge, new variabilities that were not previously described or understood. These variations result, in part, from the anisotropic exposure to high-energy protons as SMM passes through the SAA and in part to selection effects associated with the SMM observations and their resulting correlations with parameters such as latitude. As Figures 2 and 4 indicate, these effects can be quite large; showing modulations in affected lines of 50% and higher. It is clearly very important to take such variations into consideration when analyzing satellite data for sources and for source variability. These have been of considerable concern for the SMM observations of celestial sources such as the diffuse ^{26}Al and 0.511 MeV observations from the galactic plane[7,8] and the recent detection of radioactivity from SN1987A.[9]

It might appear that these concerns are unique to SMM. This is not the case. Previous high-energy missions such as HEAO-1 and HEAO-3 operated in scanning modes wherein the scan plane was always perpendicular to the Earth-Sun line. Hence, the detectors always viewed in the plane perpendicular to the Earth-Sun line rather than parallel to it. This can result in anisotropic exposures during passes through the SAA analogous to those experienced by SMM. The slow scan undertaken by those spacecraft in the view plane may have served to reduce the effect somewhat. The roll periods used were also close to a major fraction of an orbital period, which could result in a correlation between viewing direction and rigidity. It is important in the analysis of data from these and other missions, as well as in the planning of observations on future missions, to consider the possible impact of these phenomena. Clearly, the trend toward imaging detectors and rapid modulation of source and background observations will go a long way toward minimizing these effects.

ACKNOWLEDGEMENTS

The authors acknowledge the considerable assistance of Dan Messina in the analysis of the data reported here. This work was supported at NRL by NASA contract S-14513-D and the Office of Naval Research, NASA Grant NAG 5720 at UNH, and BMFT contract 010K 017-za/ws/wrk 0275:4 at MPE.

REFERENCES

1. C.S. Dyer, J.I. Trombka, S.M. Seltzer, and L.G. Evans, Nucl. Instr. Meth. 173, 585 (1980).

2. D.J. Forrest, E.L. Chupp, J.M. Ryan, M.L. Cherry, I.U. Gleske. C. Reppin, K. Pinkau, E. Reiger, G. Kanbach, R.L. Kinzer, J.D. Kurfess, G.H. Share an W.N. Johnson, Solar Physics 65, 15 (1981).

3. Share, G.H., R.L. Kinzer, M.S. Strickman, J.R. Letaw, E.L. Chupp, D.J. Forrest and E. Rieger, this conference (1988).

4. Letaw, J.R., G.H. Share, R.L. Kinzer, R. Silberberg, E.L. Chupp, D.J. Forrest, E. Reiger "Satellite Observations of Atmospheric Nuclear Gamma Radiation," submitted to Journal of Geophysical Research-Space Physics (1988).

5. Watts, J.W., T.A. Parnell, and H.H. Heckman, this conference (1988).

6. Wertz, J.R., Spacecraft Attitude Determination and Control (D. Riedel, 1978) Ed. J.R. Wertz.

7. G.H. Share, R.L. Kinzer, J.D. Kurfess, D.J. Forrest, E.L. Chupp and E. Reiger, Astrophys. J. (Lett) 292, L61 (1985).

8. G.H. Share, R.L. Kinzer, D.C. Messina, W.R. Purcell, J.D. Kurfess, E.L. Chupp, D. J. Forrest, and E. Rieger, to be published in Astrophys. J. 336 (1988).

9. Matz, S.M., G.H. Share, M.D. Leising, E.L. Chupp, W.T. Vestrand, W.R. Purcell, M.S. Strickman and C. Reppin, Nature 331, 416 (1988).

COMPARISON OF BACKGROUNDS IN *OSO-7* AND *SMM* SPECTROMETERS AND SHORT-TERM ACTIVATION IN *SMM*

P. P. Dunphy, D. J. Forrest, and E. L. Chupp
University of New Hampshire, Durham, NH 03824

G. H. Share
E. O. Hulburt Center, Naval Research Laboratory,
Washington, DC 20375

ABSTRACT

The backgrounds in the *OSO-7* Gamma-Ray Monitor and the Solar Maximum Mission Gamma-Ray Spectrometer are compared. After scaling to the same volume, the background spectra agree to within 30%. This shows that analyses which successfully describe the background in one detector can be applied to similar detectors of different sizes and on different platforms.

The background produced in the *SMM* spectrometer by a single trapped-radiation belt passage is also studied. This background is found to be dominated by a positron-annihilation line and a continuum spectrum with a high energy cutoff at ~ 5 MeV. The decay times of the major background components are determined and possible sources are discussed.

OSO-7 AND *SMM* BACKGROUNDS

The *OSO-7* Gamma-Ray Monitor (GRM) and the Solar Maximum Mission (*SMM*) Gamma-Ray Spectrometer (GRS) are almost ideal for comparing backgrounds in their respective spacecraft. Both instruments use NaI(Tℓ) scintillators for their central detectors, and, except for size, are similar in geometry and construction.

The GRM [1] (Figure 1) is a single 7.6-cm diameter×7.6-cm thick NaI(Tℓ) crystal, mounted on a photomultiplier tube. This unit is surrounded by a 4-cm thick CsI(Na) scintillator shield, except for the forward aperture, which is covered by a 0.5-cm thick CsI(Na) slab. *OSO-7* was launched in 1971 September into an eccentric orbit with an altitude excursion of 330 km to 575 km and an inclination of 33°. The GRM operated until 1972 December.

The *SMM* GRS [2] (Figure 2) is an array of seven 7.6-cm diameter×7.6-cm thick NaI(Tℓ)/photomultiplier main channel detectors and a 25-cm diameter×7.6-cm thick Cs(NaI) back detector/shield. The detectors are shielded on the sides by a 2.5-cm thick CsI(Na) annulus and on the front and back by 1-cm thick plastic scintillators. The satellite was launched in 1980 February into a nearly circular orbit at an altitude of 570 km and an inclination of 28°.

An analysis of the GRM background has been reported previously [3,4]. The orbital eccentricity of the *OSO-7* spacecraft resulted in periodic deep penetration into the Van Allen radiation belts, when the latitude of the apogee was in the South Atlantic Anomaly. This, in turn, resulted in background enhancements from activation with a period of ~ 38 days. Unlike *SMM*, which is continuously Sun-oriented, the *OSO-7* instrument section spun with a 2-s period in a plane that always

© 1989 American Institute of Physics

included the Sun. Data were accumulated separately during two 0.5-s intervals when the GRM was pointed at the solar and anti-solar quadrants. Total spectral accumulation time was 3 minutes.

GRM and *SMM* GRS background spectra are plotted in Figure 3. For this comparison no special selection of the data was attempted. The GRM data were accumulated approximately 1 year after launch. At the time the data were collected, the latitude of the apogee was farthest south (i.e., during a period of maximum activation from radiation belt penetration). Also, the axis of the detector viewing-quadrant was within 90° of the "anti-Earth" direction for all data.

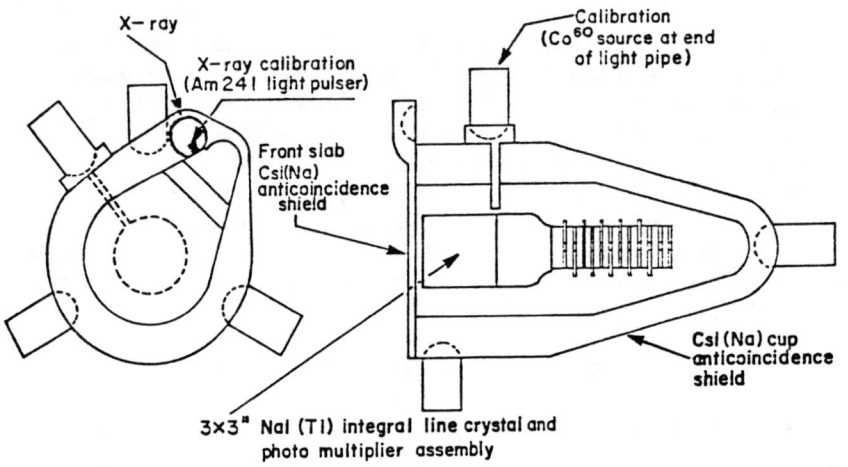

Fig. 1. Drawing of the *OSO*-7 Gamma-Ray Monitor (GRM). Also shown are an associated X-ray detector and ^{60}Co and ^{241}Am calibration sources.

The background in the *SMM* GRS is described in other papers at this Conference [5,6]. The *SMM* GRS spectrum was accumulated after $\sim$ 7 years in orbit. These data were taken in the "singles" mode (i.e., events occuring in more than one detector in the GRS array were eliminated) for a better comparison with the single *OSO*-7 detector.

As can be seen from Figure 3, the GRM and *SMM* GRS spectra show excellent agreement when scaled to the same volume (a factor of 7). The GRS background rate is only $\sim$ 30% higher than the GRM rate at energies below 800 kev. Above this energy, the agreement is even better. We conclude that accurate background models derived from one detector can be confidently applied to others if variables, such as detector materials, spacecraft orbits, etc., are "reasonably" close.

SHORT-TERM ACTIVATION EFFECTS IN *SMM* GRS

By "short-term effects," we refer to the background variations during a single orbit caused by the decay of radioisotopes produced during passage of the *SMM* spacecraft through the radiation belt. Our motivation for this study is the magnitude of the effect - variations in the background of about a factor of 5 on the time scale of one orbit. This large an effect, if ignored, can certainly compromise the analysis of solar flare or cosmic data; therefore, determining the magnitude and duration of the effect is important.

For the present analysis, we have used a single belt passage and orbit of *SMM* for "source" measurements. For corresponding "background" measurements, we have used data from an orbit 2 days later. The source and background time histories are aligned in time, using the vertical cutoff rigidity at the spacecraft as a benchmark. This procedure, to first order, provides a background with the same altitude, rigidity, and pointing aspect as the source data. A simple subtraction of the background from the source, scaled by livetime, yields a net spectrum due to the belt passage. The spectrum can then be studied in terms of time behavior and the presence of identifiable γ-ray lines.

Figure 4 shows the net spectrum for the orbit following the belt passage. The dominant features in this spectrum are: (1) an activation line at 0.515 ± 0.003 MeV and (2) a continuum which extends as high as ~ 5 MeV. Additional line features are evident, but are weaker than the line at 0.5 MeV by an order of magnitude or more. The line at 0.5 MeV is apparently due to positron annihilation, probably from positron emitters produced outside the NaI scintillators. Offset of the fitted line energy from the expected 0.511 MeV may be due to an additional nearby line feature (e.g., at 0.528 MeV from ^{115}Sb with a half-life of 31 minutes [7]). The continuum is most likely caused by β emitters produced in the scintillators.

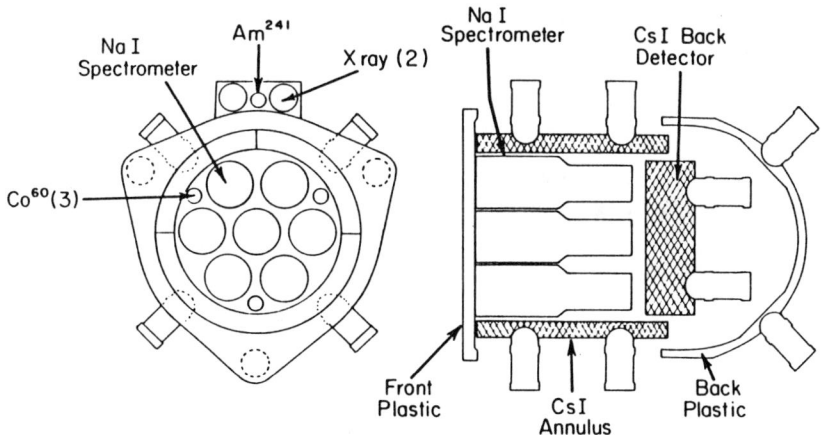

Fig. 2. Drawing of the SMM Gamma-Ray Spectrometer (GRS). Also shown are two X-ray detectors and calibration sources.

To study the time history of the annihilation line, we have plotted its intensity versus time over one orbit in Figure 5. This figure shows that the time behavior is well fit by a single exponential function. The exponential decay gives a best-fit half-life of 22.8 ± 3.1 minutes. Effects from this source, therefore, become negligible after two or more 90-minute orbits. The annihilation line could originate from a number of sources. One possible contributor of positrons is ^{52}Mn (from a metastable state with a half-life of 21.1 minutes), produced from ^{56}Fe in the steel casing of the scintillators. However, lines from excited states of the ^{52}Mn daughter are not seen at an intensity comparable to that of the annihilation line. The positron emitter ^{119}I has a half-life of 19 minutes and can be produced by spallation of I and Cs in the annulus and back shields. Its daughter produces a 0.258 MeV line, which is below the GRS threshold and thus would not be visible. However, the end-point energy of the positron is 2.4 MeV, so the CsI shields, with thresholds at 100 kev, have high vetoing efficiencies against this source. Another possible source is ^{11}C, which decays with a branching ratio of 99% via positron emission with no associated γ-ray lines. The half-life of this decay is 20.4 minutes. A potential source of ^{11}C is ^{12}C in the plastic charged-particle shield covering the front face of the SMM GRS. The end-point energy of the positron is 0.96 MeV, while the shield threshold is set at ~ 0.5 MeV. Thus, a large fraction of these positrons could come to rest and annihilate in the plastic, undetected by the shield veto electronics. The argument that the shield is a source of the annihilation line is supported by an excess shield counting rate which decays with a half-life of ~ 20 minutes after the belt passage. The shield rate is also consistent with the line intensity seen in the main channel detectors.

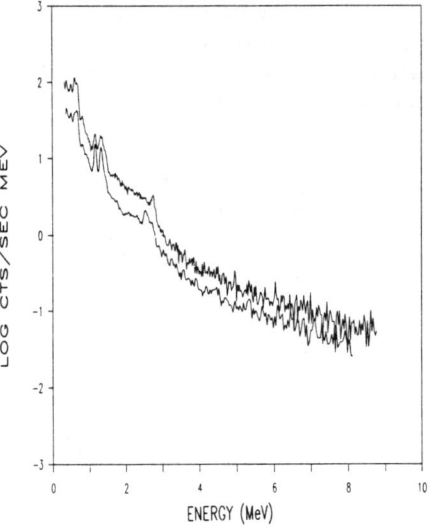

Fig. 3. Comparison of SMM GRS "singles" background spectrum (scaled by a factor of 1/7) and the OSO-7 GRM background spectrum. The scaled GRS spectrum (upper curve) is offset by a factor of 2 for clarity. The spectra agree within 30%.

The time behavior of the continuum has been analyzed in a similar way. We have divided the continuum into two energy ranges : 0.3 – 2.0 MeV and 2.0 – 5.0 MeV. This was motivated by the change in slope in the continuum near 2 MeV and the negligible background above 5 MeV. The time histories for these two ranges are shown in Figure 6. Attempts to model the decays with single exponentials gave poor fits. A model using the sum of two exponentials gave a substantially better fit. The

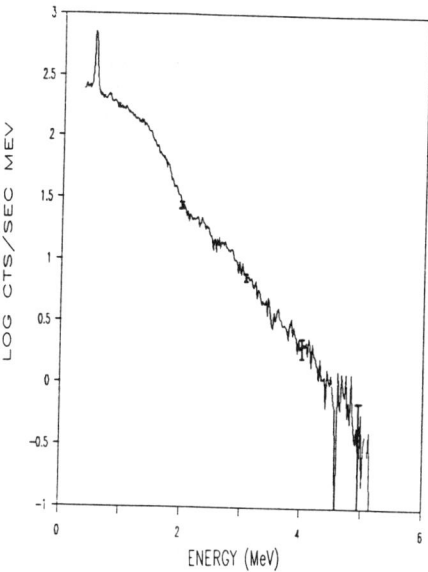

Fig. 4. Plot of the net background spectrum produced in the GRS following a passage through the trapped-radiation belt. The dominant features are a continuum and a line at 0.515 ± 0.003 MeV. Data are sampled from a single orbit. The accumulation livetime is 2822 s.

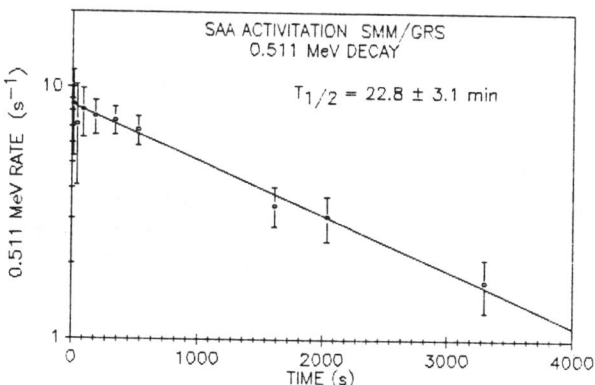

Fig. 5. Data points show the time history of the strong line at 0.5 MeV (assumed to be due to positron annihilation). Data gaps are caused by the GRS going into a calibration mode. The curve is a fit using a single exponential decay.

nominal half-lives from the two-component fits are 11.2 minutes and 40.6 minutes for the 0.3 – 2.0 MeV range, and 6.25 minutes and 71.3 minutes for the 2.0 – 5.0 MeV range. It is likely, however, that the continuum is the summation of many decays, especially spallation products of ^{127}I, including isotopes of I, Xe, and Sb.

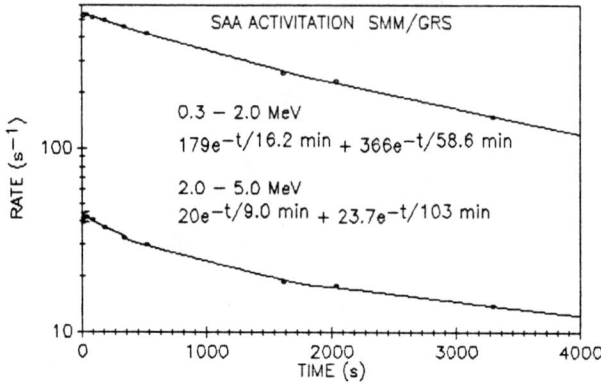

Fig. 6. Data points show the time history of the continuum in two energy ranges. The curves are fits using a sum of two exponentials as shown in the figure.

CONCLUSIONS

A comparison of the background spectra in the *OSO-7* GRM and the *SMM* GRS indicates that the backgrounds agree within a factor smaller than 2 when scaled by volume. The backgrounds, to that extent, are not sensitive to details of materials and construction of the spacecraft. Therefore, background models which successfully describe one detector should be directly applicable to similar detectors.

A study of activation of the *SMM* GRS from a single radiation belt passage shows that the background is dominated by a positron-annihilation line and a β-particle continuum. Other γ-ray lines are weaker by a factor of 10 or more. The annihilation line may be due primarily to ^{11}C produced in the plastic charged-particle shield. If this is the case, then its magnitude in other detectors would depend on the presence of a plastic shield and the setting of its threshold. The β-particle continuum does not extend above ~ 5 MeV. These components of the background are short-term in the sense that they dominate for times within tens of minutes of a belt passage, but become negligible after ~ 200 minutes.

ACKNOWLEDGEMENTS

The authors thank Mary Chupp, Karen Dowd and Ben Hersh for their help in producing this paper. This work was supported by NASA contracts NAS5-28609 at the University of New Hampshire and S-14513-D at the Naval Research Laboratory.

REFERENCES

1. P. R. Higbie, E. L. Chupp, D. J. Forrest, and I. Gleske, IEEE Trans. Nucl. Sci. NS − 19, 606 (1972).
2. D. J. Forrest, E. L. Chupp, J. M. Ryan, M. L. Cherry, I. U. Gleske, C. Reppin, K. Pinkau, E. Rieger, G. Kanbach, R. L. Kinzer, J. D. Kurfess, G. H. Share, and W. N. Johnson, Solar Phys. 65, 15 (1980).
3. P. P. Dunphy, D. J. Forrest, E. L. Chupp, and C. S. Dyer, Proc. 14th ICRC 9, 3116 (1975).
4. C. S. Dyer, P. P. Dunphy, D. J. Forrest, and E. L. Chupp, Proc. 14th ICRC 9, 3122 (1975).
5. G. H. Share, R. L. Kinzer, M. S. Strickman, J. R. Letaw, E. L. Chupp, D. J. Forrest, and E. Rieger, this volume (1988).
6. J. D. Kurfess, G. H. Share, R. L. Kinzer, W. N. Johnson, J. H. Adams Jr., E. L. Chupp, D. J. Forrest, and C. Reppin, this volume (1988).
7. C. S. Dyer, J. I. Trombka, S. M. Seltzer, and L. G. Evans, Nucl. Instr. Meth. 173, 585 (1980).

INSTRUMENTAL AND ATMOSPHERIC BACKGROUND LINES OBSERVED BY THE SMM GAMMA-RAY SPECTROMETER

G.H. Share, R.L. Kinzer, M.S. Strickman
E.O. Hulburt Center for Space Research
Naval Research Laboratory
Washington D.C. 20375

J.R. Letaw
Severn Communications Corp.
Severna Park, Md. 21146

E.L. Chupp and D.J. Forrest
University of New Hampshire
Durham, New Hampshire 03824

E. Rieger
Max Planck Inst. for Extraterrestrial Physics
Garching, FRG

ABSTRACT

Preliminary identifications of instrumental and atmospheric background lines detected by the gamma-ray spectrometer on NASA's Solar Maximum Mission satellite (SMM) are presented. The long-term and stable operation of this experiment has provided data of high quality for use in this analysis. Methods are described for identifying radioactive isotopes which use their different decay times. Temporal evolution of the features are revealed by spectral comparisons, subtractions, and fits. An understanding of these temporal variations has enabled the data to be used for detecting celestial gamma-ray sources.

INTRODUCTION

Advances in the field of gamma-ray astronomy are dependent on understanding the various sources of background which dominate the spectra measured by space instrumentation. Detailed calculations and measurements have been performed in order to identify various spectral features observed by such instruments[1-3]. In this paper we describe an analysis of the background line radiation observed by the SMM gamma-ray spectrometer during its eight-year lifetime.

The Gamma-Ray Spectrometer (GRS) on NASA's Solar Maximum Mission satellite (SMM) has been in operation since February 1980. It consists of seven 7.6 cm x 7.6 cm cylindrical NaI detectors contained within an anticoincidence shield. A drawing of the basic elements of the instrument is included in a companion paper[4] in these Proceedings. Details of the instrument are described by Forrest et al.[6] The spectrometer has an active gain stabilization system which, along with the other sub-systems of the instrument, has performed flawlessly since launch. This has enabled spectra to be accumulated over long time periods without degradation in resolution.

As a result of this stable performance, the SMM data have been used to detect diffuse Galactic gamma-ray lines from decay of ^{26}Al and from positron annihilation[7,8], and to detect ^{56}Co-decay gamma-rays from SN1987A[9]. These celestial measurements could not have been made without an understanding of the various background lines and their temporal variations. These time variations are discussed in accompanying papers[4,5] in these Proceedings. In this paper we summarize our current understanding of the different line features appearing in the background spectrum from 0.3 to 8.5 MeV. Lines from both instrumental and spacecraft background and the Earth's atmosphere are identified.

Spectra revealing both the instrumental and atmospheric lines are plotted in Figure 1. Both spectra were accumulated during SAA-free orbits from 1980 to 1983 for vertical geomagnetic cutoff rigidities between 4 and 11 GV. The upper spectrum includes data accumulated with the detector pointed within 36° of Earth center and exhibits both instrumental and atmospheric γ-ray lines and continua. The lower plot reveals the spectrum of atmospheric gamma-radiation and has been produced by subtracting a comparably accumulated spectrum, with the detector pointed away from the

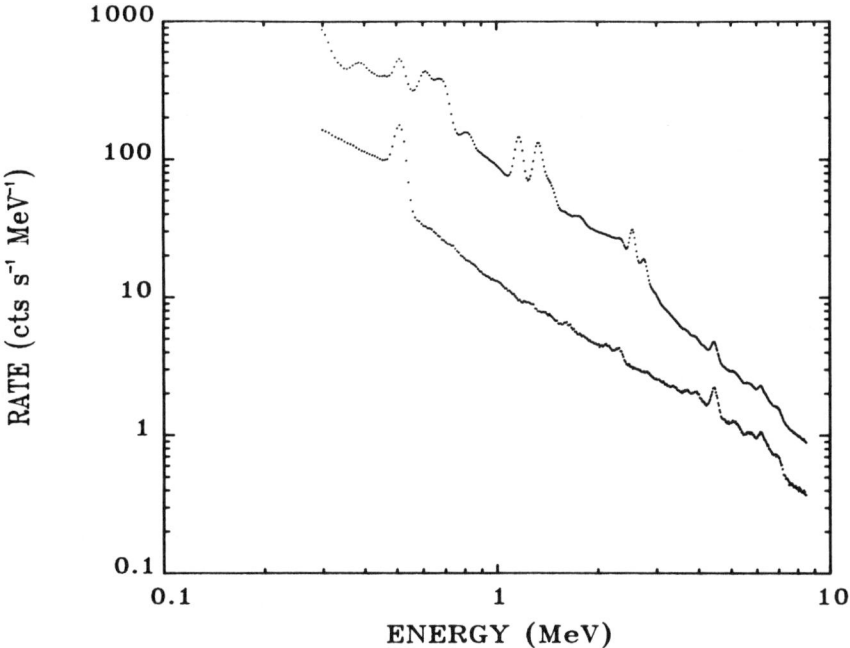

Fig. 1. Count rate spectra from the GRS revealing the sum of instrumental and atmospheric background (top) and atmospheric background (bottom).

Earth, from the upper spectrum. This normalized subtraction removes instrumental background lines and permits the atmospheric spectrum to be studied in detail[10,11]. Comparison of these two spectra indicates that instrumental continua and lines dominate the GRS spectra below about 4 MeV, even when the instrument is pointed directly at the Earth. These instrumental features are not evident above about 5 MeV where the atmospheric continuum and lines dominate the spectrum.

TEMPORAL VARIATION OF SPECTRAL FEATURES

Due to the moderate energy resolution of the GRS, 7% FWHM at 662 keV, many of the features appearing in Figure 1 are blends of more than one gamma-ray line. This is revealed by measurements indicating that the widths of the lines are broader than the instrumental resolution. We have therefore utilized the long-time base of observations to assist in the separation and identification of these lines. This is illustrated in Figure 2 which displays spectra accumulated over single 9-day intervals at four different times over the eight years that the GRS has been in operation. The spectra were accumulated for all times with the exception of orbits when SMM passed through the SAA. The strong lines appearing in the spectrum near 1.17 and 1.33 MeV ('a') are from the onboard ^{60}Co

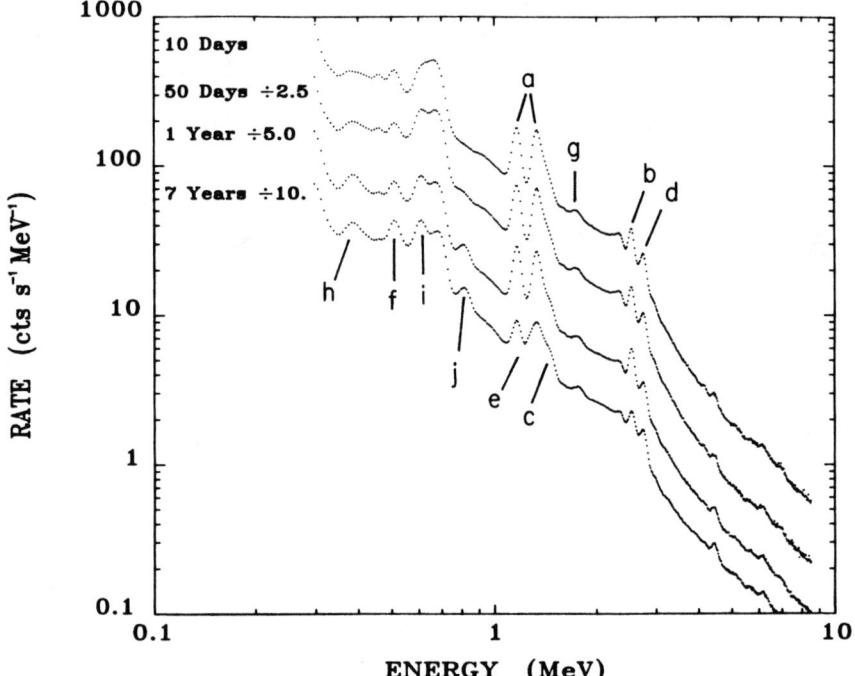

Fig. 2. GRS spectra accumulated over 9 days at different times during SMM's lifetime. Letters indicate line features discussed in detail the text.

calibration source; the sum peak near 2.5 MeV ('b') is also visible. All three lines show a marked decline in intensity with time, reflecting the 5.27 year half-life of ^{60}Co. As the intensity of the calibration source diminishes other background lines in that vicinity become more apparent.

One of these features is near 1.4 MeV ('c') and is in large part due to ^{24}Na, produced by spallation of aluminum in the housing of the instrument. This isotope also produces a line at 2.75 MeV ('d') which is apparent above the sum peak of the ^{60}Co calibration source. The time variation of the 2.75 MeV line is displayed and discussed by Kurfess et al.[5] ^{24}Na is also produced in the NaI detectors by neutron capture; because this isotope decays by emission of an electron, the narrow lines are broadened beyond recognition due to the β continuum.

Another spallation product of aluminum is ^{22}Na ($\tau_{1/2}$ 2.6 y) which produces a background line at 1.275 MeV ('e'), as well as the 511 keV line ('f') from annihilation of the positron emitted in its beta decay. The presence of the 1.275 MeV line is revealed by the increase in intensity of the valley between the two ^{60}Co lines. An increase in the intensity of the 511 keV line due to ^{22}Na is also apparent in Figure 2. We discuss time variations of the ^{22}Na lines in more depth below.

The broad feature near 1.8 MeV ('g') does not reveal any visibly striking long-term variation. However, detailed analysis of this region has enabled at least three separate line features to be identified.[7] One of these is a line at 1.786 MeV which exhibits the same build up in time as the 511 keV and 1.275 MeV lines and which is identified as the sum peak of these two lines from ^{22}Na. This same analysis has revealed a striking increase in the intensity of a line near 1.8 MeV when the Galactic center transits the instrument's aperture. This increase is attributed to a Galactic line at 1.809 MeV from the decay of radioactive ^{26}Al dispersed in the interstellar medium.[7,12]

Moving to lower energies, we notice a clear increase in intensity of a feature near 400 keV ('h'). The rate of increase and its energy suggests that the feature is due to ^{113}Sn (392 keV, $\tau_{1/2}$ 115 d) which can be produced by spallation of iodine in the central detector and CsI shield. This isotope decays by electron capture, therefore a gamma-ray line from decays in the NaI detector appears as a narrow feature with an energy equal to the sum of the transition energy and k-escape x-ray. Alternatively, a ^{113}Sn gamma ray produced in the CsI shield will exhibit a narrow line in NaI at the transition energy (392 keV).

The spectral region between 600 and 700 keV also reveals striking time variability over the early months of the mission. This is especially true of the feature near 600 keV ('i') which can be attributed to production of ^{121}Te (573 keV, $\tau_{1/2}$ 16.78 d) in the NaI detectors. The sum of the nuclear line plus the k-escape X-ray following electron capture produces a line near 605 keV.

The region between 800 keV and 900 keV also exhibits striking long-term variability. Early in the mission this region is generally featureless. There is no evidence for significant line emission near 844 and 847 keV where background features from aluminum and iron have been observed in other instruments. However, there is evidence for weak emission at 847 keV from ^{56}Mn within the first hours after exiting the SAA that was revealed in our study of radiation from SN1987A.[9] A significant feature with a center energy near 820 keV ('j') becomes apparent about a year after launch. This feature is a blend of two lines from radioactive isotopes produced in the stainless steel housing of the detector. These lines are ^{58}Co (810.8 keV, $\tau_{1/2}$ 70.8 d) and ^{54}Mn (835 keV, $\tau_{1/2}$ 312 d).

FEATURES REVEALED BY SPECTRAL SUBTRACTION

One method for assisting in the identification of line features in the GRS data is spectral subtraction. The presence of lines with half lives in excess of weeks are revealed in Figure 2. Their identifications can be confirmed by subtracting spectra obtained just after launch from those obtained at later times. An example of portions of such a difference spectrum is shown in the two parts of Figure 3. This difference spectrum was obtained by

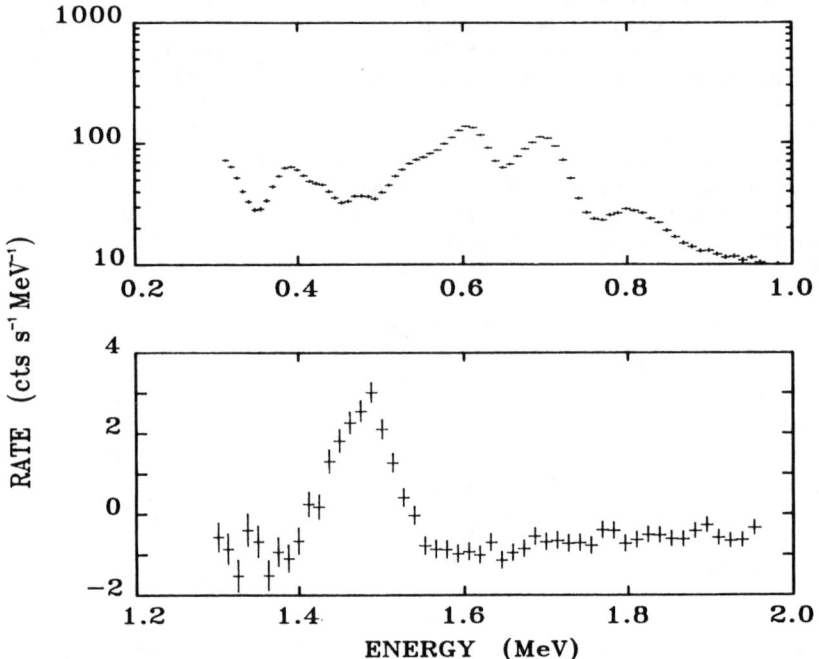

Fig. 3 The change observed in the SMM background from 2 weeks to 8 weeks after launch revealed by spectral subtraction.

subtracting a 9-day accumulation beginning about 2 weeks after launch of SMM from one accumulated beginning about 8 weeks after launch. The subtraction was performed after normalizing for livetime. There is evidence for the production of ^{113}Sn in both the CsI and NaI as revealed in the features appearing near 390 and 420 keV; some contribution from ^{127}Xe (375 keV, $\tau_{1/2}$ 36 d) produced in the NaI is also expected. Production of ^{126}I ($\tau_{1/2}$ 13.0 d) in the NaI detector is revealed by the presence of the strong line near 690 keV (666 keV plus 33 keV k-escape x-ray) and its weaker companion near 1.45 MeV (1.42 MeV plus 0.03 MeV). ^{126}I produced in the CsI (389 keV) blends with the ^{113}Sn line. The presence of ^{121}Te ($\tau_{1/2}$ 16.8 days) in the NaI is revealed in the features near 540 and 605 keV which result from the lines at 508 and 573 keV and the k-escape x-rays. The feature near 800 keV is dominated by ^{58}Co.

A different spectral subtraction reveals other line features shown in Figure 4. This spectrum was created by subtracting one

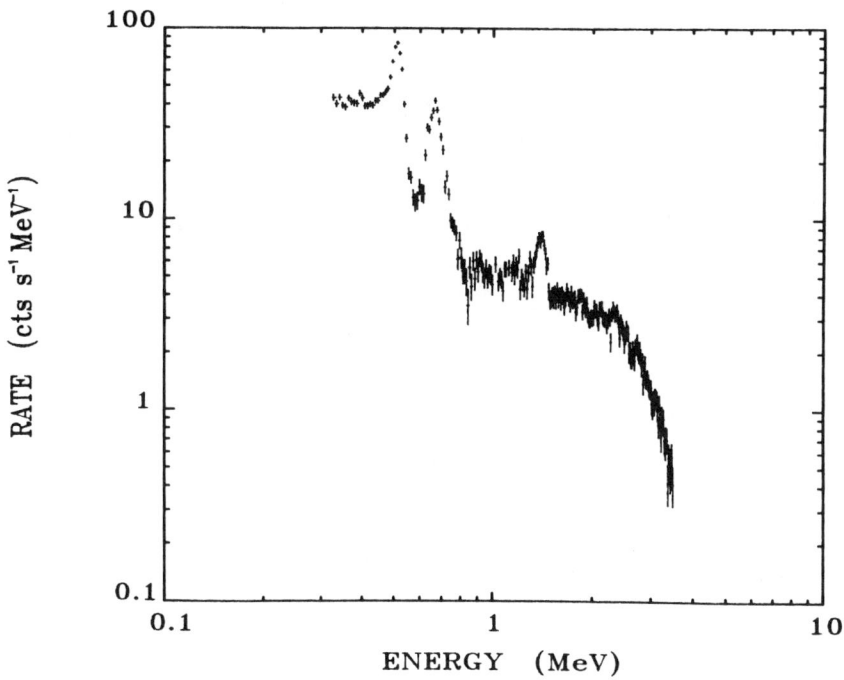

Fig. 4 Change in background spectrum in a one week period in 1987.

week accumulations separated by about a week in 1987. Once again all data outside of SAA orbits have been included. The residual lines and continuum appearing in this spectrum have half lives of a few days and appear because of the varying exposure of the detector to particles in the SAA, which has a period of about 48 days[5].

Evident at 511 keV is the positron annihilation line. The line at about 660 keV is probably a blend of ^{124}I (635 keV in NaI from electron capture, $\tau_{1/2}$ 4.15 d) and ^{132}Cs (668 keV, $\tau_{1/2}$ 6.5 d), the feature near 1.39 MeV is probably a blend of lines produced in the stainless steel shield: ^{57}Ni (1.378 MeV, $\tau_{1/2}$ 36 h) and ^{52}Mn (1.434 MeV, $\tau_{1/2}$ 5.6 d). The underlying continuum agrees well with the computed response function for various β^+-decays of ^{124}I produced in the NaI detector (see Figure 2 in reference 2).

LINE IDENTIFICATIONS FROM FITS

A complementary method for identifying lines in the SMM background spectrum employs detailed fits of line features in specific regions of the spectrum. This method was applied successfully in the analysis of SMM data which revealed the presence of ^{26}Al gamma radiation from the vicinity of the Galactic center.[7] In that analysis the blended complex of lines near 1.8 MeV was approximated by three Gaussian-shaped lines plus a power-law continuum. From the temporal variation of these three lines we were able to identify ^{119}Te, ^{22}Na, and the Galactic line as contributors to the feature (see earlier discussion).

We have also applied this technique to a study of 511 keV positron annihilation from the Galactic plane[8]. We review this technique in some detail below. A three-day spectral accumulation from 350 to 760 keV is shown in Figure 5, taken from reference 8. The solid curve is the best fit to the data with a model containing a quadratic continuum and 5 Gaussian lines. The temporal

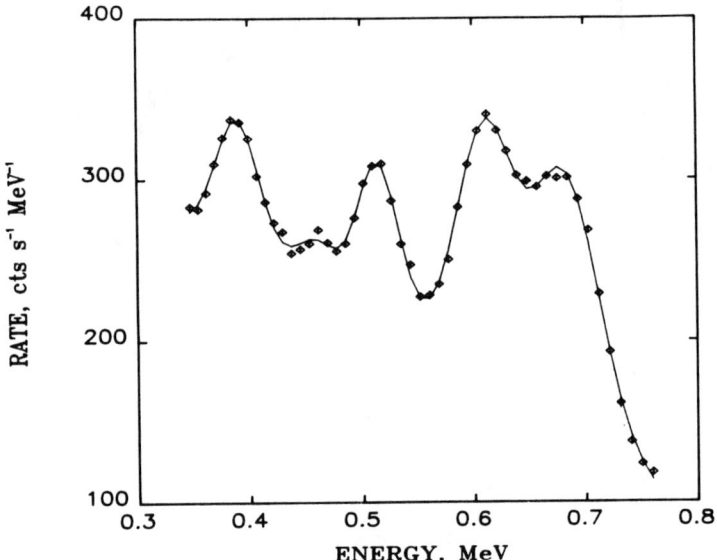

Fig. 5 Fit to background spectrum used to reveal the Galactic 511 keV line. Reprinted courtesy of the Astrophysical Journal, published by the Univ. of Chicago Press (c) 1988 (Amer. Ast. Soc.)

variations of the six intensities (continuum and 5 lines) of the model are shown in Figure 6, taken from reference 8. With the exception of the 511 keV intensities, the intensity profiles are similar to that exhibited by ^{24}Na produced by spallation of aluminum. These variations are described in detail by Kurfess et al.[5] and result from a combination of changes in altitude, exposure to the SAA particles, and the precession of SMM's orbit. In addition, the 511 keV time profile shown in panel a) exhibits a long-term build up from the production of ^{22}Na ($\tau_{1/2}$ 2.6 y) as well as an increase each year as the Galactic center transits the aperture.[8]

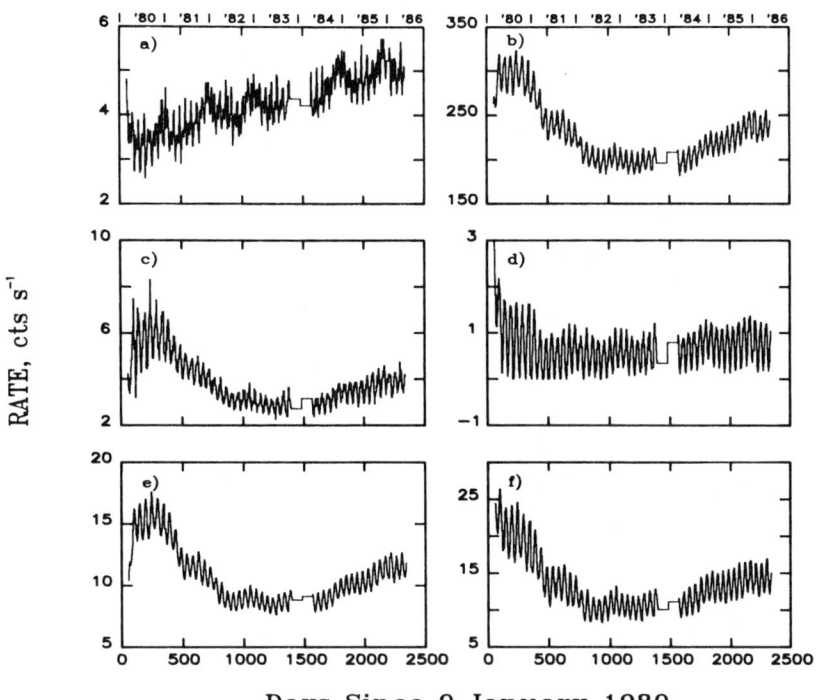

Fig. 6 Time variations of amplitudes of six spectral features: a) 511 keV line, b) continuum, c) 390 keV feature, d) 460 keV feature, e) ~610 keV feature, and f) ~670 keV feature. Reprinted courtesy of the Astrophysical Journal, published by the Univ. of Chicago Press (c) 1988 (Amer. Ast. Soc.).

Closer inspection of the individual fits reveals that the intensity of the line feature near 670 keV (Fig. 6f), primarily from ^{132}Cs ($\tau_{1/2}$ 6.5 d), exhibits a rapid decrease during the first year of the mission. This decrease is due to the change in altitude from 570 km to about 550 km and is similar to that exhibited by ^{24}Na ($\tau_{1/2}$ 15 h)[5]. The feature near 600 keV (Fig. 6e) exhibits a long-term variation with a flatter response, consistent with ^{121}Te ($\tau_{1/2}$ 16.8 days). The feature near 390 keV (Fig. 6c) exhibits a slower rise early in the mission which can be attributed to the build up of ^{113}Sn ($\tau_{1/2}$ 115 d). Superimposed on these long-term variations are the 48-day periodicities which follow the exposure to anisotropic particles fluxes in the SAA.[5] These periodicities require the presence of isotopes with shorter half-lives than ^{113}Sn near 400 keV such as ^{129}Cs (372 and 411 keV, $\tau_{1/2}$ 32.3 h) and ^{126}I. We attribute the line feature near 460 keV (Fig. 6d) to ^{132}Cs (469 keV, $\tau_{1/2}$ 6.5 d).

Perhaps the clearest example of how this fitting technique can be used to reveal and identify blended line features in complicated spectra was detailed by Leising et al.[13] in their study of ^{22}Na gamma-ray line emission from novae. The spectral range from 0.9 to 1.6 MeV (see Figure 1) was fitted by a cubic continuum function and six Gaussian lines, including a background line at 1.275 MeV in the valley of the strong ^{60}Co calibration lines. The time profile of the 1.275 MeV line is shown in Figure 7, taken from reference 13,

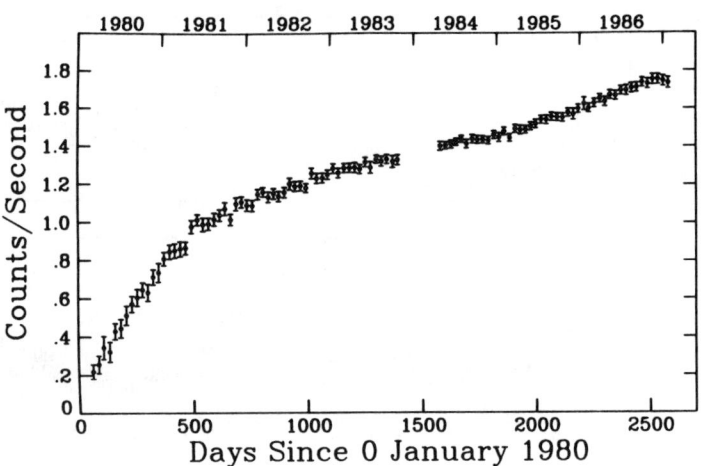

Fig. 7. The GRS count rate in a line at 1.275 MeV as a function of time in sky-viewing data. Reprinted courtesy of the Astrophysical Journal, published by the Univ. of Chicago Press (c) 1988 (Amer. Ast. Soc.).

is consistent with a radioactive isotope with a half life of 2.60 $\pm$ 0.08 years ($\tau_{1/2}$ for ^{22}Na is 2.6 y).

IDENTIFIED RADIOACTIVE ISOTOPES

In Table 1 we list preliminary identifications of radioactive isotopes from background studies with the GRS. Many of the identifications have been discussed previously in this paper.

TABLE 1
ISOTOPES IDENTIFIED IN SMM/GRS BACKGROUND

ISOTOPE	MAJOR LINES (MEV)	HALF LIFE
^{22}Na	0.511, 1.275, 1.786 (sum peak)	2.6 y
^{24}Na	1.368, 2.754, 4.12 (sum peak)	15.0 h
^{28}Al	1.779	2.24 m
^{52}Mn	0.744, 1.434	5.59 d
^{54}Mn	0.835	312.2 d
^{56}Mn	0.847	2.57 h
^{58}Co	0.811	70.8 d
^{60}Co	1.173, 1.332, 2.50 (calibration)	5.27 y
^{57}Ni	1.378	36.0 h
^{110}In	0.658	4.9 h
^{113}Sn	0.392	115.1 d
^{119}Te	0.644, 1.749	16.1 h
^{121}Te	0.507, 0.573	16.8 d
^{124}I	0.603	4.15 d
^{126}I	0.389, 0.666, 0.753	13.0 d
^{127}Xe	0.375	36.4 d
^{129}Cs	0.372, 0.411	32.4 h
^{132}Cs	0.464, 0.506, 0.668	6.5 d

ATMOSPHERIC GAMMA-RAY LINES

We have described in the Introduction how the instrumental background features produced in the GRS can be removed by normalized subtraction. Such a subtraction resulted in the lower spectrum in Figure 1 which reflects the time-averaged shape of Earth-albedo gamma radiation. This spectrum has been studied in depth and the results of this analysis have been presented elsewhere[10,11]. In this section we shall briefly describe the techniques which have been used in the analysis and list some of the line features which were identified.

The analysis utilized the instrument response function in order to determine how the incident atmospheric spectrum would appear in the detector. Our model atmospheric spectrum contained a power-law continuum and several Gaussian shaped lines. Several iterations were performed using a least squares algorithm in order to obtain the optimum parameters (i.e. line energy, width, and amplitude) which fit the data best. In Table 2 we list the best fit center energies of line features found in the spectrum along with suggested identifications. Many of the features are blends of more than one line; Doppler broadening of many of the atmospheric lines is also consistent with the data. These features can be explained by the interaction of secondary neutrons on atmospheric ^{14}N and ^{16}O.[14]

TABLE 2
ATMOSPHERIC LINE FEATURES

ENERGY (MeV)	IDENTIFICATION
0.511	positron annihilation
0.74	^{10}B, ^{14}N
1.64	^{14}N
2.14	^{11}B
2.32	^{14}N
2.78	^{16}O
3.03	^{13}C
3.40	^{14}N
3.69	^{13}C
3.91	^{13}C, ^{14}N
4.19	^{16}O
4.45	^{11}B, ^{12}C
5.12 (broad feature)	^{14}N, ^{15}N, ^{15}O
5.70 (broad feature)	^{14}N, ^{11}C
6.18	^{14}C, ^{15}O, ^{16}O
6.46	^{11}C, ^{14}N, ^{15}N
6.93 (broad feature)	^{11}B, ^{14}C, ^{14}N, ^{16}O

ACKNOWLEDGMENTS

We wish to express our thanks to Dan Messina for his assistance in the analysis and in preparation of this manuscript. We have profited from discussions with Clive Dyer, Neil Johnson, Jim Kurfess, Mark Leising, Steve Matz, and Bill Purcell. This work is sponsored by NASA contract S-14513-D at NRL, Grant NAG 5720 at UNH, and BMFT contract 010k 017-za/ws/wrk 0275:4 at MPE.

REFERENCES

1. G.J. Fishman, Ap. J. 171, 163 (1972).
2. C.S. Dyer, J.I. Trombka, S.M. Seltzer, and L.G. Evans, Nucl. Instr. Meth. 173, 585 (1980).
3. N. Gehrels, Nucl. Instr. Meth. A239, 324 (1985).
4. P.P. Dunphy, D.J. Forrest, E.L. Chupp, and G.H. Share, this volume (1988).
5. J.D. Kurfess, G.H. Share, R.L. Kinzer, W.N. Johnson, J.H. Adams, Jr., E.L. Chupp, D.J. Forrest, and C. Reppin, this volume (1988).
6. D.J. Forrest, et al., Solar Physics 65, 15 (1980).
7. G.H. Share, R.L. Kinzer, J.D. Kurfess, D.J. Forrest, E.L. Chupp, and E. Rieger, Ap. J. (Letters) 292, L61 (1985).
8. G.H. Share, R.L. Kinzer, J.D. Kurfess, D.C. Messina, W.R. Purcell, E.L. Chupp, D.J. Forrest, and C. Reppin, Ap. J. 326, 717 (1988).

9. S.M. Matz, G.H. Share, M.D. Leising, E.L. Chupp, W.T. Vestrand, W.R. Purcell, M.S. Strickman, and C. Reppin, Nature 331, 416 (1988).
10. J.R. Letaw, G.H. Share, R.L. Kinzer, R. Silberberg, E.L. Chupp, D.J. Forrest, and E. Rieger, Adv. Space Res. 6, 133 (1986).
11. J.R. Letaw, G.H. Share, R.L. Kinzer, R. Silberberg, E.L. Chupp, D.J. Forrest, and E. Rieger, to be published in J. Geophys. Res. (1988).
12. W.A. Mahoney, J.C. Ling, W.A. Wheaton, and A.S. Jacobson, Ap. J. 286, 578 (1984).
13. M.D. Leising, G.H. Share, E.L. Chupp, and G. Kanbach, Ap. J. 328 (1988).
14. J.C. Ling, J. Geophys. Res. 80, 3241 (1975).

RADIOACTIVITY INDUCED IN GAMMA-RAY SPECTROMETERS

C. S. Dyer, P. R. Truscott
Space Department, RAE Farnborough, Hampshire, GU14 6TD, England

N. D. A. Hammond, C. Comber
Scicon Ltd, Abbey House, Farnborough Road, Farnborough, England

ABSTRACT

A review is given of the data and methods used in the prediction of radioactivity induced in spaceborne gamma-ray spectrometers by the space radiation environment. The large masses of current and future detectors requires the application of particle transport codes and results from such codes are presented for scintillator detectors. Calculations and irradiation data are also presented for germanium and bismuth germanate and future requirements are outlined.

INTRODUCTION

Background produced in gamma-ray spectrometers and spacecraft material by the space radiation environment has been known to set a severe limitation on instrument sensitivity since the earliest flights. Major contributions to the background arise from the following components:

1. Direct charged particle counts.
2. Electromagnetic cascades induced by protons.
3. Nuclear de-excitation gamma-rays from proton-induced nucleonic cascades.
4. Bremsstrahlung from electrons.
5. Secondary neutron interactions in detector materials.
6. Radioactivity induced in detector and spacecraft materials by protons and secondary neutrons.

Of these components items 1 to 4 can be significantly reduced by active shielding and anticoincidencing techniques, which also diminish the contribution of item 5 if most of the secondary neutrons originate in the active detector volume. Thus radioactivity in the central detector elements remains as the most significant and difficult to remove component. Predictions of this component and its sensitivity to materials and design parameters are therefore of the utmost importance.

COMPONENTS OF THE CALCULATION

Prediction of induced radioactivity involves the following components:

1. Proton spectra for cosmic ray and inner belt environments.
2. Radiation transport of protons through spacecraft and detector materials.

© British Crown Copyright 1988

3. Spallation cross-sections from semi-empirical formulae and intranuclear cascade simulations and hence nuclide production rates.
4. Low energy neutron transport and hence capture rates.
5. Time history of irradiations and hence decay rates of radionuclides.
6. Energy-loss spectra of the decay of radionuclides within detector material computed by stochastic calculations to allow for coincidencing of cascade gamma-rays and beta-particles.

EARLY CALCULATIONS AND OBSERVATIONS

In a previous review paper[1] predictive methods were described which concentrated upon relatively small detectors of NaI and CsI carried on light spacecraft (eg OSO's 7 and 8). For such cases the spallation interactions of primary particles dominate and the predictions showed a reasonable match with observations. For heavier spacecraft, such as Apollo, observations showed significant fluxes of secondary neutrons arising from cosmic ray interactions with the spacecraft. These enhanced the background in small crystals (7 x 7 cm) of NaI which were returned to earth for low-level counting.

CURRENT CALCULATIONS FOR SODIUM IODIDE

Since that time there has been an evolution towards increasingly massive detectors carried on increasingly massive spacecraft (eg HEAO, GRO). This trend, while leading to increased shielding against trapped protons, also leads to increased production of secondaries, largely within the detector volumes themselves.

The treatment of secondary particles requires the application of particle transport codes. A number of studies have been performed for large assemblies of NaI utilising both 1-dimensional codes, such as ANISN[2], and full 3-D Monte Carlo simulations, such as HETC[3] and MORSE[4]. The components of the calculation are summarised in Table I.

TABLE I Components of the calculation

Proton environment for GRO in 450 km, 28.5° orbit in 1991
Daily fluences for E>100 MeV
Trapped protons: 1.5×10^6 cm^{-2}
Cosmic rays: 1.4×10^4 cm^{-2}

Spallation reactions vs depth + average neutron multiplicities used with ANISN for 1-D study.

Proton transport by HETC + low energy neutron transport by MORSE used in 3-D study.

The detector configurations studied are representations of the Oriented Scintillation Spectrometer Experiment (OSSE)[5] to be carried on the Gamma Ray Observatory. This detector system employs four identical telescopes, each of which comprises a central NaI/CsI phoswich (13 inch diameter, by 4 inches of NaI plus 3 inches of CsI), provided with an active NaI shield of about 3 inches thickness and a passive tungsten collimator. The geometrical representations used in the calculation are given in figure 1. A 0.3 cm layer of ^{6}LiH around the inner detector represents the ^{6}LiF-impregnated silicone rubber employed in the actual design. This material was incorporated to shield against thermal neutrons arising from hydrazine tanks and plastic scintillators carried on the spacecraft. In certain studies this thickness was increased to assess whether any reduction could be effected in the influence of the more energetic neutrons arising from the spacecraft and outer NaI shield. In the case of the 1-D study such increases were allowed to expand the detector volume, while for the 3-D study any increases were at the expense of outer NaI so as to keep the detectors within a constant envelope. Certain of the 3-D detector studies have examined an isolated detector for more ready comparison with the 1-D case. For the four detector array the local spacecraft materials are represented by 10 cm aluminium plates on two sides.

Results show the importance of secondary neutrons produced within the detector[5]. The cascade component leads to enhanced spallation while the evaporation component leads to I-128 production by neutron capture following energy loss via scattering interactions. Figures 2 and 3 show these background components as estimated from the 1-D ANISN studies and indicate that the latter component can in fact dominate the background at less than 2 MeV. The influence of increased ^{6}LiH shielding has been calculated for both the 1-D and 3-D cases and results for the trapped proton case are presented in figure 4. Results are in agreement to within about a factor of 2 and all show the influence of the moderating absorber to be counter-productive for all reasonable thicknesses. This effect is more marked for the 1-D case due to the increase in the neutron source with increasing detector volume, but is present in all cases due to the deleterious influence of the increasingly moderated neutron spectrum exceeding the benefit of the net neutron attenuation[7]. In general thermalising materials should be avoided in the vicinity of the central crystal wherever possible. Silicone rubber padding is presumably always required for shock proofing and the deleterious influence of 0.3 cm is minimal. Impregnation with ^{6}LiF has possibly reduced such influence and will remain of benefit for a thermalised external neutron source.

Full 3-D studies for incident cosmic rays are also in hand and preliminary results give spallation rates enhanced by a factor of 2.7 compared with the 1-D calculations and ^{128}I rates enhanced by a factor of 1.7. Such agreement is really very good and the general trend of increase is probably due to the improved treatment of the cascades in the coupled Monte Carlo codes. By contrast with the trapped proton case, which showed a net shielding benefit from adjacent detectors and spacecraft structure, the cosmic ray backgrounds show a small increase due to their higher energies leading to deeper penetration and greater multiplicity for secondary particle production. The importance of secondary particles should be noted. The overall radio-active background is enhanced by factors of 5 and 19 for trapped protons and cosmic rays respectively when compared with the interaction rates of primaries alone. Further studies are needed on the influence of detector orientation and particle anisotropies. The South Atlantic Anomaly protons show considerable anisotropy (Watts and Parnell, these proceedings) and this can influence the levels of induced radioactivity (Kurfess et al, these proceedings). Preliminary results show that there could be a factor of two effect for the OSSE instrument, which could make appropriate orientation during SAA passes beneficial.

CALCULATIONS FOR GERMANIUM

Spallation calculations have been extended to an isolated germanium crystal using semi-empirical cross-sections and estimated detector response functions based on a few accurately computed examples. An example prediction for trapped proton irradiation is given in figure 5. There is a need to further refine these calculations and extend the range of computed response functions. In addition full transport codes should be applied to allow for secondary particle effects including capture. There is also a need for a comprehensive set of irradiation data to be published.

CALCULATIONS FOR BISMUTH GERMANATE

This material is enjoying increasing use in high efficiency shielding and detector applications. Some preliminary assessment has been made of the most significant spallation products based on semi-empirical formulae and some examples are presented in Table II.

TABLE II Estimates of dominant nuclear production cross sections for bismuth germanate

Parent nucleus	Daughter nucleus	Maximum Cross section (mBarns)	Energy of Maximum (MeV)	Half life
^{209}Bi	^{207}Po	1065.15	30	5.7 hrs
^{209}Bi	^{205}Po	620.15	50	1.8 hrs
^{209}Bi	^{204}Po	372.77	60	3.6 hrs
^{209}Bi	^{209}Po	336.88	10	103 yrs
^{209}Bi	^{207}Bi	269.11	50	30 yrs
^{74}Ge	^{73}As	449.97	30	80 days
^{74}Ge	^{74}As	334.85	10	17.9 days
^{72}Ge	^{71}As	334.34	30	62 hrs
^{72}Ge	^{72}As	319.03	10	26 hrs
^{70}Ge	^{70}As	266.38	10	52 mins

Irradiations have also been performed on several one-inch crystals. Figure 6 presents some of the internal spectra obtained, while figure 7 presents a spectrum of escaping photons as measured by a well calibrated germanium crystal in order to enable accurate line identification and measurement of absolute activities. In addition such spectra are clearly of considerable use in predicting background to be expected in germanium spectrometers actively shielded by bismuth germanate.

FUTURE REQUIREMENTS

Some of the major requirements are as follows:

Extension of the particle transport studies on NaI and CsI to include particle anisotropies and orientation effects.

Extension of the particle transport codes to systems incorporating germanium and bismuth germanate.

Generation of a response function data base for both germanium and bismuth germanate.

Comprehensive irradiation data on germanium and bismuth germanate.

Further exploitation of background data sets such as HEAO and SMM.

Performance of controlled space experiments to monitor the background, with return of samples for ground monitoring, as in the Shuttle Activation Monitor experiments (Rester et al, these proceedings).

ACKNOWLEDGEMENTS

The 60 MeV proton beam was provided at the Variable Energy Cyclotron at AERE, Harwell, England and D Mapper and T Sanderson are thanked for their assistance and beam calibration. R Hutchings and A Skates assisted in the accumulation of the bismuth germanate spectra. The germanium counting of the bismuth germanate crystal was performed and analysed by G Colvin at Harwell.

REFERENCES

1. C S Dyer, J I Trombka, S M Seltzer, and L G Evans, Nuc Inst and Meths, 173, p585 (1980).
2. W W Engle Jr, ORNL-RSIC, CC-254 (1973).
3. K C Chandler, and T W Armstrong, ORNL-4744 (1972).
4. M B Emmett, ORNL-4972 (1983).
5. J D Kurfess, W N Johnson, R L Kinzer, G H Share, M S Strickman, M P Ulmer, D D Clayton, and C S Dyer, Adv Space Res, 3, p109 (1983).
6. C S Dyer, N D A Hammond, IEEE Trans on Nuc Sci, NS-32, p4421 (1985).
7. C S Dyer, P R Truscott, C Comber, N D A Hammond, IEEE Trans on Nuc Sci, NS-34, (1987). To be published.
8. C S Dyer, IEEE Trans on Nuc Sci, NS-31, p1061 (1984).
9. R Silberberg and C H Tsao, Ap J Supp, 25, p315 (1973).

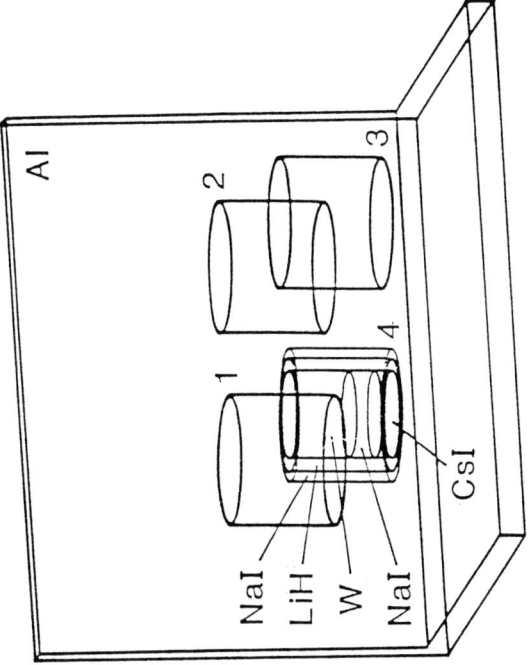

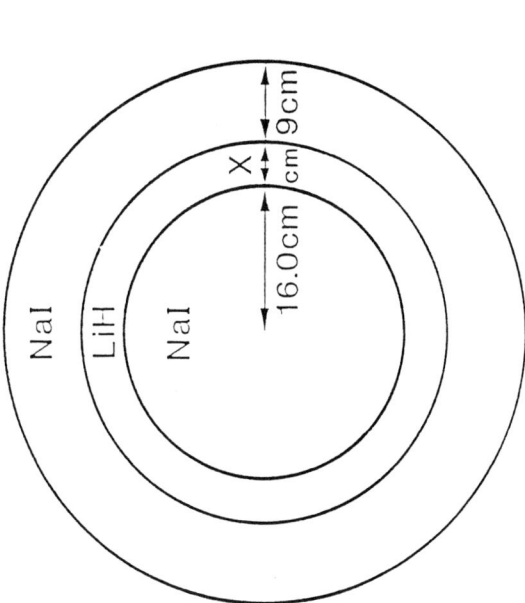

FIG 1

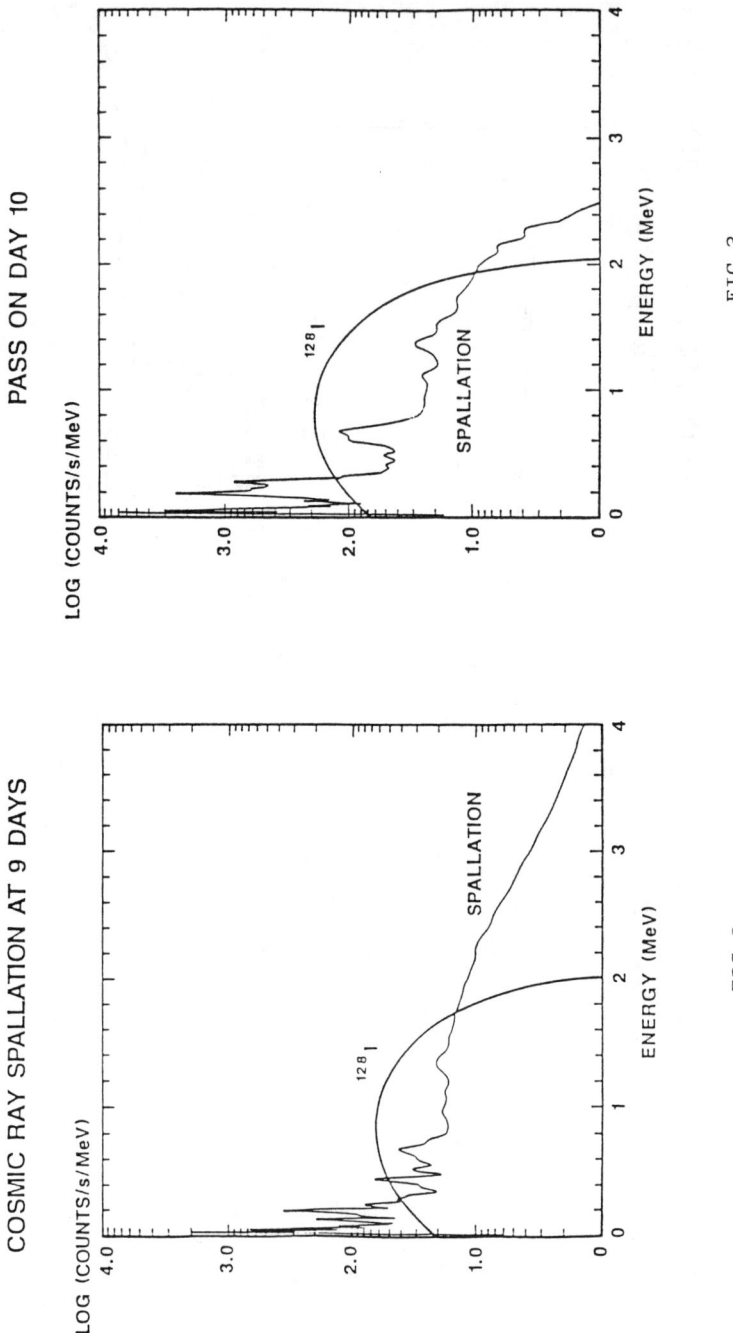

FIG 2

FIG 3

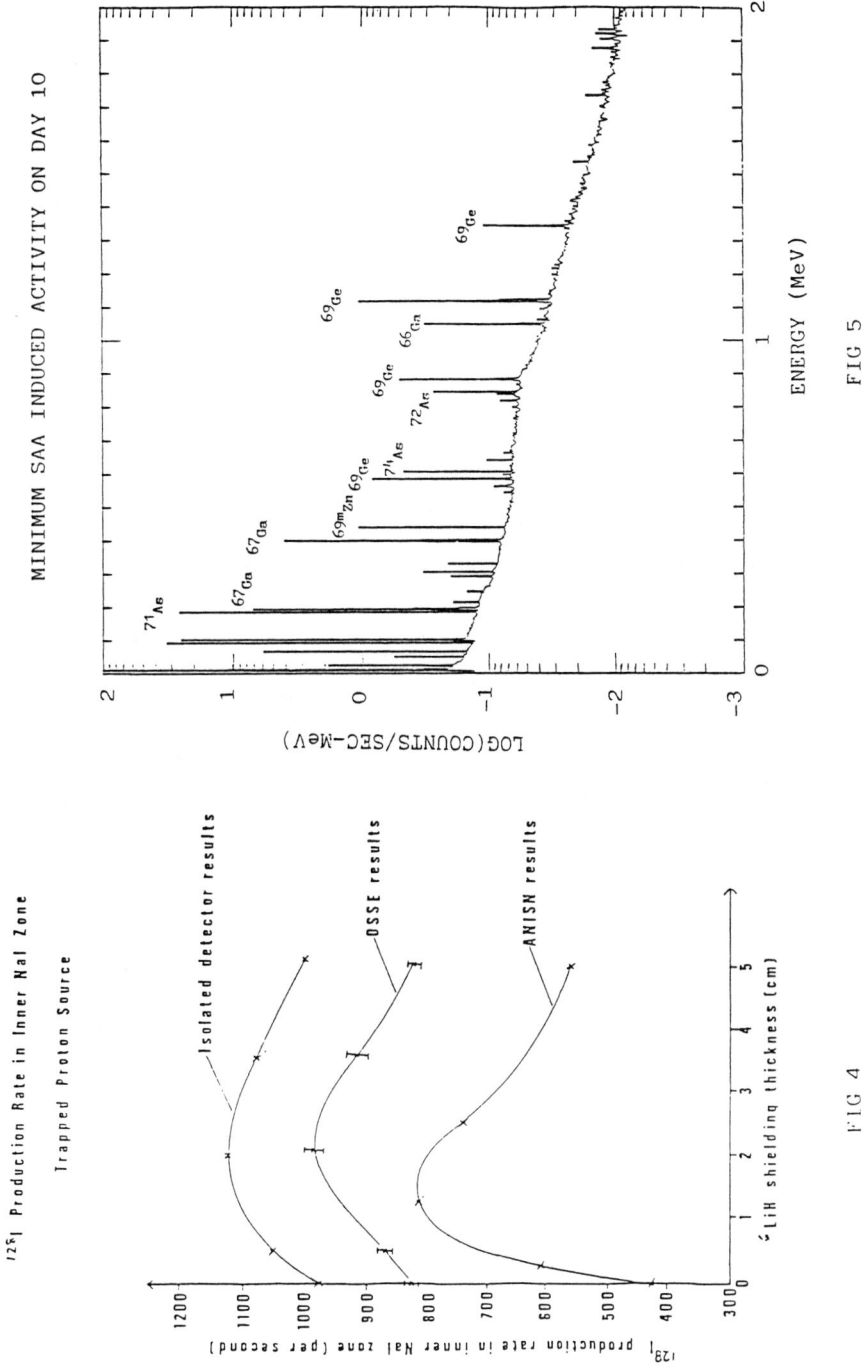

FIG 4

FIG 5

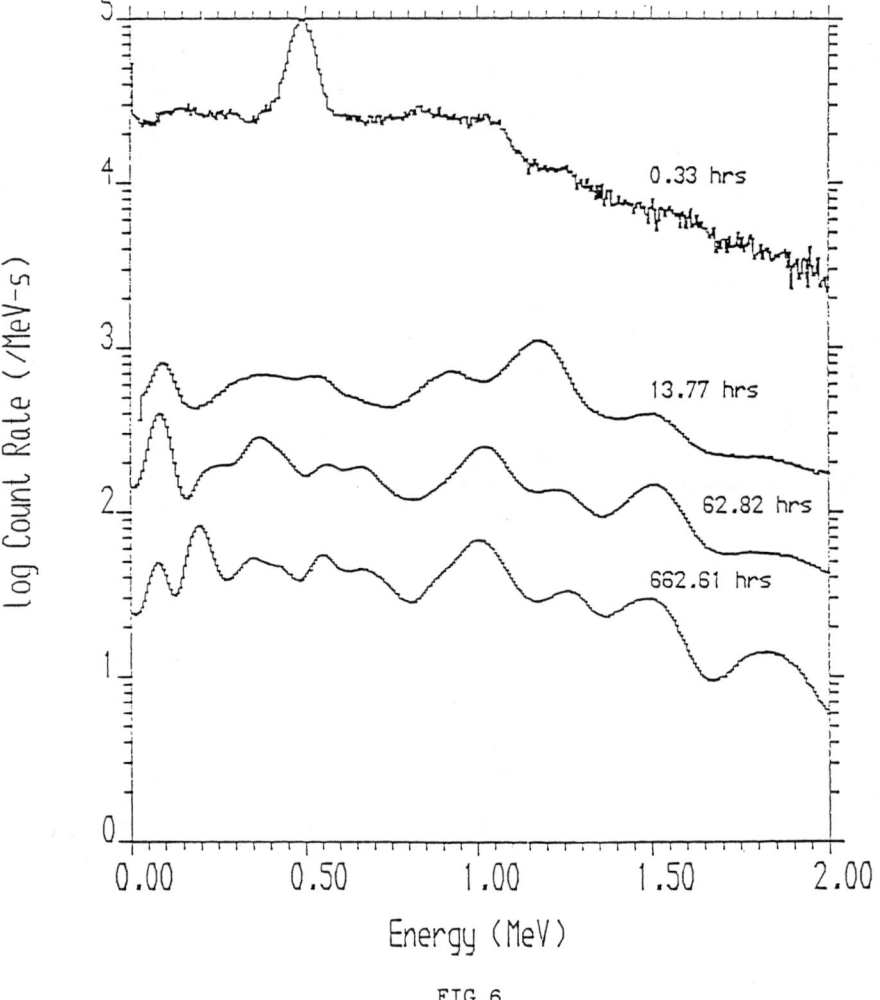

FIG 6

PRELIMINARY PEAK IDENTIFICATION FOR FIG 7

Peak No	Energy (KeV)	Probable Nuclides
1	72 to 79	K_α from Pb,Bi,Po
2	84 to 92	K_β from Pb,Bi,Po
3	175.2	As-71
4	184.6	Ga-67
5	262.3	Po-209
6	279.2	Pb-203
	286.4	Po-206
7	338.4	Po-206
8	401.3	Pb-203
9	500.0	As-71
10	511.0	β^+-decay
11	516.2	Bi-206
12	522.5	Po-206
13	537.5	Bi-206
14	703.4	Bi-205
15	807.5	Po-206
16	849.6	Po-206
17	899.1	Bi-204
18	983.9	Bi-204
	987.8	Bi-205
19	992.3	Po-207
20	1032.2	Po-206
21	1039.4	Ga-66
22	1050.6	As-72
23	1106.4	Ge-69
24	1336.2	Ge-69

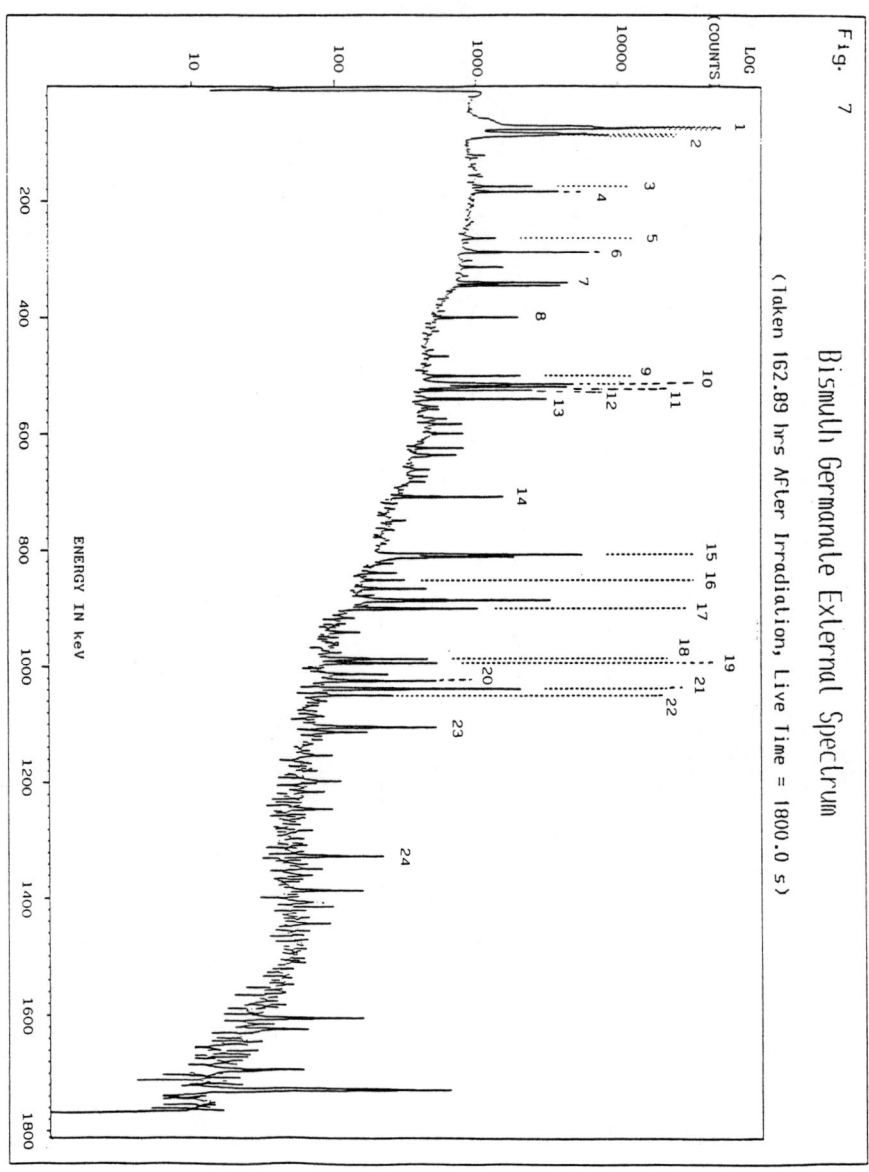

Fig. 7 Bismuth Germanate External Spectrum
(Taken 162.89 hrs After Irradiation, Live Time = 1800.0 s)

SPACE RADIATION SHIELDING ANALYSIS AND DOSIMETRY
FOR THE SPACE SHUTTLE PROGRAM

William Atwell and E. R. Beever
Rockwell International, Houston, Tx. 77058

A. C. Hardy and R. G. Richmond
NASA/Johnson Space Center, Houston, Tx. 77058

B. L. Cash
Lockheed EMSCO, Houston, Tx. 77058

ABSTRACT

Active and passive radiation dosimeters have been flown on every Space Shuttle mission to measure the naturally-occurring, background Van Allen and galactic cosmic radiation doses that astronauts and radiation-sensitive experiments and payloads receive. A review of the various models utilized at the NASA/Johnson Space Center, Radiation Analysis and Dosimetry is presented. An analytical shielding model of the Shuttle was developed as an engineering tool to aid in making premission radiation dose calculations and is discussed in detail. The anatomical man models are also discussed. A comparison between the onboard dosimeter measurements for the 24 Shuttle missions to date and the dose calculations using the radiation environment and shielding models is presented.

INTRODUCTION

The Radiation Analysis and Dosimetry (RAD) group at the NASA/Johnson Space Center (JSC) utilizes radiation environment models, shielding models, and dose exposure programs as analytical tools to provide functional, preflight, and real time support products. The trapped radiation environment models have been discussed in an earlier paper[1] at this Conference. In this paper we will concentrate on the Shuttle space radiation analysis support, shielding models and computer program interfaces, and a comparison of the onboard dosimeter measurements with the premission dose calculations.

SHUTTLE SPACE RADIATION ANALYSIS SUPPORT

ANALYTICAL TOOLS

An analytical shielding model[2] of the Space Shuttle has been developed for use with the space radiation environment models to compute dose exposure to astronauts and radiation-sensitive experiments and payloads. The shielding model, called SHLDAT, consists of approximately 2000 volume elements (geometrical shapes) that is used as an input data file with the program, EVDP (Elemental Volume Dose Program)[3]. Using an equal solid-angle, ray-tracing technique, EVDP computes the shield mass distribution for any location (dose point) in the Shuttle. Figure 1 shows the shield distribution for a particular dosimeter location, DLOC #1, using 968 rays. All materials, such as iron, copper, etc., are converted to an equivalent aluminum thickness. The amount of shielding material intercepted is computed for each ray. It has been determined that 500-1000 rays are adequate to provide a statistically significant shield distribution. Generally, 512-ray shield distributions are computed.

To compute critical body organ (skin, eye, and depth) doses to astronauts in the Shuttle, a computerized anatomical man (CAM) model[4] that was modified by Billings and Yucker[5] is used. The body organ shield distributions shown in Figure 2 have been converted into 512 shield thicknesses[6]. For extravehicular activity (EVA), spacesuit models have been developed by the RAD. These are used with the CAM to compute EVA crew doses.

The proton dose program uses the shield distributions and the proton spectrum external to the spacecraft or spacesuit to calculate dose at various locations. Using the "straight-ahead" approximation both absorbed dose (rad) and dose equivalent (rem) are calculated. The rad-to-rem conversion is based on the LET (linear energy transfer) dependent quality factor, Q, defined by the International Commission on Radiological Protection (ICRP) Report 26[7]. The major contributors to dose inside the Shuttle are the protons in the South Atlantic Anomaly (SAA) and galactic cosmic radiation.

Electron dose calculations are made utilizing a modification to the Shieldose[8] program. Again, the shield distribution and external radiation environment are used as input to make the dose computations. The Q for electrons is assumed to be unity. Due to the inherent bulk shielding of the Shuttle, electron and the associated bremsstrahlung doses have been found to be negligible, but for the thinly-shielded EVA spacesuits that is not the case. Electrons can contribute a significant portion of the total dose during extravehicular activity.

Since we do not presently have the programmatic capability to compute galactic cosmic radiation (GCR) doses in complex, three-dimensional shielding such as the Shuttle, the work of Silberberg, et al,[9] is used for GCR dose estimates. For 28.5 deg inclination missions, we have been using 4.5 mrad/day and 8 mrem/day. For 57 deg inclination missions, the GCR absorbed dose rate is 8 mrad/day with a dose equivalent of 17 mrem/day. These are solar minimum "worst case" values.

SUPPORT PRODUCTS

The RAD group provides functional, preflight, and real time support products. Various groups in the Shuttle program office have required shield distributions and dose estimates to Shuttle systems such as the avionics for use in single-event upset studies. The inertial measuring units (IMU) were upgraded and required RAD support. At the JSC-center level, RAD support has been utilized by the Crew Systems Division for the development of a new spacesuit design. Within the NASA agency, the RAD has supported, at the headquarters level, advanced programs such as the Space Station and has provided a variety support products to other NASA centers.

Preparatory to each Shuttle mission, preflight support products are provided in the form of crew dose exposure and dose calculations at each dosimeter location for post flight comparisons. These products assist in flight planning and support the Radiological Health group at JSC, who maintain crew health records. For mission specific flights, the RAD provides support to experimenters and radiation-sensitive payloads, for example, radiation dose exposures to photographic film. In addition, reports are prepared and submitted for radioactive payload safety assessment.

A three-man, real time mission support is provided by the RAD for every Shuttle flight. This support consists of monitoring the nominal radiation environment, and daily crew dose exposure reports are submitted to the Flight Surgeon. Frequent solar activity status reports are received from the National Oceanic and Atmospheric Administration (NOAA). NOAA provides solar data and information on active regions, solar flares, magnetic activity, and proton emission. A contingency situation exists if it is determined that the radiation environment is enhanced. This situation can arise if a solar particle event or a nuclear detonation occurs.

ANALYTICAL MODELS & PROGRAMS INTERFACE

Figure 3 is flow diagram of the analytical models and computer programs interface used to provide the nominal radiation environment support. For a given Shuttle mission, a mission trajectory computer tape is received that contains the following mission orbital parameters: mission elapsed time (1-min intervals), spacecraft longitude, latitude, and altitude. Using the appropriate magnetic field model, the orbital parameters are converted to B,L-space, where B (gauss) is the strength of the geomagnetic field at the particular trajectory point and L (earth radii) is the physical distance to the field line. For each B-L point, the instantaneous proton and electron flux are determined using the AP-8 proton[10] and AE-8 electron[11] models. Proton and electron spectra are computed for the mission and are used with the desired shield distributions as input to the dose programs for both intravehicular and extravehicular dose calculations. As mentioned earlier, we are limited in the GCR area. We are looking to the scientific community to eventually supply us with an operational GCR program.

MEASUREMENTS VERSUS DOSE CALCULATIONS

Both active and passive dosimeters that are calibrated premission are flown on each Shuttle mission. The passive dosimeters are thermoluminescent, TLD-100. Each crewman wears a passive dosimeter during the mission. In addition, six TLD's are placed at fixed locations in the spacecraft - three on the flight deck and three in the mid deck. The active dosimeters are ionization chambers and are stored in a pouch in an assigned stowage locker. These dosimeters are used to support EVA and could be read out periodically during a mission if an enhanced radiation environment is suspected. These read outs would assist the RAD Group in determining the degree of radiation enhancement for accurate assessment of the hazard. Upon completion of the mission, the dosimeters are read by the Radiation Dosimetry Laboratory personnel at JSC.

The previous chart (Figure 3) showed the calculational techniques used to make dose calculations. Proton and electron spectra are generated premission and are used with the six TLD shield distributions to compute the expected mission dose at each dosimeter location. Due to the unpredictable movement of the astronauts inside the Shuttle, it is difficult to make accurate premission crew dose calculations. Previous flight experience has shown that crew dose averages are approximately 10% higher than dosimeter location #1, which is the most heavily shielded of the six fixed locations. During the Shuttle era, we have been using the solar minimum models (AP-8 MIN and AE-8 MIN) and more recently adjusting the dose calculations to include the solar cycle effects. Figure 4 shows a plot of the normalized Zurich sunspot number (jagged line), an 84-month (7 yr) sine wave, and the time after solar maximum that the Shuttle mission was flown.

Figure 5 shows a comparison of the dosimeter measurements at location #2 with the dose calculations as a function of altitude for several Shuttle missions that flew at a 28.5 deg inclination. The upper solid line is the dose calculations for solar minimum, and the lower solid line is for solar maximum. The dose calculations shown by the "triangles" include the solar cycle modulation effect and 4.5 mrad/day for the GCR dose contribution. Similar results were obtained for the other five dosimeter locations and for the five Shuttle missions that flew at higher inclinations (one mission at 49.5 deg inclination and four missions at 57 deg inclination).

SUMMARY

The NASA/Johnson Space Center, Radiation Analysis and Dosimetry group utilizes a number of models and computer programs to calculate space radiation dose exposures to astronauts and radiation-sensitive payloads and equipment. In addition, functional, preflight, and real time support products are provided by the Group. Onboard dosimetry is used to monitor the space environment in real time and provides a means for comparing the premission dose calculations.

REFERENCES

1. E. G. Stassinopoulos, This Conference (1987).
2. W. Atwell and E. R. Beever, Rockwell-Houston Internal Memorandum (1979).
3. S. C. Hamilton and B. Liley, Rockwell International Document No. SD 76-SA-0184-1 (1976).

4. Paul G. Kase, Technical Report No. AFWL-TR-69-161, Air Force Weapons Laboratory, Kirtland Air Force Base, NM (1970).
5. M. P. Billings and W. R. Yucker, McDonnell Douglas Astronautics Report MDC G4655 (1973).
6. William Atwell, Rockwell-Houston Internal Memorandum (1975).
7. ICRP, International Commission on Radiological Protection. Publication 26, Ann. of ICRP 1, (3), Pergamon Press, Oxford, England (1977).
8. S. Seltzer, U.S. Department of Commerce, National Bureau of Standards, NBS Technical Note 1116 (1980).
9. R. Silberberg, C. H. Tsao, J. H. Adams, Jr., and J. R. Letaw, Adv. Space Res. 4, No. 10, 143-151 (1984).
10. D. M. Sawyer and J. I. Vette, NSSDC/WDC-A-R & S-76-06, National Space Science Data Center, NASA/Goddard Space Flight Flight Center, Greenbelt, MD (1976).
11. J. I. Vette, NASA/Goddard Space Flight Center, Greenbelt, MD, private communication to W. Atwell, Rockwell International, Houston, TX (1984).

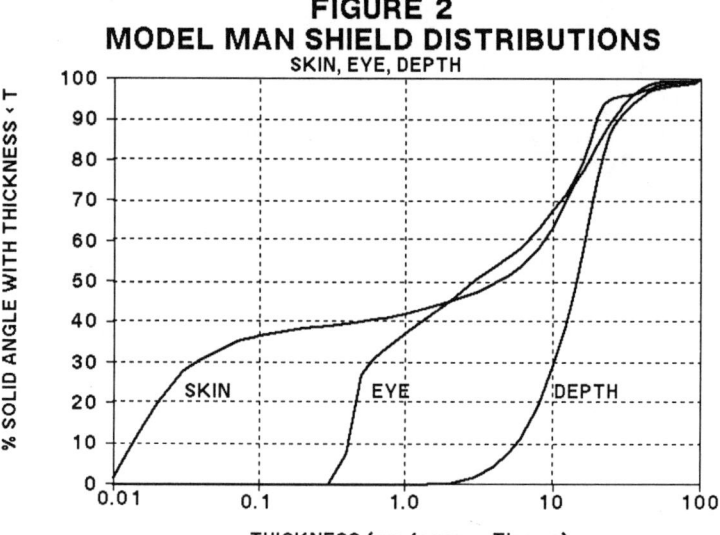

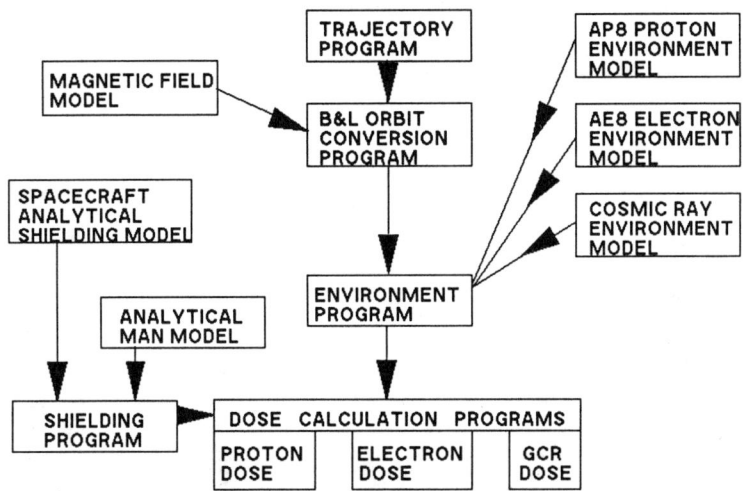

FIGURE 3
SPACE RADIATION ANALYSIS
NOMINAL ENVIRONMENT ANALYTICAL MODELS/PROGRAMS INTERFACE

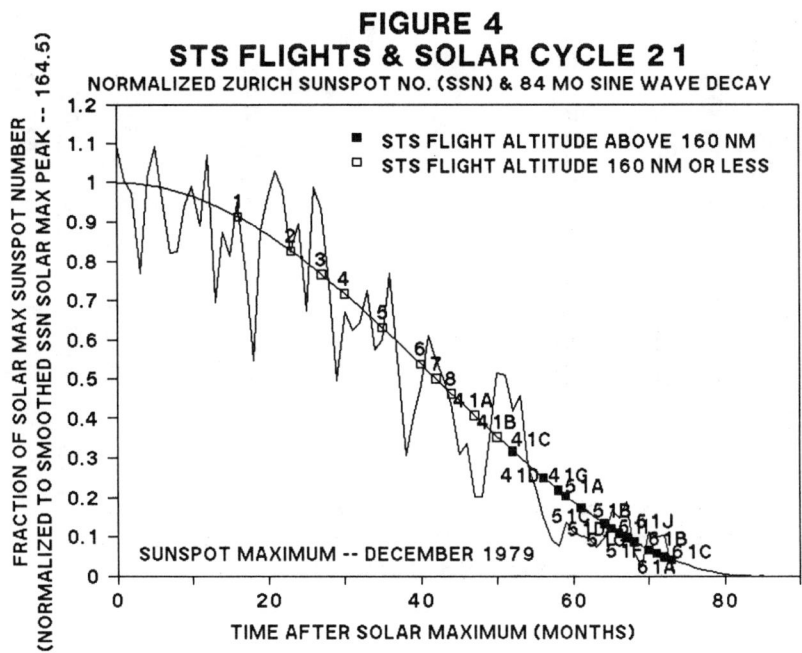

FIGURE 4
STS FLIGHTS & SOLAR CYCLE 21
NORMALIZED ZURICH SUNSPOT NO. (SSN) & 84 MO SINE WAVE DECAY

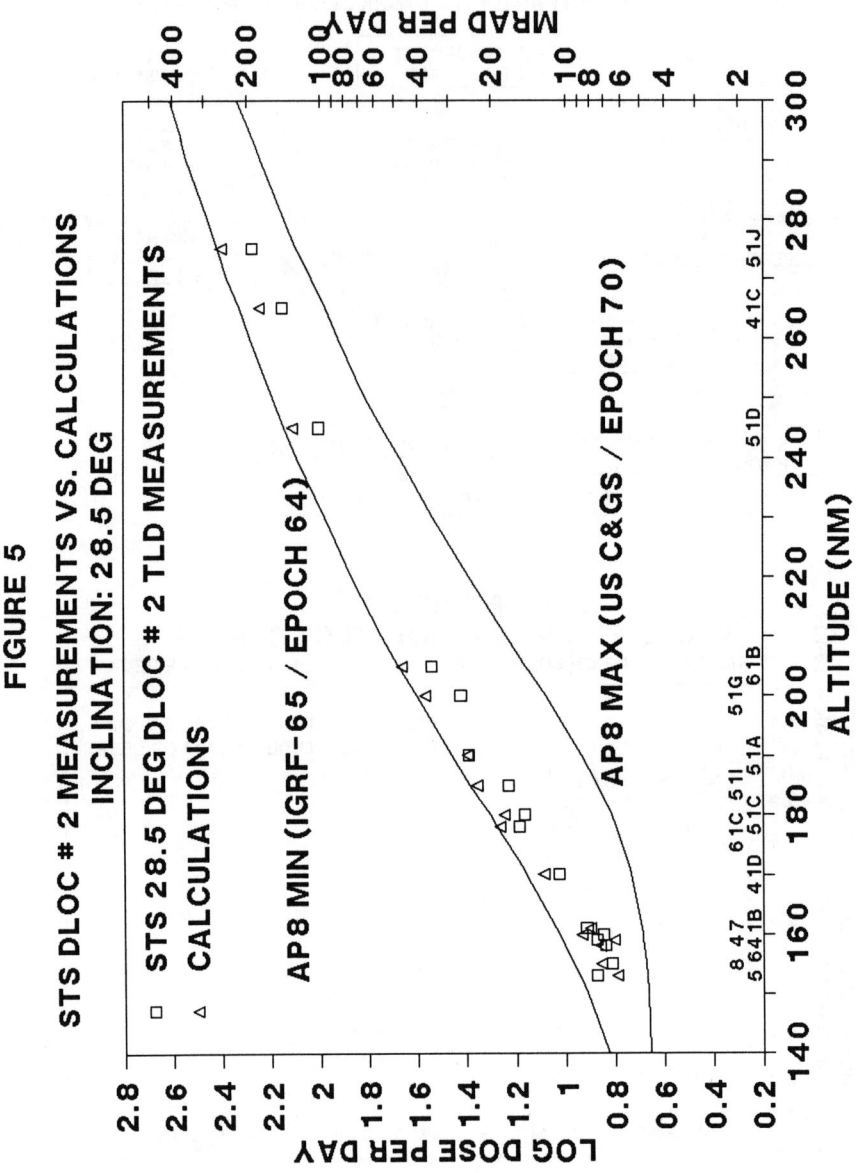

THE RADIATION DOSE IN A MOLNIYA-TYPE ORBIT

J.B. Blake and J.E. Cox
Space Sciences Laboratory
The Aerospace Corporation
P.O. Box 92957
Los Angeles, California 90009

ABSTRACT

A pair of dosimeters aboard satellites in a Molniya orbit have made measurements of the radiation dose under 0.69 gm/cm^2 of aluminum. The measured dose was substantially less than that predicted by the NASA AE-8 and AP-8 models. The cause of the difference cannot be determined with certainty. One possibility is that AP-8 is in agreement with the observations and that AE-8 leads to a substantial overestimate of the dose due to electrons.

INTRODUCTION

Measurements have been made since August 1983 of the radiation dose encountered by satellites in a Molniya-type orbit. The Molniya orbit is named for the Soviet COMSATS which utilize this orbit. The inclination of the orbit, 63°, is selected such that the argument of perigee does not change. The orbital period, just under 12 hours, is selected such that apogee "hangs" above the same two places on the Earth every day. The period requirement fixes the semi-major axis of the orbit of course and, in order to maximize the "hang time" over the region of interest, perigee height is as low as possible commesurate with orbital lifetime requirements.

The orbital parameters result in a variety of magnetospheric environments for the satellite ranging from the Southern auroral zone, the equatorial regions of the inner zone with its energetic protons and electrons, the outer zone energetic electrons, the high-latitude plasma sheet, the magnetosheath and, at times, the interplanetary medium. The dosimeter makes a direct measurement of the dose in silicon in a slab geometry under 100 mils of aluminum shielding. The geometry of the spacecraft installation is such that approximately π solid-angle is not heavily shielded; the other 3π is heavily shielded.

In this paper data from two independent measurements will be presented and the implications discussed.

DOSIMETER

The dosimeter, built by the Space Sciences Laboratory of The Aerospace Corporation, uses technology proven by flight aboard many USAF and NASA satellites. Each dosimeter consists of two separate, single-detector units. Only one unit is operated at a time; the other serves as a flight spare.

Each dosimeter uses a small, silicon, surface-barrier detector that is in the shape of a circular disc. The dosimeter directly measures the ionization in the silicon disc caused by the natural radiation - that is, the radiation dose. A particle passing through the active volume of the detector produces a number of electron-hole pairs which is directly proportional to the amount of energy deposited in the detector. The charge is collected and results in a pulse at the amplifier output. Use is made of the fact that the area as well as amplitude of individual pulses is directly proportional to the energy which a particle deposits in the active volume of the silicon disc. An integral discriminator, with a threshold corresponding to deposited energy of 66 keV, is used to analyze the pulse-height spectrum of signals produced by protons, electrons and photons (primarily bremsstrahlung). A threshold of 66 keV was selected on the basis of detector noise present when the entire package is operated at the planned temperature of less than 40° C. A gated integrator accepts pulses whose amplitudes exceed the value of the gating threshold of 66 keV. Every pulse meeting the gating criteria is integrated and the resulting charge (area) deposited in a storage capacitor which in general already contains an accumulated charge from preceding signals. Whenever an added increment of charge causes the total charge on the capacitor to exceed 7.6 MeV equivalent of deposited energy, the capacitor is discharged in 7.6 MeV equivalent increments. For each discharge increment, a logic pulse is produced at the integrator output and counted in a storage register.

Counts from the registers are read out once every thirty-two seconds by the telemetry system. The registers are never re-set unless the power is turned off, therefore the number of counts read out at any given time represents the total energy deposited in the active volume of the silicon during the entire mission. The data system contains the storage registers where counts from the dosimeter are accumulated, and shift registers for temporary storage of data during readout. It also serves as the interface with the spacecraft telemetry system.

The dosimeter storage register has a capacity of 36 bits. The counts are accumulated directly in binary code, so the total capacity of each dose-monitor channel is 2^{36} counts, or 6.87×10^{10} counts. Each count corresponds to 2.7×10^{-6} rads as determined by calibration with radioactive sources. The maximum dose capacity is 1.85×10^5 rads. The maximum and minimum measurable doserate is determined by maximum discharge pulse rate (10^5 pulse/sec) and the system leakage current (1 pulse/500 sec) respectively. The maximum measurable dose rate is 0.27 rads/sec. The minimum detectable dose rate is 5.4×10^{-9} rads/sec (0.17 rad/yr).

RESULTS

The dosimeters were flown aboard two satellites. The dosimeters are labeled, for reasons irrelevant to the present discussion, as PL03 and PL04.

Figure 1 gives the daily dose as measured by PL03 from August 1983 to August 1985. A large variation in the daily dose can be seen. Note the three very large peaks late in 1984. These are the largest daily doses seen to date. These peaks are separated by the synodic period of the Sun, and are the result of a high-speed solar-wind stream impinging upon the Earth's magnetosphere.

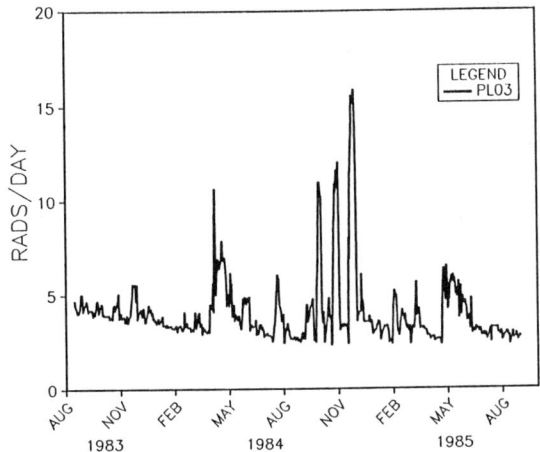

Figure 1. The radiation dose measured by PL03 is shown as a function of time between August 1983 and August 1985.

Figure 2 shows the dose as measured by PL03 and PL04 from February 1985 to February 1986 on the same plot. Similar structure is seen in both measurements but the magnitude of the dose is quite different in the two cases. Note also that the "quiet-time" level in PL03 was flat at the end of the time period whereas for PL04 it was decreasing. These differences were unexpected.

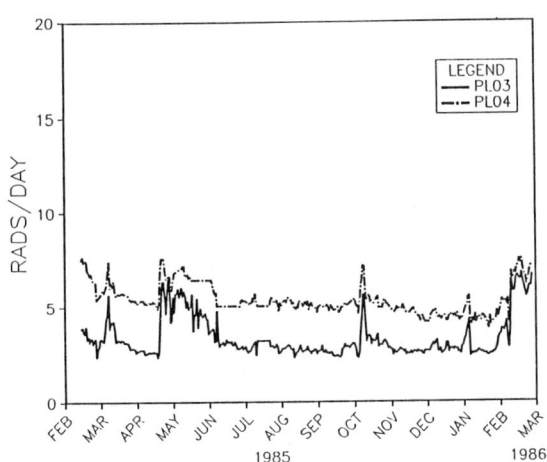

Figure 2. The radiation dose measured by PL03 and PL04 is shown as a function of time between February 1985 and February 1986.

Figure 3 shows the run of data from February 1985 until February 1988. This plot shows that the dose from PL04 continued to decline whereas the dose from PL03 began to increase and by late 1986 they became identical.

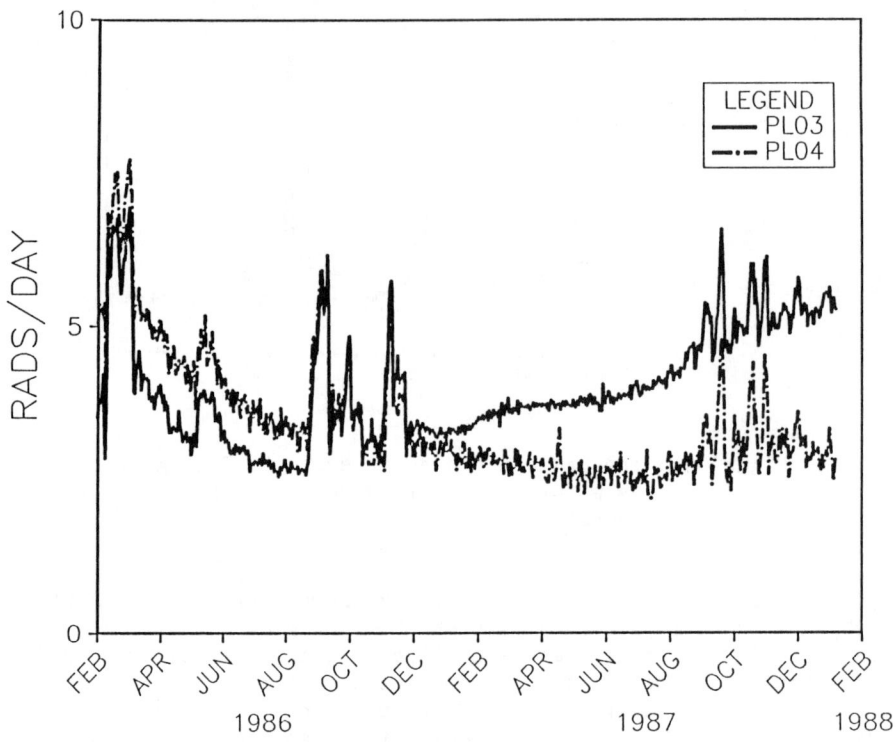

Figure 3. A plot similar to Figure 2 for the time period between February 1985 and February 1986.

In Figure 4 the height of perigee is plotted vs. the dose difference between PL04 and PL03. Note that, when the perigee altitudes were identical, the measured daily doses were essentially the same. It is not a longitude effect since the orbital planes were close together, and fixed relative to each other. This comparison shows that the initial difference observed in the data from the two dosimeters was due to exposure to a different environment and not due to a difference in the dosimeters themselves. This figure shows that the height of perigee has a strong effect upon the dose experienced by a Molniya satellite.

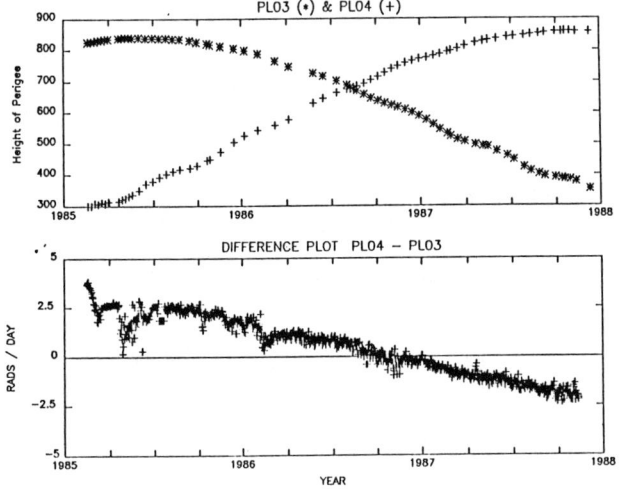

Figure 4. The height of perigee of the two satellites, and the dose difference is plotted a a func- of time.

Figure 5 showing the entire data set for PL03 clearly shows the variation of the quiet-time flux with perigee height.

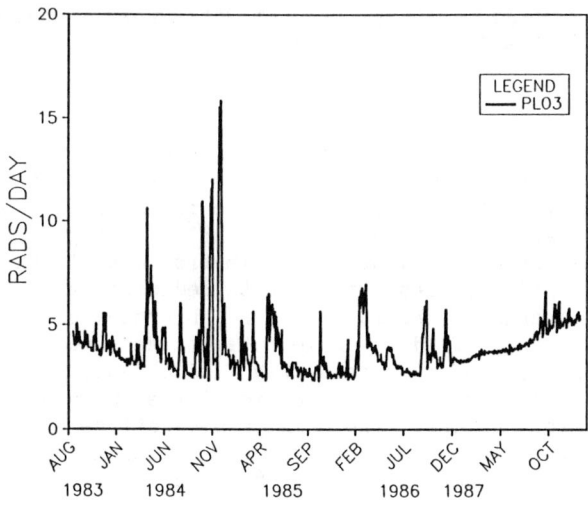

Figure 5. The entire run of data presently available from PL03 is shown.

MODEL COMPARISON

In Table 1 the calculated dose from the Shielddose code using the NASA AE-8 and AP-8 models as input is compared with the observations for perigee heights of 341nm and 857nm. The actual location of the orbital planes of the two satellites was used in the calculations; they differed only by 11°. The measured dose has been multiplied by two because only ~ 1/2 of the semi-infinite shield was exposed to the ambient environment.

Table I. Calculated/Measured Dose Comparison

341nm Perigee			857nm Perigee		
Electron & Brems	Proton	Measured	Electron & Brems	Proton	Measured
14.3	6.9	8.8	21.1	4.5	5.44
21.2			21.1		

The actual measured doses are multiplied by 2 because of the geometry of the dosimeter, see above.

Note that the calculated dose is more than a factor of two larger than the measured dose, that electrons provide the largest contribution to the total dose, and that the electron dose is larger at 857nm than at 341nm whereas the proton dose is larger at 341nm than at 857nm.

DISCUSSION

Why the large difference between the measured and calculated radiation dose? Gussenhoven et al.[1] have made measurements of the radiation dose in the DMSP orbit which is circular at 450nm and at a 96° inclination. They found that the outer zone plus inner zone electron dose in the DMSP orbit is substantially less than that predicted by the NASA model. For a spherical shield with a thickness of 0.55 gm/cm^2 of aluminum, they find a calculated to measured ratio of ~ 6, and for a 1.55 gm/cm^2 shield they find a ratio ~ 9.

Thus we are led to consider the possibility that the disagreement between the Molniya predictions and measurements is due to an overestimate of the electron dose by the NASA models. Remember that the Molniya shield is 0.69 gm/cm^2 aluminum in a semi-infinite slab geometry.

Let f be the convection factor to bring the measured and observed dose into agreement. Then from Table 1:

$$14.3 f + 6.9 = 8.84,$$
$$21.1 f + 4.5 = 5.44.$$

These equations give f = .0814. Thus the overprediction of the NASA model is 12.3 for the Molniya orbit under the assumption that the discrepancy we observe is due to the NASA electron models alone.

CONCLUSIONS

1) The dose measurements made in a Molniya orbit suggest that the NASA AE-8 model substantially overpredicts the dose under a ~ 0.69 gm/cm^2 aluminum shield. Note that these measurements do not prove that this is the case, more complex differences between prediction and measurement are possible.
2) The number of large storms, which substantially add to the radiation dose, varied greatly in number and intensity over the mission to date.

REFERENCE

1) M.S. Gussenhoven, E.G. Mullen, R.C. Filz, D.H. Brantegun, and F.A. Hauser, IEEE Trans. Nucl. Sci., NS-34, 676, 1987.

ACKNOWLEDGEMENT

This work was supported at The Aerospace Corporation by the U.S. Air Force Systems Command's Space Division under Contract F04701-85-C-0086.

THE HEAO 3 BACKGROUND: SPECTRUM OBSERVED BY A LARGE GERMANIUM SPECTROMETER IN LOW EARTH ORBIT

Wm. A. Wheaton, A. S. Jacobson, J. C. Ling, W. A. Mahoney, and L. S. Varnell
Jet Propulsion Laboratory 169-327, California Institute of Technology
Pasadena, CA 91109

ABSTRACT

We present a 50-day (1,436,000 live second) accumulation of the background spectrum observed by the high-resolution germanium gamma-ray spectrometer flown on HEAO 3 in low earth orbit, with a tabulation of approximately 130 observed line energies, count rates, and tentative identifications.

INTRODUCTION

High-resolution astronomical gamma-ray spectroscopy using cooled germanium spectrometers holds great promise in the 50 keV to 10 MeV range where nuclear lines are prominent. The sensitivity of an instrument of this type is, for a given observing time, determined by two factors: the effective collecting area, and the background. Of these, the effective area is generally straightforward to compute by means of Monte Carlo photon transport calculations. It can also be measured directly on the ground with calibrated radioactive gamma-ray sources. The background, due to the great number and complexity of the contributing processes, is very difficult to compute accurately[1]. It also cannot be meaningfully measured on the ground, as the space radiation environment is very different due to the dominance of high-energy cosmic rays and their accompanying secondary radiations and the associated activation.

The Jet Propulsion Laboratory High-Resolution Gamma-Ray Spectrometer Experiment flown on NASA's Third High Energy Astronomy Observatory (HEAO 3) in 1979 and 1980 was the first large germanium spectrometer flown in space to be dedicated to astronomy, and is the largest yet placed in orbit. The background spectrum it observed is of interest both for the analysis and understanding of data from current experiments and for the design of new ones. The principal purpose of this paper is to present a long background accumulation from HEAO 3 showing well over 100 discrete lines together with fitted line energies, rates, and tentative identifications.

INSTRUMENT AND ANALYSIS

The JPL spectrometer on HEAO 3 consisted of four independent p-type high-purity germanium sensors which operated in the energy range from 45 keV to about 10 MeV. The total volume of the germanium detectors was about 400 cm^3; the total effective area was about 75 cm^2 at 100 keV. Each detector had its own independent signal chain, 8192-channel pulse height analyzer, and telemetry. Every good event was tagged with time, pulse height, and status bits. Telemetry capacity allowed a maximum of 15.6 events per second to be individually transmitted to Earth per detector, compared to the typical on-orbit background rate of 10 s^{-1} in the 45 keV to 10 MeV range. A common cryostat contained the four Ge detectors, which were cooled to 92 K by a two-stage methane/ammonia cooler, and surrounded by a 6.6 cm thick CsI active anticoincidence shield. Holes drilled in the CsI collimated the detector fields-of-view to 30° (FWHM) at 45 keV, increasing at high energy. The experiment was launched with HEAO 3 on 1979 September 20 into a 500 km, 43.6° inclination orbit, and was fully

© 1989 American Institute of Physics

operational from shortly after launch until cryogen exhaustion on 1980 June 1. During its normal operation, the spacecraft spun every 20 min about an axis pointed at the sun, causing the instrument to scan a great circle perpendicular to the ecliptic. Thus every six months a complete survey of the sky was obtained. More complete information about instrumental details appears in References 2 and 3.

The data presented here were obtained during the period 1979 September 23 through 1979 November 11. The p-type detectors, in the normal electrode configuration[4] used, were subject to radiation-induced resolution degradation with time[3]. However, during this immediate post-launch period, radiation damage was small so that the detector resolutions were near their pre-launch values, which averaged 3.1 keV (FWHM) at 1460 keV; differences among the individual detectors and their dependence on time are detailed in Reference 3.

Variations in detector gain, of the order of one percent during the mission, and among the four detectors, were corrected before the data from the four were combined. The PHA channel number corresponding to each of a number of the stronger lines, with known energies, was determined by the line fitting program HYPERMET[5] and the best-fit gain and offset for each detector entered into a table used for reference by other analysis programs. The spectra from the four separate germanium detectors were combined by reducing ("normalizing") all to a standard nominal PHA system of 1.2200 keV per channel, with zero offset. This is the binning shown in the figures. However, the best possible energy resolution was obtained if the detectors were analyzed individually prior to such gain normalization. For this reason, Table I is based on detector number 4 only, taken during 1979 October 23 to November 11, when its gain and offset were exceptionally stable.

Each event was telemetered with five accompanying flag bits: "simultaneous" (S, 2 bits), "veto" (V, 1 bit), and "window" (W, 2 bits). Of these, S indicated the number of germanium detectors with simultaneous energy loss above the lower level discriminator (45 keV) for each event, V indicated the presence of a shield veto, and W counted the number of shield segments with simultaneous energy loss in a window bracketing 511 keV; this was to allow the analysis of events with annihilating positrons. For normal gamma-ray astronomy data analysis, S = 1, V = W = 0 is appropriate, and the spectrum shown in Figures 1 and 2 was collected with this flag condition. However, for the purposes of the line identifications and analyses of Table I, V and W were ignored. (Because the on-board veto function of the shield normally suppressed V=1 events, for the data given here, only the difference in the W condition is actually significant.) The main effect of this is to approximately double the rate in the 511 keV annihilation line, from 0.065 s^{-1} in the figures and data used for analysis to 0.131 s^{-1} in the table due to the acceptance of β^+ annihilation in which a single 511 keV photon reached the shield. Counting rates in this line for various flag conditions are given by Mahoney, Ling, and Jacobson[6]. About a dozen other line rates with possible coincident β decay from spallation products in the shield are also somewhat affected; these are indicated in the notes to the table.

The data were selected to eliminate possible transmission errors and precipitating particle events. The detectors were turned off during South Atlantic Anomaly (SAA) passages, and data 600 s thereafter were also excluded, but Earth-looking data were included.

BACKGROUND SPECTRUM

Figure 1 presents the resulting overall spectrum on a log/log plot, to show the global continuum and its relation to the stronger lines. The spectrum decreases below 100 keV due to absorption of the diffuse cosmic background and atmospheric albedo fluxes by the dead layers on the detector end faces. From about 100 keV to 1 MeV the

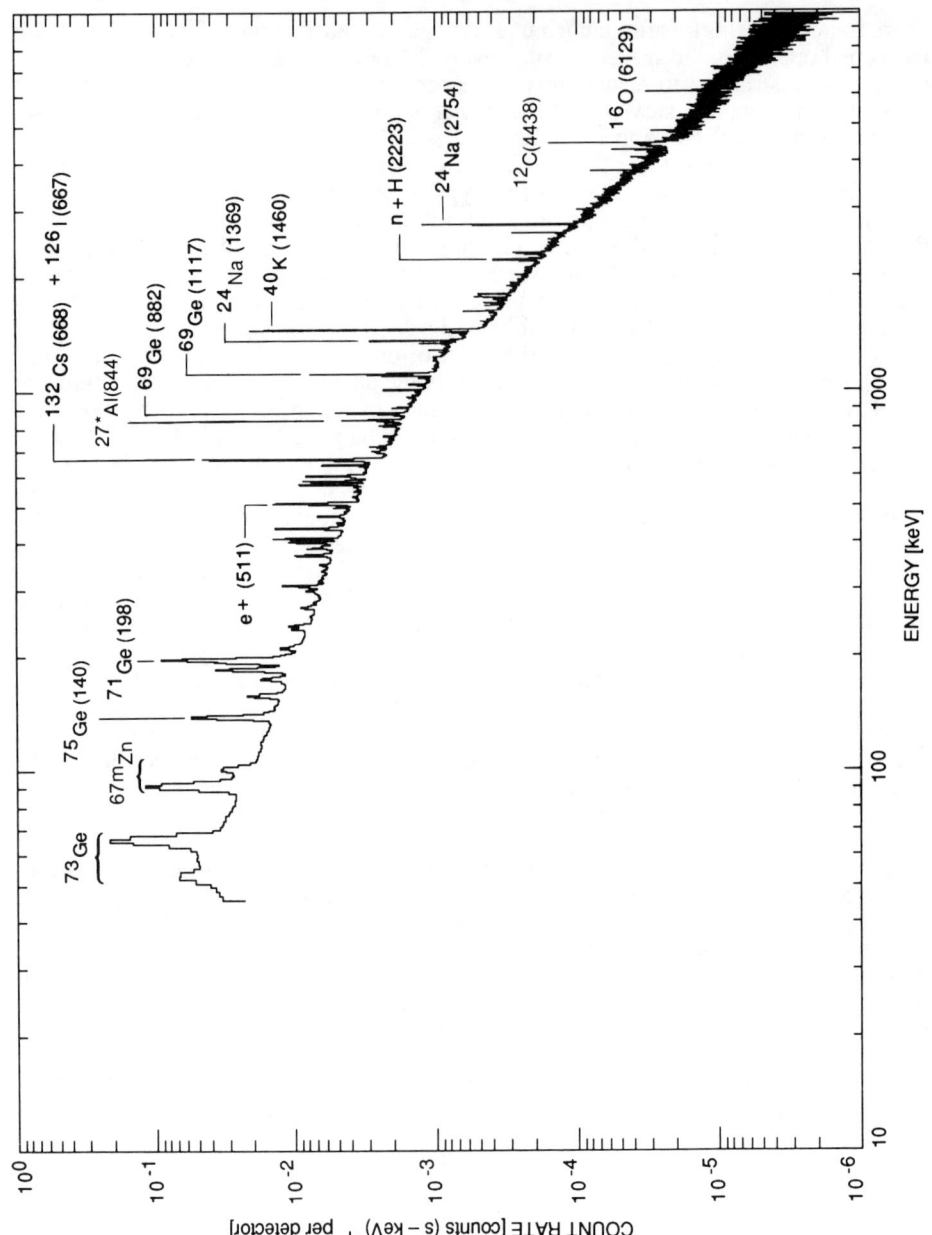

Figure 1 (opposite) : HEAO 3 50-day background spectrum plotted on a log/log scale to show overall shape. Spectra from the four separate detectors have been combined as described in the text. Thus the instrument total was four times higher. The data were accumulated from 1979 September 23 to 1979 November 11. Both earth- and sky-looking data are included.

underlying continuum may be coarsely described as 1/E in form; above 1 MeV the continuum steepens to nearly $1/E^3$. The total counting rate was about 10 s^{-1} in each of the four detectors. Figure 2 shows the same spectrum on an expanded linear scale so that the wealth of detail revealed by the high-resolution detectors can be seen, along with weaker lines invisible in Figure 1. In addition to the many strong lines that are apparent, there are several triangular features due to inelastic neutron scattering[7]. One producing the noticeable edge at 693 keV is especially clear.

Table I gives a tabulation of the lines observed. We have not attempted to apply a strictly uniform criterion of statistical significance (based on HYPERMET line fits) for the inclusion of line-features in the table, as there are other considerations which affect our confidence in the reality of the many features observed. However, the significance of otherwise unidentified lines is typically about 5 σ or above. Column 1 gives the observed energy as determined by HYPERMET. Here and elsewhere, errors in the last place of decimals are indicated by parentheses. The HEAO 3 errors shown are statistical, and make no allowance for systematic effects which may produce larger energy shifts in particular situations. The 92/102 keV line pair, discussed further below, is an example of a case where an associated continuum may produce a systematic inaccuracy in the HYPERMET energy determination. After small corrections for signal-chain non-linearities, we believe the final gain determinations are accurate to ≤ 0.1 keV below 1 MeV, increasing to possibly as much as 1 keV near 10 MeV. Some very broad lines, such as the carbon 4.4 MeV feature with width about 100 keV, were not detected by the HYPERMET line survey. Their rates were estimated separately, from the four-detector, 50-day sum.

Columns 2 - 4 give, where a firm or tentative identification exists, the laboratory energy, the parent isotope, and the half-life, respectively. Identifications have been based on energy agreement with compilations of gamma-ray line energies[8], plausibility considering the materials of the instrument and their likely spallation products, and the existence of a state long-lived enough to escape the presumed veto of the exciting primary particle, as well as life-time estimates from the observed orbital variations in count rates[9, 10]. They are not due to detailed calculation from first principles using the dosage and cross-sections, beyond rough consistency estimates. In several cases a consistency check is possible based on the presence and intensity of other lines in the spectrum; such cases are indicated in the notes to Table I. Column 5 gives the fitted count rate in the line for one detector as determined by HYPERMET, with its (statistical) error.

DISCUSSION

Mahoney, Ling, and Jacobson[3] give approximate cosmic-ray, van Allen, and neutron fluences encountered by HEAO 3. They estimate that the fluence of high energy protons due to cosmic rays was approximately 6 x 10^4 cm^{-2} per day, accurate to about a factor of two, with a mean energy of about 4 GeV. They give the fluence of approximately 100 MeV protons as 2.5 x 10^5 cm^{-2} per day, about 80% due to the SAA, the rest cosmic-ray secondaries. The fluence of neutrons with energies > 1 MeV they estimate as about 1.5 x 10^4 cm^{-2} per day. Of the lines in Table I, most are neutron activation or spallation products of germanium, of the Cs and I in the shield, or of other materials used in the construction of the cryostat or near the detectors, such as aluminum and stainless steel. Table II contains a partial inventory of cryostat materials. Most of the remaining lines are associated with the natural radioactivity of residual uranium and thorium in the instrument.

Distinctive spectral features can arise from the summing of sequential transitions from internal radioactivity when the lifetime of the intermediate level is comparable to the time constant of the shaping amplifier (1 µs for HEAO 3) or to the time that the gate of the pulse analyzer is open. For example, the 0.499 s level at 66.7 keV in ^{73}Ge decays by emission of a 53.4 keV gamma ray or by internal conversion to the 2.95 µs level at 13.3 keV. If the 2.95 µs level decays within the analysis time of the electronics, the two transitions will be summed and a peak at 66.7 keV results. If the decay is long compared to the analysis time, the 13.3 keV emission will not contribute and a line at 53.4 keV will be recorded. Intermediate decay times result in partial summing, filling in the region between the two peaks. In the HEAO 3 spectrum, the area of the 53.4 keV peak may have been reduced by the lower level discriminator (LLD) which was set at about 50 keV. Similar internal summing can occur for the 9.1 µs level at 93.3 keV in ^{67}Zn. The level can be populated by the electron capture decay of ^{67}Ga with the prompt de-excitation of the vacancy in the K shell of the Zn atom. If the 9.1 µs level decays within the analysis time, the K binding energy of 9.66 keV will be added to the 93.3 keV transition and a peak at 102.98 keV results. When the decay time is long, only the 93.3 keV transition is analyzed, the 9.66 keV de-excitation energy being below the LLD. Partial summing fills in the region between the two peaks.

The strong line at 1460 keV was due to two calibration sources of enriched ^{40}K placed on the front of the shield on opposite sides of the collimator, so that photons from them had to leak through the shield to reach the detectors. Compton scattering out the aperture of 1460 keV photons from these sources produces the broad hump near 1240 keV. The strong feature at 667 keV, which is absent or weak in balloon-borne instruments, originates in long-lived states of cesium and iodine which require weeks to build up to equilibrium. The 511 keV annihilation line results from a large number of instrumental and environmental sources, including β+ decays in the germanium and nearby materials[1], leakage from the atmospheric positron annihilation line[6], and of course a contribution from cosmic sources[11].

ACKNOWLEDGEMENTS

We thank Guenter Riegler and John Bradley for development of the analysis software used in the accumulation, and Gerry Share for discussions. The work described in this paper was carried out by the Jet Propulsion Laboratory, California Institute of Technology, under contract with the National Aeronautics and Space Administration.

REFERENCES

1. N. Gehrels, Nuc. Instr. and Meth. A 239, 324, (1985).
2. W. A. Mahoney, J. C. Ling, A. S. Jacobson, and R. M. Tapphorn, Nuc. Instr. and Meth. 178, 363, (1980).
3. W. A. Mahoney, J. C. Ling, and A. S. Jacobson, Nuc. Instr. and Meth. 185, 449, (1981).
4. R. H. Pehl, N. W. Madden, J. H. Elliot, T. W. Raudorf, R. C. Trammell, and L. S. Darken, IEEE Trans. Nucl. Sci. NS-26, 321 (1979).
5. G. W. Phillips and K. W. Marlow, Nuc. Instr. and Meth. 137, 525 (1976).
6. W. A. Mahoney, J. C. Ling, and A. S. Jacobson, J.G.R. 86, 11098, (1981).
7. R. L. Bunting and J. J. Kraushaar, Nuc. Instr. and Meth. 118, 565-572 (1974).

8. U. Reus and W. Westmeier, Atomic Data and Nuclear Data Tables, 29,1, (1983).
9. Nuclear Data Sheets, produced by The National Nuclear Data Center, J. K. Tuli editor, Academic Press.
10. Evaluated Nuclear Structure Data File, maintained by the National Nuclear Data Center, Brookhaven National Laboratory, Upton, NY 11973; for information contact J. K. Tuli or S. Pearlstein.
11. G. R. Riegler, J. C. Ling, W. A. Mahoney, W. A. Wheaton, J. B. Willett, A. S. Jacobson, and T. A. Prince, Ap. J. (Letters) 248, L13, (1981).

Figure 2 (following): Same data as Figure 1, on an expanded linear scale to show spectral details. a: 0 - 300 keV; b: 300 - 600 keV; c: 550 - 1000 keV; d: 1 - 2 MeV; e: 2 - 4 MeV; f: 4 - 10 MeV.

(a)

Figure: Count rate spectrum showing peaks labeled ^{73}Ge {66.4}, ^{67m}Zn {92.1}, 102.0, 53.5, ^{75}Ge (139.8), and ^{71}Ge (198), with energy axis from 0 to 300 keV and count rate axis from 0 to 0.25 counts (s·keV)$^{-1}$ per detector.

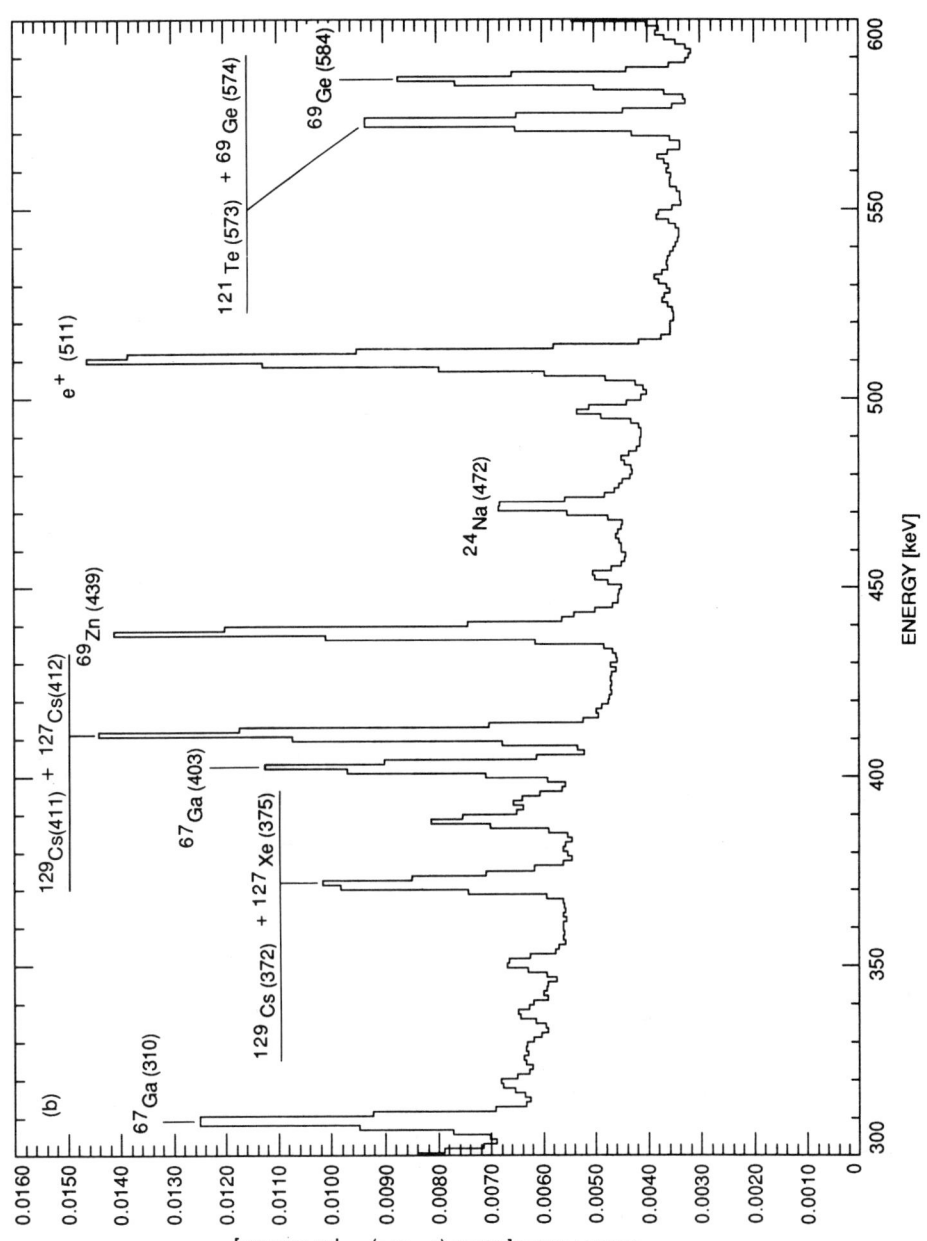

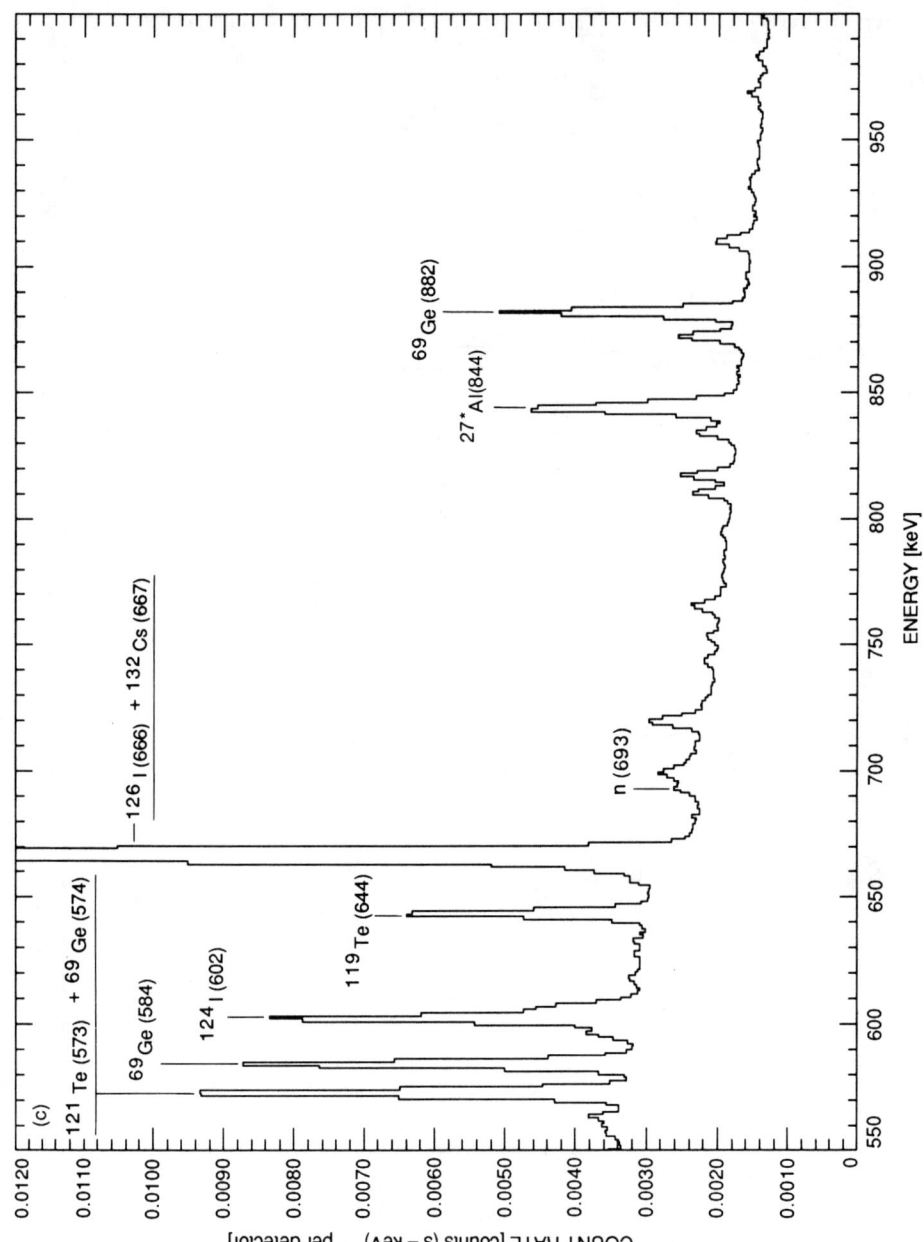

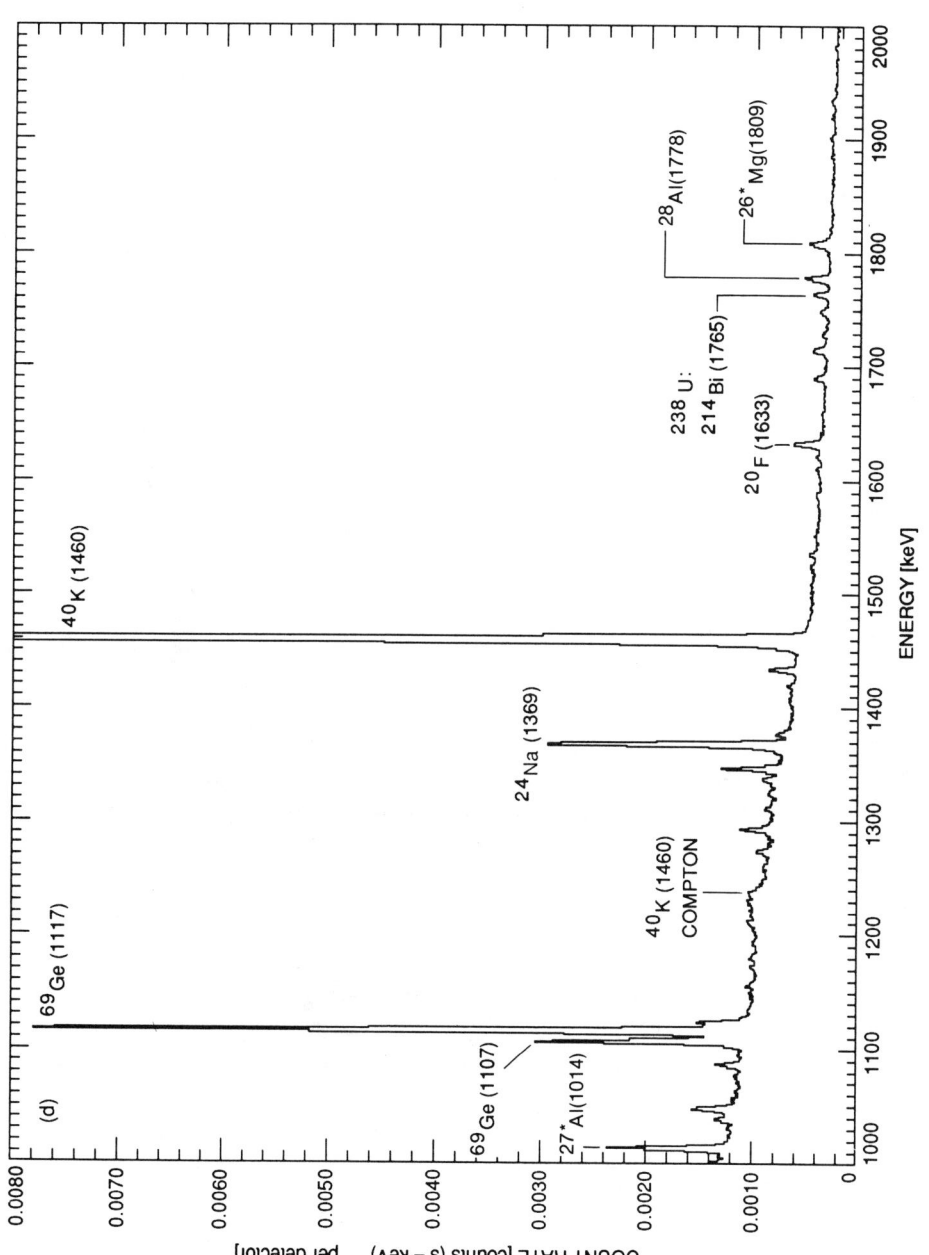

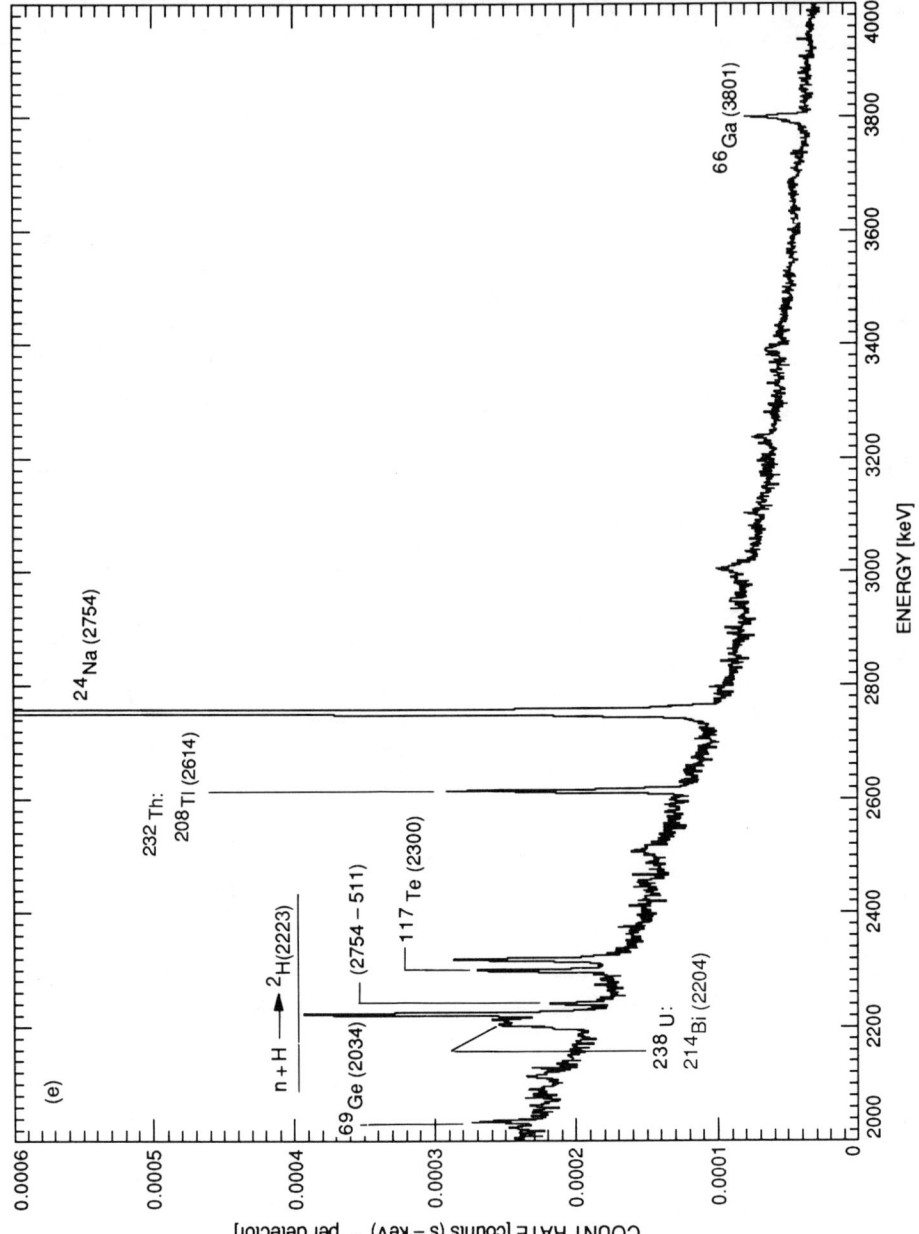

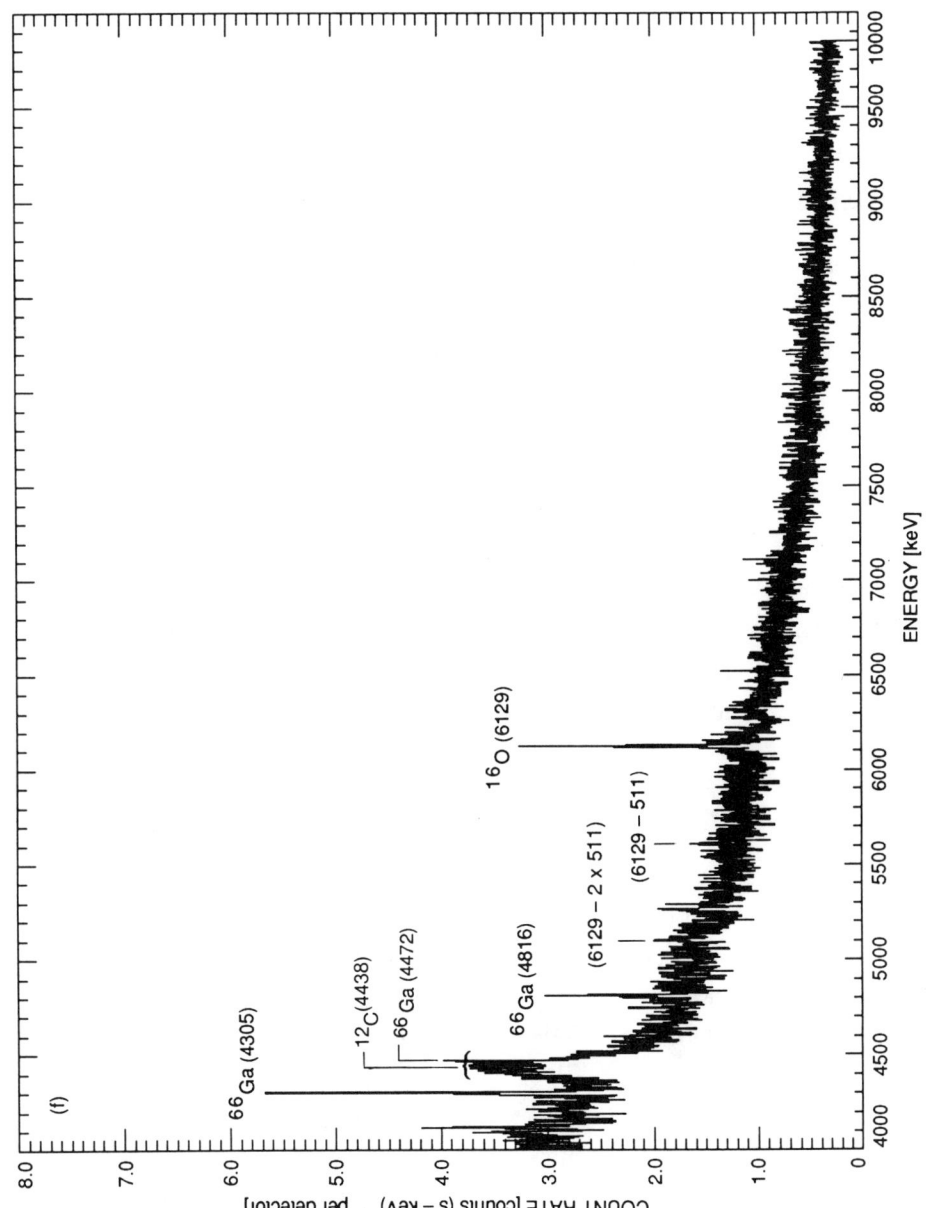

Table I

HEAO 3 BACKGROUND LINES

Energy (keV)		Parent			
Measured	True	Isotope	Halflife	Count Rate[a] [s^{-1}]	Notes
53.54(8)				0.0993(56)	b
	53.437(9)	^{73}Ga	4.86(3) hr		
		^{73m}Ge	0.499(11) s		
		^{73}As	80.30(6) d		
66.43(6)				0.546(12)	b, c
	66.700(20)	^{73}Ga	4.86(3) hr		
		^{73m}Ge	0.499(11) s		
		^{73}As	80.30(6) d		
92.07(7)	93.32(2)	^{67m}Zn	9.1(1) us	0.345(15)	b
101.98(29)				0.0412(87)	
	100.927(5)	^{67}Ga	3.261(1) d		b, e
	102.982(5)	^{67}Ga	3.261(1) d		e
119.38(16)	?	?	?	0.00297(43)	
124.32(25)				0.00237(42)	
	124.75(4)	^{65}Ga	15.2(2) m		e
	124.7(2)	^{127}Cs	6.25(10) hr		
139.80(5)	139.68(3)	^{75m}Ge	47.7(5) s	0.13896(79)	
143.28(10)	143.582(3)	^{57}Co	271.65(13) d	0.00789(47)	e
148.77(17)	148.9(2)	^{123}Xe	2.08(2) hr	0.00288(43)	
158.92(6)	158.97(5)	^{123}I	13.2(1) hr	0.03292(49)	d
175.06(10)	174.93(2)	^{71m}Ge	78 ns	0.0159(13)	
	174.954(5)	^{71}As	65.28(15) hr		
176.77(16)	?	?	?	0.0089(13)	
185.90(5)				0.0853(12)	
	184.58(2)	^{67}Ga	3.261(1) d	[0.017]	d, f
	186.0(1)	^{71}As	65.28(15) hr		e
188.23(15)	188.43(3)	^{125}Xe	16.9(2) hr	0.0108(10)	d
194.15(5)	194.24(2)	^{67}Ga	3.261(1) d	0.08530(91)	d, e
198.29(5)	198.40(4)	^{71m}Ge	20.3(1) ms	0.2870(18)	

Energy (keV) Measured	Energy (keV) True	Parent Isotope	Halflife	Count Rate^a[s^{-1}]	Notes
202.79(10)	202.860(10)	^{127}Xe	36.4(1) d	0.01086(61)	d
212.13(7)	212.189(27)	^{121}I	2.12(1) hr	0.01448(57)	d
216.16(22)	?	?	?	0.00273(55)	
238.63(7)	238.626(5)	^{212}Pb	10.643(12) hr	0.00795(30)	^{232}Th, g
243.30(6)	243.40(4)	^{125}Xe	16.9(2) hr	0.01158(29)	d
257.63(14)	257.5(1)	^{119}I	19.1(4) min	0.00252(24)	d
271.17(8)	271.241(10)	^{44m}Sc	2.442(4) d	0.00749(30)	
274.29(15)	273.44(1)	^{128}Ba	2.43(5) d	0.00321(24)	
280.20(12)	?	?	?	0.00258(23)	
300.77(13)	300.20(2)	^{67}Ga	3.261.1(1) d	0.00514(43)	d
	300.087(10)	^{212}Pb	10.643(12) hr	[0.0005]	^{232}Th, f, g
309.87(8)	309.86(2)	^{67}Ga	3.261(1) d	0.02402(62)	d, e
319.83(22)	320.11	^{51}Cr	27.704(4) d	0.00295(37)	
329.16(31)	329.00(20)	^{69}Ge	39.05(10) hr	0.00193(35)	d, e, h
338.28(24)	338.7(4)	^{228}Ac	6.15(2) hr	0.00227(34)	^{232}Th, g
351.27(19)	351.992(62)	^{214}Pb	26.8 min	0.00321(35)	^{238}U, g
	350.5	^{21}F	4.32(3) s		
372.03(9)	371.918(2)	^{129}Cs	32.35(10) hr	0.01474(65)	
374.66(19)	374.991(12)	^{127}Xe	36.4(1) d	0.00499(60)	d
388.56(7)	388.633(11)	^{126}I	13.02(7) d	0.01941(57)	h
393.44(18)	393.57(3)	^{67}Ga	3.261(1) d	0.00420(43)	d
402.82(7)	403.23(3)	^{67}Ga	3.261(1) d	0.01852(54)	d, e
411.69(6)	411.490(2)	^{129}Cs	32.35(10) hr	0.03197(61)	
	411.9	^{127}Cs	6.25(10) hr		
438.58(6)	438.634(18)	^{69m}Zn	13.76(2) hr	0.03313(33)	
442.86(7)	442.873(9)	^{128}Cs	3.62(2) min	0.00685(21)	h
453.77(13)	453.83(5)	^{125}Xe	16.9(2) hr	0.00218(20)	d
464.65(20)				0.00143(21)	h
	463.05(4)	^{228}Ac	6.15(2) hr	[0.0005]	^{232}Th, g
471.96(8)	472.3(1)	^{24m}Na	20.21(14) ms	0.01042(32)	
476.27(26)	?	?	?	0.00147(19)	
484.60(20)	?	?	?	0.00130(20)	
497.20(9)	498.0(5)	^{115}Sb	31.8(2) min	0.00667(27)	
510.98(6)	511.0034(14)	e$^\pm$ annihilation		0.1305(19)	j, k

Energy (keV) Measured	Energy (keV) True	Parent Isotope	Halflife	Count Rate[a] [s^{-1}]	Notes
526.39(25)	526.577(10)	^{128}Cs	3.62(2) min	0.00158(30)	
532.03(31)	531.9(3)	^{121}I	2.12(1) hr	0.00120(29)	d
548.13(40)	548.945(8)	^{129}Cs	32.35(10) hr	0.00139(39)	
563.72(34)				0.00216(41)	
	563.72(10)	^{69}Ge	39.05(10) hr	[0.0011]	d, e, f
573.05(6)				0.02938(56)	
	573.139(11)	^{121}Te	16.78(35) d	[0.021]	d, f
	574.11(10)	^{69}Ge	39.05(10) hr	[0.005]	d, f
584.47(6)	584.48(10)	^{69}Ge	39.05(10) hr	0.01959(55)	d, e
596.57(32)	595.90(8)	^{74}As	17.79(5) d	0.00234(38)	
602.53(6)	602.72(4)	^{124}I	4.15(3) d	0.02186(57)	
606.87(15)	607.00(8)	^{74}As	17.79(5) d	0.00608(40)	
	609.312(10)	^{214}Bi	19.9(4) m		^{238}U, g
644.00(8)	644.01(4)	^{119}Te	16.05(5) hr	0.01380(49)	d
667.33(6)				0.1946(25)	
	666.331(12)	^{126}I	13.02(7) d	[0.050]	d, f
	667.5(1)	^{132}Cs	6.475(10) d	[0.150]	d, f
699.65(47)	699.85(6)	^{119}Te	16.05(5) hr	0.00147(35)	d
720.05(23)	719.7(7)	^{117}Te	62.(2) min	0.00483(43)	d
744.04(35)	744.214(5)	^{52}Mn	5.591(3) d	0.00102(20)	
753.46(29)	753.819(13)	^{126}I	13.02(7) d	0.00123(20)	d, h
766.32(27)	?	?	?	0.00192(23)	
795.48(20)	795.0(2)	^{228}Ac	6.15(2) hr	0.00159(17)	^{232}Th, g, h
810.81(12)	810.757(21)	^{58}Co	70.78(7) d	0.00262(16)	
817.90(9)	817.868(21)	^{58}Co	70.78(7) d	0.00385(15)	e
834.68(12)	834.827(21)	^{54}Mn	312.20(7) d	0.00300(17)	
843.81(7)	843.76(3)	^{27*}Al	33.(2) ps	0.01437(30)	
	843.22(10)	^{66}Ga	9.40(7) hr	[0.00033]	d, e, f
846.81(14)	846.78(6)	^{56}Fe	6.7(2) ps	0.00495(30)	
872.68(20)	871.98(10)	^{69}Ge	39.05(10) hr	0.00385(22)	d
882.28(6)	882.35(10)	^{69}Ge	39.05(10) hr	0.01361(34)	d, e
909.83(21)	911.07(7)	^{228}Ac	6.15(2) hr	0.00217(19)	^{232}Th, g
931.30(82)	931.800(50)	^{116}Sb	15.8(8) min	0.00057(20)	
968.82(49)	968.8(3)	^{228}Ac	6.15(2) hr	0.00101(23)	^{232}Th, g

Energy (keV) Measured	Energy (keV) True	Parent Isotope	Halflife	Count Rate^a[s^{-1}]	Notes
983.85(49)	983.501(2)	^{48}V	15.976(3) d	0.00068(18)	
1000.74(62)	?	?	?	0.00061(17)	
1014.36(20)	1014.43(4)	^{27*}Al	1.48(13) ps	0.00585(23)	
1039.15(45)	1039.29(10)	^{66}Ga	9.40(7) hr	0.00072(15)	d
1048.66(20)	1048.95(10)	^{66}Ga	9.40(7) hr	0.00236(17)	d, e
1087.25(38)	1087.1(1)	^{68}Ga	68.33(9) min	0.00113(21)	
1107.55(8)	1106.77(10)	^{69}Ge	39.05(10) hr	0.00797(31)	d
1117.40(6)	1117.14(10)	^{69}Ge	39.05(10) hr	0.02968(47)	d, e
1124.32(17)	1124.50(3)	^{65}Zn	241.1(2) d	0.00265(22)	d, e
1156.78(35)	1157.002(3)	^{44}Sc	3.927(8) hr	0.00063(12)	h
1229.28(36)	1229.6	^{118m}Sb	5.00(1) hr	0.00053(12)	
1238.54(49)	1238.11(1)	^{214}Bi	19.9(4) min	0.00052(12)	^{238}U, g, l
1274.20(35)	1274.55(4)	^{22}Na	2.602(2) y	0.00062(10)	
1293.20(15)	1293.538(40)	^{116}Sb	15.5(10) min	0.00167(11)	
1311.60(40)	1312.087(3)	^{48}V	15.976(3) d	0.00045(10)	
1337.48(46)	1336.60(10)	^{69}Ge	39.05(10) hr	0.00053(12)	d
1346.99(12)	1346.97(10)	^{69}Ge	39.05(10) hr	0.00261(15)	d, e
1368.63(6)	1368.53(4)	^{24}Na	15.030(3) hr	0.01111(25)	
1377.81(31)	1377.59(4)	^{57}Ni	35.99(12) hr	0.00081(14)	
	1377.669(14)	^{214}Bi	19.9(4) min		^{238}U, g
1434.27(36)	1434.056(12)	^{52m}Mn	21.1(2) min	0.00158(25)	h
	1434.082(10)	^{52}Mn	5.591(3) d		
1460.83(6)	1460.75(6)	^{40}K	1.277(8)x10^{9}y	0.09519(80)	
1508.85(82)	1509.228(17)	^{214}Bi	19.9(4) min	0.000251(89)	^{238}U, g
1535.47(42)	1536.21(10)	^{69}Ge	39.05(10) hr	0.000488(90)	e
1588.86(69)	1587.9(4)	^{228}Ac	6.15(2) hr	0.000236(90)	^{232}Th, g
1611.49(75)	?	?	?	0.000306(85)	
1633.30(25)	1632.6(8)	^{20}F	11.00(2) s	0.00153(12)	
1690.59(19)	1691.02(4)	^{124}I	4.15(3) d	0.001565(95)	h
1715.84(25)	1716.4(7)	^{117}Te	62.(2) min	0.001019(88)	d
1731.02(71)	1732.135(6)	^{24}Na	15.030(3) hr	0.000375(75)	n
1748.92(42)	1749.65(8)	^{119}Te	16.05(5) hr	0.000460(78)	d
1764.71(30)	1764.50(2)	^{214}Bi	19.9(4) min	0.000655(76)	^{238}U, g

Energy (keV) Measured	Parent True	Isotope	Halflife	Count Rate[a][s⁻¹]	Notes
1778.70(21)	1778.70(17)	^{28}Al	2.2405(3) min	0.001255(84)	
	1779.2	^{58}Mn	65.3(7) s		
1809.23(25)	1808.65(7)	^{26*}Mg	0.49 ps	0.001234(87)	^{26}Na, g, k
1934.27(86)				0.00027(12)	
	1933.4(2)	^{69}Ge	39.05(10) hr	[0.00008]	d, e, f,
2036.47(45)	2034.02(20)	^{69}Ge	39.05(10) hr	0.000307(46)	d, e
2203.38(57)	2204.21(4)	^{214}Bi	19.9(4) min	0.000420(63)	^{238}U, g
2212.66(43)	2211.	^{27}Al	27.7(14) fs	0.000653(63)	
2224.03(21)	2223.351(46)	^{2}H		0.001655(74)	
	2225.327	^{116}Sb			
2242.74(31)	2243.139(6)	^{24}Na	15.030(3) hr	0.000811(69)	m, h
2300.23(33)	2300.0(7)	^{117}Te	62.(2) min	0.000724(80)	d
2319.20(28)	?	?	?	0.001037(91)	
2375.84(121)		^{65}Ga	15.2(2) min	0.000217(58)	
2500	?	?	?	0.0003	o
2614.49(43)	2614.533(13)	^{208}Tl	3.053(4) min	0.001185(76)	^{232}Th, g
2754.22(16)	2754.142(6)	^{24}Na	15.030(3) hr	0.00763(15)	
2763.4(14)	2761.76(20)	^{66}Ga	9.40(7) hr	0.000124(71)	d, e
3007.2(19)	3004.	^{27}Al	58.(5) fs	0.000135(46)	o
3236.1(20)	3238.92(10)	^{66}Ga	9.40(7) hr	0.000066(35)	d, e
3392.0(29)	3390.98(10)	^{66}Ga	9.40(7) hr	0.000064(30)	d, e
3798.6(7)	3801.13(10)	^{66}Ga	9.40(7) hr	0.000441(50)	d, e
4094.9(21)	4096.02(10)	^{66}Ga	9.40(7) hr	0.000040(26)	d, e
4118.8(15)	4122.672(40)	^{24}Na	15.030(3) hr	0.000068(26)	c
4294.7(5)	4295.70(20)	^{66}Ga	9.40(7) hr	0.000076(14)	d
4303.3(16)	4305.36(20)	^{66}Ga	9.40(7) hr	0.000232(68)	d, e
4438	4438.01(31)	^{12}C	42 fs	0.001023(40)	p
4472.8(9)	4471.67(15)	^{66}Ga	9.40(7) hr	0.000120(29)	d, e
4812.8(24)	4816.25(15)	^{66}Ga	9.40(7) hr	0.000092(29)	d, e
5107	5107.163(43)	^{16}O	17 ps	0.000031(20)	n, p
5616.0(17)	5618.167(43)	^{16}O	17 ps	0.000102(28)	m
6128.9(11)	6129.170(43)	^{16}O	17 ps	0.000172(27)	

Notes to Table I:

a   Count rate in one detector (detector 4) ignoring V and W flags (except as noted under p below). Total (four detectors) instrument rate was approximately four times higher.

b   Complex peak structure. (See text.)

c   Sum peak.

d   Rate consistent with one or more other line rate in the same decay scheme.

e   Energy shown includes k-edge energy as follows:

Co	(Z = 27)	7.7114 keV
Ni	(28)	8.3324
Cu	(29)	8.9805
Zn	(30)	9.6608
Ga	(31)	10.367
Ge	(32)	11.104
As	(33)	11.867

f   Rate estimated from other line rates observed in same decay scheme, or from comparisons of proton interaction cross sections. The latter estimates are believed accurate to about a factor of two.

g   Half-life of immediate parent shown.

h   Rate significantly affected by shield window flag W; see text.

j   If flags are required to be $V = 0$, $S = 1$, $W = 0$, the normal condition for science analysis, then rate is 0.067 counts s^{-1}. See text.

k   Cosmic contribution included in line rate.

l   Rate may be affected by presence of ^{40}K Compton edge.

m   Single escape.

n   Double escape.

o   Broad.

p   Rate estimated from data of Figures 1 and 2.

Table II

HEAO 3 Cryostat Materials

Element	Estimated Mass (kg)
Al	1.735
Fe	1.552
Cr	0.438
Ni	0.238
Ag	0.080
Mn	0.046
Si	0.033
Mg	0.017

ON-ORBIT OBSERVATIONS OF SINGLE EVENT UPSET IN
HARRIS HM-6508 RAMs: AN UPDATE

J. B. Blake
Space Sciences Laboratory, The Aerospace Corporation,
Los Angeles, CA 90009

R. Mandel
Lockheed Missles and Space Corporation,
Sunnyvale, CA 94088

ABSTRACT

The observed single-event-upset rate of Harris HM-6508 RAMs in a low polar orbit is presented. These data were acquired during a four year period from 1983 through 1986.

INTRODUCTION

The single event upset (SEU) phenomenon has continued to be of great interest to designers of spaceflight hardware. Although a great deal of ground testing has been and continues to be carried out, quantitative on-orbit measurements are very limited. Space measurements are a key part of the effort to ensure that the ground-based testing yields accurate predictions of on-orbit upset rates. Blake and Mandel have published[1] data from two years of flight observations from a subsystem consisting of 384 Harris HM-6508 1K RAMs. This paper is an update of that study; the results of 2560 days of observation are presented.

THE EXPERIMENT

The Harris HM-6508 1K x 1 RAMs are in a satellite subsystem in low, polar orbits. The memory module used in the subsystem containing the RAMs consists of three printed circuit cards with each card containing eight 2K byte memory hybrids for a total of 48K bytes. Thus each memory hybrid contains 16 HM-6508 RAM chips.

On a regular basis all but 256 bytes of the 48K bytes are examined for bit errors. Two different techniques are used for detecting bit errors. The first technique, a memory check sum, is capable of automatically detecting all single bit and some double bit errors which occurred within a page of memory. A memory page consists of 256 bytes. Memory check-sum tests are performed approximately every 90 minutes. To detect a multiple error or to determine the exact location of the bit error within the page, the entire contents of the memory is dumped and compared to the load file. Memory dumps are normally performed once a month, or immediately after the check sum routine detects an error. Once the location of the error is found, the correct value is reloaded into the memory. After the memory is reloaded, the contents of the memory location in question is verified in order to determine if the error was a soft error generated by an SEU or a hard error generated by a part failure or cosmic-ray induced latchup.

© 1989 American Institute of Physics

RESULTS

A total of 234 SEUs were observed during 2560 days of observation. The average upset rate per day thus is:

$$(2.62 \pm 0.17) \times 10^{-7} \text{ upsets/bit day}.$$

The distribution of upsets as a function of time is given in Figure 1.

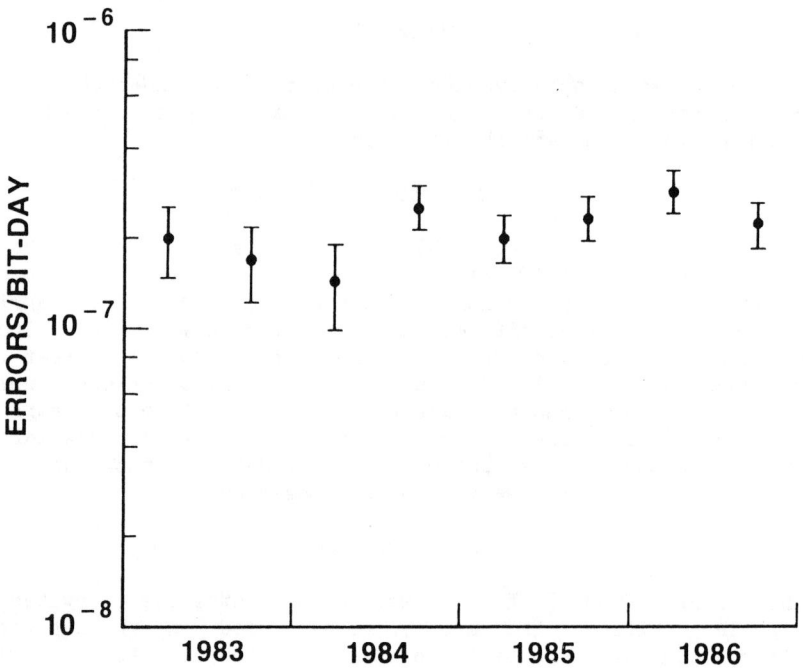

Figure 1. Measured upset rate as a function of time

The data have been grouped into six month bins as a compromise between time resolution and counting statistics. Solar cycle modulation can be seen to be modest at best; a horizontal straight line is not a bad fit to the data. A computer code written by Adams[2] has been used to predict the on-orbit upset rate. The RAM input data are based upon accelerator testing of the HM-6508 RAMs; the data were discussed in Blake and Mandel[1]. Some results of the calculation are given in Figure 2 for shielding thicknesses of 1, 2 and 5 gm/cm^2 of aluminum.

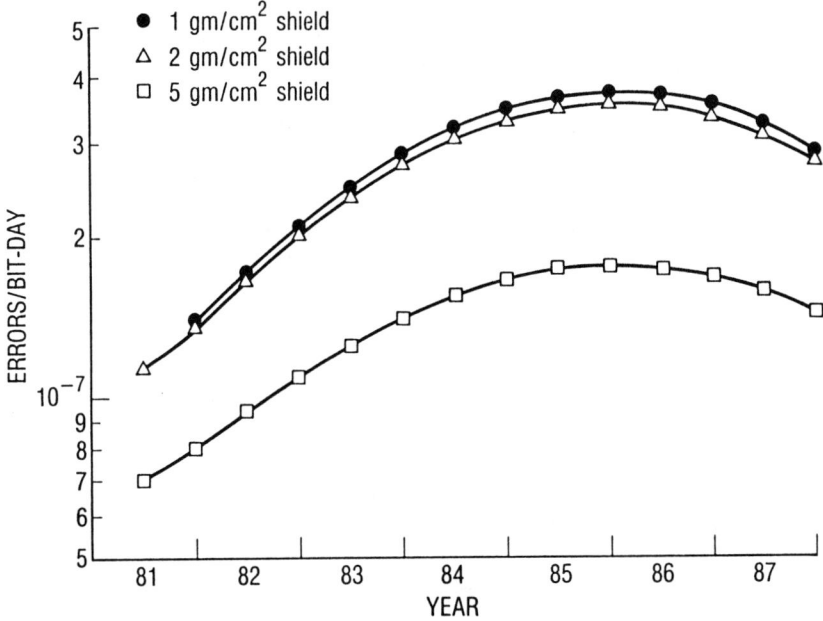

Figure 2. Calculated upset rates as a function of shield thickness

It can be seen that the observed SEU rate, given above, is consistent with a shielding thickness of a few gm/cm^2 which is a reasonable value for the satellite in question. The predicted solar-cycle modulation may be greater than that observed, although the issue is unclear given the counting statistics.

Multiple events were observed in 14% of the upsets. The multiplicity distribution is given in Table I.

Table I. SEU Multiplicity

Number of Events	Multiplicity
202	1
27	2
3	3
1	4
1	All bits on chip

The relative location of the upsets in a multiple event is interesting. In many of the events the multiple errors were adjacent, including the 4-fold error[1].

However in eight cases the errors were not even on the same board. The obvious question arises: are these multiple SEU events really two independent events which occurred between verifications. As discussed in Blake and Mandel[1], the probability of such an accidental event is less than once per decade given the observed upset rate. Perhaps these separated multiple events were due to a single cosmic ray which creates a shower in the satellite vehicle due to a nuclear interaction. An understanding of multiple events will be an important consideration in the effective implementation of EDAC (error detection and correction) techniques in spaceflight hardware.

REFERENCES

1. J. B. Blake and R. Mandel, <u>Proceedings IEEE Trans. Nucl. Sci.</u>, <u>33</u>, No. 6, 1616, 1986.

2. J. H. Adams, Jr., private communication, 1985.

V. Detectors and Experimental Programs

THE SPACE RADIATION ENVIRONMENT AT 840 KM

E.G. Mullen, M.S. Gussenhoven and D.A. Hardy
Air Force Geophysics Laboratory, Hanscom AFB, MA 01731

ABSTRACT

The Defense Meteorological Satellite Program (DMSP) F7 satellite, launched in November, 1983, carries a dosimeter provided by the Air Force Geophysics Laboratory. The dosimeter uses planar silicon detectors behind four thicknesses of aluminum shielding to measure both radiation dose and high energy particle fluxes in the space radiation environment at 840 km. Energy thresholds in the detectors are set to distinguish low (electron), high (proton), and very high (>40 MeV) energy particle depositions. The dosimeter returns accurate, high-time-resolution dose measurements. Maps of the radiation dose (electron and proton) at 840 km are presented and compared to the NASA models. Maps of the very high energy deposits which can produce Single Event Upsets (SEUs) in microelectronic components are also presented. Characteristics of energetic particles that enter the polar cap regions during solar particle events are discussed and compared to inner belt proton and cosmic ray background levels. Included is an analysis of two of the largest solar proton events since launch of the satellite, those of 16 February, 1984, and 26 April, 1984.

I. INTRODUCTION

A sensitive volume in living tissue can be affected by energy deposition from high-energy particles to such a degree that a person's performance can be degraded or impaired. As such satellite-borne humans are subject to radiation effects as they are carried through naturally occurring or artificially-produced regions of radiation in space. The effects can be produced in three general ways, by total dose, dose rate, and single event upsets (SEUs). As the frequency and duration of manned space flights increase, we need to improve our ability to predict and model the space radiation environment to ensure the safety and maintain the performance capabilities of the crew members.

There are essentially three near-Earth radiation regions: a) the inner radiation belt region populated mainly by stably-trapped high energy protons; b) the outer radiation belt region populated mainly by trapped, but highly variable fluxes of electrons; and c) the polar regions populated mainly by steady galactic cosmic rays and infrequent but high intensity solar proton event particles. Low altitude vehicles in high inclination orbits encounter parts of all the various radiation environments, the extent depending on the altitude of the satellite.

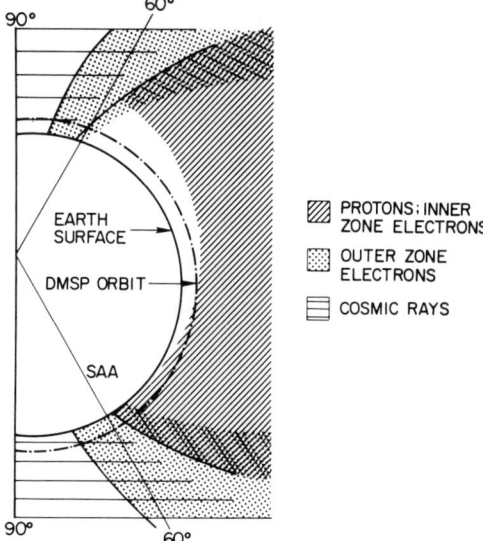

Fig. 1. Schematic diagram of the DMSP polar orbit passing through the low altitude extensions of the major radiation regions. The shading code for the various regions is given on the right.

DMSP/F7 is a three-axis stabilized satellite in a sun-synchronous orbit whose orbital plane is the 1030-2230 local time meridian. The spacecraft altitude is 840 km, its period is 101 minutes, and its inclination is 98.7°. Figure 1 schematically illustrates the three high energy particle populations as they are encountered by the DMSP satellite. The offset of the Earth's

© 1989 American Institute of Physics

dipole is exaggerated to illustrate the low-altitude extension of the proton belts. The stable inner zone proton belt (cross-hatching) reaches low altitudes in the region of the South Atlantic Anomaly (SAA) due to the Earth's offset dipole. The SAA is limited in longitude and latitude, and as the Earth rotates underneath the satellite, approximately 10 of 14 of the daily orbits encounter portions of the SAA. The outer zone electrons (dotted region) are located in two high latitude rings at 840 km, one in each hemisphere. The outer zone fluxes are highly variable. The polar caps (horizontal lines) are regions of direct entry for high energy particles both galactic cosmic rays and solar particles; however, under certain conditions, the higher energy cosmic rays and solar particles can penetrate to 840 km at any latitude.

Of the three regions of high energy particles, two are considered most hazardous to man. They are the SAA and the polar regions during solar particle events. Significant dose in these regions is produced by high-energy protons and heavy ions, which are difficult, or impossible, to shield. The outer zone electrons are lower in energy and can be effectively shielded; however, they are highly variable in intensity and can produce hazardous dose rates behind minimal shielding, such as would be the case for an astronaut only protected by a space suit. The human effects of the higher energy heavy ion particles are not yet fully understood, although it is known that they can produce light flashes in the eyes and may be able to upset certain brain functions, similar to a SEU in a microelectronic chip.

The DMSP dosimeter is extremely versatile, returning information on electron and proton dose, electron and proton flux, and nuclear star events, all with 4 second time resolution and in 4 integral energy steps. It is our intention here to show the versatility of the instrument, and to present results from the major analysis efforts undertaken to date using the dosimeter data. These efforts include both short-term and empirical model results of dose and nuclear star events, as well as case studies of major solar particle events. The dose model results are compared to the NASA model predictions. The short-term results are presented to indicate the range over which the environment can deviate from the models (both ours and NASAs), and where the deviations are likely to be most important. Because of the capability of the dosimeter, we are able to present our direct dose and dose rate measurements separately for electrons and protons.

II. SSJ* INSTRUMENT DESCRIPTION

The DMSP/F7 dosimeter measures the radiation dose from both electrons and protons occurring behind four different thicknesses of aluminum shielding. Additionally, information is provided on the differential and integral fluxes of electrons and protons at energies above the thresholds defined by the shields, and on the number of nuclear star events in each detector. The basic measurement technique is the determination of the amount of energy deposited in a simple solid-state detector from particles with sufficient energy to penetrate the shielding. A lower limit cutoff of 50 keV is set for measuring energy deposition in each detector. Each of the four detectors is mounted behind a hemispheric aluminum shield. The aluminum shields are chosen to provide electron energy thresholds for the four sensors of 1, 2.5, 5, and 10 MeV, and for protons of 20, 35, 51, and 75 MeV. The 1 MeV threshold sensor has a detector area of .051 cm^2, and the remaining three each have areas of 1.00 cm^2. Particles that penetrate the shield and bremsstrahlung produced in the shield will deposit energy in the device producing a charge pulse. The charge pulse is shaped and amplified. The pulse height is proportional to the energy deposition in the detector, and the dose is proportional to the sum of the pulse heights.

Energy depositions in the range between 50 keV and 1 MeV are used to calculate the low linear energy transfer (LOLET) dose; depositions between 1 MeV and 10 MeV provide the high linear energy transfer (HILET) dose; energy depositions above 75 MeV in detector 3 and above 40 MeV in detectors 1, 2, and 4 are counted as very high linear energy transfer (VHLET) events. The LOLET dose (which we will call electron dose below) results primarily from electrons, high energy protons (above approximately 100 MeV incident), and bremsstrahlung. The HILET dose (which we will call proton dose below) comes primarily from protons below about 100 MeV incident and above the dome threshold for each detector. The integral flux is proportional to the number of energy depositions counted. The VHLET counts (which we will call "star counts" below) come from a.) high energy nuclear (mostly proton induced) interactions inside and /or near the sensitive device volume, b.) direct energy deposition by heavier cosmic rays, or c.) direct energy deposition by protons that have long path lengths in the detectors. These will be referred to respectively as a.) nuclear stars, b.) cosmic ray events and c.) direct deposit proton events below. (The 'star'

TABLE 1

DMSP/F7 J* DOSIMETER CHARACTERISTICS

Dome	Aluminum Shield (gm/cm^2)	Dome Thresholds		Area (cm^2)	Detector Thickness (microns)	Threshold		
		Electron (MeV)	Proton (MeV)			LOLET (MeV)	HILET (MeV)	VHLET (MeV)
1	0.55	1	~20	.051	398	.05-1	1-10	>40
2	1.55	2.5	35	1.000	403	.05-1	1-10	>40
3	3.05	5.0	51	1.000	390	.05-1	1-10	>75
4	5.91	10.0	75	1.000	384	.05-1	1-10	>40

description originally comes from the array of emulsion tracks observed when a high energy proton interacts with a nucleus producing secondaries and a recoiling fragment.) Thus, five separate outputs are obtained from each of the four hemispherically shielded detectors: LOLET (electron) dose, LOLET (electron) flux, HILET (proton) dose, HILET (proton) flux, and VHLET flux (star counts). (A complete description of the instrument can be found in Gussenhoven et al.[1]) A summary of the detector properties and their shielding is given in Table 1, and a summary of the dome shielding effectiveness for protons is given in Table 2.

TABLE 2

DOME SHIELDING EFFECTIVENESS FOR PROTONS

INTERNAL PROTON ENERGY (MeV)	EXTERNAL PROTON ENERGY (MeV)			
	DOME 1	DOME 2	DOME 3	DOME 4
0	20.0	35.0	51.0	75.0
8	21.8	37.0	52.5	75.2
10	23.4	37.8	53.0	75.7
20	30.0	43.0	56.8	78.5
40	45.0	55.0	68.0	87.5
100	103.2	110.0	119.0	124.2
1000	1001.	1003.	1005.	1010.

III. DOSE MEASUREMENTS AND MAPS

During quiet times when there are no high energy solar protons present, the high energy proton fluxes are measured by DMSP/F7 dosimeter only in the SAA. A survey of the satellite's daily encounters of the SAA is shown in Figure 2. Here the average flux count rates for protons >75 MeV are plotted as a function of universal time (UT) in hours for 9 November 1984. The averaging interval is one minute. Each spike in the count rate indicates the crossing of a portion of the SAA as the Earth rotates underneath the DMSP orbit. Between each crossing, the satellite passes over the two polar caps and the equatorial region opposite the SAA. During these intervals, only background counts are detected. Similar spatial distributions occur for the lower energy proton flux counts and the proton dose counts.

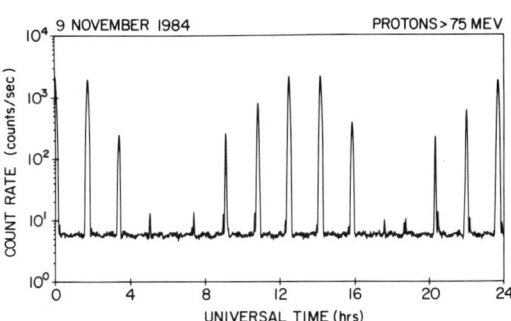

Fig. 2. Proton flux counts for energies greater than 75 MeV, measured by DMSP on 9 November 1984 when no solar event was in progress. The spikes occur during crossings of the South Atlantic Anomaly.

The depth-dose spectra for the electron, proton and total dose, obtained from a single traversal through the heart of the SAA on February 3, 1985, are given in Figure 3. The spectra are hard (highly energetic), with the dose decreasing only somewhat more than a factor of two over the entire shielding range and becoming nearly asymptotic for the thickest shielding. The proton dose is approximately twice the electron dose. The total dose behind the thinnest shield is .38 rad(Si) and behind the thickest shield is .17 rad(Si). For the DMSP satellite, the total dose per day from the SAA is nearly 2 rad(Si) behind the thinnest shield and 1 rad(Si) behind the thickest shield.

To make a more quantitative estimate of the dose accumulated per day from the SAA we constructed an empirical DMSP dose rate map for 840 km altitude. Dose count-rates from individual passes were accumulated for a full year (from 1 November, 1984 to 31 October, 1985) in bins five degrees by five degrees wide in corrected geomagnetic latitude and longitude coordinates. Average count rates were calculated for each bin, and contours of constant dose drawn. The dose rate map for the >75 MeV HILET channel (with 5.91 g/cm^2 aluminum shielding) is shown in Figure 4. The map is in corrected geomagnetic coordinates. The contours are in rad(Si) per day. As can be seen, the only significant proton dose occurs in the SAA where the contours range from 0.1 to 10 rad(Si) per day.

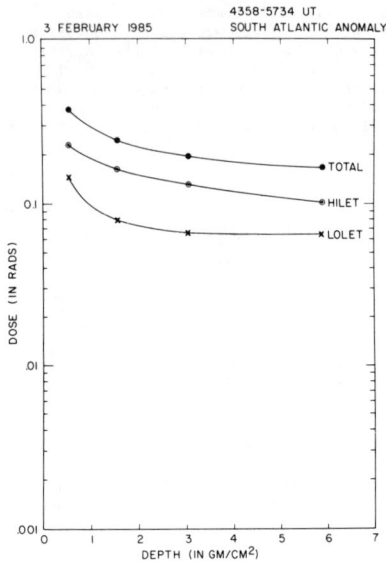

Fig. 3. Depth-dose spectrum for a DMSP/F7 crossing of the heart of the South Atlantic Anomaly on 3 February, 1985.

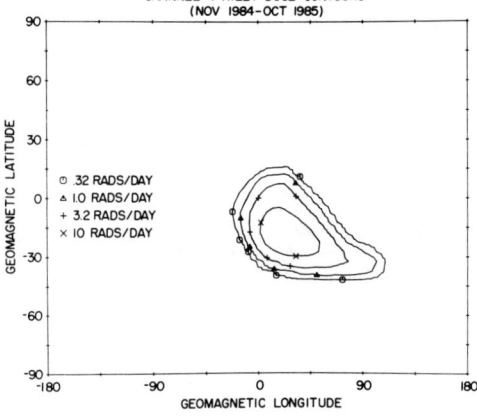

Fig. 4. Contours of constant dose-rate for the > 75 MeV proton dose detector, plotted in corrected geomagnetic latitude and longitude coordinates.

The outer zone electrons are mostly found at magnetic latitudes between 50° and 70° at 840 km. The extent of these zones is shown in Figure 5, which is the DMSP dose rate map for electron detector 1 (behind 0.55 g/cm² shielding). The outer zone electron dose is very evident and appears as two high latitude bands. The SAA is also apparent; however, in the SAA the dose results from a combination of electrons above threshold energies (>1 MeV) and protons with energy greater than ∼100 MeV.

To show how the DMSP measured dose compares with the NASA model values, Table 3 gives a listing of individual day and yearly average proton and electron dose/day obtained from DMSP and the NASA model predictions for the same orbit for both solar maximum and solar minimum options. The first three measured values result from summing the DMSP dose counts for the three individual days: 9 November 1984, 26 November 1984, and 3 February 1985, respectively. There were no solar proton events on these days. The average DMSP values result from running nine successive days of DMSP/F7 orbits through the DMSP dose rate map values and taking the average accumulation. The standard deviation varied from ±8% in the first proton detector up to ±10% in the highest energy detector. The standard deviations are listed just below the average DMSP values. Similarly the DMSP/F7 satellite orbit was run through appropriate NASA radiation belt models for solar maximum and solar minimum conditions with the dose calculated at the center of a solid aluminum sphere of

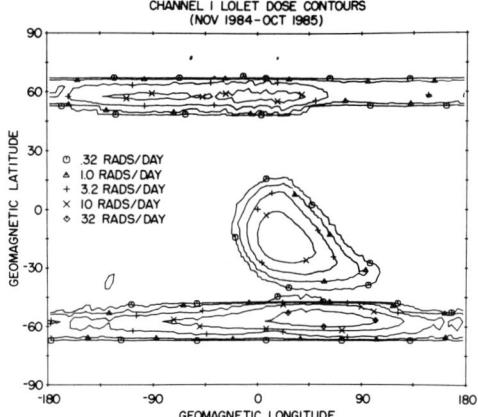

Fig. 5. Contours of constant dose-rate for the >1 MeV electron dose detector, plotted in corrected geomagnetic latitude and longitude coordinates.

thickness equal to the DMSP dosimeter shielding values. The NASA model values for solar minimum and maximum are given at the bottom of the table. (Details of the NASA models can be found in Singley and Vette[2], Teague and Vette[3], and Sawyer and Vette[4], and problems associated with radiation belt modelling can be found in Vette et al.[5])

TABLE 3

COMPARISON OF NASA MODEL AND DMSP DOSE VALUES

DETECTOR NO	PROTONS RAD(Si)/DAY				ELECTRONS RAD(Si)/DAY			
	1	2	3	4	1	2	3	4
DAY 11/09/84	1.20	0.83	0.63	0.48	1.61	0.35	0.28	0.27
DAY 11/26/84	1.22	0.77	0.57	0.44	13.02	0.50	0.27	0.28
DAY 2/03/85	1.18	0.80	0.59	0.46	6.31	0.35	0.26	0.26
AVERAGE	1.21	0.83	0.63	0.49	2.51	0.34	0.26	0.27
STANDARD DEVIATION	±8%	±9%	±9%	±10%	±2%	±9%	±9%	±9%
NASA SOLAR MIN	1.28	0.70	0.49	0.20	10.83	0.58	0.03	0.00
NASA SOLAR MAX	0.89	0.47	0.33	0.18	11.62	1.28	0.07	0.00

SOLAR MIN = AP8MIN, AE5MIN, AE17LO
SOLAR MAX = AP8MAX, AE6MAX, AE17HI
NASA MODEL PROPAGATED THROUGH A SOLID ALUMINUM SPHERE

For the proton dose, the DMSP individual day and DMSP average values are approximately the same. They all agree to within 10%, which is the same variation found in the day to day orbital runs. This indicates that the proton radiation belts are very stable over the one-year duration of the data acquisition, as expected. The solar minimum NASA model values when compared to the near-solar minimum DMSP values for proton dose are only slightly higher for the thinnest shielding and are only slightly lower for the remaining thicknesses. The agreement is felt to be remarkably good given the differences in measurement techniques and solar cycles, and indicates the long term stability of the inner belt.

For the electron dose, the directly measured daily dose varies greatly, from 1.6 to more than 13 rad(Si) behind the thinnest shielding. For higher shielding the dose is relatively constant. The average DMSP model value is 2.51 rad(Si) for the thinnest shielding. This indicates the high variability of the outer zone electrons, over time periods much less than one year. Furthermore, the comparison of the DMSP model and NASA model values is poor. The dose calculated for the DMSP orbit from the NASA models is higher by factors of 4 and 1.8 for Channels 1 and 2, respectively. For the last two channels, the DMSP directly-measured and model electron dose values are nearly constant, while the NASA model values fall to zero. This behavior results from bremsstrahlung effects in the DMSP measured values which would not be predicted from electron fluence in the NASA model values. While it is possible that on occasion (i.e., 26 November, 1984) the daily measured electron dose in Detector 1 can be higher than that predicted by the NASA models, the Detector 2 measured values have not been observed to be as large as the NASA model values. We conclude that the NASA models for electrons are both too high in intensity and too hard in spectral shape.

Comparing total dose behind the thinnest shielding for the DMSP/F7 orbit on a yearly basis, the NASA models give values that are too high by somewhat more than a factor of 3 for the 1984 period. The electron contribution to the total dose is the major source of the discrepancy. Below we will more closely examine the highly variable electron dose from the DMSP data set.

In order to quantify the high variability in outer zone electron fluxes we constructed a daily averaged flux count rate using the region in space where the maximum fluxes occur. Figure 5 illustrates that the maximum occurs in the southern hemisphere in a rectangle centered near -55° geomagnetic latitude and 55° geomagnetic longitude. The region lies far enough below the SAA to avoid contamination of the flux and dose measurements from protons. All flux count rates which fell in this rectangular region were averaged for one day intervals to show the variability of the outer zones. In Figure 6 the average flux count rate for electrons with energy >2.5 MeV is plotted as a function of day number for 1984, the first full year of the dosimeter operation. The day numbers are marked off in 27-day solar rotation units. The average count-rates vary over more than two orders of magnitude, and high fluxes can persist for many days on end. The 27-day recurrence of outer zone electron enhancements is particularly evident in the three largest events in the last quarter of the year.

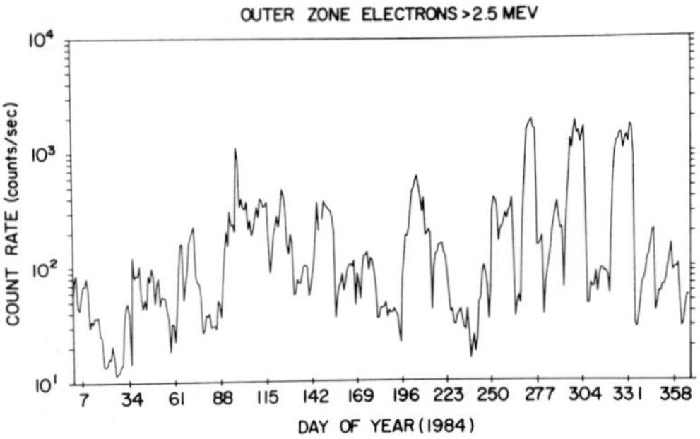

Fig. 6. Daily average count-rates for the >2.5 MeV outer zone electrons plotted as a function of day number for 1984.

IV. STAR COUNT MEASUREMENTS AND MAPS

To perform meaningful statistical analyses of the star count (VHLET) particles, the data must be separated into appropriate regions of space. For this study, the data are summed over all longitudes in broad latitudinal ranges designated as the North Pole (NP) from 40° N to 90° N, Middle Latitude (ML) from 15° N to 40° N, South Atlantic Anomaly (SAA) from 55° S to 15° N and South Pole (SP) from 55° S to 90° S.

Figure 7 shows the average daily star count rates for dosimeter detectors 1 and 2 for 1984 plotted versus day of the year in each of the 4 latitudinal bins. Similarly Figure 8 shows the count rates for detectors 3 and 4. The sharp peaks in the polar cap regions are due to solar particle events. Only 3 solar particle events of any magnitude occurred in 1984. The flare events seen in the polar regions do not penetrate into the mid-latitude or SAA regions. Aside from these peaks, the data are extremely stable over the year and continue to be stable through 1985 and 1986 where we also have plotted the data (not shown). The 1984 through 1986 data were all collected during solar minimum conditions.

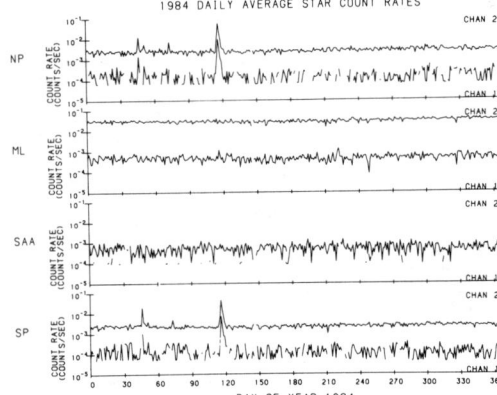

Fig. 7. 1984 daily average star channel counts for detectors 1 and 2.

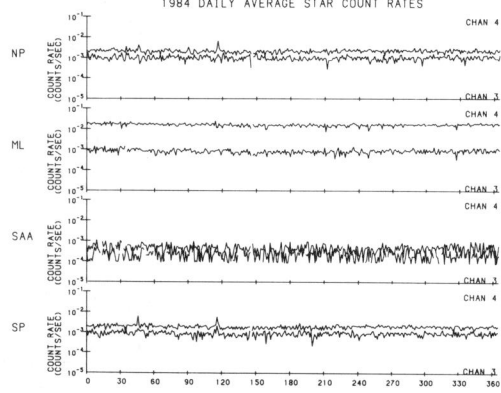

Fig. 8. 1984 daily average star channel counts for detectors 3 and 4.

Figures 9 and 10 show the occurrence frequency of star counts in map format for detectors 2 and 3 of the dosimeter respectively. (Maps for detectors 2 and 4 and additional information on the star counts are provided in Mullen et al.[6]) Shown in gray scale are the average star count rates (counts/s) for the period from November 1984 through October 1985. The average has been calculated for each one degree geographic latitude and longitude box. The gray scale extends from 10^{-3} counts per second to greater than .05 counts per second in logarithmic intervals. The gray scale code is displayed to the right of the pictures. Small holes (data gaps) in the maps occur because the satellite orbit is synchronized with the Earth's rotation resulting in incomplete coverage of all 1° by 1° bins. The maps provide information on the particle populations producing the high energy deposits which will be discussed below.

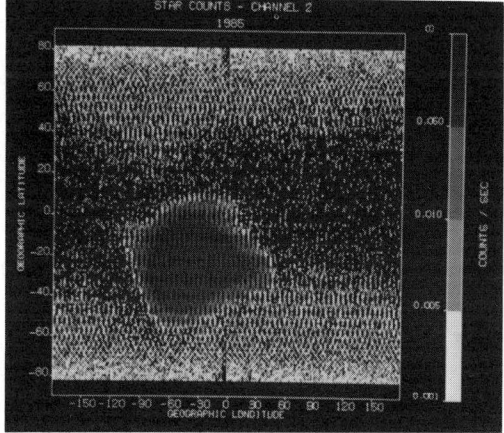

Fig. 9. Gray scale plot of star count rates for detector 2.

The general features pictured in both Figures 9 and 10 may be summarized as follows: a.) highest count rates occur in the region of the South Atlantic Anomaly where the inner radiation belt particles extend down to DMSP altitudes; b.) next highest count rates occur in the polar regions where solar protons and heavy ions have direct access along magnetic field lines down to lower altitudes; and c.) scattered count rates occur at all locations. This means that the highest LET particles might be expected any place in an 840 km orbit but the highest probability is in the region of the SAA, where it is proton dominated.

Information on the characteristics of the particles producing the counts can be obtained by statistically examining individual counts and count rates and the ratios of counts in the different detectors. To avoid mixing regions having different characteristics, all the statistics were done in the 4 latitude bins discussed above. Table 4 gives the total VHLET counts and average yearly count rates. Approximately 312 equivalent full days of data between November 1984 and October 1985 were used to calculate the yearly averages. During this period, no solar proton events were seen in the data. Table 5 gives the yearly average count rate ratios for these 4 regions for each detector.

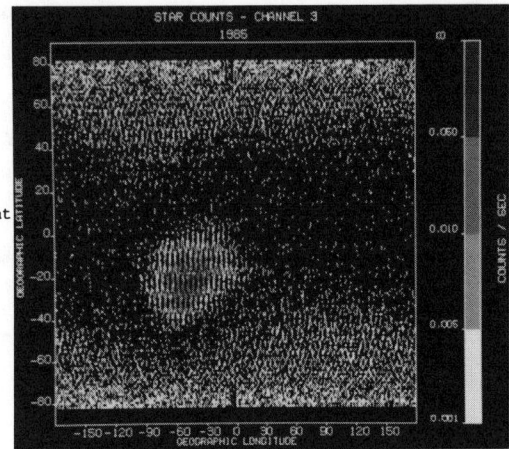

Fig. 10. Gray scale plot of star count rates for detector 3.

TABLE 4

TOTAL STAR COUNTS AND COUNT RATES

TOTAL COUNTS

LATITUDE BIN	DET 1	DET 2	DET 3	DET 4
NP	1164	19813	8408	17884
ML	121	2100	893	2015
SAA	5889	335641	10714	201530
SP	897	15329	6615	13979

COUNT RATES (COUNTS/SEC)

LATITUDE BIN	DET 1	DET 2	DET 3	DET 4
NP	1.5E-04	2.6E-03	1.1E-03	2.3E-03
ML	3.1E-05	5.3E-04	2.3E-04	5.1E-04
SAA	5.3E-04	3.0E-02	9.6E-04	1.8E-02
SP	1.7E-04	2.8E-03	1.2E-03	2.6E-03

TABLE 5

DETECTOR STAR COUNT RATE RATIOS IN LATITUDE BINS

DETECTOR RATIOS

LATITUDE BIN	3/1	2/1	2/4
NP	7.22	17.0	1.11
ML	7.38	17.4	1.04
SAA	1.82	57.0	1.67
SP	7.37	17.1	1.10

As mentioned above, star counts can be produced directly by proton energy deposition, directly by cosmic rays, and/or indirectly by nuclear star events in or near the detectors. In order for the protons to directly produce a pulse in the detector, they must have sufficient path length in the detector and sufficient energy to deposit energy above the threshold level (≈ 40 MeV or ≈ 75 MeV depending on the detector). Because the detectors are planar, they have small lateral dimensions such that there is only a very narrow angular and energy range of particles that can produce pulses by direct deposition. In detectors 1 and 3 no pulses can be produced by direct proton energy deposit since the total path length is too short for any energy proton to produce a pulse of the required size to be counted. In detectors 2 and 4, only protons incident between approximately 87.5° and 90° with external energies between approximately 55 MeV and 58 MeV for detector 2 and between approximately 85 MeV and 88 MeV for detector 4 can produce pulses of the required magnitude to be counted. Protons below these energies do not have sufficient energy to produce VHLET size pulses, and protons above these energies will pass through the detectors without depositing sufficient energy. Particles at 0° incidence must have a mass of oxygen or greater to produce a VHLET pulse. Particles with masses less than oxygen (such as helium) must be incident at an angle less than the normal to produce a pulse. The differences in the shielding, detector area, and response characteristics of the 4 detectors will be used to gain insight into the properties of the particles creating the pulses.

Since detectors 1 and 3 do not measure any counts due to direct energy deposition by protons, the counts in these two detectors are produced only by nuclear star events and direct cosmic ray events with appropriate masses and incident angles. Star counts in the Middle Latitude (ML) region are almost entirely due to extremely high energy cosmic rays since the geomagnetic cutoff of the Earth's magnetic field prohibits normal solar protons from directly entering into this region of space. We also know that the star counts in the South Atlantic Anomaly (SAA) region are due almost entirely to protons from the inner belt trapped particle population. In Table 5 the ratios of star counts of channels 3 to 1, 2 to 1 and 2 to 4 are listed. By comparing ratios from the polar regions (NP and SP) to the ML and SAA ratios, it is evident that the polar region ratios are the same as (within statistical error) those of the middle latitude region. We conclude that since the ratios are the same, the counts are produced by the same type of particles, namely cosmic rays. This is not unexpected since high energy cosmic rays have their most direct access into the near-Earth environment down the open magnetic field lines in the polar regions.

If we assume that the numbers of high energy deposits produced by particles of the same energy is directly proportional to the area of the detectors, we can use the ratio of star counts from detectors 1 (40 MeV threshold) to those from detector 3 (75 MeV threshold) to get a first order feel for the relative upset susceptibility in space for materials (tissues) that might have different sensitivity thresholds. The ratio of the area of detector 1 to detector 3 is 19.6. The 3/1 channel ratio for the poles is ∿7.3 and for the SAA is 1.82. This indicates that the relative upset susceptibility of materials with a 40 MeV threshold to materials with a 75 MeV threshold is approximately 3 times as great in a cosmic ray environment and 11 times as great in the South Atlantic Anomaly if the materials are proton sensitive.

We can separate nuclear stars from direct deposits in detector 2 by comparing the data from detectors 1 and 2. We know that detector 1 responds only to nuclear stars and detector 2 responds to both nuclear stars and direct proton deposition in the region of the SAA. The number of nuclear stars (statistically speaking) in detector 2 should differ from those in detector 1 primarily by the relative detector area factor of ∿19.6. The shielding difference between the two detectors however, reduces this factor somewhat. From Table 5 it can be seen that factor is ∿17.4 for the cosmic ray dominated middle latitude region. For the proton dominated SAA region, the factor could be more or less than the 17.4 depending on the shape of the proton spectrum, but could not be significantly different. Subtracting the estimate of the detector 2 nuclear stars (determined by multiplying detector 1 by 17.4) from the total detector 2 counts in Table 4 shows that for the SAA region the direct deposit protons in detector 2 are, on the average, approximately a factor of 2 greater than the nuclear star counts.

Since detectors 2 and 4 differ only in their shielding thickness, their areas and detector thresholds being the same, shielding effectiveness can be estimated by examining the counts ratio of these two detectors. In the mid-latitude and polar regions where the higher energy cosmic rays dominate, the shielding effectiveness is only 10% or less. In the SAA where the protons dominate, the shielding effectiveness is much greater because the energetic proton spectrum is softer. Even in the SAA region, there are major differences in the hardness of the spectra as a function of position. Figure 11 shows the ratio of average star count rates for detector 2 to detector 4 for the period from November 1984 through October 1985 in one degree geographic latitude and longitude boxes. The ratios are gray scale coded to indicate ranges from 1 to ∞ in logarithmic type intervals. The gray scale code is displayed to the right of the figure. The

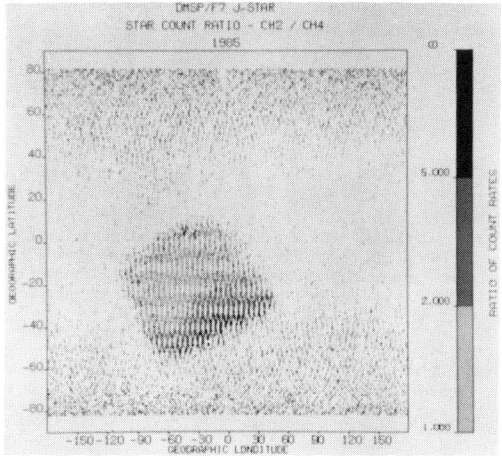

Fig. 11. Gray scale plot of star count ratios of detector 2 to detector 4.

gradual shift in the ratio across the South Atlantic Anomaly region is due to the magnetic field strength (L-Shell) variation. Higher energy particles are able to diffuse further inward across magnetic field lines, thus reaching lower L values and producing a harder spectrum at the lower L values.

V. THE SOLAR PARTICLE EVENTS OF 16 FEBRUARY AND 25-29 APRIL, 1984

The two largest solar particle events during 1984 occurred in February and April. Figures 12a and 12b are survey plots of the DMSP flux count data for protons >75 MeV (the highest energy channel) on the peak days of the two events: 16 February and 26 April, respectively. They are in the same format as Figure 2. In each plot, three types of counts can be identified: background counts; the systematically occurring SAA flux counts that stood alone in Figure 2; and the large flux levels across each of the polar regions. The latter are solar protons and heavy ions which have direct access to low latitudes in the polar regions. On 16 February there is a sudden onset of solar particles in the caps at ~09 UT. The fluxes fall by an order of magnitude within 4 hours, after which there is a steady exponential decay of the particles that continues into the 17th. At the beginning of 26 April the high energy precipitation into the polar caps was already in progress. This event had a much slower buildup and decay. The widths of the flux peaks in the polar regions show that during solar proton events significant dose can be received at polar orbiting vehicles. Even in these survey plots one can discern significant spatial and/or temporal variations in the cap fluxes. They are particularly large at the beginning of the 16 February event.

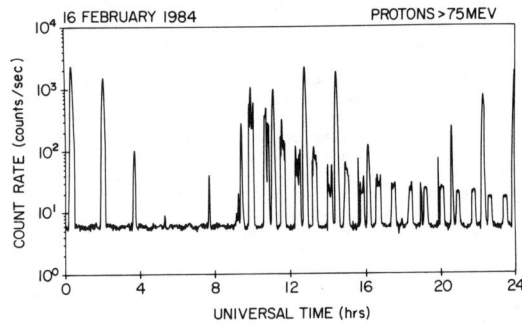

Fig. 12a. Daily plot of one-minute-averaged flux count rate for protons with energy greater than 75 MeV on 16 February, 1984.

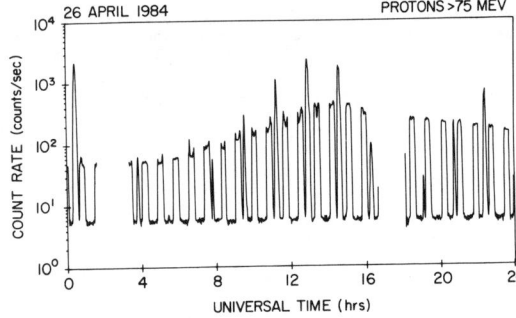

Fig. 12b. Daily plot of one minute averaged flux count rate for protons with energy greater than 75 MeV on 26 April, 1984.

To show the kind of variability that exists across the polar cap during solar events, the paths of two polar crossings on 16 February are shown in Figure 13. Here the satellite track is plotted in corrected geomagnetic latitude (MLAT)-magnetic local time (MLT) coordinates. The length of the black lines vertical to the track represents the count rate in the lowest energy proton channel (>20 MeV). These plots show a) the low latitude cutoffs (<60° MLAT) of the solar particles; b) that the particles fill both the regions normally occupied by the outer zone electrons and the polar cap; c) that relatively deep troughs occur in the flux levels which are located differently in the northern and southern hemispheres, but are near the magnetospheric cusp. Because the variations in electron and proton fluxes (not shown) are so similar in the 16 February event, it is very likely the case that the "electron" count rates contain a significant contribution from the >100 MeV protons.

339

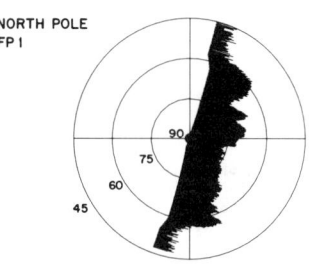

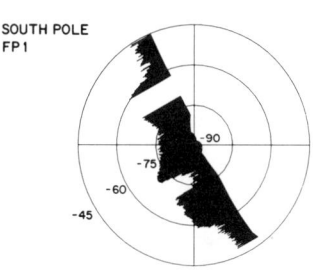

DMSP - F7 J*   FEBRUARY 16, 1984
                UT START 35090

Fig. 13. Polar plots of the >20 MeV protons during the peak of the 16 February event. The coordinates are corrected geomagnetic latitude and geomagnetic local time.

Since the shape of the spectra are very important in assessing damage during solar events, the proton flux counts are used to find an average power law spectrum for protons for each flare. For the 16 February data, at the peak of the proton flux the optimal spectral fit for the counts averaged across the polar cap is:

16 February (peak): $j(E) = 86(\pm 2) \times \{20/E\}^{1.8}$ p/cm^2-s-MeV,

where E is in MeV. Under the assumption of complete particle isotropy in the upward hemisphere at DMSP altitudes and integrating from 10 MeV to infinity, the directional integral flux from the equation above is 298 p/cm^2-s-sr. The average spectrum for 17 February, using the cap data for the entire day is:

17 February: $j(E) = 0.85 (\pm 0.05) \times \{20/E\}^{2.7}$ p/cm^2-s-MeV.

Because the solar event in April is so slowly varying we list the spectra determined by averaging the cap data for each day from 25 - 28 April. They are:

```
24 April:     Background levels
25 April:     j(E) = 5.6  (+0.4) x {20/E}^3.0  p/cm^2-s-MeV
26 April:     j(E) = 116  (+4.0) x {20/E}^3.8  p/cm^2-s-MeV
27 April:     j(E) = 6.8  (+0.4) x {20/E}^4.3  p/cm^2-s-MeV
28 April:     j(E) = 18.  (+2.0) x {20/E}^4.0  p/cm^2-s-MeV
```

The deviation listed in parenthesis is the uncertainty in the fitting procedure and does not include the instrument's inherent uncertainties. From the power law fits to the flux counts we draw the following conclusions: a) Throughout the February event the spectra were much harder than any of those occurring in the April event. b) The February event softened in time. c) The April event softened over the first three days, then increased in intensity and became slightly harder.

Figures 14a and 14b are the depth-dose spectra for single polar cap crossings during the peaks of the two events. The time intervals were approximately the same for the crossings (21.3 min on 16 February, and 19.2 min on 26 April). In each plot, the electron dose, the proton dose, and their sum are plotted separately for each of the four domes as a function of aluminum shielding thickness (in mass per unit area).

We compare these to the depth-dose spectrum for the SAA crossing (Figure 3) which took 23 min. The depth-dose spectrum for the peak flux on 16 February, Figure 14a, shows that the total dose behind the minimum shielding is 0.2 rad(Si), or somewhat less than in the SAA. The dose source is mainly protons, the electron contribution being less than 25% of the total. The

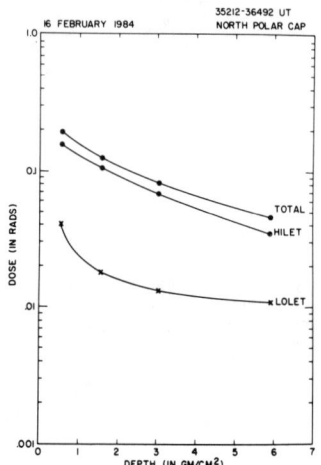

Fig. 14a. Depth-dose spectrum for a DMSP/F7 crossing of the north polar cap at the peak of the solar particle event on 16 February, 1984.

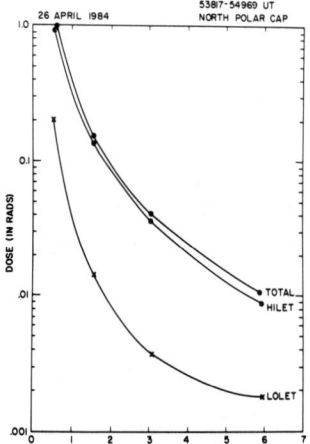

Fig. 14b. Depth-dose spectrum for a DMSP/F7 crossing of the north polar cap at the peak of the solar particle event on 26 April, 1984.

spectrum is falling slowly with increased shielding, but faster than in the SAA indicating a softer (less energetic) particle population. Shielding is still effective at 5.91 gm/cm^2, where the total dose is reduced to 0.05 rad(Si). Since this event decays quickly, the maximum dose behind the minimum shielding is less than 2 rad(Si) over a period of less than one day.

The depth-dose spectrum across the northern polar cap for the peak of the solar particle event on 26 April is shown in Figure 14b. More than 83% of the total dose comes from protons. The spectrum is considerably softer than that in the SAA or at the peak of the 16 February event. It is, however, much more intense at low shielding values. For .55 gm/cm^2 aluminum shielding, 1.2 rad(Si) are received. This falls by two orders of magnitude to 0.012 rad(Si) at maximum shielding (5.91 gm/cm^2) where the dose is approximately an order of magnitude less than encountered in the SAA or in the 16 February event. Shielding remains effective throughout the spectrum. The dose per polar crossing remained high for several days, and in the first three days the total dose exceeded 25 rad(Si) for .55 gm/cm^2 of aluminum shielding. This is more than four times the dose received in the same time period and behind the same shielding from the SAA.

The dosimeter star data can also be used to study solar proton events. Here we examine the entire periods of the February and April 1984 solar particle events seen in Figures 1 and 2. The peaks of the events were on 16 February 1984 and 26 April 1984. Since the events are only seen in the polar regions, only the polar region data is used. The two polar cap regions were added together to get better statistics for the event days. Table 11 gives the daily average of star channel count rates for the 4 detectors in counts per second times 10^{-4} for the periods 15-17 February 1984 and 24-30 April 1984. Also given are the 1985 average count rates for comparison as a normal background level and the differences from the 1985 averages for the days of interest. The count rates were determined only from the polar regions: between 40° and 90° North Latitude and 55° and 90° South Latitude. Data equatorward of 55° South Latitude were excluded from the table to prevent contamination from the SAA. It is obvious that during the events the number of energetic particles that can produce 40 MeV pulses (detectors 1, 2, and 4) can go up significantly from factors of approximately 2 to 50 depending on shielding thickness. It is also evident that with a shielding equivalent of 3 gm/cm^2 (detector 3), there is less than a factor of 2 increase in particles that can produce a 75 MeV pulse. Subtracting the 1985 average from the star counts on the flare days gives the flare produced numbers shown in the bottom of Table 11. Comparing these back to the 1985 averages shows that in no cases did the flares produce more pulses in detector 3 than the averages, and only at the peak of the events were there more particles seen in detector 4. This says that neither particle event had a spectral hardness at the highest energies greater than the natural cosmic ray background. The softness of the event spectra probably accounts for the fact that unusually high upsets have not been seen on satellites during flare periods to date.

Table 11 Star Count Rates

Count Rates (counts/sec x 10^{-4})

DETECTOR	1985 Ave.	Feb 84 15	16	17	Apr 84 24	25	26	27	28	29	30
1	1.6	0.6	12.0	1.8	1.3	9.6	83.0	26.0	9.6	7.9	2.2
2	27.0	28.0	170.0	38.0	27.0	87.0	500.0	190.0	91.0	38.0	27.0
3	11.0	12.0	14.0	8.8	12.0	9.6	18.0	12.0	8.5	7.1	10.0
4	24.0	22.0	53.0	26.0	20.0	25.0	62.0	29.0	22.0	18.0	19.0

Flare Produced Differences from 1985 Averages

1		--	10.4	0.2	--	8.0	81.4	24.4	8.0	6.3	0.6
2		1.0	143.0	11.0	--	60.0	473.0	163.0	64.0	11.0	--
3		1.0	3.0	--	1.0	--	7.0	1.0	--	--	--
4		--	29.0	2.0	--	1.0	38.0	5.0	--	--	--

VI. DISCUSSION AND CONCLUSIONS

The dosimeter on the DMSP/F7 satellite has been shown to be a very versatile instrument which can measure energetic particles of different types over a wide dynamic range. It can measure the dose effects of electrons and protons separately, and gives accurate measurements over short time periods. The dosimeter also can distinguish those very high LET particles that can produce SEU type behavior in sensitive materials.

From the data we see that dose accumulated in the SAA is extremely stable, showing no measurable variation from day to day. The SAA contributes 2 rad(Si)/day behind .55 g/cm^2 aluminum shielding. Approximately 60% of the total is dose from protons with energy less than 100-200 MeV, and the remaining 40% is LOLET dose, from electrons, high energy protons (>100-200 MeV) and bremsstrahlung. The depth-dose spectrum for the SAA is very hard, falling only by 60% for shielding thicknesses between .55 and 5.91 gm/cm^2. For non-solar particle event periods, the SAA provides most of the proton dose at DMSP altitudes. The NASA solar minimum proton model for DMSP altitudes predicts a proton dose which is within statistical error of the dose measured on DMSP.

The dose from the outer zone electrons provides the remainder of the daily dose when there are no solar particles from solar proton events. The dose per day from the outer zone electrons is highly variable and large, from 1 to more than 10 rad(Si)/day for .55 gm/cm^2 aluminum. The average electron dose rate determined from the DMSP dose rate model for the year 1984-85 is 2.5 rad(Si)/day. This is four times less than that predicted by the NASA models for solar minimum from electrons alone. The electron depth-dose spectrum is very soft, falling to a near-constant bremsstrahlung level by 3.05 gm/cm^2 aluminum.

Dose effects from the two largest solar particle events in 1984 were measured. The event that occurred on 16 February, 1984, lasted less than a day, had a reasonably hard spectrum, and led to an accumulated dose of less than 2 rad(Si) behind .55 gm/cm^2 aluminum, which is about the same as the SAA-accumulated dose for one day. The event that occurred on 25-29 April, 1984, had a very soft spectrum compared to the SAA and the 16 February spectra. The accumulated dose, however, for a three-day period was larger than that for the SAA behind .55 gm/cm^2 of shielding (25 rad as compared to 6 rad), these levels could prove harmful for an astronaut performing EVA during this period.

For short flights in polar orbit, our results show that a three-day mission at high inclination and at 840 km altitude during moderately large solar particle events could lead to a dose deposition of 30 rad(Si) or more behind minimal spacecraft skin thicknesses which approximate .5 gm/cm^2 aluminum. This level may be sufficient to damage the ocular nerve unless shielded (M.A.

Shea, private communication, 1984). For manned space flights in the polar regions, care must be taken in planning and carrying out activities where minimally shielded astronauts could be exposed to dangerous radiation levels.

For the highest LET particles, the probability of seeing effects in proton sensitive materials is greatest in the region of the SAA. For materials not proton sensitive, the highest probability is in the polar regions. Counts observed outside the polar cap and SAA regions are almost entirely due to higher energy cosmic rays that penetrate beyond the magnetic cutoff region. Although these particles are statistically small, they can appear anywhere in an 840 km orbit.

In conclusion, accurate, short-term models of dose rate are needed for future space missions that are manned for long periods of time, or that rely on new microelectronic technologies. The NASA models projected to low altitude give orbital dose values which are too high by a factor of four during solar minimum. The DMSP proton and electron models, constructed for solar minimum conditions, show that the discrepancy is due to an overestimate of the electron dose. The conservative NASA model predictions can impose more stringent shielding conditions than are necessary. Models of dose in the polar regions during solar proton events do not exist because of the high temporal and hardness variability of the particle fluxes. Nevertheless, significant dose levels can be accumulated during these events and must be taken into account in orbital planning. This can most likely be done on a "worst case" and probability of occurrence basis. It should be remembered that for satellites in near-Earth orbit, the lifetime total dose is mostly accumulated in short bursts and not at a steady constant rate. This leaves much of a satellite system's time in a relatively benign environment during which many of the dose effects can anneal out of certain sensitive materials.

REFERENCES

1. Gussenhoven, M.S., R.C. Filz, K.A. Lynch, E.G. Mullen and F.A. Hanser, Space Radiation Dosimeter SSJ* for the Block 5D/Flight 7 DMSP Satellite: Calibration and Data Presentation, Air Force Geophysics Laboratory, Hanscom AFB, MA, AFGL-TR-86-0065, March, 1986.

2. Singley, G.W. and J.I. Vette, A Model Environment for Outer Zone Electrons, NASA Goddard Space Flight Center, Greenbelt MD, NSSDS 72-13, December, 1972.

3. Teague, M.J. and J.I. Vette, A Model of the Trapped Electron Population for Solar Minimum, National Space Science Data Center, NASA Goddard, NSSDC-74-03, 1974.

4. Sawyer, D.M. and J.I. Vette, AP-8 Trapped Proton Environment for Solar Maximum and Solar Minimum, National Space Science Data Center, NASA Goddard Space Flight Center, Greenbelt MD, WDC-A-R&S 76-06, December, 1976.

5. Vette, J.I., K.W. Chan, and M.J. Teague, Problems in Modeling the Earth's Trapped Radiation Environment, Air Force Geophysics Laboratory, Hanscom AFB, MA, AFGL-TR-78-0130, 1978.

6. Mullen, E.G., M.S. Gussenhoven, K.A. Lynch and D.H. Brautigam, DMSP Dosimetry Data: A Space Measurement and Mapping of Upset Causing Phenomena, IEEE Transactions on Nuclear Science, to be published December 1987.

THE COSMIC RADIATION EFFECTS AND ACTIVATION MONITOR

C. S. Dyer, A. J. Sims, R. J. Hutchings
Space Department, RAE Farnborough, Hampshire GU14 6TD, England

D. Mapper, J. H. Stephen, J. Farren
Instrumentation & Applied Physics Division, AERE Harwell, OX11 0RA, England

ABSTRACT

The Cosmic Radiation Effects and Activation Monitor (CREAM) has been designed to monitor real-time LET spectra together with mission-integrated particle fluences and radioactivity, with the aim of improving the predictive models used to define the aerospace environment. In this paper the experiment will be described together with the expected environments and ground calibration data.

INTRODUCTION

The linear energy transfer (LET) spectrum of space radiation is of considerable significance in quantifying single event upset rates in microelectronics as well as the radiobiological hazard to man in space. Such spectra are highly dependent upon geomagnetic location, physical shielding and interplanetary and solar conditions. Predictive codes[1] have been produced based on cosmic ray measurements combined with Stormer cut-off theory etc, but further measurements are required, both in a variety of satellite orbits and in the upper atmosphere, where the penetration of energetic particles can reach supersonic transport and transatmospheric vehicles. In addition induced radioactivity, as well as providing a monitor for particle fluxes, produces background in gamma-ray spectrometers. The CREAM experiment is intended to actively monitor LET spectra as a function of time and geomagnetic location, while accumulating integrated information on particle fluences and induced radioactivity.

This experiment was originally designed as a Shuttle middeck experiment for deployment by the first UK astronaut and has been modified for use on free-flyer spacecraft, as well as on aircraft, in order to explore the fullest range of locations where cosmic rays are experienced. The current hiatus in spaceflight opportunities means that the first experiments will explore the upper atmosphere using supersonic aircraft.

AIMS

The aims of the CREAM package are as follows:

- To improve space environment and radiation shielding models used in the prediction of upset rates and induced radioactivity.

- To obtain real-time energy deposition spectra in silicon for a number of locations afforded various amounts of physical shielding.

© 1989 American Institute of Physics

© British Crown Copyright 1988

- To obtain total dose, high-Z particle fluence, and neutron spectra accumulated in the same locations.
- To monitor the radioactivity induced in materials employed in gamma-ray detectors.

DESCRIPTION

The experiment as initially conceived is summarised in Table I.

TABLE I CREAM experiment description

- STS NOMINAL ORBIT 28.5°, 160 nm
- ACTIVE DETECTOR ARRAY
    - 16 pin diodes to give effective area of 9.6 cm^2.
    - 8 channels of pulse-height analysis.
    - LET range 1.5 to 5×10^4 MeV/(gm cm^{-2}).
    - Battery power plus 64 kbyte solid-state memory.
    - 5 minute time bins for mapping variations over geomagnetic coordinates.
- PASSIVE DETECTORS (Four of each type, deployed one per location)
    - Cosmic ray track detectors containing 215 cm^2 sheets of CR-39, Kapton plus ^{7}LiF thermoluminescent dosimeters.
    - Neutron activation detectors containing 98 cm^2 foils of Au, Ti, Ni plus ^{6}LiF thermoluminescent dosimeters.
- PASSIVE GAMMA-RAY DETECTOR
    - 7.5 cm cylindrical NaI(TI) crystal.

In this form continuous operation of the active monitor requires a battery change approximately once per day and the consequent payload occupies the majority of a standard middeck locker. In orbit the passive packages will be located at a number of locations (minimum of three) afforded a range of shielding. These locations include maximum shielding (air lock) and minimum shielding (middeck wall) situations. The active monitor will then be relocated at these various locations on a daily basis. The passive sodium iodide crystal will remain in the locker throughout and the level of radio-activity will be monitored as soon as possible after landing. Besides yielding important information on background rates in gamma-ray spectrometers, important information on secondary neutron fluences will be forthcoming.

The battery-powered version allows full flexibility with respect to location and operation but is somewhat bulky. A more compact version has been produced to operate from a middeck 28V DC supply and in this case the total experiment can stow in a half-locker. In this configuration it is hoped to share flights with the Shuttle Activation Monitor (Rester et al these proceedings) in order to maximise the data yield from the two complementary experiments. A 28V version has also been configured in an ARINC avionics crate to allow aircraft flights to explore the upper atmosphere.

ACTIVE BOX

The active detector comprises three boards as illustrated in the overall block diagram of Figure 1. The analogue board contains the sensor array comprising ten PIN diodes each 1 cm^2 in area. An ionising particle produces electron-hole pairs which are rapidly collected by the 60V reverse-bias field. A charge amplifier is used to integrate the charge and give a proportionate output voltage pulse. One of two possible amplifications is selected depending on the pulse amplitude thus allowing coverage of the full LET range of interest. The digital board 'B' features 8-bit analogue to digital conversion and channel assignment via a 80C81 microprocessor into a 64K x 8 bit non volatile RAM. Board 'C' is responsible for power conversion and the version illustrated is for battery operation. A back-up battery is brought into operation if the main voltage falls below 8 volts and sustains operation during main battery changes.

The detector has been calibrated using Americium-241 alpha particles (LET 600 MeV cm^2 g^{-1}) and Californium-252 fission products (LET 4.3 x 10^4 MeV cm^2 g^{-1}), and a sample spectrum is given in Figure 2.

ENVIRONMENTS

The Shuttle middeck area affords a variety of shielding environments and the maximum and minimum distributions are illustrated in Figures 3 and 4 (Atwell et al, these proceedings). The influence of shielding on the LET spectrum is illustrated in Figures 5 and 6, which have been produced using the methods described in Ref 1. Based on these calculations a prediction has also been made of the accumulated counts in 5 minute time bins for some sample LET channels for both STS orbits of 28.5° and 57° inclination, as well as for an atmospheric vehicle at 60000 ft and 2 GV rigidity cut-off, and the results are presented in Table II.

TABLE II   Predicted count rates

Channel No.	LET Range MeV/(gm cm^{-2})	Counts per 5 min STS 28.5°	STS 57°	SST 60000 ft, 2 GV
1	1-3	220	975	200
2	3-10	28	124	10
3	10-30	0.2	20	3
4	30-100	1.3	6	0.3
5	100-300	1.2	5	0.07
6	300-1000	0.3	1.3	0.006
7	1000-3000	0.13	0.5	5 x 10^{-4}
8	>3000	1.2 x 10^{-5}	0.02	3 x 10^{-5}

FUTURE DEVELOPMENTS

With the current hiatus in spaceflight opportunities the first flights of the CREAM active package will be on supersonic aircraft probably commencing in Spring 1988. Shuttle flights with SAM are anticipated in late 1989 or early 1990. Meanwhile the active package is being modified with a view to flights on free flying spacecraft where the environment monitoring will assist in the interpretation of anomalies and device performance. In this form it will be combined with active total dose monitors such as RADFETs[2] and will hence be called CREDO (Cosmic Radiation Effects and Dosimeter). In parallel, improved forms of LET spectrometer are being studied employing co-incidencing between diodes in various geometries in order to provide higher resolution angular and energy-loss information.

It is hoped that once the full range of flight opportunities is realised the LET environment will have been extensively explored from sea level, through the upper atmosphere to LEO and thence GEO.

REFERENCES

1. J H Adams, Jr., R Silberberg, and C H Tsao (1981), NRL Memorandum Report 4506.
2. A Holmes-Siedle, L Adams, S Marsden, and B Pauly, IEEE Trans on Nuc Sci, NS-32,6,4425, Dec 1985.

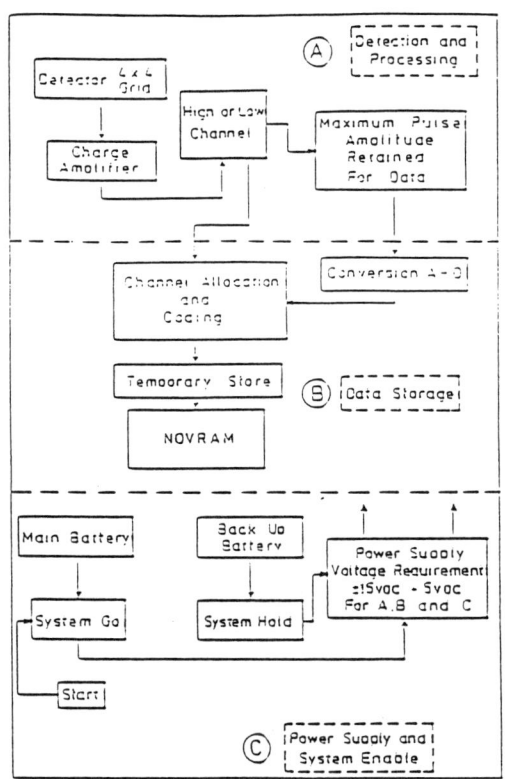

FIG 1   OVERALL BLOCK DIAGRAM

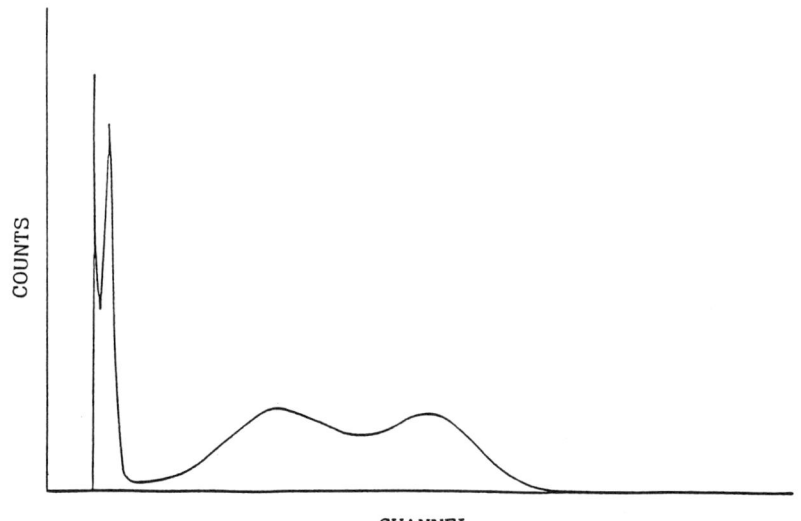

FIG 2   Am-241 AND Cf-252 SPECTRA ON PHOTODIODE UNDER SAME GAIN

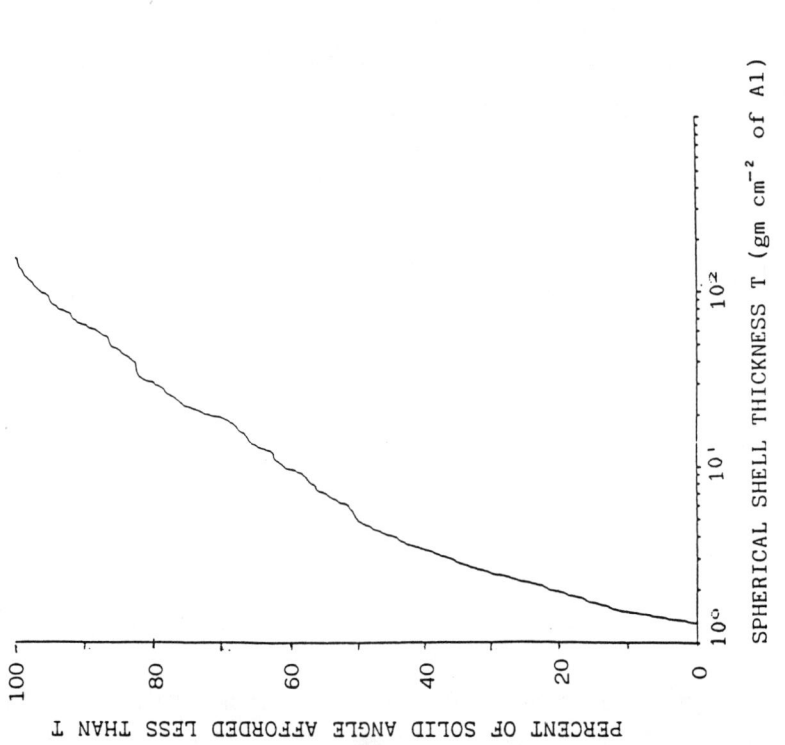

FIG 3 STS SHIELDING DISTRIBUTION AT MINIMUM (PORT WALL OF MIDDECK ABOVE HATCH)

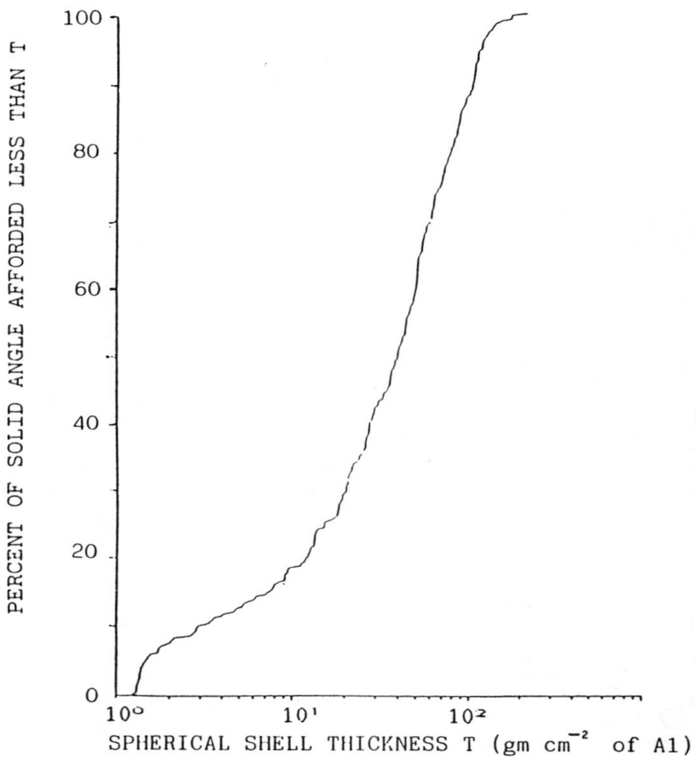

FIG 4 STS SHIELDING DISTRIBUTION AT MAXIMUM (MIDDECK SIDE OF AIRLOCK)

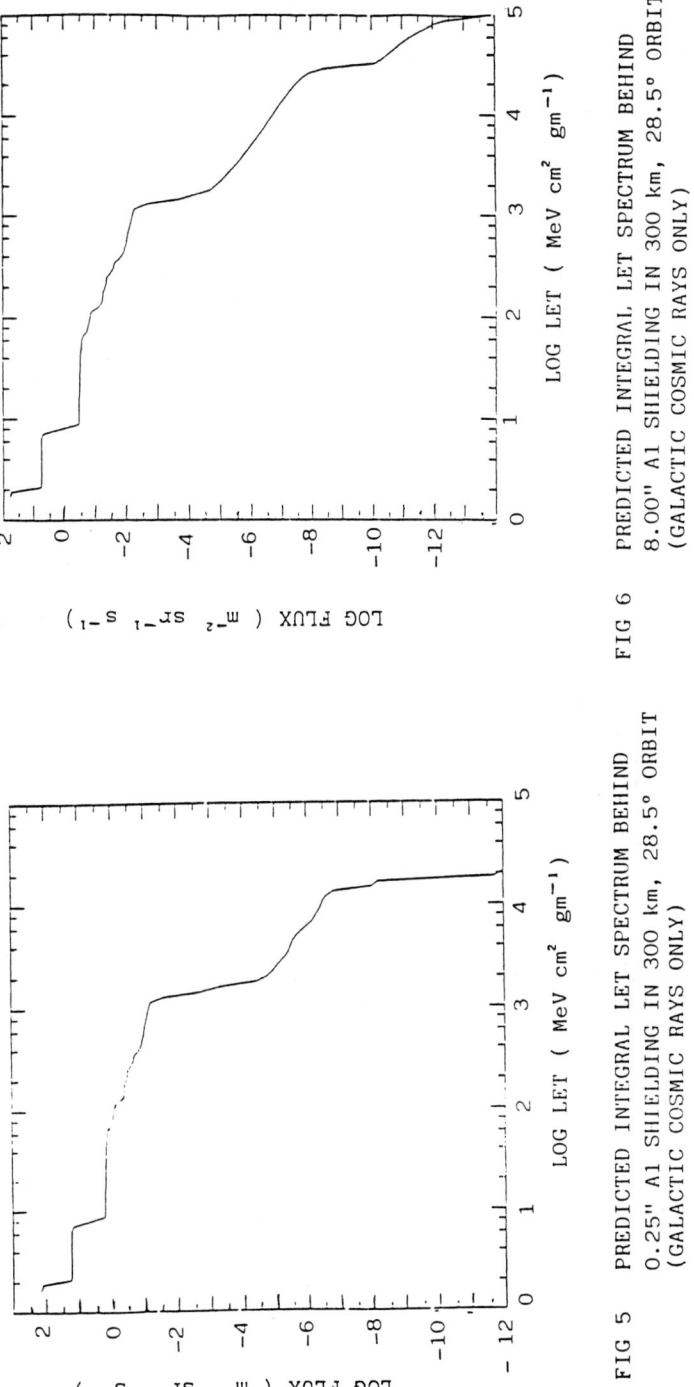

FIG 5 PREDICTED INTEGRAL LET SPECTRUM BEHIND 0.25" Al SHIELDING IN 300 km, 28.5° ORBIT (GALACTIC COSMIC RAYS ONLY)

FIG 6 PREDICTED INTEGRAL LET SPECTRUM BEHIND 8.00" Al SHIELDING IN 300 km, 28.5° ORBIT (GALACTIC COSMIC RAYS ONLY)

Characterization of Space Radiation Environment in Terms of the Energy Deposition in Functionally Important Volumes*

L.A. Braby, N.F. Metting, W.E. Wilson, and C.A. Ratcliffe
Pacific Northwest Laboratory, Richland, WA 99352

INTRODUCTION

Ionizing radiation can have effects which seem out of proportion to the amount of energy deposited. An energy deposition which will raise the temperature of a body only a small fraction of a degree can cause the death of an animal or the permanent failure of a complex electronic system. Even smaller energy depositions can result in potentially lethal tumors in animals or loss of information in a computer. Furthermore, equal amounts of energy deposited by different types of radiation or even by the same charged particle at different velocities result in significantly different degrees of effect. A goal of radiological physics is to understand how these small energy depositions can produce such dramatic and varied effects. Closer inspection of both electronic and biological systems suggests that the damage is a consequence of crucial alterations of small components of the overall system. The disruption of a computer may begin with the radiation induced alteration of the charge stored in a very small circuit element. Tumors are frequently initiated by the rearrangement of a small amount of DNA in a single cell. Thus, to understand these important effects, we need to study the interaction of radiation with very small volumes of matter.

In order to understand these local radiation effects we must deal with the spectrum of charged particles at the position where the damage occurs. This spectrum results from the transport of the incident radiation through the surrounding medium and the target structure itself. One needs a great deal of information in addition to the incident radiation spectrum to calculate that charged particle spectrum at any point. The atomic composition, size, and relative position of each component of the entire system, and all of the relevant cross-sections for the interactions of the incident radiation and their secondaries with these components must also be known. The charged particle spectrum, when calculated, can be used to calculate dose (the average energy imparted per unit mass) at the point of interest. However, we do not yet know how to convert this information into reliable predictions of the probability of different effects. We do know that the initial products are the same for all ionizing radiations. The same types

*Work supported by the Office of Health and Environmental Research (OHER) U.S. Department of Energy under Contract DE-AC06-76RLO 1830.

of ionizations and excitations are produced by electrons and alpha particles. These radiations do however differ in the average distance between these products, and other characteristics generally referred to as the "track structure."

Since the damage which initiates detrimental effects occurs in a small site (semiconductor junctions, or biological cell nuclei), these differences in spatial distribution of ionization may be the relevant factor controlling the effectiveness of different radiations. When the appropriate cross section data is available Monte Carlo methods can be used to simulate the positions of all ionizations and excitations produced by a typical charged particle. This calculated track structure must interact with the biological or electronic entity in which it occurs to produce the effect. However, we do not know the mechanisms of this interaction and thus cannot specify which characteristics of the charged particle track are responsible for the relevant damage. From track structure we can obtain the spectrum of energy deposition in small volumes which may be relevant to the processes of concern. This has led to a new approach to dosimetry, one which emphasizes the stochastic nature of energy deposition in small sites, known as microdosimetry.

MICRODOSIMETRIC QUANTITIES

The energy, ϵ, imparted in a site by the passage of a charged particle is a stochastic variable depending on the energy loss processes and the path length through the site. It is convenient to use the quantity lineal energy, y, equal to the energy imparted divided by the mean chord length of the site, ℓ, $y = (\epsilon/\ell)$ which is the stochastic equivalent of the average quantity linear energy transfer (LET). One can also use the quantity specific energy, z, the energy deposited in a site by one or more charged particles divided by the mass of the site. This quantity is clearly analogous to the dose and is offered[1] as the basis for an alternative definition of dose.

In those situations where the continuous slowing down approximation is valid, the energy deposited in a small site is equal to the stopping power times the chord length. For spherical sites, where the chord length distribution has a simple triangular shape, the distribution of lineal energy will also be triangular (Figure 1). The mean value of y will be approximately proportional to the LET. Typical values of z for a single particle passing through a 7 μm diameter sphere are 0.004 Gray (0.4 Rad) for ^{60}Co irradiation and 0.2 Gy (20 Rad) for 1 MeV neutrons. The dose is equal to the average specific energy when an event occurs times the probability of an event occurring. Thus at a dose of 0.01 Gray (1.0 rad) of ^{60}Co gamma rays a cell nucleus will experience an average of 2.5 energy deposition events. But at the same dose of 1 MeV neutrons only about one of every 20 cell nuclei will experience any energy deposition. However, those that do will receive the

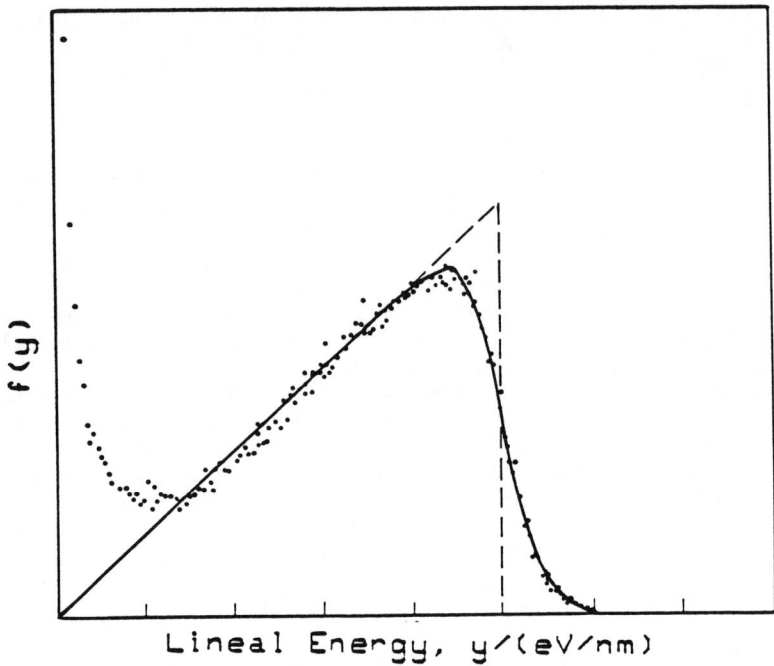

Fig. 1. Lineal energy in a spherical site. The dashed line is the chord length distribution for a sphere, the points are data for 5 MeV alpha particle irradiation of a wall-less proportional counter simulating a 1 μm diameter sphere, and the solid line shows the effect of energy loss straggling.

same amount of damage as the average of nuclei in a sample exposed to 20 rad. The simple relationship between lineal energy, track length distribution and LET makes it possible to calculate the LET distribution from a measured lineal energy distribution. However, it has been shown[2] that the continuous slowing down approximation is not a valid approximation for determining energy deposition in most small sites for most charged particles. Three factors contribute to significant additional variations in the distribution of energy imparted. They are the range of the charged particles, the energy loss straggling, and the delta ray effects. If the size of the site is comparable to or larger than the range of charged particles, some of the charged particle tracks will begin or end in the site. In this situation, the mean path length in the site is less than the mean chord length for the site, and the energy deposited in the site is less than that which would be predicted based on the chord length distribution and the stopping power of

the particles. Thus, if one measures the energy imparted and
calculates the linear energy transfer, the result will be less than
the actual LET for the charged particles producing these events.
Energy loss straggling is the consequence of the fact that charged
particles lose energy in discrete interactions with the atoms they
pass. These interactions are statistically independent and
consequently the number of interactions and the energy deposited in
a short segment of path is a random variable. For small sites or
sparsely ionizing irradiations (electrons and very fast heavy
charged particles) the variation in the number of ions for that
path length can exceed the variation in path length through the
site. Even for 5 MeV alpha particles crossing a 1 micrometer
diameter site the energy loss straggling is noticeable and results
in a rounding of the peak of the ideal track length distribution.
Some of these energy transfer interactions along the path of the
charged particle produce secondary electrons with enough kinetic
energy to ionize atoms themselves. These energetic secondary
electrons, known as delta rays, carry energy away from the ion
track. If the site diameter is less than or comparable to the
maximum range of the delta rays, they can carry significant amounts
of energy outside the site in which the initial ionization
occurred. Similarly, charged particles which do not cross a
specific site may deposit some energy in it. In the 5 MeV alpha
particle distribution (Figure 1), the large number of events with
low values of lineal energy are the result of these delta rays.

If we adopt the suggestion[3] of using a small spherical volume
as a phantom for a cell, we soon realize that all cells do not
experience the same energy deposition nor can that energy
deposition be reliably described by reference to the LET. As a
result of these factors, it has recently been suggested[4] that the
quality of radiation be expressed in terms of lineal energy rather
than LET. However, such a definition requires a choice of the site
size to be used in evaluating lineal energy. If we could specify
the chemical and biological mechanisms responsible for biological
effects or the specific characteristics involved in electronic
upsets, we might be able to determine the optimum site size to use
in specifying radiation quality. However, microcircuit features
vary in size depending on manufacturing techniques and we have no
good indication of target site size that is universally applicable.
In the biological case the data suggests that formation of initial
biological damage depends on the energy concentration in volumes a
few nanometers in diameter, but that the biological response to
that damage depends on the number of these lesions occurring in
something nearly as large as the cell nucleus. Consequently we can
not specify an optimum site size and the recommended site has been
chosen on the basis of ease of measurement. That choice can be
partially justified on the basis that y is a slowly varying
function of site size for most radiations. Furthermore, the ratio
of y in a 10 nanometer site to y in a 1 micrometer diameter site
for protons is nearly constant at 0.6 from 5 MeV to 20 MeV and

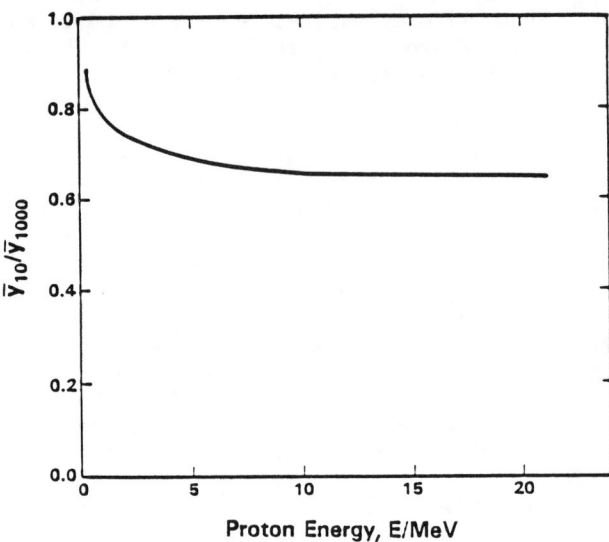

Fig. 2. The ratio of lineal energy in a 10 nm diameter site to lineal energy in a 1 μm diameter site as a function of proton energy, determined by Monte Carlo track simulation.

probably beyond (Figure 2). Below 5 MeV the ratio gradually increases as the delta ray path length decreases, resulting in fewer delta-rays escaping from the 10 nanometer site. This figure shows that nearly 40% of the energy deposited in small sites by high energy protons is transferred via delta rays. Monte Carlo calculation of energy deposition by electrons, (Figure 3), shows that the spectrum of events in 10 nanometer sites is nearly independent of the initial energy of the electrons. Thus the number of small volumes receiving delta ray events of any particular size can be related to y measured in a 1 μm or smaller diameter site. The only limitation to this is that the range of the primary ions must be large relative to the site diameter.

Thus for most radiations (except very low energy charged particles and photons which do not penetrate biological systems and are, therefore, not a problem, and for neutrons below 100 keV) the arbitrary choice of volume for the specification of quality factor as a function of lineal energy appears to be justified for radiation protection purposes, the only area where quality factor is defined. A similar choice may be appropriate for estimating effects in electronic systems if the geometry of the sensitive regions is not well known.

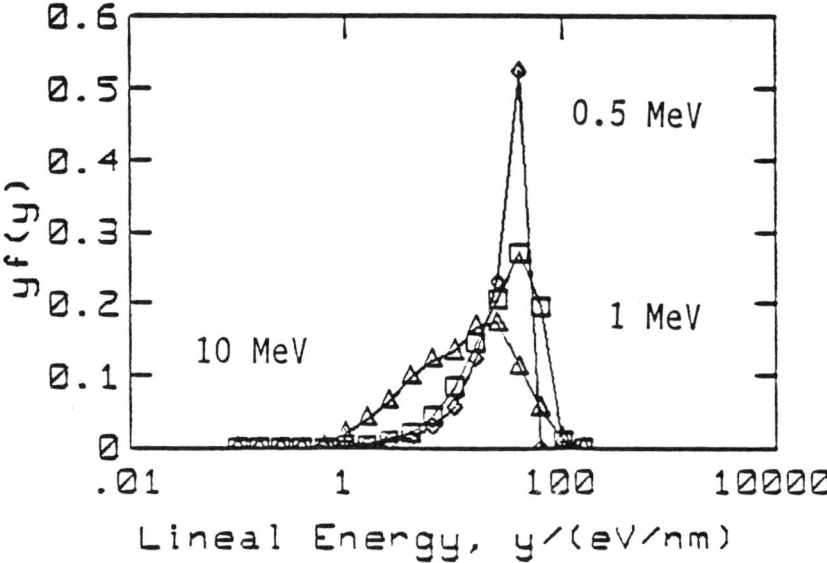

Fig. 3. Calculated lineal energy distributions for all of the interactions of electrons with 10 nm diameter sites as the initially monoenergetic electrons slow down in water vapor.

MEASUREMENT TECHNIQUES

The advantage to using lineal energy for a 1 micrometer diameter site is that this quantity can be easily measured. Measured lineal energy distributions can be used to evaluate LET distributions, either by applying simplifying approximations or by a detailed deconvolution of the chord length and LET distributions. Thus an instrument which measures lineal energy can be used to evaluate the risk imposed by a particular radiation using either LET or y as the variable in this specification of quality factor.

Lineal energy can be measured using relatively simple low pressure gas proportional counters. When the product of the path length, the density of the medium, and the stopping power are equal for media of different density the energy depositions will be the same. Thus a 10 mm diameter detector filled with gas whose specific gravity is 1/10,000 will experience the same energy depositions as a 1 micrometer diameter site in unit density material of the same composition. Proportional counters of this size and density are easily constructed. By using a tissue equivalent gas mixture or a simple hydrocarbon such as propane and applying a suitable potential difference to the anode, gas gains of around 1000 can be achieved. The major limitation of this technique is that the ions which are produced very near the anode, inside the gas avalanche radius, receive less gain than those produced further out. As the simulated site size is decreased, the gas avalanche region becomes a larger fraction of the total detector volume. For simulated sites less than about 0.3 micrometers in diameter, this results in a undesirable loss of detector resolution. Another small problem is that if the detector utilizes a solid wall (as opposed to a wire grid) some branching tracks will result in one large event rather than two smaller events. This does not affect the total dose measurement, but produces a slight shift toward larger events when solid walled detectors are used.

A typical electron track across a 1 micrometer diameter site results in an energy deposition of about 1000 eV, or the production of about 30 ion pairs. In order to properly resolve the spectrum of these events the electronic system should be sensitive to the production of about 3 ion pairs. This requires a charge-sensitive preamplifier with electronic noise less than about 200 electrons RMS. Furthermore, a heavy ion recoil can deposit several hundred thousand eV in this site size. Thus the data acquisition system must have a dynamic range of about 4 orders of magnitude. This requirement can be met by use of high quality preamplifiers and shaping amplifiers and a multichannel analyzer with logarithmic channel width (or two special multi-channel analyzers with different amplifier gains).

The original microdosimetry systems were cumbersome collections of laboratory equipment and used detectors which required a continuous gas flow (or frequent refilling) to maintain satisfactory gas gain. However, recent technological advances have eliminated the need for frequent gas changes by eliminating the sources of gas contamination. Similarly, customized electronic systems have been built which are very compact and require a minimum of operator control. These systems range from the nearly pocket size instruments being designed as alarm detectors for radiation workers[5] to a prototype approximately the size of a shuttle mid-deck locker developed to demonstrate the feasibility of characterizing the radiation quality inside the space shuttle in real time.[6] This latter instrument, (Figure 4), utilizes two detectors, one of which simulates a 2 micrometer diameter volume, the other simulates a 100 micrometer diameter volume. The larger simulation was designed to detect very high energy ions (HZE particles) which may produce qualitatively different biological effects, while the smaller is used to determine the dose and quality factor of all of the radiations. Two amplifiers plus analog-to-digital converters are used to process the data from the 2 micrometer diameter detector while only 1 ADC is required from the 100 micrometer detector since it only needs to respond to large events. A microcomputer controls the system, receives data from the individual 128 channel multi-channel analyzers, calculates the dose and dose equivalent according to the algorithm similar to that proposed in ICRU 40,[4] and displays the results in near real time.

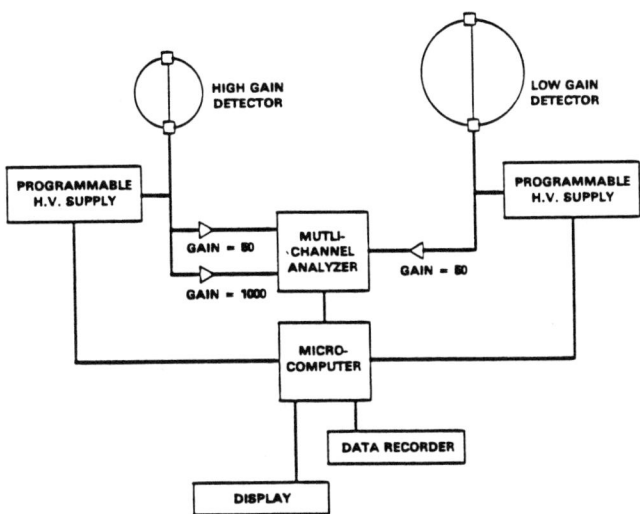

Fig. 4. Block diagram of a fully automated microdosimetry system which calculates dose and dose equivalent during the measurement.

The computer also controls calibration sources built in to each detector and adjusts the anode voltage through a digital-to-analog converter controlled power supply to place the calibration peak in the specified channel. The total instrument requires less than 3.5 watts of electrical power and the detector has operated for over a year without requiring gas purification. This system has been tested using high energy ions from the BEVALAC and also with monoenergetic neutrons produced by a Van de Graaff accelerator. The measured values of dose and dose equivalence have generally been within $\pm 7\%$ of the values obtained using fluence measurement and standard conversion techniques. This prototype instrument demonstrates that microdosimetry measurements are practical in almost any environment. The calculation discussed previously indicates that the microdosimetry data provides a reasonable means of estimating the effectiveness of different radiations.

REFERENCES

1. ICRU, Radiation Quantities and Units, (International Commission of Radiation Units and Measurements, Bethesda, Report 33, 1980).

2. A. M. Kellerer and D. Chmelevsky, Concepts of Microdosimetry-I. Quantities. Radiat. Environ. Biophys. $\underline{12}$, (1975), p. 61.

3. V.P. Bond and M.N. Varma, A Stochastic, Weighted Hit Size Theory of Cellular Radiobiological Action (Radiation Protection Proceedings of the Eighth Symposium on Microdosimetry, Edited by J. Booz and H.G. Ebert, Commission of the European Communities), p. 423, (1983).

4. ICRU, The Quality Factor in Radiation Protection, (International Commission of Radiation Units and Measurements, Bethesda, Report 40, 1986).

5. L.W. Brackenbush, J.C. McDonald, G.W.R. Endres, and W. Quam, Mixed Field Dose Equivalent Measuring Instruments, Radiation Protection Dosimetry, 10, p. 307 (1985).

6. L.A. Braby, C.A. Ratcliffe, and N.F. Metting, A Portable Dose Equivalent Monitor Based on Microdosimetry, (Radiation Protection Proceedings of the Eighth Symposium on Microdosimetry, Edited by J. Booz and H.G. Ebert, Commission of the European Communities), p. 423, (1983).

The GRAD High-Altitude Ballon Flight Over Antarctica

G. Eichhorn
Space Astronomy Laboratory
University of Florida, Gainesville, FL 32609

R.L. Coldwell, F.E. Dunnam, and A.C. Rester
Institute for Astrophysics and Planetary Exploration
University of Florida, Alachua, FL 32615

J.I. Trombka and R. Starr
NASA Goddard Spaceflight Center, Greenbelt, MD 20771

G.P. Lasche
DARPA/NMO, Arlington, VA

Abstract

The Gamma Ray Advanced Detector(GRAD) constists of a n-type germanium detector inside an active bismuth-germanate Compton and charged particle shield with additional active plastic shielding across the aperture. It will be flown on a high altitude balloon at 36 km altitude at a latitude of 78° S over Antarctica for observations of gamma radiation emitted by the radioactive decay of ^{56}Co in the Supernova SN1987A, for assessment of the performance of bismuth-germanate scintillation material in the radiation environment of near space, for gathering information on the gamma-ray background over Antarctica, and for testing fault-tolerant software.

© 1989 American Institute of Physics

Introduction

The GRAD detector is a high resolution n-type germanium gamma ray detector with an active bismuth germanate (BGO) Compton and charged particle particle shield. It was designed to be deployed in the cargo bay of the Space Shuttle on flight 62A from Vandenberg AFB in 1986. The main objectives of this mission were:

a) investigation of the behavior of the n-type germanium detector and the BGO shield in a high radiation environment,

b) evaluation of the performance of the microprocessor system and its fault-tolerant software system in a high radiation environment,

c) measurements of gamma radiation from astronomical sources (galactic center).

After the Challenger accident the GRAD experiment was not scheduled to fly before 1990.

On 24 February 1987 the Supernova SN1987A was discovered in the Large Magellanic Cloud. This was an ideal opportunity for us to use the flight-ready GRAD detector for astronomical observations of major importance. It was decided to adapt the GRAD system for a flight on a high altitude balloon. This places the detector high enough above the atmosphere to enable it to detect gamma radiation from celestial objects. To also achieve the technological objectives, other constraints had to be met. (a) The payload had to be in a high radiation environment. (b) The mission duration had to be well above the 6-8 hours achievable with regular balloon flights (e.g. from Australia). The only launch site that could provide this environment was Antarctica. In a concentrated effort of all involved agencies (University of Florida, NASA Goddard Space Flight Center, National Science Foundation, Air Force Geophysics Lab, Defense Advanced Research Projects Agency, Air Force Space Division, Office of Naval Research, Aerospace Co., New Mexico State University), the GRAD system was adapted to interface with the balloon system, the gondola and telemetry system were designed and built, the power system for the gondola was designed and built, and the logistics for the Antarctic expedition were developed.

Mission Requirements

The major objective of this mission is to measure gamma radiation from the Supernova SN1987A. According to (1) the main line emission to be expected from a Supernova should be produced by the radioactive decay of ^{56}Co at energies of 847 keV and 1238 keV. This radiation should first penetrate the shell produced during the explosion approximately one year after the first detection. This fortunately coincided with the available launch window in Antarctica of January/February, the austral summer.

A balloon flight from Antarctica has the unique advantage that the observation target is continuously above the horizon and is only slowly moving with respect to the sun (~1°/day). At our flight latitude of 78°S the Supernova moves in elevation between 57° and 81° daily. Since the field of view of the GRAD detector is 52° FWHM (see 2) no provisions for movement in elevation had to be made.

Since the emission from the Supernova is expected to be small compared with the background produced in the high radiation environment, the accurate determination of this background is essential for the success of the mission. To achieve this an azimuth pointing system was developed which enabled us to rotate the instrument with reference to the sun. The instrument controller is programmed to step continuously through an observation sequence unless commanded otherwise from the ground. This sequence consists of one 1 hour measurement pointing at the Supernova followed by 11 measurements of 5 minutes every 30° in azimuth. With this measurement sequence it is hoped that the background can be characterized accurately enough to extract the Supernova data.

<u>Detector</u>

Fig. 1 shows a cross-section of the detector (see 2). The detector is a high purity n-type germanium crystal 5.5 cm in diameter, 5.7 cm high with a counting efficiency of 30%. The detector is mounted through an aluminum cooling rod on a liquid nitrogen dewar with a capacity of 37 l. This provided a nominal maximum holding time of 21 days. The resolution of the detector after final integration in flight configuration was measured as 2.3 keV at 1332 keV.

The detector is surrounded by the active BGO shield. The Compton suppression in flight configuration was measured as approximately 80%. The entrance aperture is covered with a plastic scintillator optically coupled to the BGO shield to reject charged particles entering the detector through the aperture. The BGO shield gives the detector a directional sensitivity with a field of view of 52° FWHM.

The experiment is controlled by a Motorola 68000 based microprocessor system (3). This system integrates the data over 29 second intervals and transmits the integrated spectra to the balloon telemetry system for downlink. It also accumulates the data in on-board memory during times when the balloon is out of telemetry range. These accumulated data will then be downloaded on command to a ground station or a mobile station based on an airplane. The microprocessor system also controls the gondola pointing system. It provides the means of either stepping through the preprogrammed observation sequence or positioning the gondola by commands received through the balloon telemetry system.

During adaptation of the GRAD system to the balloon, special efforts were made to eliminate as much iron around the detector as possible. This is necessary since (pn)-reactions produce ^{56}Co from ^{56}Fe. Even small amounts of ^{56}Co produced in this fashion, especially if it is close to the detector, can make the background subtraction very difficult.

Gondola

The gondola is shown schematically in Fig. 2. It is a pyramid shaped structure with a footprint of 2.75x2.15m and a height of 3.65m. The total gondola weight in flight configuration will be less than 1000 kg. Since this flight is scheduled to last for a maximum of 21 days (the maximum holding time of the LN_2 dewar), solar cells were chosen as the main power supply. These are augmented by rechargeable batteries to supply power during times when the solar panels are not pointing towards the sun. The gondola is configured such that the detector will have the Supernova in its field of view when the solar panels are pointing at the sun. This provides the most power during the preselected automatic observation program.

To locate the balloon after it drifts out of telemetry range an ARGOS satellite positioning system will be installed. The position of the ARGOS transmitter is measured by two satellites in polar orbits. This will provid accurate enough position information to locate the balloon from an aircraft during data relays flights.

Expected Flight Profile

The balloon is to be launched in January, 1988 from the Ross Ice Shelf near McMurdo Station, Antarctica. The amount of time the balloon is within range of the telemetry station at McMurdo will depend on the wind conditions at float altitude. It should be between 12 and 24 hours. If the weather pattern is as expected for this time of the year, the balloon should circle the South Pole at an approximately constant latitude. It is planned to fly to the balloon with an LC-130 transport plane after it is out of telemetry range to downlink the stored data as often as such a plane is available through NSF. The LC-130 will be equipped with special antennas and a mobile receiving and command station. The position of the balloon will be determined through the ARGOS tracking system. After termination of the flight every effort will be made to recover the payload in order to count the induced activity in the detector and the BGO shield.

Summary

The Supernova 1987A provides a unique opportunity to utilize the GRAD detector. The fact that the experiment will be conducted near the South Pole with its associated high radiation background required special efforts to suppress this background. These included:

- a) Elimination of any unnecessary material within the BGO shield. Only aluminized mylar was used to cover the BGO shield on the inside. This is possible since BGO as opposed to NaI is not hygroscopic.
- b) Covering the entrance aperture with a plastic scintillator, coupled to the BGO shield to reject charged particles entering the detector.
- c) Removal of all iron form the inside of the shield and constructing the gondola entirely of aluminum. This included replacing all steel bolts and nuts by brass or titanium bolts and nuts. The detector itself also was disassembled and all iron parts replaced.

With these precautions it is hoped that the Supernova gamma rays and the background can be separated.

Acknowledgement

This research is supported by the Advanced Research Project Agency and the Space Test Program of the Department of Defense and is monitored by the Office of Naval Research under Grant No. N00014-87-G-1259 and by the National Science Foundation under Grant No. DPP-8715809.

References

1. N. Gehrels, C.J. MacCallum, M. Leventhal, "Prospects for Observations of the Nucleosynthetic Gamma-ray Lines and Continuum from SN1987A", Ap.J.,320:L19-L22, 1987.

2. A.C. Rester, R.B. Piercey, G. Eichhorn, J. McKisson, D.W. Ely, and H.M. Mann, "The GRAD Gamma Ray Spectrometer", IEEE Trans. on Nuclear Sci., Vol. NS-33, No. 1, PP. 732-734, 1986.

3. R.B. Piercey, J. McKisson, G. Eichhorn, and A.C.Rester, "The GRAD Data Acquisition and Control System", IEEE Trans. on Nuclear Sci., Vol. NS-33, No. 1, pp. 866-869, 1986.

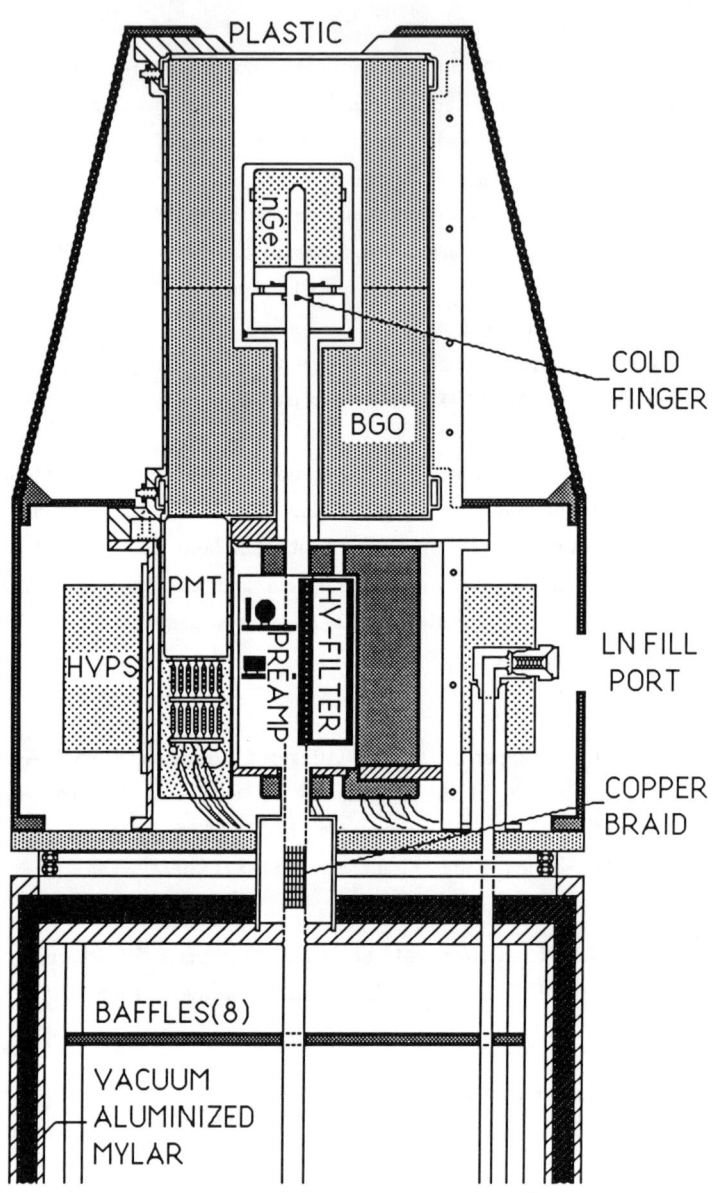

Fig. 1: GRAD Detector Cross Section

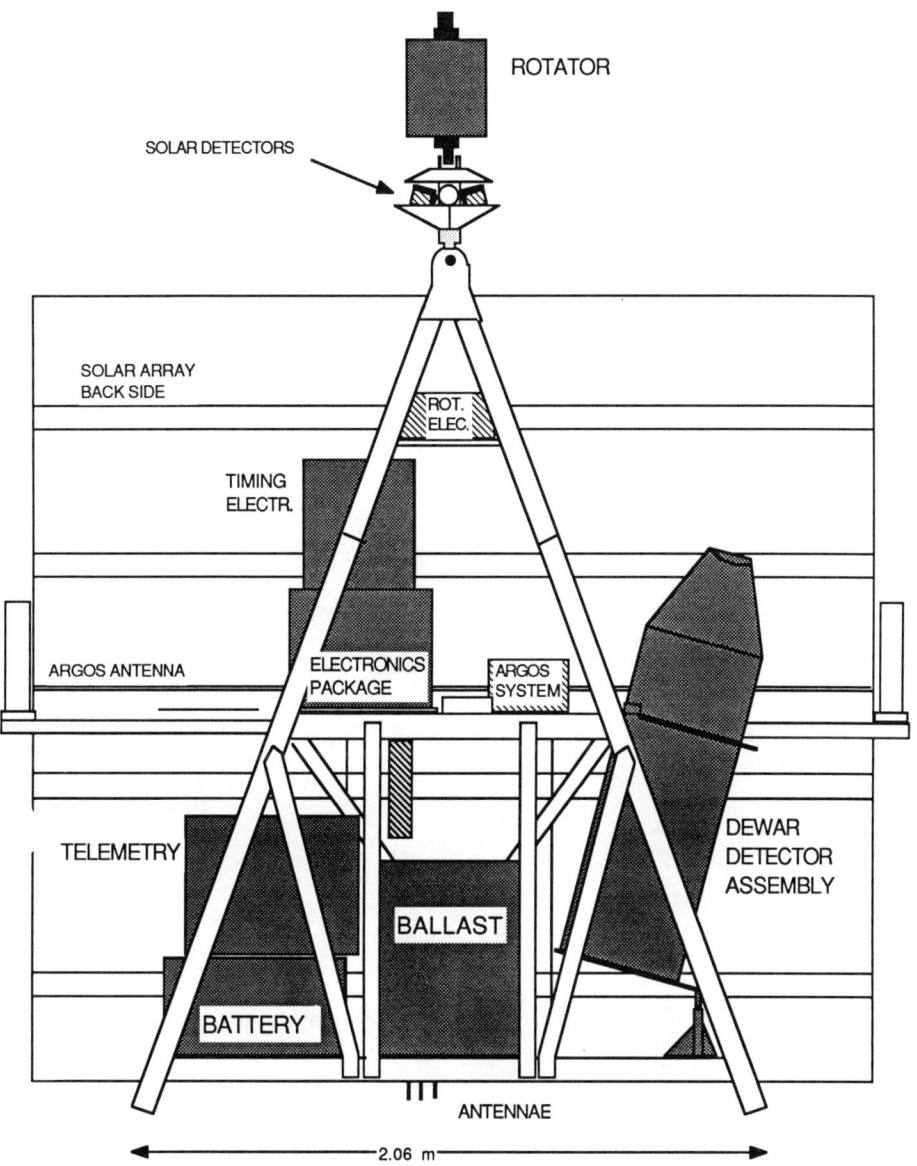

Fig. 2: GRAD Balloon Gondola Side View

VI. Biological Effects

BIOPHYSICAL ASPECTS OF HEAVY ION INTERACTIONS IN MATTER*

Walter Schimmerling, Mervyn Wong, Bernhard Ludewigt, Mark Phillips, Edward L. Alpen, Patricia Powers-Risius and Randy J. DeGuzman

Research Medicine and Radiation Biophysics Division
Lawrence Berkeley Laboratory, University of California, Berkeley, CA 94720

Larry W. Townsend and John W. Wilson

NASA/*Langley Research Center, Hampton, VA 23665*

ABSTRACT

The biological effects of high energy, high charge nuclei (HZE particles) occupy a central role in the management of space radiation hazards due to galactic cosmic rays. For the energy range of interest, the mean free path for nuclear interactions of these heavy ions is comparable to the thickness of the material traversed, and a significant fraction of stopping particles will undergo a nuclear reaction with the nuclei of the stopping material. Transport methods for HZE particles are dependent on models of the interaction of man–made systems with the space environment to an even greater extent than methods used for other types of radiation. Hence, there is a major need to validate these transport codes by comparison with experimental data. The basic physical properties of HZE particles will be reviewed and illustrated with the results of nuclear fragmentation experiments performed with 670A MeV neon ions incident on a water absorber and with measurements of multiple Coulomb scattering of uranium beams in copper. Finally, the extent to which physical measurements yield radiobiological predictions is illustrated for the example of neon.

INTRODUCTION

An understanding of the various ways in which the space radiation environment interacts with the systems we place in it is required in order to predict space radiation hazards, design strategies for avoiding unacceptable deleterious effects on personnel and instruments, and interpret scientific and engineering data collected in space. This understanding depends on models of the multiple levels of interaction between space and systems to a degree that may not be immediately apparent.

An attempt to indicate this dependence schematically has been made in Fig. 1, where the dashed blocks separate components attributed to space from components attributed to the systems used in space exploration. The boxes and the words in quotation marks are intended to indicate the models of the actual component that are used to describe it; they may be thought of as denoting computer programs written to

---

*Supported in part by the Public Health Service of the U.S. Department of Health and Human Services under Grant No. CA23247 awarded by the National Cancer Institute and in part by the National Aeronautics and Space Administration under Grant No.L22395A and conducted at the Lawrence Berkeley Laboratory (Department of Energy contract DE-AC03-76SF00098 to the University of California).

© 1989 American Institute of Physics

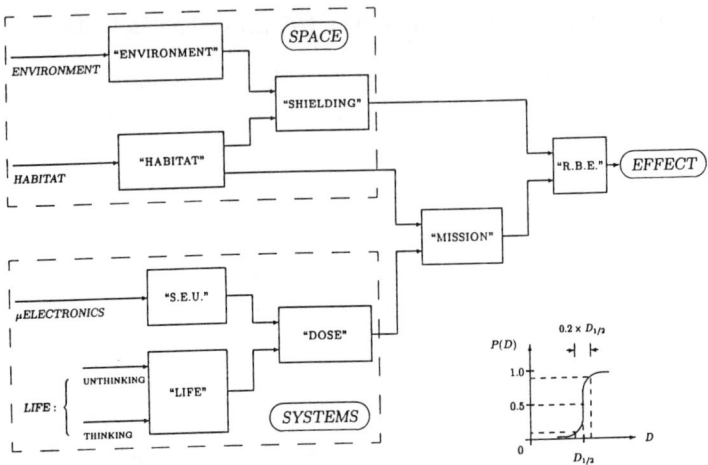

Figure 1: Schematic diagram of the interaction of space with systems to produce an effect, indicating how the problem is mediated by models at all levels. The dose effect curve in the lower right hand corner indicates the steepness of typical dose response probabilities.

calculate a restricted set of characteristics. Some of these restrictions are engineering compromises, but others are due to fundamental limitations in our knowledge.

The actual radiation environment is not known at every point of space and at any time, but depends in a complicated way on astrophysical phenomena that cannot always be predicted accurately (*e.g.*, solar storms). The description of this environment is contained in several different models, denoted by the box labelled "environment". The quotation marks indicate that this is the representation that we use for our discourse about the environment, and that it is only valid under restricted circumstances. Similar comments apply to "habitat", where a simplified description of *e.g.*, space vehicles provides the actual data used to evaluate the way in which the habitat modifies the environment; for radiation purposes, we refer to models of this interaction as "shielding", where the quotation marks indicate that model approximations have been made (*e.g.*, by neglecting various aspects that may not be important under certain circumstances, such as backscattering, or by making assumptions about time dependence and averaging).

For the purpose of understanding the radiation environment, the systems used in space are also described in restricted terms. The behavior of microelectronics circuits is most often analyzed in terms of single-event upsets. The aspects of "life" that are selected for consideration are also limited: we consider cancer induction, cell transformation, DNA damage. The effects on both personnel and instruments will be arbitrarily collected under the concept of "dose", a shorthand term for many quantities affecting radiation effects, most often approximated by a single number.

The interaction of space and systems is indicated in Fig. 1 as occurring in the box labelled "mission" which itself is also a model of the activities of personnel. The shielding filters the "quality" of the space radiation environment, depending on the mission parameters, in ways that are only approximately understood and that have been denoted by "R.B.E." to indicate relative biological effectiveness.

Finally, a schematic diagram of a dose–effect curve (probability of effect as a function of dose) has been plotted in the lower right–hand corner of Fig. 1 to make the point that most dose–effect curves are relatively sharp and that a certain degree of accuracy is required whenever the "dose" is high enough that the probability of an effect is not negligible. Quality factors, by arbitrarily defining the contents of the RBE box, attempt to position the operating point on the dose–effect curve sufficiently to the left in order to avoid having to characterize the "dose" very accurately. This makes quality factors cost effective only when the cost of transporting extra shielding, abbreviating tours of duty and taking evasive action does not exceed the cost of obtaining accurate information on the actual exposures and their biological or engineering significance (the "quality" of the radiation).

In the discussion that follows, the biophysical aspects of high–energy, high charge nuclei (HZE particles) of the types and energies present in galactic cosmic rays, will be considered. The biological effects of HZE particles are likely to be of critical importance in the management of space radiation hazards wherever the geomagnetic cutoff does not provide a natural protection.

Only recently have transport methods been developed to estimate the radiation fields produced by relativistic heavy nuclei incident upon thick layers of absorbers of different composition.[1] We begin by presenting an abbreviated description of some aspects of the physics of nuclear interactions of HZE–particles that are of importance in understanding the limitations of such transport codes. Based on the limited present state of the developing field of heavy–ion physics, it is clear that these HZE–transport methods need to be validated by comparison with experimental data; accordingly, an experimental program was started at the Lawrence Berkeley Laboratory several years ago. We report some initial results of fluence measurements of fragment yields as a function of depth in an absorber, as well as some initial results of multiple scattering measurements and of their comparison with theoretical calculations. Finally, some of the limitations in our ability to adequately predict biological effects will be illustrated by one particular case, for which details have been published elsewhere.[2]

SOME BASIC ASPECTS OF HEAVY ION PHYSICS

The properties of high energy nuclei that are most important from the point of view of their penetration into matter are their charge, given in terms of the number of protons, z; the mass, in terms of the total number of nucleons, A; and the velocity, β, in units of the speed of light c, or the kinetic energy per nucleon, ϵ.

The interaction of HZE particles with matter differs from that of an assemblage of protons and neutrons because the packing of z protons and $(A-z)$ neutrons into a nucleus of radial dimensions $\sim A^{1/3} \times 10^{-13}$ cm gives rise to correlations and to coherent effects. The most important of these for HZE–particles is manifest as the coherent z^2– dependence of the stopping power. As a consequence, the average ionization of HZE particles as a function of penetration into materials has a Bragg peak similar to that of protons, but the **range** in matter of HZE particles is shorter than that of other charged particles with the same energy per nucleon ϵ by a factor A/Z^2 and is narrower due to the reduction in energy straggling associated with the statistics of greater stopping power.

The mean free path for nuclear interactions is shorter than that of protons and neutrons by a factor of order $A^{2/3}$, itself a reflection of the "blackness" of a nuclear target. As a consequence, the mean free path for nuclear interactions of these heavy ions is comparable to their range in matter for most situations of practical interest, and the strong variation of ϵ as a function of material thickness cannot be neglected by HZE transport calculations.

Nuclear interactions of HZE particles are usually classified into three categories: **projectile** fragmentation, **target** fragmentation and **central collisions**.[3,4] Projectile and target fragmentation are isotropic in their respective rest frames, but the kinematics of relativistic particles results in collection of all projectile fragments into a gaussian distribution, sharply peaked forward in the laboratory with a width of the order of one degree.[5] As a consequence, the straight–ahead approximation long popular in transport calculations is much better in the case of HZE particles than in the case of protons and neutrons at the same ϵ, especially for the projectile fragments with mass closest to that of the projectile.

Finally, a significant number of HZE interactions result in more than one high–energy particle emerging from the reaction. The **multiplicity** of HZE reaction products is related to the impact parameter of the reaction, *i.e.*, to the extent to which the collision is central and involves momentum transfers large enough to break up the re-acting nuclei or parts of them. Contrary to the case of *e.g.*, neutron–induced spallation or fission, multiple particles emerge from HZE reactions with velocities comparable to that of the primary; their contribution to the particle fluence, even at forward angles, can be significant (of the order of $\sim 10\%$ of the reaction cross section). This is likely to have a significant impact on calculations of HZE transport through very thick materials or shielding geometries other than slabs, where the "protonization" and subsequent "neutronization" of the radiation field becomes important.

In order to characterize the radiation field to which personnel may be exposed, it is thus necessary to know the energy spectrum, the angular distribution, and the multiplicity of each type of secondary particles. In the cases of interest here, where these reactions can take place anywhere in a thick absorber, it is necessary to know these quantities as a function of particle energy not only for all particles incident upon a thickness of material, but for all the particles produced inside the material as well.

In addition to the nuclear scattering angle θ_N, it is necessary to take into account the multiple Coulomb scattering of particles in the absorber.[6] This is a statistical process characterized by a projected root-mean-square angle,[7] θ_{rms}, that is comparable to θ_N at the higher energies, and becomes very large in the stopping region.

The relevant characteristic linking the nuclear interaction phenomena to beam quality, *i.e.*, the relative biological effectiveness of heavy ion interactions, is the spatial distribution of energy deposition in the irradiated tissue. A schematic diagram of the result of a Monte Carlo calculation of electron energy deposition density of two overlapping boron fragments is shown in Fig. 2.[9] The very high energy deposition

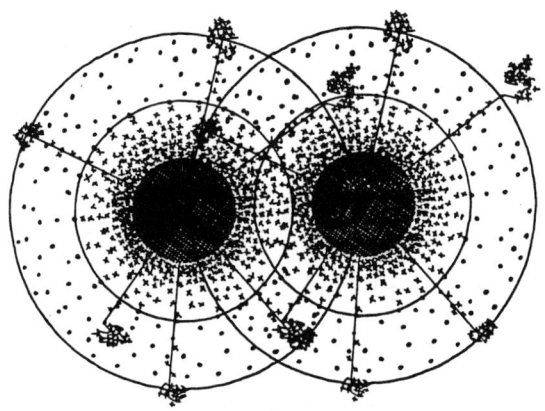

Figure 2: Cross sectional view of two overlapping $^{10}_{5}$B tracks produced in the splitting of a high–energy $^{20}_{10}$Ne–nucleus. (logarithmic polar coordinates)

density region around the trajectory of each track ("core") has a diameter of $\sim$ 60 A. The low–density energy deposition region ("penumbra"), extends to the outermost envelope, which corresponds to 10^5 Å. Energy is deposited by high–energy δ–rays. The dot density is proportional to the energy deposition density. This characteristic is also not known accurately for all cases of interest. Theoretical calculations and experimental measurements are difficult and for this reason the track structure characteristics are often approximated by the linear energy transfer (LET). This quantity is conceptually equal to the average energy deposited locally per unit track length, but it is in practice often approximated by the average stopping power, which is the linear energy loss off the traversing particle.

A calculation from first principles of the biological endpoint or relative effect requires an evaluation of the fluences of different particle species present at any point in tissue, their spatial distribution and the microscopic distribution of energy losses suffered by each particle along its track. This approach is severely limited by the complexity of the required calculations, by the paucity of knowledge on nuclear interactions of heavy ions, and by the limitations of models and theories of biological effect. The development of theories in nuclear physics, atomic physics, and radiation biology and chemistry that would make it possible to make such predictions is necessary. At the present time, such theories must be approximated by models and these models must be validated by experimental measurements. Some of these are described below.

FLUENCE MEASUREMENTS

Nuclear fragmentation experiments were performed with 670A MeV neon ions incident on twelve different thicknesses of water. A schematic diagram of the detector used for this experiment is shown in Fig. 3. In this figure, the beam traverses a water column set to different thicknesses. The velocity β of each particle emitted from the water column is obtained from the time of flight between two $\sim$ 200 μm - thick silicon semiconductor detectors T_1 and T_2 placed $\sim$ 1 m apart. The finite energy loss ΔE of

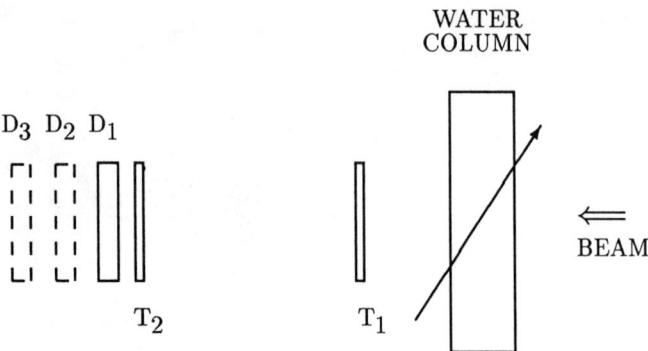

Figure 3: Schematic diagram of the detector used to measure fragments yields.

each particle was measured in the ~ 3 mm - thick D_1 detector (the presence of further detectors whose information is not included in the present discussion is indicated by the dashed outlines corresponding to D_2 and D_3 in Fig. 3).

For each particle, an effective charge is calculated as

$$Z^2 \sim (\Delta E/\rho \Delta x)/[(1/\rho) \cdot (dE/dx)_p], \tag{1}$$

i.e., as the ratio between the average stopping power of the particle in detector D_1 and the stopping power of a proton with velocity β.

Using Eq. 1 as a particle identification algorithm, fluence spectra (*i.e.*, number of particles per unit area) were obtained for particles with charges ranging from $z = 4$ (Be) to $z = 10$ (the primary neon and its isotopes), as a function of water absorber thickness. At each depth, spectra in z^2, velocity and linear energy transfer (LET) were generated.

Typical examples of LET spectra obtained for an incident 670A MeV neon beam at two depths of water (entrance and Bragg peak) are shown in Figs. 4 and 5. At the entrance to the Bragg curve the radiation field consists only of the incident beam and beam contaminants. These are present at the $\sim 1\%$ level and arise from the residual material in the beam line (vacuum windows, beam flattening lead foils, etc.). The LET spectra are narrow, even for the fragments since Fig. 4 includes the contributions due to detector resolution. Near the Bragg peak, the spectra are broadened considerably. For the primary neon, the broadening of the spectrum is due mainly to energy straggling. For the fragments, the minimum LET corresponds to the maximum energy of the fragment, which is the energy of fragments produced at the entrance to the absorber; the maximum LET corresponds to minimum energy fragments, which are produced at the exit of the absorber.

Figure 6 shows how the LET spectrum of neon varies as a function of depth: energy straggling for these beams is minimal and the spectrum does not become significantly broadened until almost the end of the particle range.

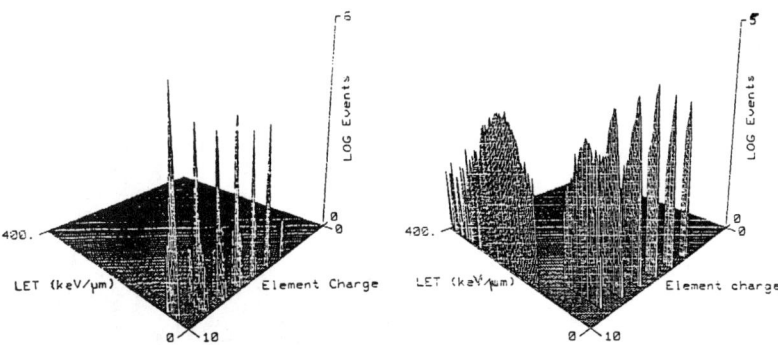

Figure 4: LET–spectra of incident neon beam and beam contaminants (note the semilogarithmic event scale).

Figure 5: LET–spectra of neon and fragments produced after 32 cm of water on which neon was incident at 670 MeV/A (note the semilogarithmic event scale).

MULTIPLE COULOMB SCATTERING

An adequate understanding of the angular distribution of radiation is required for the description of geometry–dependent factors in radiation transport and for an evaluation of the accuracy of "straight ahead" approximations. The elastic component of greatest interest for transport calculations is due to multiple Coulomb scattering (the inelastic component, not discussed here, is mainly important for very thick targets, where nuclear reaction products predominate). We have initiated a systematic investigation of heavy–ion multiple Coulomb scattering. Measurements were made using beams of 600A MeV Ne, Fe and U, incident on targets of water, Al, Cu, and Pb.

The experimental method makes use of position sensitive detectors (PSDS) to determine the spatial trajectories of individual particles, before and after target scattering. This method avoids physical collimation of the beam; collimator edge scattering and nuclear fragmentation effects are therefore eliminated. In addition, the method is independent of the beam divergence.

Four sets of $\sim$ 800–μm thick, 44–mm diam (PSDS) and their associated preamplifiers were fabricated. The position of a traversing particle is determined by charge division in a resistive sheet on the side of the detector facing the beam. The first detector in each set is used to measure the horizontal (x) position and the second detector is used to measure the vertical (y) position. The incident–particle trajectory and the deflected–particle trajectory are both reconstructed in three dimensions from the measured particle coordinates.

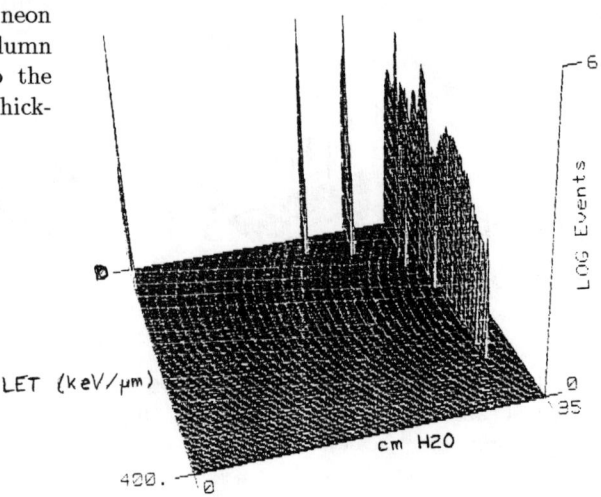

Figure 6: LET–spectra of neon as a function of water column thickness, normalized to the beam intensity at zero thickness.

The experiments with uranium offer an extreme test of how far multiple Coulomb scattering calculations based on proton (z=1) physics could be extrapolated. The data were compared with acceptance calculations. Fig. 7 shows the fraction of beam intercepted by increasing circular areas in the last x–detector. The open circles are the data for the straight–through beam and the filled circles are the data for a 2.7-mm diam Cu target. The curves (both dashed and full) are the results of the acceptance calculation performed using the PROPAGATE program,[10] which has no free parameters; the extraction energy, the measured beam profiles at the entrance to the detectors, and the measured properties of the materials in the beam were used in the calculation. Agreement between the integral calculation and experiment is at the 10–15% level.

The PSDS make possible differential measurements of the multiple scattering deflection angle. Fig. 8 shows preliminary results for the projected angular distributions for uranium scattering, both with and without a 2.7–mm Cu target in place.

The differential shapes of these distributions have been studied using a Monte Carlo simulation of the experiment, taking into account the particle energy loss and intrinsic spatial resolution of each detector, as well as detector multiple scattering. A "standard" description of multiple Coulomb scattering was used in the Monte Carlo simulation with the rms multiple scattering angle given by:[8]

$$\sigma_{proj} = \frac{14.1 \text{ MeV}/c}{p\beta} z_{in} \sqrt{\frac{t}{X_0}} \left[1 + \frac{1}{9} \log_{10}\left(\frac{t}{X_0}\right)\right] \qquad (2)$$

with energy loss taken into account by dividing this gaussian standard deviation σ by a factor $(1 - (x/R_0))^{\frac{1}{2}}$, and where x is the material thickness and R_0 is the particle residual range.

Preliminary results comparing the differential distributions for the uranium experiment with a Monte Carlo simulation also indicate agreement between theory

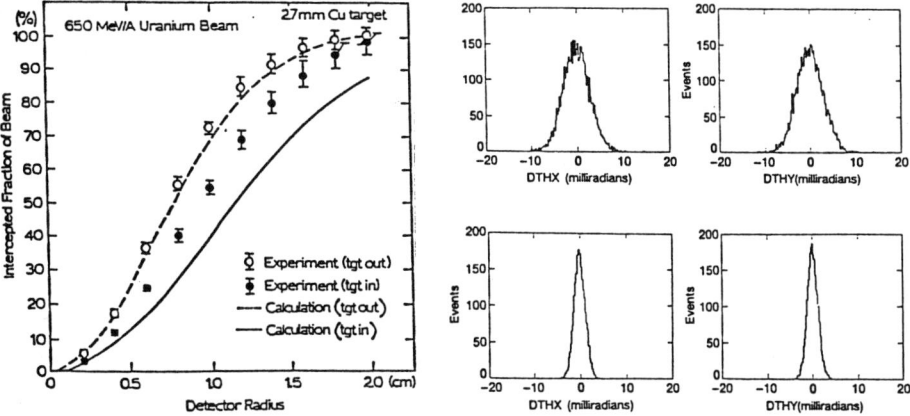

Figure 7: Acceptance for 600A MeV U beam as a function of last x–detector radius.

Figure 8: Projected angular distributions for uranium scattering with (upper panels) and without (lower panels) a 2.7-mm Cu target.

and experiment to within 15%.

RADIOBIOLOGICAL IMPLICATIONS – AN EXAMPLE

To examine the validity of the hypothesis that biological effects can be predicted based on particle fluence at depth and on the observed dependence of RBE on LET, the central axis fluence measurements for neon were evaluated in terms of their relative biological effectiveness (RBE) for testes weight loss, a tissue model that measures the killing effectiveness for early spermatogonia in the mouse testes.[11] The RBE as a function of LET, obtained by Alpen and Powers–Risius,[2] is shown in Fig. 9.

From the LET spectrum of each particle at each depth of water, a dose D(z) was calculated from the measured fluence spectrum of each particle species, and a biological "dose equivalent" $D_e(z)$ was obtained by weighting the fluence spectra with R(L) at each value of L as follows:

$$R.B.E. = \frac{\sum_z \int_L D_e(L)dL}{\sum_z \int_L D(L)dL}, \quad (3)$$

where

$$D_e(L) = R(L)D(L)$$

and

$$D(L) = \varphi(L)L.$$

The RBE for the mixed radiation field is calculated as the ratio $\overline{RBE} = \sum D_e(z)/\sum D(z)$ using the data of Fig. 9 to interpolate between the plateau data

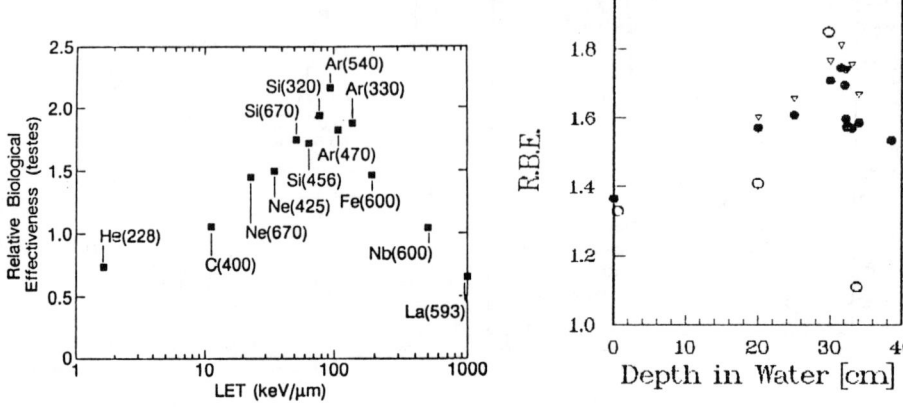

Figure 9: Relative biological effectiveness for spermatogonial cell survival, labelled with the particle species and nominal energy per nucleon used to obtain each data point

Figure 10: RBE for spermatogonial cell survival vs. water column thickness for 670A MeV neon nuclei. ●: calculated, ○: measured, ▽: calculated using neon only.

of pure beams. In collaboration with Alpen and Powers-Risius, measurements of spermatogonial cell survival were made in the mixed radiation field of the neon beam at four depths in water to obtain experimental values of $\overline{RBE}$ for comparison with the data calculated using the physical measurements and the RBE values for pure beams at different values of LET.

The results are shown in Fig. 10, for cell survival measured in the mixed radiation field produced by the incident neon at depth and for the calculated RBE based on measured fluences and RBE from Fig. 9. The RBE calculated using only the neon fluence is also shown to illustrate the contribution of secondaries due to one projectile at one energy.

The difference between the RBE calculated neglecting the fragment contribution (open triangles) and the values calculated including the fragment fluences becomes important beyond the range of the particle. The fluence measurements did not include particles with $z \leq 3$. These particles will tend to lower the actual RBE in the mixed radiation field and would result in RBE values closer to those actually measured at 20 cm and at 33 cm; at $\sim$ 30 cm, however, the observed discrepancy would be even greater. Furthermore, the H and He nuclei cannot fully account for the discrepancy observed in the Bragg curve tail at $\sim$ 33 cm.

In order to account for the observed discrepancies between RBE interpolated as a function of LET and the measured values of RBE, it is necessary to consider the track structure of the beams used for the RBE determination. Using the homogeneous track model of Chatterjee and Schaeffer,[9] at 1 μm from the track core, the radial energy density is found to be ∼ 15% greater in the mixed radiation field than in the pure beam used to obtain RBE in the LET–range of the mixed beam.

There is evidence that cell inactivation depends on the local production of one or more lesions requiring $\geq$ 300 eV each,[12] and that these lesions may be double DNA strand breaks.[13] In the latter case, the amount of energy required to produce such breaks, at high LET , can be estimated to be 520 eV.[14] The specific energy deposition at maximum RBE for spermatogonial cell killing, is 600 – 800 eV/$(\mu m)^3$ at 1 μm from the track core of HZE-particles. In a radio–sensitive sytem like the spermatogonial cells, where the exponential survival curve indicates a lack of repair, the variations in RBE due to different energy densities in the penumbra of heavy ions reflect the statistics of lesion production and there would seem to be qualitative agreement with such an understanding of the physics responsible for biological effect.

CONCLUSIONS

The experimental results illustrate the physical characteristics of radiation that need to be known in order to make predictions of biological effects necessary for management of radiation hazards in space. In the case of HZE–particles, a measurement of the fluence spectra of all particles present in the radiation field at depth in matter enables track structure information to be applied to the prediction of biological effects with an accuracy of the order of a few percent. Multiple Coulomb scattering describes that part of the elastic component of the angular distribution of nuclear interaction products which predominates at the smallest angles and is of greatest interest for transport calculations. For geometrical effects involving materials with thickness less than the incident particle range, current knowledge of multiple Coulomb scattering seems to be sufficient for calculations that are accurate at the ∼ 15% level. The experimental data base on which the discussion of HZE particles is based is very sparse, mainly because laboratory beams of HZE particles have become available relatively recently and the field of relativistic heavy ion physics is still under development. In view of the ability of adequate physical characterization of these particles to provide useful biological information, it is desirable to expand this data base with accurate and comprehensive experimental measurements. Such a program is currently in progress at the Lawrence Berkeley Laboratory.

REFERENCES

1. J.W. Wilson, L.W. Townsend, H.B. Bidasaria, W. Schimmerling, M. Wong and J. Howard,*Health Phys.* **46**,1101 (1984).

2. W. Schimmerling, E. L. Alpen, P. Powers–Risius, M. Wong, R. DeGuzman, and M. Rapkin *Radiat. Res.* **112**, 436(1987).

3. A.S.Goldhaber and H. H. Heckman *Ann. Rev. Nucl. Sci.***28**,161 (1978)

4. R. Babinet, *Ann. Phys. Fr.*, **11**,113(1986)

5. J. P. Wefel, J. M. Kidd, W. Schimmerling, and K. G. Vosburgh *Phys. Rev.* **C19**, 1380-1392 (1979).

6. G. Moliére *Z. Naturforschg.* **3a**, 78-97 (1948).

7. J.D. Jackson *Classical Electrodynamics*, 2nd. ed., (John Wiley and Sons, New York) 1975.

8. M. Aguilar-Benitez *et al.*(Particle Data Group) *Phys. Lett.* **170b**, 45 (1986).

9. A. Chatterjee and H.J. Schaefer *Rad. and Environ. Biophys.* **13**, 215-227 (1976).

10. W. Schimmerling, M. Rapkin, M. Wong, and J. Howard, *Med. Phys.* **13**, 217(1986).

11. E. L. Alpen, P. Powers-Risius, and M. A. McDonald *Radiat. Res.* **88**, 132 (1980).

12. D.T. Goodhead, R.J. Munson, J. Thacker and R. Cox *Int. J. Radiat. Biol.* **37**,135 (1980).

13. D. L. Dugle, C.J. Gillespie and J. D. Chapman *Proc. Nat. Acad. Sci. USA* **73**,809 (1976).

14. R. Roots and K. C. Smith *Int. J. Radiat. Biol.* **26**,467 (1974).

DELAYED EFFECTS OF PROTON IRRADIATION IN MACACA MULATTA
(22-YEAR SUMMARY)

D. H. Wood, K. A. Hardy, A. B. Cox, Y. L. Salmon
M. G. Yochmowitz and R. E. Cordts.
U. S. Air Force School of Aerospace Medicine, Brooks AFB, TX 78235

ABSTRACT

Lifetime observations on a group of rhesus monkeys indicate that life expectancy loss from exposure to protons in the energy range encountered in the Van Allen belts and solar proton events can be correlated with the dose and energy of radiation. The primary cause of life shortening is nonleukemic cancers. Radiation also increased the risk of endometriosis (an abnormal proliferation of the lining of the uterus in females). Other effects associated with radiation exposures are lowered glucose tolerance and increased incidence of cataracts. Calculations of the relative risk of fatal cancers in the irradiated subjects reveal that the total body surface dose required to double the risk of death from cancer over a 20-year post exposure period varies with the linear energy transfer (LET) of the radiation. The ability to determine the integrated dose and LET spectrum in space radiation exposures of humans is, therefore, critical to the assessment of lifetime cancer risk.

INTRODUCTION

The commitment of the U. S. government to manned lunar exploration in the 1960s was made at a time when the radiation hazards of such a venture were largely undefined. To help fill this urgent requirement, the U. S. Air Force School of Aerospace Medicine, with the support of the National Aeronautics and Space Administration, conducted a series of experiments to determine the acute and long term effects of the types of radiation that could be encountered in extended manned space flights. Adolescent rhesus monkeys of similar age and of both sexes were exposed to single, total body doses of one of several types and energies of radiation, including protons, electrons, and X-rays. The acute effects of these radiations have been described in a series of reports[1-7]. A select population of the original group of exposed animals and their age matched controls was set aside for long term observation of possible delayed effects.

EXPERIMENTAL DESIGN

The exposure data for all groups in the lifetime study are summarized in Table I.

Table I Simulated Space Radiation Exposures in Rhesus Monkeys

TYPE	ENERGY (MEV)	DOSE RANGE (CGY)	DATE	MALES	FEMALES
PROTON	32	280-560	JUL 64	6	6
PROTON	55	25-600	APR 65	50	22
PROTON	138	210-650	JAN 65	19	13
PROTON	400	50-600	MAR 65	28	27
PROTON	2300	56-560	OCT 65	21	25
X-RAY	2	446-716	MAR 64	15	17
ELECTRON	2	900-1500	NOV 69	5	7
ELECTRON	1.6	1000-1500	MAY 68	0	12
PROTON (9:1)	10 & 100	300-1200	APR 69	17	11

The exposure conditions have been described in the earlier reports. During the proton exposures, the beam characteristics precluded simultaneous irradiation of all body surfaces; therefore, to simulate as closely as possible the multidirectional exposure in space, the animals were confined in rotating wire mesh cylinders aligned perpendicular to the beam. The cylinders revolved at 2 rpm, and could be raised or lowered to expose the upper or the lower half of the body. Depending upon the source, the exposure rates varied from 12.5 to 100 cGy/min. The protons represent the most significant radiation hazard in solar particle events. The availability of controlled data on proton irradiation in a primate is important because previous guidelines have been derived from data on populations exposed to other types of radiation. Conversion of nuclear fission exposure data to space radiation risk estimates requires the use of estimated quality factors to correct for the relative biological effectiveness of the different radiations. The simulation of space radiation theoretically eliminates the requirement for a quality factor, and the nonhuman primate model reduces the uncertainty of extrapolation of the animal data to humans. The delayed effects study began with 301 irradiated and 57 sham irradiated animals. The animals have been individually housed under the same environmental conditions since the study began. No animals are sacrificed, except for humane considerations. All animals that die receive a thorough post-mortem examination.

RESULTS

<u>Mortality and Life Shortening.</u> In the original acute experiments, sufficient numbers of animals were lethally irradiated to permit reasonable estimates of the $LD_{(50)}$. The effect of radiation type and energy on the $LD_{(50)}$ is illustrated in Table II.

Table II LD$_{(50)}$ in Rhesus Monkeys*

RADIATION TYPE	TISSUE PENETRATION	LD$_{(50)}$
32 MeV Proton	$\leq$ 1.0 cm	1600 cGy
55 MeV Proton	$\leq$ 2.5 cm	1150 cGy
138 MeV Proton	total	520 cGy
400 MeV Proton	total	500 cGy
2300 MeV Proton	total	400 cGy
2.0 MeV X-ray	total	670 cGy

Our data indicate that the LD$_{(50)}$ for protons of sufficient energy to penetrate the whole body is, within the limits of the experimental error, similar to that for high energy x-rays. The LD$_{(50)}$ for monkeys appears to be only slightly greater than that reported for humans. As expected, the LD$_{(50)}$ is greatest for those radiations that do not uniformly penetrate all blood forming organs.

In those animals assigned to the long term study, the mortality, as represented by Kaplan-Meier product-limit estimates is related to both dose and energy (Figure 1.)

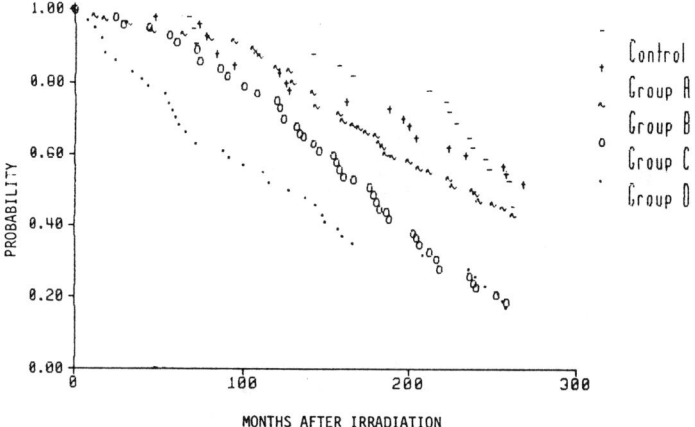

Fig. 1. Survival probability related to time after irradiation in four dose-energy combinations. Group A: 32-55 MeV, 25-280 cGy. B: 138-2300 MeV, 25-280 cGy. Group C: 138-2300 MeV, 360-800 cGy. Group D: 32-55 MeV, 360-800 cGy.

The high dose animals in both energy groups had a similar

---

*Modified after Krupp (1976)[8]

survival probability at the 22 year point, but the low energy (high LET) exposed monkeys had higher mortality in the first ten year period, while the high energy exposed animals experienced greater mortality in the later phase of the study. Recent accelerated mortality in the control group has closed the gap between the exposed and nonexposed animals. Our data show that the median age at death for nonirradiated rhesus monkeys in our colony is approximately 24 years.

The charts in Figure 2 show causes of death in the irradiated and control populations. Neoproliferative disease, including both tumors and endometriosis, accounts for a considerably greater fraction of the total deaths in the irradiated animals than in the controls.

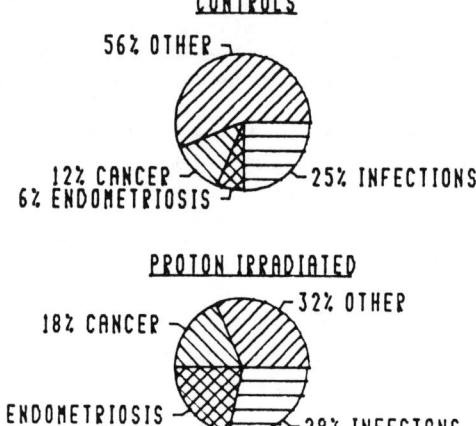

Fig. 2. Primary causes of mortality in control and irradiated monkey populations during a 20-year postirradiation period.

The mortality from nonleukemic cancers is the principal reason for the loss of life expectancy in the irradiated animals. Estimates of the loss of life expectancy, calculated by the Cohen-Lee method[9], have been computed for a 20-year post-irradiation interval. The Cohen-Lee formula for estimating life expectancy loss from a single factor uses the death rate from all factors in a reference population as its' basis of comparison. To normalize data from reference populations that have different average life spans, such as monkeys and humans, a correction factor based on the ratio of the average life expectancies of the two populations may be used. For monkey-to-human conversions, the factor is 2.5, based on a human/monkey life expectancy ratio of 75/30. The $LD_{(50)}$ studies suggest that the two species do have similar dose-response relationships. As we have reported previously[10], endometriosis is a significant cause of mortality in irradiated female monkeys, but it is not normally a life threatening condition in humans. Therefore, estimates excluding the contribution of endometriosis were calculated. These estimates did not differ appreciably from those including the endometriosis because a significant number of controls also died of endometriosis. Life expectancy loss associated with proton dose and energy is shown in Figure 3.

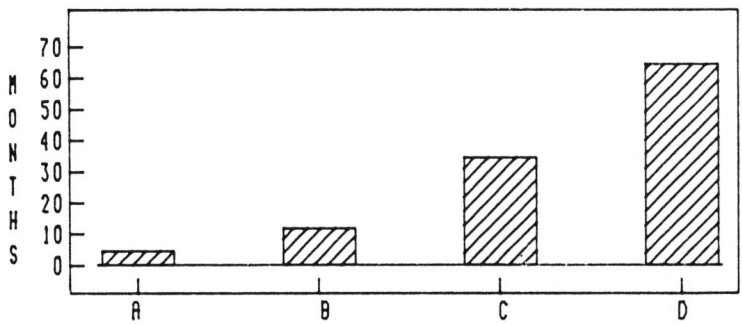

Fig. 3. Life expectancy loss over a 20-year interval related to both dose and energy. Group A: 32-55 MeV, 25-280 cGy. Group B: 138-2300 MeV, 25-280 cGy. Group C: 138-2300 MeV, 360-800 cGy. Group D: 32-55 MeV, 360-800 cGy.

The results reflect the greater biological effectiveness of the 55-MeV protons with the low energy-high dose combination resulting in a 73-month or 31% life expectancy loss for a 20-year period.

As demonstrated in Figure 4, life expectancy loss from proton radiation within the energy range encountered in space, can be fitted to an exponential curve.

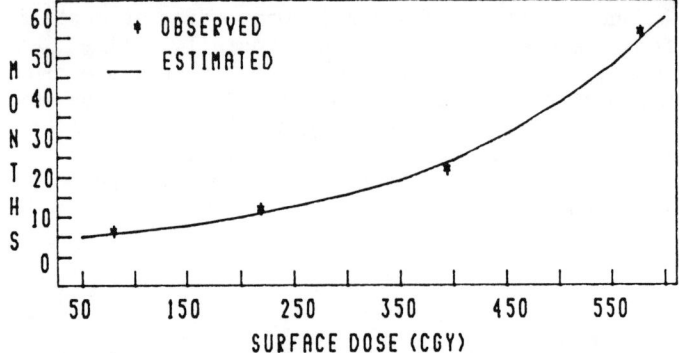

Fig. 4. Observed life expectancy loss (ignoring the effect of endometriosis) at four weighted mean doses of proton irradiation in monkeys at 20 years after irradiation. The curve of estimates is derived from least squares analysis.
($y = 3.06 \times e^{.0052x}$)

The calculated life expectancy loss per unit of radiation of all types is shown in Figure 5. This saddle shaped curve can be used to estimate life shortening from the doses recorded by passive personal dosimeters where the type and energy of the radiation is undetermined. Note that for 55-MeV protons and 2-MeV X-rays, the estimates are constants (3.6 and 1.7 days per cGy).

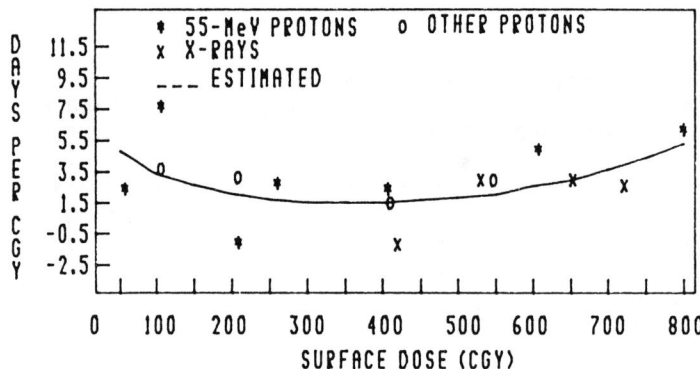

Fig. 5. Observed life expectancy loss (days per cGy) from all types of experimental radiation in monkeys. The curve was derived by least squares analysis ($y = .000025x^2 - .02x + 5.2$).

Risk of Fatal Cancer. Three control and 35 irradiated animals have died of cancer. This translates into an excess mortality of

2.15 x 10-8 fatal cancers per year per cGy. We have expressed annual relative risk as the ratio of the number of cases per monkey year in the irradiated versus control populations. A value of one indicates equality between the groups, while a value of less than one indicates a lower incidence in the irradiated population. Because the number of years at risk is a random variable, we have not attempted to estimate relative risk confidence limits. The relative risks for the major causes of death are compared in Figure 6. The effect of dose and energy is again apparent. The combination of low energy and high dose results in the greatest relative risk for all major causes of death except endometriosis. The location of the uterus makes it vulnerable only to the deeply penetrating high energy radiations. The overall incidence of endometriosis in the irradiated animals is slightly more than 50%, including non-fatal cases, compared to 14% in the controls. As indicated in Figure 6, all dose groups had a greater susceptibility to endometriosis than the controls (Fischer's exact test, p<.01).

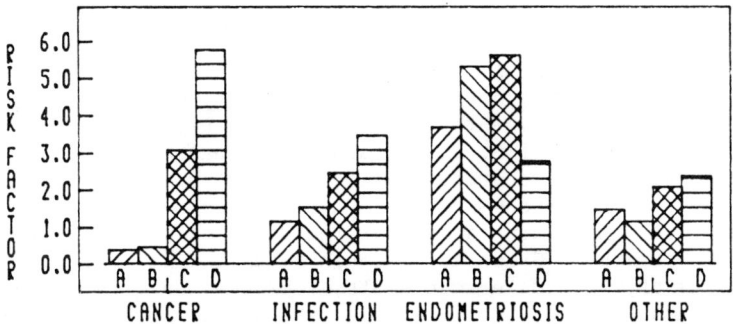

Fig. 6. Annual relative risk of death from major causes in four dose-energy combinations. Group A: 32-55 MeV, 25-280 cGy. Group B: 138-2300 MeV, 25-280 cGy. Group C: 138-2300 MeV, 360-800 cGy. Group D: 32-55 MeV, 360-800 cGy.

The relative risk of cancer death is also related to both dose and energy. The risk from high energy radiation is presented in Figure 7. The estimated dose to double the relative risk of fatal cancer is 342 rem, assuming a quality factor of 1.0 for conversion of cGy to rem. Of greater interest is the risk associated with lower energy protons, which constitute the majority of the Van Allen Belt protons and solar flare particles.

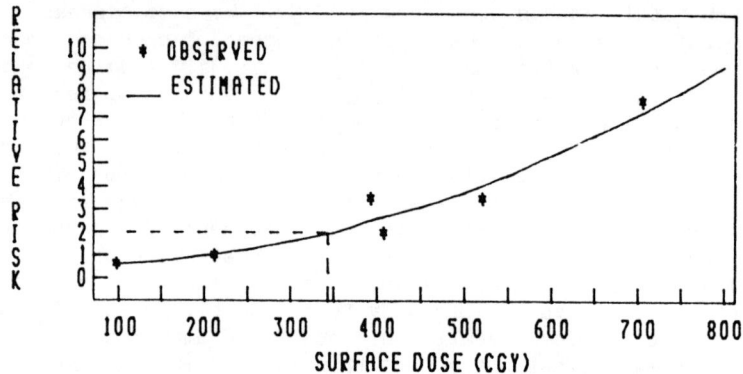

Fig. 7. Annual relative risk of fatal cancer observed at six weighted mean doses of low LET (138-2300 MeV proton or 2 MeV X-ray) radiation in monkeys. The curve fits a quadratic equation. ($y = .000145x^2 + .0063x + .521$).

Although the rate of all cancers is significantly higher in the exposed subjects, only one type of cancer has occurred in sufficient numbers to suggest that it is associated with a particular type of radiation. Nine fatal brain cancers of similar morphology (Grade IV astrocytoma or glioblastoma multiforme) have occurred in the 55-MeV proton group. These tumors occur spontaneously in humans with an annual incidence of approximately one in 20,000 persons. Glioblastomas can occur at any age, but are seen predominantly in middle-aged adults. Experience with tinea capitis patients treated with X-rays has shown that low-LET brain irradiation, in doses as low as 150-200 rem, results in a statistically significant increase in the incidence of intracranial neoplasms[11]. The magnitude of this risk compared with that of our monkeys is given in Table III.

TABLE III Brain Tumor Risk in Irradiated Humans and Monkeys

GROUP	POPULATION	TUMORS	YEARS	DOSE (GY)	TUMORS/GY/YR	RELATIVE RISK
Human Tinea Capitis Patients (100KV X-Ray Exposures)	10902	8	200,000	1.4	1.4×10^{-6}	
Rhesus Monkeys (55-MeV Proton Exposures)	72	9	960	3.5	1.3×10^{-4}	94x

According to the United Nations Scientific Committee on the Effects of Atomic Radiation (UNSCEAR)[12], radiation induced tumors, with the exception of leukemia, generally have mean latencies of 20 years or greater in the human. One might expect this to be proportionately less (35/70) in the monkey. Only one case of leukemia has occurred in the irradiated monkey population. This case was seen at 14 months post-irradiation in a male exposed to 100 cGy of 400 MeV protons.[5] Observation of 72 animals exposed to 55-MeV protons, revealed that the dose to double the cancer risk was 245 cGy (Fig. 8).

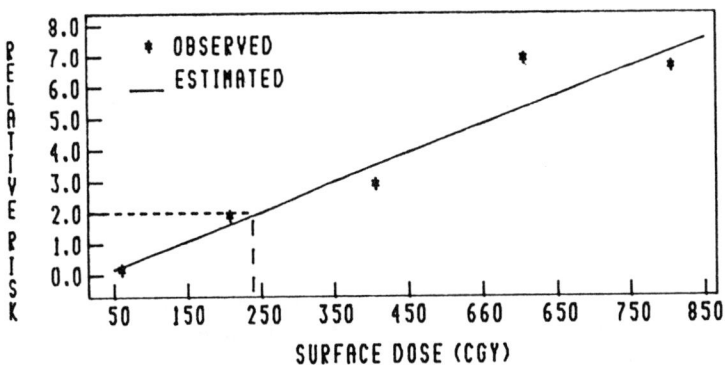

Fig. 8. Annual relative risk of fatal cancer observed at five weighted mean doses of high LET (55-MeV) proton irradiation in monkeys. The estimates best fit linear equation (y=.0092x - .266).

Cataracts. Eye examinations are now performed annually. The observations in 1986 and 1987 have revealed an interesting trend in the development of late cataracts. There appears to be an increase in the number and severity of lens opacities which corresponds to the late phase of radiation cataract induction reported in rabbits and dogs[13]. If the trend proves to be significant, it will markedly increase the value of the data from the lower mammalian species in predicting the dose response relationship of radiation induced cataracts in humans. Cataracts are an example of a nonstochastic, or threshold effect, in which the physical manifestation of the injury is not apparent until a minimum exposure has been exceeded. For a number of years, the apparent threshold in our monkeys has been approximately 2.0 Gy. The final figure and its variability with type, dose and energy of radiation will not be known until the late phase results are complete.

Diabetes. The average blood glucose clearance rate (glucose tolerance) in the human population declines as a function of age.

In some individuals the condition progresses to the point of clinically apparent diabetes mellitus. Data from the colony has suggests that the same is true for monkeys[14]. Monkeys that received radiation of sufficient energy to penetrate the whole body had lower glucose tolerance than age-matched controls in a single survey in 1981. The age-matched controls were less glucose tolerant than younger nonirradiated animals (Table III).

Table III Incidence of Diabetes (1964-1981)

RADIATION TYPE	CASES/AT RISK	%	MEAN GLUCOSE CLEARANCE*
32-55 MeV Proton	2/84	2.4	2.57 %/min
138-2300 MeV Proton	5/32	4.5	2.20 %/min
10 & 100 MeV Proton	0/28	0.0	2.55 %/min
Electron	0/24	0.0	2.71 %/min
2.0 MeV X-ray	5/32	15.6	1.93 %/min
Untreated (19-yr-old)	0/57	0.0	2.57 %/min
Untreated (10-yr-old)* 1981 Survey	0/10	0.0	3.27 %/min

Because diabetes is a metabolic disease associated with aging, the results are of interest in the consideration of the effect of total body irradiation on the overall aging process. So far, the only cases of clinically apparent type II diabetes (glucose clearance of less than 1%/min.) have occurred in irradiated animals.

CONCLUSIONS

Effective radiation protection measures for space crews require both the use of dose limitation and the ability to assess quantitatively the risk of potentially hazardous activities. Limits assure that risk remains within tolerance when doses can be forecast with reasonable certainty. This has value in planning crew rotations and redeployment as well as maximum career exposures. The data from our experimental animal studies, together with human cancer incidence figures compiled by the NAS-BIER[15] and with new estimates of the doses at Hiroshima and Nagasaki[16], support the new tentative recommendations of the NCRP Panel 75 for lowering the individual career exposure limits[17]. Interim doses should also be scaled down, because experience by NASA has shown that astronaut careers may extend to a least ten years instead of the five years assumed in the 1970 Space Science Board recommendations[18]. Until the relative sensitivity of human and monkey females to radiation induced endometriosis can be established, we recommend that the maximum career cumulative dose for radiation in women of child bearing age not be allowed to exceed 25 rem. In some critical missions, dose limitation must be disregarded in favor of risk versus gain decisions. Both life shortening and cancer risk can be forecasted if accurate information on the type, energy, dose, and dose rate is known. Life expectancy loss provides a means for assessing radiation risk relative to other environmental hazards. Within the dose range of 50-800 cGy, total body surface radiation, an average life expectancy loss from all types of space radiation has been found to be 2 to 5 monkey-days or 5 to 12.5 person-days per cGy. Assuming a 200 rem career exposure and ignoring dose rate effects for a "worst case" estimate this translates into an average life expectancy loss of 1000 to 2000 days. Equivalent non-radiation hazards are: being a coal miner (1100 days); being 30% overweight (1300 days); having heart disease (2100 days). The risk is less than that of a cigarette smoking male (2250 days) but greater than that of a cigarette-smoking female (800 days)[9]. The association of brain tumors with 55 MeV protons emphasizes the importance of accurate determination of the type and linear energy transfer spectrum of the radiation in assessing the risk of delayed effects. These measurements are also important in evaluating the threat from the heavy ion component of galactic cosmic radiation. The lack of information on the physiological and behavioral effects of heavy ions is a major void in our understanding of the space radiation threat. When the contribution of these heavy ions to the radiation environment is determined, realistic ground-based simulations can be conducted to evaluate their potential for undesirable late effects.

REFERENCES

1. G. V. Dalrymple, I. R Lindsay, J. J. Ghidoni, H. Kundel, E.T Still, R. Jacobs and I. L. Morgan, Radiat Res. 28, 406 (1966).
2. G. V. Dalrymple, I. R Lindsay, J. J. Ghidoni, J. D. Hall, J. C. Mitchell, H. L. Kundel and I. L. Morgan, Radiat. Res. 28, 471 (1966).
3. G. V Dalrymple, I. R. Lindsay, J. J. Ghidoni, J. C. Mitchell and I. L. Morgan, Radiat. Res. 28, 507 (1966).
4. I. R. Lindsay, G. V. Dalrymple, J. J. Ghidoni, J. C Mitchell, and I. L. Morgan, Radiat. Res. 28, 446 (1966).
5. A. M. Siegal, H. W. Casey, R. W. Bowman, and J. E. Traynor, Blood 32, 989 (1968).
6. H. W. Casey, P. S. Coogan and J. E. Traynor. Radiat. Res. 39, 634 (1969).
7. J. E. Traynor, and H. W. Casey. Radiat. Res. 47, 143 (1971).
8. J. H. Krupp, Radiat. Res. 67, 244 (1976).
9. B. L. Cohen, and I. Lee, Health Physics 36, 707 (1979).
10. D. H. Wood, M. G. Yochmowitz, Y. L. Salmon, R. L. Boster, and R. L. Eason. Aviat. Space Environ. Med. 54, 718 (1983).

11. B. Modan, H. Mart, D. Baidatz, R. Steinitz and S. G Levin. Lancet 1, 277 (1974).
12. United Nations Scientific Committee on the Effects of Atomic Radiation (UNSCEAR), U. N. Publication E. 77. ix. 1, (1977).
13. J. T. Lett, A. B. Cox and A. C. Lee. Adv. Space Res. 6 (11) 295, (1986).
14. Y. L. Salmon, M. G. Yochmowitz and D. H. Wood. USAFSAM-TR-84-7, (1984).
15. National Research Council, Report of the Committee on the Biological Effects of Ionizing Radiation: The Effects on Populations of Exposure to Low Levels of Ionizing Radiation (BEIR Report). National Academy of Sciences, Washington, D. C. (1980).
16. U. S. Department of Energy. Reevaluation of the Dosimetric Factors, Hiroshima and Nagasaki. DOE Symposium Series 55, DOE Technical Information Center, Washington, D. C. (1982).
17. W. K. Sinclair, Adv. Space Res. 6(11), 295 (1986).
18. Space Science Board, Radiation Protection Guides and Constraints for Space-Mission and Vehicle-Design Studies Involving Nuclear Systems. National Academy of Sciences (1970).

RESPONSE OF CARAUSIUS MOROSUS TO SPACEFLIGHT ENVIRONMENT

G. Reitz, H. Bücker, R. Facius, and G. Horneck
DFVLR-Institute for Aerospace Medicine, 5000 Köln 90, FRG

W. Rüther
University of Marburg, 3550 Marburg, FRG

R. Beaujean
University Kiel, 2300 Kiel 1, FRG

W. Heinrich
University of Siegen, 5900 Siegen 21, FRG

ABSTRACT

Already during the early biosatellite program, a synergistic action of radiation and spaceflight factors - probably microgravity - was observed in disturbances of development of Drosophila larvae, such as chromosome translocations and body anomalies[1]. Radiation was applied by an onboard source of gamma-radiation[2]. The synergism was supposed to be due to an increase in chromosome breakage followed by a loss or exchange of genetic information.

A striking finding, that could not be explained at that time, was a high frequency of genetic damage, such as lethal mutations, chromosome crossovers, and chromosome translocations, in the "flight-control" larvae which were not exposed to the onboard radiation source. These effects could have been caused by either HZE particles of cosmic radiation or microgravity or both parameters together.

A different approach to separate pure radiobiological effects from disturbances produced by microgravity or by combined actions of both parameters was applied in our experiment which uses the Biostack concept[3] - biological objects arranged between nuclear track detectors - and the 1 g reference centrifuge of the BIO-RACK[4] in combination. For the investigations Carausius morosus eggs of different stages of development were exposed in the German D1 mission and allowed to continue their development during spaceflight. After retrieval, hatching rates, growth kinetics and anomaly frequency were determined for the different test samples. Radiation data were obtained from the detectors used in the Carausius experiment and from dosimeters located in several additional positions inside BIORACK.

The early stages of development turned out to be highly sensitive to single hits of high LET cosmic ray particles as well as to the temporary exposure to microgravity during their development. In some cases, the combined action of radiation and microgravity even amplified the effects exerted by the single parame-

© 1989 American Institute of Physics

ters. Microgravity exposure leads to a reduced hatching rate. However, it cannot be excluded that cosmic background radiation or low LET HZE particles are also causally involved in damage observed in the microgravity samples. A synergistic action of HZE particle hits and microgravity was established in the unexpectedly high frequency of anomal larvae. Neither cosmic radiation nor microgravity alone produced an effect of similar extent nor was this extremely high anomaly rate reached by adding up the effects of the two parameters.

INTRODUCTION

The experiment is part of a radiobiological space research program to obtain experimental data on the biological effectiveness of the structured component of cosmic radiation during spaceflight[1]. Using the Biostack concept, to localize precisely the trajectory of individual heavy ions of cosmic radiation relative to the biological objects under consideration and to correlate the physical data of the particles relative to the biological effects along their path, it was demonstrated that single heavy ions can severely damage individual biological test systems. This has been proven for a variety of dormant systems, such as bacterial spores, animal eggs and embryos and plant seeds[5-10]. Such in-flight data, in addition to a dosimetric record for each spaceflight mission[11], are required as basic information for a quantitative assessment of the potential radiation hazard to man or any biological specimen in space, especially during future long duration or repeated spaceflights and in view of plannings for the space station.

To reach conclusive results on the biological response to single parameters of space, special correlation methods are required. During the 7 day German D1 mission, in the ESA facility BIORACK the influence of HZE particles of cosmic radiation and/or microgravity was studied on eggs of the stick insect Carausius morosus. The use of the Biostack sandwich method[3] in combination with an in-flight 1 g reference centrifuge[4] allowed to separate the biological changes caused by HZE particle hits from those caused by microgravity or the combined action of these two factors of space.

MATERIALS AND METHODS

Biological System

The stick insect Carausius morosus is bred in the laboratory on ivy leaves. The female produces 1-2 eggs per day and has a life span of approximately one year. In the eggs (1.8x2.7 mm), kept at 18-22°C and at high relative humidity (70-80 %), larval development lasts for 75-105 days until hatching takes place.

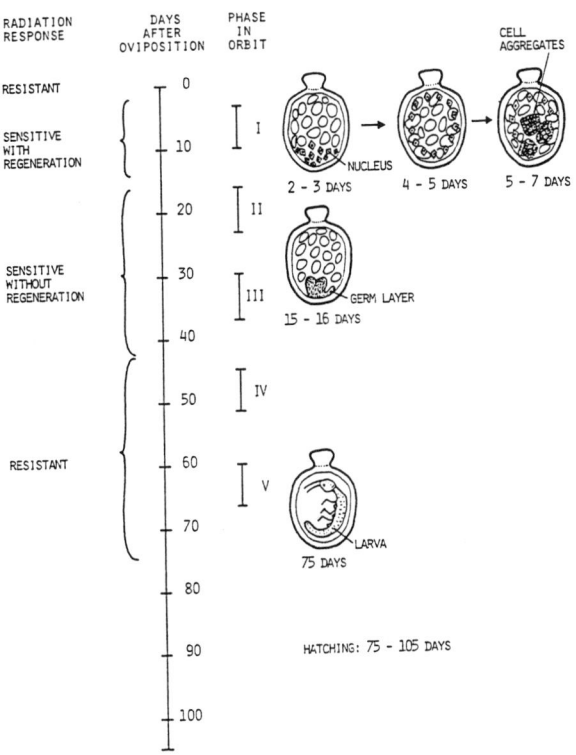

Fig. 1. Characterization of the 5 stages of development (I through V) of Carausius morosus used in the spaceflight experiment.

The embryo passes the following stages, differing in radiation sensitivity and regenerative capacity:

Undeveloped eggs: radiation resistant. Center of cleavage and differentiation is oriented at the pole opposite to the egg lid.

1-14 days after oviposition : radiation sensitive with regeneration. This stage covers the multiplication of the cleavage nuclei, the blastula stage, the ventral aggregation of cells, and finally, the formation of the germ layer.

16-43 days : radiation sensitive without regeneration. This stage starts with the delimitation of the head limbs and the segmentation of the thorax. Growth and differentiation of the embryo continues. The head region is always oriented in

43-75 days	the ventral, the abdominal in the dorsal area. At the end of this stage the limbs are differentiated in femur, tibia, and tarsus and the residual yolk is consumed. : radiation resistant. In this phase, the organs of thorax and abdomen are completed and the larva streches itself until the head is positioned under the egg lid, ready for hatching.

Experimental Set-up

In the experiment, eggs of five different stages of development (stages I-V) were used (Fig. 1). Their development continued when in orbit. The samples were accommodated in the BIORACK 22°C incubator, either at microgravity in a type 2 aluminum container (100 x 66 x 66 mm^3) or on the 1 g reference centrifuge in two type 1 aluminum containers (90 x 42 x 22 mm^3)[4].

Each of the containers accommodated a stack consisting of monolayers of Carausius morosus eggs in perpex sheets sandwiched between different nuclear track detectors (cellulose nitrate, CR 39). Thermoluminescence dosimeters (TLD) were mounted at the top and the bottom of each stack. The eggs were mounted in the following way. On perpex sheets (3 mm in height) with holes of 3 mm in diameter, scotch tape was attached to the bottom side. The eggs were inserted into the holes and pressed onto the tape. It was proved in simulation tests that under these conditions, development continued normally. Each egg was separately numbered. The total number of eggs was 153 in each layer of type 2 container and 96 in each layer of type 1 container. The type 2 container accommodated 14 layers with eggs and each type 1 container 5 layers. The layers, each carrying eggs of a separate stage of development, were arranged as follows:

Table I Arrangement of Layers of Carausius Morosus Eggs in the BIORACK Containers

Development Stage	No. of Layers in Container	
	Type 1	Type 2
I	1 (192)*	2 (306)
II	1 (192)	4 (612)
III	1 (192)	4 (612)
IV	1 (192)	2 (306)
V	1 (192)	2 (306)

*Total number of eggs flown, in brackets

From photographs, taken before and after flight, it was confirmed that the eggs in the holes had not changed their position.

The fully assembled stack was marked with two "reference" notches. This procedure permits to correlate the coordinates of the different layers after disassembly.

Flight Protocol

The experiment was exposed to spaceflight conditions for 7 days during the Spacelab D1 mission inside the 22°C incubator of the BIORACK. By use of the in-flight 1 g reference centrifuge in combination with the Biostack sandwich method it was achieved to differentiate between the effects of the different parameters of space as listed in Table II.

In addition, a parallel mission simulation experiment was conducted on ground.

Postflight Evaluation

Table II Space Parameters Applied to the Different Biological Samples During Spaceflight and Ground Control

HZE particle hit	Background radiation	g-value	
+	+	$10^{-3} - 10^{-6}$	
−	+	$10^{-3} - 10^{-6}$	spaceflight
+	+	1	
−	+	1	
+	negligible	1	ground
−	negligible	1	control

After the space mission, the stacks were disassembled. Photographs were taken from each egg layer. Thereafter, the eggs were immediately removed from the perpex sheets for further development in separate chambers. After hatching the larvae were isolated in cages to follow their growth individually until adolescence and sexual maturity.

Cellulose nitrate (CN) plastic detectors were etched in 6 normal NaOH at 50°C for 2 hours, CR 39 plastic detectors at 80°C for 24 hours. The latent heavy ion tracks were thereby developed into conically shaped, microscopically visible etch tracks. CN detectors were scanned manually under a light microscope. CR 39 detectors were scanned with an automatic system [12]; all detected tracks thereafter were visually controlled.

TLDs were evaluated with the Harshaw 2800 TLD-analyzer at standard settings. Calibration was done with a Co^{60} standard source of the Gesellschaft für Strahlenforschung (GSF) in Munich. Absorbed doses from medium energy neutrons up to about 1 MeV neutrons were determined from difference measurements between Li^7

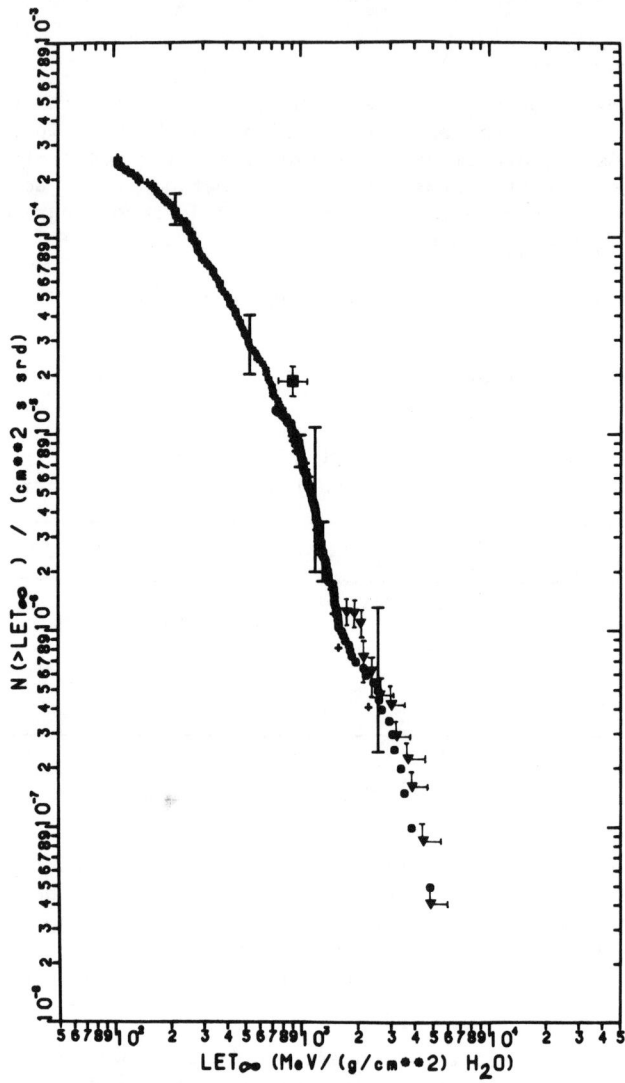

Fig. 2. Preliminary integral LET spectra of the D1 experiment
▼ CN, Group of Kiel
■ CN, Group of DFVLR
+,○ CR 39, Group of Siegen

(TLD-700) and Li^6 (TLD-600) enriched LiF chips. The TLD-700 response was used as measurement of the received average dose. Eggs hit were determined from the adjacent track detectors and the photographs of the egg layers. For this purpose, only those HZE particles were considered that produced visible tracks in the adjacent detectors both above and below the egg layer under investigation. By this method, particles of a linear energy transfer (LET) above 1 GeV cm^2 g^{-1} were selected. Therefore, an egg was determined to be "hit", if it was located in the trajectory of an HZE particle of LET > 1 GeV cm^2 g^{-1}. For these particles, for which no quality factor can be defined, dramatic effects were found in previous Biostack experiments[13, 14].

A complete LET spectrum is given in Fig. 2^{11}. From this spectrum the total fluence of particles with LET > 100 MeV cm^2 g^{-1} can be calculated to about 150 particles cm^{-2}. So each of the residual so-called "non-hit" eggs had experienced several hits of particles with LET < 1 GeV cm^2 g^{-1}. All eggs were exposed in addition to the low LET component of the radiation such as electrons, protons, and electromagnetic radiation of at most 130.6 mrad + 3.6% and to neutrons of at most 12 mrad with an uncertainty of more than 30%.

RESULTS

Classification of the eggs

The eggs were classified in 25 categories according to their stage of development and the spaceflight parameter or combination of parameters concerned. The total number of eggs in each category is given in Table III.

According to their history of exposure, the eggs were classified in the following groups:

Group FH: This group includes eggs that were kept in microgravity environment during the mission and that were hit by an HZE particle of high LET (>1 GeV cm^2 g^{-1}).

Group F: This group comprises the residual "non-hit" eggs of the microgravity samples.

Group FGH: This group includes all eggs kept on the 1 g reference centrifuge during the mission, that were hit by an HZE particle of high LET (>1 GeV cm^2 g^{-1}).

Group FG: This group comprises the so-called "non-hit" eggs on the 1 g reference centrifuge.

Group C: The ground control eggs, that were kept in the simultaneously running mission simulation experiment at KSC, are represented in this group.

Of development stages II and III, aliquots of group "F" and "FH" were fixed after hatching for subsequent histological examination. These results will be published later on.

Table III Total Number of Eggs of the Different Categories According to Their History of Exposure, of Larvae Hatched, of Anomal Larvae Detected, and of Adults 150 Days after Hatching (For explanation of the categories see text)

Stage of Development	Group	Number of			
		Eggs	Larvae	Anomal Larvae	Adults
I	F	244	166	1	95
	FH	62	40	7	25
	FG	88	72	0	46
	FGH	8	6	0	4
	C	306	254	0	165
II	F	248	117	0	69
	FH	58	21	12	15
	FG	168	139	0	100
	FGH	24	19	0	15
	C	306	257	0	187
III	F	258	201	10	111
	FH	48	27	15	18
	FG	176	144	1	81
	FGH	16	12	2	6
	C	306	239	0	147
IV	F	261	193	0	143
	FH	45	33	1	19
	FG	175	133	0	98
	FGH	17	14	0	10
	C	306	230	0	162
V	F	263	208	0	155
	FH	43	36	2	26
	FG	165	137	0	96
	FGH	27	21	0	18
	C	306	245	0	188

Hatching

At day 75-105 after oviposition, hatching took place. The numbers of larvae hatched in each group are given in Table III. The mean values and their standard deviations of the hatching rates of the ground control samples were (83+2)% for stage I, (84+2)% for stage II, (78+3)% for stage III, (75+3)% for stage IV, and (80+2)% for stage V. Similar values were obtained for the flight samples of late stages of development (stages IV and V). The mean hatching rates of the flight samples of groups F, FH, FG, and FGH of stage IV amounted to (74+3)%, (73+8)%, (76+4)%, and (82+14)%, respectively. For stage V the corresponding hatching rates were (79+3)%, (82+8)%, (83+3)%, and (78+11)%, respectively. This shows, that from day 45 after oviposition on, an interim exposure to spaceflight conditions for 7 days did not disturb embryonic development. This late stage of development, when organogenesis has been completed, was already known from laboratory experiments to be relatively radiation resistant.

The hatching rate was significantly reduced, if the early stages of development were exposed to microgravity (group F); it decreased to (68+3)% for stage I and to (47+4)% for stage II. An additional hit by an HZE particle (group FH) further reduced the hatching rate to (65+7)% for stage I, (36+7)% for stage II, and (56+8)% for stage III. Eggs of stage II were the most sensitive ones, as well to microgravity as to cosmic radiation plus microgravity.

Table IV Levels of Significance of Differences Between Groups in Rates of Hatching and of Anomalies (Fisher-Yates test. Only values >0.9 are indicated, values <0.9 are given by a dash, values >0.999 are given as 1) (For explanation of groups see text)

Stage of development	Group	Hatching				Anomalies			
		C	F	FG	FGH	C	F	FG	FGH
I	F	1		0.982		—		—	
	FH	0.999	—		—	1	1		—
	FG	—							
	FGH	—		—					
II	F	1		1		1			
	FH	1	(0.856)		0.999	1	1		1
	FG	—							
	FGH	—		—					
III	F	—		—		0.999		0.971	
	FH	0.998	0.998		—	1	1		0.963
	FG	—							
	FGH	—		—		0.998		0.984	
IV	F	—		—					
	FH	—	—		—	(0.875)	(0.859)		—
	FG	—							
	FGH	—		—					
V	F	—		—					
	FH	—	—		—	0.984	0.979		—
	FG	—							
	FGH	—		—					

No significant reduction in hatching frequency was observed for eggs, that were kept on the 1 g reference centrifuge, regardless of the stage of development during the mission and a hit of an HZE particle. Early stages of group FGH, these are "hit" eggs on the centrifuge, showed slightly reduced hatching rates; however, the difference was not significant. Table IV gives the levels of significance obtained from Fisher-Yates test[15].

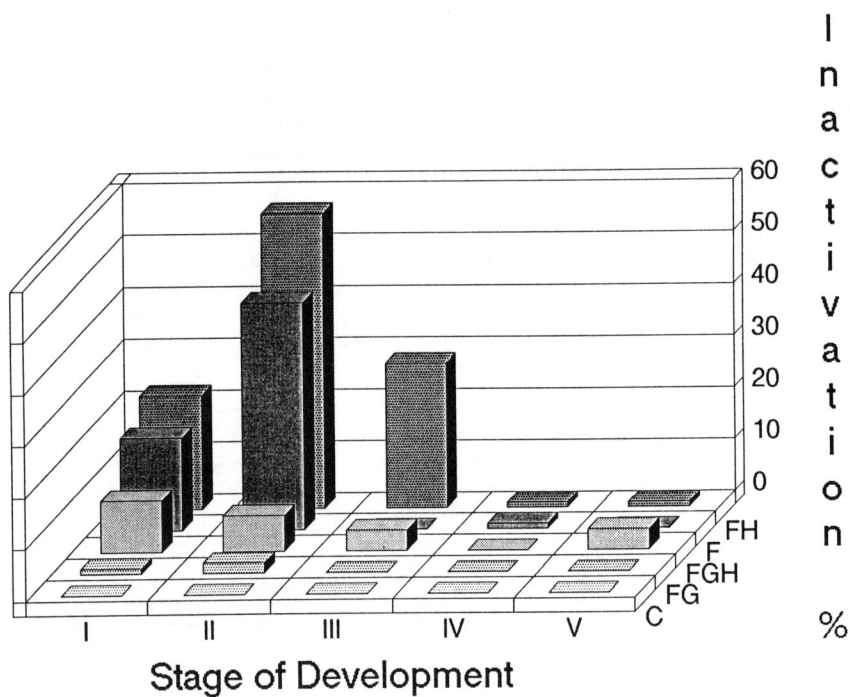

Fig. 3. Inactivation of Carausius morosus eggs, exposed to spaceflight conditions in different stages of development
F   = flight samples in microgravity
FH = flight samples in microgravity, hit by an HZE particle.
FG = flight samples on the 1 g centrifuge
FGH = flight samples on the 1 g centrifuge, hit by an HZE particle
C   = ground control

From the hatching data it is seen, that development of Carausius morosus is disturbed, if early stages of development, up to an age of approximately 40 days after oviposition, are exposed to microgravity. Combined action of microgravity and HZE particles of cosmic radiation seems to aggravate the damage. Both facts are clearly demonstrated in Fig. 3, in which the damage (inactivation of eggs) is plotted for the different stages. From laboratory experiments with heavy ions of accelerators this early developmental phase is known to be highly radiation sensitive[16].

Malformations

A substantial portion of larvae hatched from flight samples showed body anomalies, such as deformation of abdominal segments, of extremities or antennae. No malformations were observed in larvae of the ground control groups though the sample sizes of these groups were larger than of any flight group (Table III). The rate of anomalies determined as ratio of anomal larvae to total number of larvae hatched is given in Fig. 4, the levels of significance according to Fisher-Yates test are given in Table IV.

The high frequency of malformations in the group FH, comprising eggs "hit" under microgravity is striking. Anomalies were induced in all stages of development. From eggs, that were in stages II or III during the mission, a hit by an HZE particle in microgravity resulted in more than 50 % anomal larvae.

Stage III turned out to be the most sensitive one with respect to induction of anomalies. Only in this stage, anomalies were also induced by HZE particles in eggs, that were kept on the 1 g reference centrifuge ((17+11)%). In this stage the eggs are a very sensitive detector, since they are radiation sensitive without the capacity to regenerate. Unfortunately, the sample size of group FGH was very small due to the limited space available on the centrifuge. More date from future space missions are required to examine the statistical significance of this observation.

In eggs of stage III anomalies were also observed in the so-called "non-hit" individuals, whether kept in microgravity ((5+1)%) or on the 1 g reference centrifuge ((0.6+0.5)%). "Non-hit" eggs from all other developmental stages, did not produce any anomal larvae, neither those kept in microgravity (except eggs of stage I with (0.6+0.5)% anomalies) nor those on the 1 g reference centrifuge.

In stage III of development, the combined action of an HZE particle of cosmic radiation and microgravity (group FH) induced more malformations than one would expect from the sum of the effects of HZE particles (group FGH) and microgravity (group F) alone. This result is a clear indication of a synergistic effect of microgravity and radiation. A synergism of microgravity and radiation is also obviously effective in all other stages of development, in which the single parameters alone did not induce any malformations or only to a negligible extent.

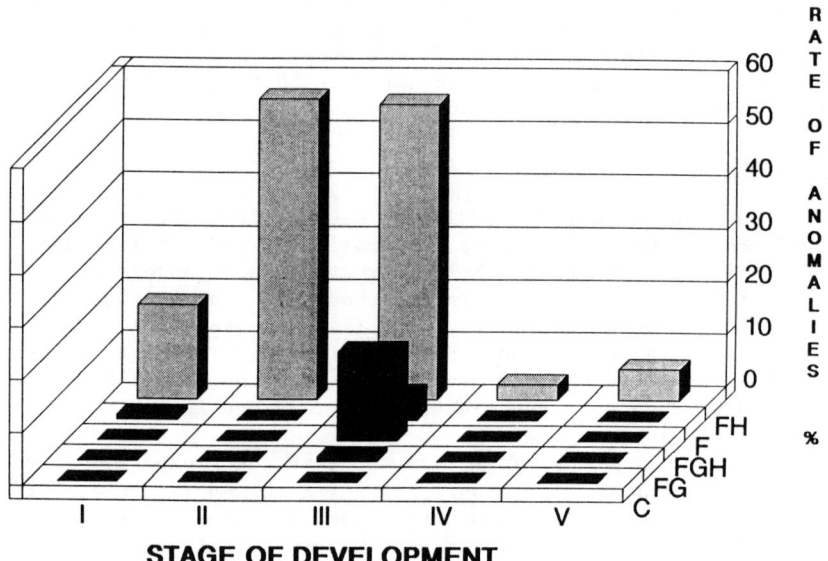

Fig. 4. Rate of anomalies of larvae hatched from eggs of Carausius morosus, that were exposed to spaceflight conditions in different stages of development
F = flight samples in microgravity
FH = flight samples in microgravity, hit by an HZE particle.
FG = flight samples on the 1 g centrifuge
FGH = flight samples on the 1 g centrifuge, hit by an HZE particle
C = ground control

Regarding a potential effect of microgravity alone, one has also to bear in mind, that each of the so-called "non-hit" eggs had obtained several hits of HZE particles of lower LET; therefore, even in the "non-hit" eggs, anomalies might be caused by cosmic radiation.

DISCUSSION

The first time, it was possible by use of the combination of the Biostack concept[1] and the 1 g reference centrifuge of BIO-RACK[9] to separate pure radiobiological effects from disturbances produced by microgravity or by combined actions of both parame-

ters. The biological process under investigation was the embryonic development of the stick insect Carausius morosus and its further development up to an adult organism. The following causalities could be established:
(1) HZE particle hits cause body anomalies, if applied during the stages of organogenesis.
(2) Microgravity leads to disturbances of early embryogenesis, resulting in reduced hatching rates.
(3) Combined action of HZE particles and microgravity probably aggravates the damage to early embryogenesis.
(4) HZE particles and microgravity act synergistically on eggs of all developmental stages, resulting in high rate of body anomalies of the larvae.

In this experiment, the potentiation of the biological response, if an HZE particle hits the system under microgravity conditions, was unequivocally proved in respect to body anomalies of the larvae. Neither cosmic radiation nor microgravity alone produced an effect of similar extent nor was this extremely high anomaly rate reached by adding up the effects of the two parameters. To understand the mechanism responsible for this potentiation of the biological effect, further studies in space are required. Especially, it needs to be investigated whether in the radiobiological chain of events certain processes, such as repair of radiation damage, are disturbed by microgravity.

REFERENCES

1. I.I. Oster, in: The experiments of Biosatellite II, ed. J.F. Saunders, NASA SP-204 (1971), p. 41.
2. J.F. Saunders, O.E. Reynolds, and G.D. Smith, in: The experiments of Biosatellite II, ed. J.F. Saunders, NASA SP-204 (1971), p. 1.
3. H. Bücker, Life Sci. Space Res. 12, 43 (1974).
4. A.F.L. Soons, ESA Bulletin #31, 46, (1982).
5. W. Rüther, E.H. Graul, W. Heinrich, O.C. Allkofer, R. Kaiser, and P.Cüer, Life Sci. Space Res. 12, 69 (1974).
6. G. Horneck, R. Facius, W. Enge, R. Beaujean, and K.-P. Bartholomä, Life Sci. Space Res. 12, 75 (1974).
7. H. Planel, J.P. Soleilhavoup, Y. Blanquet, and R. Kaiser, Life Sci. Space Res. 12, 85 (1974).
8. R. Facius, H. Bücker, G. Horneck, G. Reitz, and M. Schäfer, Life Sci. Space Res. 17, 123 (1979).
9. D.D. Peterson, E.V. Benton, M. Tran, T. Yang, M. Freeling, L. Craise, and C.A. Tobias, Life Sci. Space Res. 15, 151 (1977).
10. H. Bücker, G. Horneck, R. Facius, G. Reitz, M. Schäfer, J.U. Schott, R. Beaujean, W. Enge, E. Schopper, W. Heinrich, J. Beer, B. Wiegel, R. Pfohl, H. Francois, G. Portal, S.L. Bonting, E.H. Graul, W. Rüther, A.R. Kranz, U. Bork, K. Koller-Lampert, B. Kirchheim, M.E. Starke, H. Planel, and M. Delpoux, Science 225, 222 (1984).

11. G. Reitz, H. Bücker, R. Beaujean, W. Enge, R. Facius, W. Heinrich, and E. Schopper, Adv. Space Res. $\underline{6}$, #12, 107-113 (1986).
12. W. Trakowski, B. Schöfer, J. Dreute, S. Sonntag, C. Brechtmann, J. Beer, H. Drechsel, and W. Heinrich, Nucl. Instr. Meth. $\underline{225}$, 92-100 (1984).
13. H. Bücker and G. Horneck, in: <u>Radiation Research</u>, eds. O.F. Nygaard, H.J. Adler, and W.K. Sinclair, Academic Press, New York (1975), p. 1138.
14. H. Bücker and R. Facius, Acta Astronautica $\underline{8}$, 1099 (1981).
15. M. Kendall and A. Stuart, The Advanced Theory of Statistics, 4th ed., Vol. 2, Charles Griffin & Co., London, 1979.
16. H. Höffken, Radiobiologische Untersuchungen an Eiern von Carausius morosus (indische Stabheuschrecke) nach Einrwirkung technisch erzeugter HZE-Partikel, Thesis, (1984), University Marburg.

THE PROTONS OF SPACE AND BRAIN TUMORS:
I. Clinical and Dosimetric Considerations

G. V. Dalrymple, W. A. Nagle, A. J. Moss, L. A. Cavin,
J. R. Broadwater, E. L. McGuire, and C. S. Eason
John L. McClellan Memorial Veterans Hospital and The University of
Arkansas for Medical Sciences, Little Rock, AR, 72205

and

J. C. Mitchell, K. A. Hardy, D. H. Wood, Y. A. Salmon,
and M. G. Yochmowitz
The USAF School of Aerospace Medicine, Brooks, AFB, TX, 78235

ABSTRACT

Almost 25 years ago a large group of Rhesus monkeys were irradiated with protons (32-2300 MeV). The experiments were designed: 1) To estimate the RBE of protons, **per se** , and 2) To provide some estimate of the hazards of the radiation environment of space. The initial results showed the RBE to be about 1.0 for acute radiation effects (mortality, hematologic changes, etc). The colony has been maintained at Brooks AFB, TX since irradiation. The survivors of 55 MeV proton irradiation have developed a very high incidence of Glioblastoma multiforme, a highly malignant primary brain tumor. These tumors appeared 1-20 yrs after surface doses of 400-800 rads. Reconstruction of the dosimetry suggests that some areas within the brain may have received doses of 1500-2500 rads. More than 30 radiation induced Glioblastomas have been reported in human patients who had received therapeutic head irradiation. The radiation doses required to induce Glioblastoma were of the same order of magnitude as required to induce Glioblastoma in the Rhesus monkey.

INTRODUCTION

In the early 1960' s a study was performed at the USAF School of Aerospace Medicine, Brooks, AFB, in collaboration with NASA. Part of the effort was concerned with the potential hazards of the radiations of space to the space traveller. Prior to this project, relatively little information was available relative to the total body effects of protons, the radiation of greatest concern in space.[1] The initial project considered several aspects of the problem.

First , protons have a finite range in tissue; they give up increasing amounts of energy per unit path length until the end of the range. At this point they produce considerable local ionization, the Bragg peak.[1]

Second , the space proton environment is comprised of a spectrum of energies.[2] There are more "softer" components (**i.e.**, lower energy) than higher energy protons. This means that the occupants of the spacecraft would receive a relatively higher radiation dose to the skin and

© 1989 American Institute of Physics

sub-cutaneous tissues than more deeply placed organs.

Third, the anticipated biological effect would reflect a convolution of these factors.[2]

Fourth , only a finite range of proton energies are available from existing Cyclotrons and other accelerators.

Fifth , suitable animal models were needed to obtain data in the most efficient manner.

THE ORIGINAL EXPERIMENTS

While Rhesus monkeys were used as the primary animal model for the original irradiations (1954-1966), some studies with mice were also performed. The results of these initial efforts have been published as a supplement to **Vol 28** of RADIATION RESEARCH (June 1966). The proton energies and sources used were: 1) 32 and 55 MeV protons--The Oak Ridge Isochronous Cyclotron (ORIC), Oak Ridge National Laboratory, 2) 138 MeV protons--Harvard University, 3) 250 and 400 MeV protons--the University of Chicago, and 4) 2300 MeV protons--Brookhaven National Laboratory. Additional irradiations, to simulate the "Solar Flare" proton spectrum were performed using the Langley Cyclotron, Langley AFB, VA.

The observations were directed toward estimation of the RBE of the protons for causing acute radiation effects. The measurements included: 1) Mortality (for estimation of LD_{50} statistics), 2) Hematologic changes, including depression of WBC and platelet concentrations plus evaluation of ^{59}Fe Ferrokinetics, 3) Evaluation of several physiologic and biochemical responses, and 4) Histopathology. The proton induced responses were compared with responses after 2 MeV X-irradiation and ^{60}Co γ irradiation. A very great deal of work can be summarized very briefly. The estimated RBE for all of the variables studied was essentially 1.0. This meant that the anticipated biologic effects for protons would be about the same as seen with the far more widely studied electromagnetic radiations.

THE CHRONIC/DELAYED EFFECTS OF PROTON IRRADIATION

Fortunately, the monkeys which survived the acute effects of proton irradiation have maintained in a modern Primate Colony at Brooks AFB, Tx. During the two decade interval from irradiation to the present, at least two major findings relative to chronic/delayed effects have been reported-endometriosis and the induction of Glioblastoma multiforme. We are concerned with the latter phenomenon. Krupp[3] and Yochmowitz, Wood and Salmon[4] published reports of the colony status at 9 and 17 years after irradiation. Approximately 8% of all of the delayed mortality was a consequence of Glioblastoma multiforme. This, of itself, is highly unusual because brain tumors are very rare in the Rhesus monkey.[3,4] Of greater interest is the fact that all of the Glioblastomas were induced by surface doses of 400 to 800 rads of 55 MeV protons. No Glioblastomas have been observed in any of the controls, any of the animals irradiated with X-rays or at any other proton energies.

GLIOBLASTOMA MULTIFORME

Glioblastoma multiforme is a highly malignant primary brain tumor, of glial origin. In humans, the average survival time after diagnosis is about 1 year, in spite of the most aggressive therapy.[5] Frequently, the first symptom of Glioblastoma is that of a psychiatric disorder. Dementia, impairment of judgement, etc, may the the first manifestation of the disease.[6] For some patients, various forms of psychiatric management may be tried before the presence of a space-occupying mass in the brain is evident. Even the advent of new diagnostic modalities such as CT scanning and MR imaging plus newer chemotherapeutic agents, the prognosis is as bleak as ever.

More than 45 cases of radiation induced intra-axial CNS tumors(of these 75% are Glioblastoma multiforme) have been reported in humans.[7] These patients had previously received therapeutic irradiation of the head for three basic reasons: 1) Definitive treatment of a neoplasm of the Central Nervous System(CNS), 2) Prophylactic irradiation of the head of patients with Acute Lymphoblastic Leukemia (ALL) to prevent CNS recurrence, and 3) Patients who had received head irradiation for treatment of benign infectious disease of the scalp (such as Tinea capitis -"ringworm"). The time required for induction of tumor ranged from about 5.5 yrs for the patients with ALL to more than 31 yrs for the patients who received scalp irradiation with low energy X-rays for treatment of Tinea capitis. The time required for induction of tumor after definitive treatment of CNS neoplasia (craniopharyngioma, pituitary tumors, meningioma, etc) was about 14 yrs.

Determination of the radiation dose required to induce tumor is more difficult. For the Rhesus monkeys, they were given single dose , total-body irradiation. With the exception of the patients with Tinea capitis , the patients received fractionated dose radiation therapy. Consequently, we have used the Ellis concept of the Nominal Standard Dose (NSD) as a means to unify the findings.[8] Ellis stated the NSD as:

$$NSD = (Dose)(N^{-0.24})(T^{-0.11})$$

Where **Dose** is the Total dose used for the entire treatment, **N** is the number of fractions and **T** is the time, in days, over which the treatment was given. The NSD can be considered to represent a single dose equivalent of a course of radiation therapy. It has units of **RET** (**R**adiation **E**quivalent **T**herapeutic). We have found NSD values to be about 950 RET for induction of Glioblastoma in a patient with ALL and about 1275 RET for the induction of tumor in 10 patients with CNS neoplasia.

A prior study by Wakisaka, **et al** reported the induction of Glioblastoma multiforme in a group of monkeys head-only irradiated with high energy X-rays .[9] Of these animals, three sets of 4 monkeys each were irradiated with 1000, 1500, and 2000 rads. The 2000 rad animals all died of acute effects. None of the 1000 rad group showed Glioblastoma. Of the 1500 rad irradiated monkeys, two of the 4 developed Glioblastoma multiforme.

Originally we thought that the maximum dose to the brain of the 55 MeV proton irradiated monkeys was approximately equal to the surface doses.[10] A later re-construction of the dosimetry, suggests that the protons may have produced "hot spots" such that the maximum dose could have been as high as 1500-2500 rads, for those monkeys which developed Glioblastoma (K. Hardy --USAF School of Aerospace Medicine, Personal Communication).

Other CNS tumors **have** been observed after proton irradiation in primates (D.H.Wood--USAF School of Aerospace Medicine, Personal Communication). Although fewer in number than the Glioblastomas, a total of 8 extra-axial tumors (meningiomas, etc) have been observed. Of considerable interest is the fact that these have **not** been seen after 55 MeV proton irradiation. Instead, one was seen in a control and the others after irradiation with 2-MeV X-rays and protons of 138 MeV energy or higher. These tumors, in general, have required a longer induction time than the Glioblastomas. These findings also parallel our experience with human patients.[7] More than 200 cases of radiation induced extra-axial tumors have been reported (these are mostly meningiomas).

CONCLUSIONS

At present, we believe the evidence suggests that ionizing radiation, **per se**, may be the explanation for the findings--not necessarily some specific "toxic" property of 55 MeV protons. We believe, however, that considerable additional investigation will be necessary to verify (or deny) this hypothesis. Such studies are in progress at the USAF School of Aerospace Medicine and at the John L. McClellan Memorial Veterans Hospital, Little Rock, AR. A dose of about 1000-1500 rads (single dose equivalent) seems to be necessary to induce Glioblastoma. Higher doses may cause death by acute effects. Lower doses may be less capable of inducing this type of tumor. The observation of so many Glioblastomas after 55 MeV proton irradiation may be a consequence of this proton's capability of delivering a very high brain dose while not causing so much somatic radiation injury that the animals would die from acute bone marrow effects. These protons have a range of about 2.5 cm in soft tissue. When this range is superimposed upon the size of the Rhesus monkey, it is noted that the dose to the bone marrow is relatively reduced.[10] To be certain, the monkeys which had been irradiated with 400-800 rads of 55 MeV protons showed evidence of bone marrow injury. The magnitude of this injury, however, was not enough to cause death. As a result, they survived the acute effects and lived long enough to develop brain tumors.

REFERENCES

1. G. V. Dalrymple and I. R. Lindsay, Radiat. Res. ,28, 365 (1966).
2. G. V. Dalrymple, J. J. Ghidoni, H. L. Kundel, T. L. Wolfe, and I. R. Lindsay, Radiat. Res., 28 548 (1966).
3. J. H. Krupp, Radiat. Res., 67, 244 (1977).
4. M. G. Yochmowitz, D. H. Wood, and Y. L. Salmon, Radiat Res, 102, 14 (1985).
5. K. A. Kelly, J. M. Kirkwood, and D. S. Kapp, Cancer Treatment Rev., 11,1 (1984).
6. A. G. Donald, C.N. Still, and J. M. Pearson, S Med J., 65, 1006 (1972).

7. G. V. Dalrymple, L. A. Cavin, J. R. Broadwater, and J. L. McGuire, submitted for publication.
8. F. Ellis, Clin Radiol., 20, 1 (1969).
9. S. Wakisaka, T. L. Kemper, H. Nakagaki, and R. R. O'Neill, Fukoka Acta Med. , 73, 585 (1982).
10. J. C. Mitchell, G. V. Dalrymple, G. H. Williams, J. D. Hall, and I. L. Morgan, Radiat Res, 28, 390 (1966).

THE PROTONS OF SPACE AND BRAIN TUMORS
II. Cellular and Molecular Considerations

W.A. Nagle, A.J. Moss Jr, and G.V. Dalrymple
McClellan Veterans Hospital, and University of
Arkansas for Medical Sciences, Little Rock, AR 72205
and
A.B. Cox, J.F. Wigle, and J.C. Mitchell
USAF School of Aerospace Medicine
Brooks AFB, TX 78235

ABSTRACT

An increased incidence of highly malignant gliomas, termed glioblastoma multiforme has been observed in Rhesus monkeys irradiated with 55 MeV protons, and in humans treated with therapeutic irradiation to the head. The results suggest a radiation etiology for these tumors. In this paper, we review briefly some characteristics of glioma tumors, and summarize the genetic changes associated with malignant gliomas in experimental animals and in humans. The genetic abnormalities include cytogenetic alterations, and changes in the structure and expression of specific oncogenes. We discuss the potential for these genetic changes to contribute to several putative mechanism leading to aberrant growth stimulation and, ultimately, to tumorigenesis. In addition, we review briefly some recent data concerning the molecular nature of radiation-induced somatic cell mutation and oncogene activation, and discuss the significance of these results for the radiation etiology of malignant gliomas. Finally, some implications of these results are discussed in relation to human radiation exposure in space.

INTRODUCTION

In a collaborative study between the US Air Force School of Aerospace Medicine and the National Aeronautics and Space Administration (NASA), Rhesus monkeys irradiated with high energy protons showed a high incidence of brain tumors, in particular, of malignant gliomas (see Hardy et al,[1] and Dalrymple et al,[2] this volume). Moreover, Dalrymple et al [2] summarized the availabl clinical data that document the occurrence of human malignant gliomas in individuals who had received therapeutic radiation to the head. The results observed with both monkeys and humans show that an increased incidence of malignant gliomas appeared to result from radiation exposure.

In recent years, the genetic changes associated with neoplastic transformation have begun to be identified. In particular, studies with the retroviral "oncogenes" (cancer-causing genes) have now identified several of their cellular

© 1989 American Institute of Physics

counterparts, the so called "proto-oncogenes" (recently reviewed by Bishop).[3] The proto-oncogenes comprise a subset of cellular genes that function in the normal processes of cellular proliferation and differentiation. These processes are controlled by a complex pattern of growth signal generation and transduction within the cell: a) Signals typically are initiated by the interaction of discrete mitogens ("growth factors") with specific "receptors" found on and within particular target cells; and b) The initiating signal often results in an intracellular cascade of biochemical events that propagates the signal to the nucleus, to effect the expression of specific genes. Several proto-oncogenes now have been shown to code for different components of these signal-transducing pathways (see Weinstein et al.[4] for a recent review). Moreover, most tumor cells exhibit characteristics, for example a reduced requirement for exogenous growth factors, thought to result from a breakdown in these pathways.

PATHOLOGY AND INCIDENCE OF MALIGNANT GLIOMAS

The term "gliomas" describes central nervous system tumors derived from the neuroglia, a collection of several cell types of embryonic neuro-ectodermal origin that provide structural and metabolic support for the neurons. Malignant gliomas are thought to rise from the astrocytes, the largest cellular component of the neuroglia, and to progress through a process of continual de-differentiation, i.e. gradual loss of glial cell properties.[18] In the Kernohan graded system of classification, astrocytic tumors range from the grade I astrocytomas, which appear cytologically benign, to the astrocytomas, grades II through IV, which exhibit a progression of malignancy, characterized by increasing amounts of vascular proliferation, tumor cell mitoses, and intratumor necrosis.[6] Generally, astrocytomas of grades III and IV are designated as the glioblastoma multiforme, because of their distinguishing heterogeneous morphology.[7] The biology of glioblastoma multiforme has been described in detail.[7]

Although the incidence of gliomas is relatively low (representing about 2% of all human malignancies),[5] these tumors comprise the majority of human primary intracranial neoplasms, whereas tumors originating from the neurons are extremely rare.[6] The "spontaneous" occurrence of gliomas is highest in two age intervals: during childhood, years 3-12, and during later life, years 50-70,[6] and at both times shows a sex ratio of 3:2, males to females.[5] Typically, the pediatric tumors differ markedly in histology and behavior from their counterparts in adults;[6] for purposes of this paper, only the data obtained with adult tumors is included.

ETIOLOGY OF MALIGNANT GLIOMAS

Both chemical and viral agents have been implicated in the etiology of malignant gliomas in experimental animals. For example, the carcinogenic alkylating agent N-ethyl-N-nitrosourea (ENU), when administered to pregnant rats on about the 18th day of gestation, was found to result in a high incidence of gliomas among the offspring [reviewed in 8 & 9]. Administration of ENU at earlier times, e.g. days 13-15 of gestation, resulted in the development of brain tumors mostly of neuronal origin.[74,75]

Many species of mammals, including primates, developed gliomas after treatment with avian oncoviruses,[10] primate retroviruses,[11] and human polyoma virus.[12] The avian Rous sarcoma virus RSV), which produces only meningiomas in chickens,[10] was found to induce brain tumors of mostly glial origin when innoculated intracerebrally into neonatal or adult Fischer rats.[13] Moreover, malignant astrocytomas with morphologies similar to human GM were observed after intracerebral innoculation of newborn marmosets with simian sarcoma virus (SSV),[14] and of adult owl monkeys with human JC polyoma virus.[12]

A radiation etiology for malignant gliomas is less well documented. The results obtained with irradiated monkeys, together with the available human clinical data, are reviewed elsewhere in this volume [1,2] and will not be discussed here.

Ionizing radiation has been used in conjunction with ENU for the induction of brain tumors in rats, however radiation generally was observed to lower the ENU-induced tumor frequency.[15] A more recent study,[16] in which groups of pregnant rats received radiation only, failed to demonstrate radiation-induced brain tumors in the surviving offspring, where the whole embryos were irradiated with 500-1500 rads of X-rays on the 3rd - 13th day of gestation (no ENU). In contrast, Knowles [17] reported a 12 percent incidence of glial-derived tumors in rats given radiation treatment only, but in this case, whole body irradiation was administered to neonates 24-40 hours after birth. The observation of similar histologies for both the radiation- and ENU-induced gliomas, suggested that the progenitor cells exhibited greatest susceptibility to the two carcinogenic treatments at different times during development.[17]

GENETIC CHANGES ASSOCIATED WITH GLIOMAS

Analysis of the chromosomal composition (karyotype) of human gliomas has revealed a complex, yet distinctly non-random, pattern of abnormalities.[19,20] In general, the karyotypes from lower grade astrocytomas were near-diploid in composition, while cells from higher grade tumors (e.g. the glioblastoma multiforme) showed more heterogeneity.[20] In detailed studies of a total of 27 human malignant gliomas, Bigner et al [19] observed frequent structural abnormalities of chromosome #9, gains or structural abnormalities of #7, losses of #10, #22 and the sex chromosomes, and a high proportion of cells (about 50%) with double minute (DM) chromosomes (The DMs are extrachromosomal elements that, lacking centromeres, are randomly segregated during mitosis. The DMs generally consist of amplified DNA sequences, i.e. multiple copies per cell of a specific gene or DNA sequence. Gene amplification usually results in overproduction of the protein products of the amplified gene(s). Gene amplification is thought to represent a manifestation of a mutational event, and has been the subject of several recent reviews.)[76-78]

A pattern of chromosome abnormalities similar to that found by Bigner et al [ref #19, discussed above], also was reported by Rey et al,[79] in a study of 34 human malignant astrocytomas. In addition, Rey et al [80] analyzed 22 low grade gliomas, including nine astrocytomas grades I-II, and nine oligodendrogliomas, and found chromosome alterations similar to those in the high grade gliomas.

Thus, the available cytogenetic data show that changes within a subset of human chromosomes, presumably involving a relatively small number of genes, are consistent features of malignant gliomas. Some gliomas showed only gains in chromosome #7, losses in #10, and the presence of DMs, in an otherwise near-diploid karyotype.[118] Therefore, Bigner et al [118] suggested that these three features may represent early chromosomal alterations in the development of gliomas. Yet, no single chromosomal abnormality has been identified in all malignant gliomas. Moreover, the pattern of chromosomal abnormalities characteristic of malignant gliomas appears to be shared by the lower grade gliomas, as well as by brain tumors of non-glial origin, the meningiomas. Saadi et al [81] studied the chromosomal composition of 45 meningiomas, brain tumors that generally appear histologically benign,[6] and found a non-random pattern of changes very similar to those described in the malignant gliomas. Consequently, it is not yet clear whether the cytogenetic abnormalities observed represent the common involvement of certain genes early in tumorigenesis,[80] or merely secondary events, i.e. events associated with the progression, rather than the induction, of brain tumors.[20]

Identification of the specific genes that are altered in malignant glioma cells is the subject of current investigations [reviewed in reference #82]. For example, it is known that chromosomes 7 and 22 contain the proto-oncogenes c-erbB and c-sis, respectively; c-erb-B codes for the epidermal growth factor receptor (EGFR),[35] while c-sis produces the beta chain of the platelet derived growth factor (PDGF).[21,22] Normal neuroglia cells typically express functional receptors for both the PDGF and the epidermal growth factor (EGF) (see below), suggesting that glial cells normally respond to both mitogens. Thus, two genes involved in growth control are normally expressed in neuroglia cells, and are carried on a subset of chromosomes that frequently exhibit abnormalities in human malignant gliomas. Moreover, changes in the structure or the expression of the resident genes often accompany chromosomal rearrangement.[23] In fact, alterations in the expression of both proto-oncogenes, c-erbB and c-sis, have been identified in human gliomas.

Several groups of investigators studied the expression of the sis gene in cultured cell lines derived from human malignant gliomas. Eva et al [83] found that three of five tumor cell lines, including the glioblastoma line A172,[100] expressed sis-related mRNAs. Pantazis et al [90] also detected high levels of the 4.4 kilobase (kb) c-sis-related mRNA in A172 cells. Moreover, Nister et al [28] and Pantazis et al,[90] using glioma cell lines U-343 MGa and A172, respectively, found that both cell lines grown in vitro secreted a PDGF-like molecule with immunological and structural features similar to authentic PDGF. Similarly, Lens et al [29] found that two rat glioma cell lines, obtained from ENU-induced tumors, produced and excreted a mitogenic substance that behaved functionally like PDGF. Moreover, the glioma cells contained high levels of the c-sis messenger RNA (mRNA), whereas neuroglia cell cultures from untreated animals and cultures of untransformed glia cells from ENU-treated rats (premalignant cells) neither expressed detectable levels of c-sis mRNA or produced PDGF-like activity.[29] These results show that in several different cases, malignant glioma cells exhibited abnormal expression of the c-sis proto-oncogene, and produced a mitogenic substance similar, if not identical, to PDGF.

Libermann et al [53] first reported that human brain tumors expressed higher than normal levels of the epidermal growth factor receptor (EGFR). Subsequent work by Libermann et al [54] identified amplification of the c-erbB gene (also called HER, from Human Epidermal growth factor Receptor) in four out of ten human malignant gliomas that overexpressed the EGFR. These observations were extended by the data of Wong et al,[55] who analyzed malignant gliomas from 63 patients: 24 tumors that showed substantial amplification (> 8 copies per cell) of HER, also exhibited enhanced expression (at least 10-fold higher levels) of the EGFR mRNA.[55] Thus, overexpression of the EGFR was found in about 40 percent of the human gliomas studied, and was always accompanied by amplification of the HER gene. Bigner et al [84] found that the majority of malignant gliomas containing amplified DNA, carried the amplified sequences in the form of DMs. A minority of gliomas exhibited EGFR overexpression in association with structural abnormalities of chromosome #7, but trisomy of #7 appeared not to be related to HER amplification.[84]

Genetic analysis of human gliomas has identified changes in the structure or expression of other genes as well. These include: rearrangement and amplification of c-myc [85] and N-myc,[55] and amplification of a new gene, termed "gli",[86, 55] the function of which remains unknown.

POTENTIAL MECHANISMS OF GROWTH ACTIVATION IN NEUROGLIAL CELLS

The cytogenetic and genetic changes reviewed above indicate that alterations within specific genes, leading to the deregulation of normal growth control pathways, may play a role the development of gliomas. The morphologic appearance of glioblastoma multiforme, described as "an ensemble of ...

variations", a collection of heterogeneous cellular phenotypes
and associated proliferating vasculature,[7] is consistent with
a process of activated, but deregulated, growth. When the affected
genes code for mitogenic growth factors or their receptors,
several different schemes of aberrant growth regulation may
result. These include: autocrine and paracrine stimulation,
alterations in the structure or expression of growth factor
receptors, and the loss of responsiveness to growth inhibitors.
(A few examples are discussed below. These examples were chosen
because of their particular significance for neuroglial cells.
For detailed reviews, see references 18, 20 and 24.)

First, the tumor cells may exhibit "autocrine"
stimulation.[25] The genes coding for specific growth factors
may become "activated" in cell types which ordinarily do not
produce these factors. If the cell that experiences activation
of growth factor genes also expresses the appropriate receptors,
the result is a closed loop of positive growth stimulation.[26]
For example, Heldin et al [27] found that normal, as well as
malignant, glial cells expressed high affinity receptors for PDGF.
Several reports (discussed above) have documented the production
of PDGF-like material by different cell lines derived from human
malignant gliomas. Thus, these glioma cells possessed the
capacity for self-stimulation, by autocrine activation of their
endogenous PDGF receptors. In the case of the ENU-induced
gliomas,[29] analysis of the c-sis genes in the glioma cells
failed to detect either amplification or gross structural
abnormalities, and tests for overexpression of several other
oncogenes were negative.[29] The results suggest that the
inappropriate expression of the c-sis gene, presumably as a
consequence of ENU treatment, established an autocrine loop
of growth stimulation that was associated with the transformation
of normal glial cells into glioma cells. This hypothesis is
supported by recent reports [87,91; reviewed in #18] suggesting
that the necessary and sufficient transforming property of the
simian sarcoma virus (whose active oncogene, v-sis, is a
rearranged version of c-sis [reviewed in reference #18]),
is the production and secretion of PDGF-like material by cell
types that also express the PDGF receptor.

A second form of aberrant growth stimulation is the so-called
"paracrine" mechanism. A given cell may produce and secrete
mitogenic substances that stimulate the growth of other cells
(or other cell types) in the surrounding microenvironment,
providing the exposed cells express the appropriate receptors.
For example, a paracrine mechanism may be involved in the proliferation of small blood vessels (angiogenesis) often seen
accompanying, and presumably contributing to, the growth of
malignant gliomas.[24] Freshney et al.[30] found that extracts from
cultured glioma cell lines showed enhanced angiogenic capacity, as
compared to extracts from cultures of normal glia cells. However,
the identity of the putative angiogenic substances was not
established. As discussed above, several glioma cell lines were

shown to synthesize PDGF-like material, but vascular endothelial cells normally do not express receptors for PDGF.[93] However, it is known that different growth factors may interact in a synergistic way. For example, the insulin-like growth factor I (IGF I) and PDGF can interact to stimulate DNA synthesis in 3T3 fibroblast cells,[32] and wound healing of the skin in pigs.[33] Normal neuroglial cells produce IGF-I.[31] Thus, expression of IGF-I by normal neuroglial cells, together with the production of other mitogens by glioma cells, could result in paracrine stimulation of neighboring vascular endothelial cells. The transforming growth factor α, which is produced by a variety of tumor types (see below), has been shown to be a particularly potent angiogenic factor in vivo.[88,89] However, data relative to the production of TGF-α by human gliomas are not yet available.

The paracrine stimulation mechanism may function also in the reverse direction. Some vascular endothelial cells, when stimulated to divide, produce a PDGF-like mitogen.[34] Using cDNA clones for both the A and B chains of PDGF (PDGF isolated from human platelets appears to be a heterodimer of an A and B chain, bound together by interchain disulfide bonds; see reference #18 for a review), Tong et al [92] detected mRNA transcripts for both the A and B chains in normal human umbilical vein endothelial (HUVE) cells and the A172 line of human glioma cells, and in addition, found evidence for independent regulation of the synthesis of the two chains in HUVE cells. The significance of the independent expression of the two chains of PDGF is not yet appreciated. However, the v-sis gene product is homologous to a homodimer of PDGF B-chains, and of course is mitogenic.[18] Recent data indicate that either homodimer, A-A or B-B, as well as the heterodimer, recognize the PDGF receptor, and serve to initiate mitogenic responses.[93] Moreover, the dimerization of chains may itself not be required for receptor recognition. Giese et al,[94] using site-directed mutagenesis techniques to generate specific, altered forms of the v-sis gene product (homologous to the B-B homodimer), showed that the only cysteine residues required for receptor binding and mitogenic function were the four residues which participated in intrachain disulfide bonds, not the cysteines that participated in the interchain bonding. Therefore, the monomeric B-chain molecules (and perhaps the A-chain monomers as well) exhibited full PDGF-like activity.

Thus, the complex pattern of normal growth control, mediated by the timely production of specific growth factors for specific target cells, may be subverted by the establishment of aberrant autocrine and paracrine loops of growth stimulation: a) a transformed glial cell may begin to produce PDGF and other mitogens; b) these mitogens in turn stimulate the proliferation of tumor cells, and of vascular endothelial cells, which form new blood vessels (angiogenesis); c) the activated endothelial cells produce PDGF-like material, which in turn stimulates additional tumor cell division.[24] The cells that participate in the autocrine and paracrine loops thereby gain a selective growth advantage, because their continued proliferation becomes ever more autonomous, i.e. not dependent on exogenous growth factors from outside their immediate microenvironment.

Aberrant growth stimulation can occur by mechanisms other than the autocrine/paracrine pathways. Changes in the genes which code for the growth factor receptors, can lead to a) structural changes in the receptors, and b) changes in the number of specific receptors. In the former case, the structural changes may cause the receptor to function in a continual state of "activation", independent of the presence or absence of the exogenous ligand (the specific growth factor).[26] An example is the viral oncogene v-erbB, an altered form of the cellular proto-oncogene c-erbB, which codes for the EGF receptor (EGFR).[35] The EGFR is a 1186 amino acid transmembrane protein with several functional domains: an extracellular domain that includes the EGF-binding region; a membrane-spanning region; an intracellular, protein tyrosine kinase domain (which includes an ATP-binding site), and an intracellular carboxy-terminal region, with autophosphorylation sites thought to be important in regulation of the tyrosine kinase activity of the receptor protein.[42] Correspondingly, the v-erbB gene appears to represent a double truncated version of c-erbB [reviewed in reference #36]; as a result of the two deletions in the oncogene, the protein product of v-erbB lacks most of the extra-cellular domain (including the EGF-binding region), as well as the terminal 32 amino acid sequence, containing an autophosphorylation site (the tyrosine at residue #1173) at the carboxy-end of the receptor. Studies with chimeric forms of the EGFR [37] have associated the 32 amino acid terminal deletion with the transforming activity of the altered receptor protein. As a consequence of the loss of the terminal tyrosine autophosphorylation site, which appears to function as a competitive inhibitor with other intracellular tyrosine-containing substrates,[95] receptor function becomes deregulated. The tyrosine kinase activity of the altered receptor may be constituitively "turned on", or may recognize different substrates for phophorylation.[36] In the case of the double-truncated v-erbB protein (lacking also the EGF-binding sequences), the receptor function also is independent of the presence or absence of EGF. Similar structural alterations, i.e., truncations of carboxy-terminal amino acid sequences containing a tyrosine autophosphorylation site, with or without an accompanying truncation of a portion of the extracellular, ligand-binding domain, have been identified in other oncogene-coded receptor-analogs within the tyrosine kinase family of receptors [reviewed in reference #38]. These oncogenes include: v-ros, which codes for a receptor closely related to the human insulin receptor (HIR) (38); v-fms, whose product is an altered form of the receptor for the macrophage colony-stimulating factor (CSF-1) (39); v-src, the Rous sarcoma virus oncogene, whose cellular cognate (c-src) codes for a tyrosine kinase receptor (40) for which no ligand has yet been identified; and of course, v-erbB.

Most of the recent data suggest that the protein tyrosine kinase activity of the tyrosine kinase family of receptors is required for signal-transducing function. Chou et al.[56] and Honegger et al.[57] each used the technique of site-specific mutagenesis to inactivate the ATP-binding domain of the tyrosine kinase portion of the HIR and EGFR proteins, respectively. In both cases, the mutant receptors exhibited normal ligand-binding activity, but were unable to carry out auto- and other substrate-phosphorylation of tyrosine residues, and did not mediate postreceptor events, such as stimulation of DNA synthesis. Thus, the data from the tyrosine kinase class of GF receptors suggest that catalytic activity of the kinase is necessary for signal transduction beyond the receptor itself, and that deregulated kinase function is often associated with cell transformation.

The alterations that distinguished v-erbB and related oncogenes (see above), from the corresponding cellular proto-oncogenes, were shown to consist of gene deletions or structural rearrangements, resulting ultimately in the production of truncated protein products with diminished or altered functions. However, "activation" of proto-oncogenes coding for growth factor receptors in glial cells has been associated with at least two other mechanisms, point mutation and gene amplification.

The oncogene neu was identified by transfection of NIH 3T3 cells with DNA extracted from several different cell lines originating from rat neuro/glioblastoma tumors.[43] The tumors were isolated from rats exposed prenatally, on the 15th day of gestation, to the alkylating agent ENU. Schechter et al [43] found neu to be partially homologous to c-erbB, the gene coding for the EGFR. Subsequent work showed that neu coded for a trans-membrane protein, with overall features similar to the EGFR,[44] but human chromosome mapping studies [45,48] confirmed that neu in fact was derived from a separate gene, distinct from c-erbB. The human cognate of the rat neu oncogene has been designated c-erbB-2,[46] or HER2 (for its close resemblance to the Human Epidermal growth factor Receptor gene).[45] Bargmann et al recently identified the ENU-induced, activating alteration of neu to be a single point mutation: a thymine residue at nucleotide position #2012 in neu replaced the normal adenine at this position in the cellular cognate, resulting in the substitution of glutamic acid for the normal valine at residue #664, i.e. in the putative transmembrane region, of the 1260 amino acid receptor protein molecule.[47] The identical point mutation was found in the activated neu oncogene isolated from four separate tumor cell lines, derived from independently-induced neuro/glioblastomas.[47] No additional structural alterations of the neu oncogene-derived product were observed; furthermore, the tumor cells expressing activated neu produced only low levels of the altered receptor-like protein.[4] The data suggested that this particular mutation was strongly associated with the induction of neuro/glial tumors in the offspring of pregnant rats treated with

ENU, and that the putative membrane-spanning region of this receptor-like protein played a fundamental role in the function, as well as the structure, of the molecule.[47] The c-erbB-2 gene was recently shown [49] to be expressed in rat brain during normal development, specifically in day-14 and -16 embryos, but not in later stage embryos, nor in postnatal or adult rats. The physiological ligand for c-erbB-2/ HER2 has not been identified.

Overexpression of genes coding for growth factor receptors also has been associated with cell transformation in vitro. For example, Di Fiore et al.[50] and Hudziak et al.[51] both found that the normal coding sequence for the HER2 (c-erbB-2) gene, when highly amplified (> 50 copies per cell) [51] or overexpressed (transfected into NIH-3T3 cells in expression vectors that contained powerful viral promoter sequences),[50] converted NIH-3T3 cells to the transformed phenotype. The identical HER2 sequences, present at lower copy numbers or expressed at lower levels, did not transform NIH-3T3 cells.[51,50] Similarly, overexpression (again in NIH-3T3 cells, driven by viral promoters) of the normal HER gene was accompanied by cell transformation,[52,96] but HER was less effective than HER2, at similar levels of gene expression.[50,96] As expected, transformation of the NIH 3T3 indicator cells by overexpression of the normal, untruncated product of HER, was EGF-dependent.[52,96] However, cells expressing the highest levels of the EGFR were found to be responsive to lower EGF levels than normal cells.[96] Thus, overexpression of a normal gene was associated with the transformed phenotype in the indicator cells. As discussed above, overexpression of the EGFR was frequently found in human malignant gliomas, and was reported to be a "hallmark" of human squamous cell carcinomas.[58] In contrast, amplification of HER2 in human tumors has been observed most frequently in adenocarcinomas of secretory epithelial cells [reviewed in reference #49]. Thus, the products of the HER2 and HER genes may play similar roles in the genesis of tumors of secretory and squamous epithelial cells, respectively.[49]

Under some conditions, overexpression of a growth factor receptor may result in either autocrine or paracrine activation.[59] An increased number of receptors can make a cell more competitive for available ligand, and could lead to constituitive stimulation at ligand concentrations lower than otherwise required. For example, Di Fiore et al.[96] showed that marked overexpression of the EGFR by transfected cells made these cells responsive to lower levels of EGF than normal cells.[96] The EGF is ubiquitous in the human body. The level detected in cerebrospinal fluid, approximately 0.2 ng/ml, is about 10-100-fold lower than the levels found in some other fluids.[97] Nevertheless, overexpression of the EGFR, for example by an altered neuroglial cell, possibly could result in active stimulation of the cell even at the low level of EGF found in the brain.

Overexpression of the EGFR, in particular, has additional potential in a scheme of autocrine or paracrine stimulation, because the EGFR can respond to a second ligand, TGF-α.

Transforming growth factor alpha (TGF-α), a member of the EGF-like family of polypeptides, was originally identified by its abilities to bind the EGFR, and thus to compete with the usual ligand, authentic EGF.[60] The EGF-binding material was initially termed "sarcoma growth factor (SGF)", because it was produced by retrovirus-transformed tumor cells, and it exhibited mitogenic properties for other cell types.[60] Subsequently, SGF was found to induce the transformed phenotype in appropriate indicator cells in vitro, and thereafter was called TGF.[61] Only later was it discovered that crude preparations of TGF actually contained two transforming growth factors, TGF-α and TGF-β[98], and that these factors sometimes were detectable in non-neoplastic, adult tissues, as well as in tumors.[103] A structural similarity between TGF-α and EGF is thought to account for the ability of TGF-α to bind to the EGFR. The amino acid sequence of rat TGF-α showed 33 and 44 percent homologies with mouse and human EGF, respectively, but more importantly, all three proteins were found to exhibit a conserved pattern of potential interchain disulfide bonds, which suggested similarities in the configurations of the three polypeptides.[62] Moreover, Massague[63] reported that the EGFR exhibited about the same affinity for TGF-α and EGF; therefore, the EGFR probably represents the physiological receptor for both ligands.

The physiological role of TGF-α is less certain, although recent data suggest a possible function during prenatal development. Twardzik and his collaborators detected a mitogen with properties similar to TGF-α in extracts of 12 and 13 day mouse embryos,[64] and later determined that maximum expression of the TGF-α activity occurred on day 7, with lower levels observed on day 13.[65] Similarly, Lee et al,[66] using a cloned TGF-α cDNA, found that TGF-α mRNA levels were highest during days 8-10 in the rat embryo, and declined to barely detectable levels by day 18. In general, production of TGF-α has been associated with transformed cells, tumors, and fetal materials;[68] however, low levels of TGF-α mRNA were detected in some adult rat tissues, including brain, liver and kidney.[67]

TGF-α expression has been detected in human tumors and in cell lines derived from human tumors. Smith et al.[73] found both expression of TGF-α mRNA and overexpression of the EGFR in all five cell lines tested, derived from human pancreatic carcinomas. Moreover, the pattern of anchorage-independent growth in these tumor cell lines suggested that TGF-α functioned as a "superagonist" of EGF, i.e. TGF-α induced anchorage-independent growth at concentrations 10-100-fold lower than observed with EGF. Hermansson et al and Betsholtz et al. [both cited in reference #18] detected TGF-α activity in fresh human tumor tissue (unspecified tumor types), and in 11/16 glioblastoma and sarcoma tumor cell lines tested, respectively. Derynck et al.[69] measured the highest level of TGF-α mRNA synthesis in four different tumor types: renal carcinomas, squamous carcinomas, mammary carcinomas, and in solid tumors of neuroectodermal origin, such as melanomas.

TGF-α mRNA was not detected in the one glioma tumor, and one glioma cell line tested, nor in any normal human tissue from adults, in normal placenta, or in cell lines derived from normal tissue. Derynck et al.[69] also observed that several tumor-derived cell lines, shown to produce substantial levels of TGF-α mRNA, failed to release TGF-α into the culture medium. The TGF-α activity remained associated with the cells, and could be detected upon cell lysis.

From analysis of the cDNA's which code for rat[67] and human[70] TGF-α, as well as for mouse [71,72] EGF, the predicted amino acid sequences for each of these genes suggest the initial synthesis of large, precursor protein molecules that themselves may serve as membrane receptors (reviewed in ref #60). In the case of rat TGF-α, the precursor polypeptide appears to consist of 159 amino acids, with the active 50-AA TGF-α sequence contained in the putative extracellular domain of the molecule.[67] Thus, a specific proteolytic cleavage of the precursor is thought to be required in order to release the active form of TGF-α. If the mechanism of production and release of TGF-α as described above is confirmed, autocrine stimulation in TGF-α-producing cells would require at least three separate activities: synthesis of functional TGF-α precursor, proteolytic release of active TGF-α, and the presence of functional receptors (EGFRs).

Rosenthal et al.[99] tested the ability of TGF-α to contribute to in vitro transformation. They transfected established, non-transformed rat fibroblast cells with a mammalian expression vector containing the TGF-α precursor coding sequence, under the control of a viral promoter. They found that the transfected cells produced and secreted a TGF-α-like material, and were converted to the transformed phenotype.[99]

The production and secretion of TGF-α by cell types that also express the EGFR, meet the criteria of autocrine stimulation. In addition, synthesis and release of TGF-α within the microenvironment of another cell found to overexpress the EGFR, e.g. an altered neuroglial cell, would be expected to result in growth stimulation of the glial cell by the paracrine mechanism. The high proportion (40-50 percent)[55,84] of human malignant gliomas found to overexpress the EGFR suggests that the establishment of autocrine/paracrine stimulation pathways may be involved in the development of gliomas. To date, only two ligands, EGF and TGF-α, have been identified for the EGFR. The role of each ligand, particularly TGF-α, in a proposed autocrine/paracrine loop is yet to be determined.

The establishment of positive loops of autocrine stimulation, such as in the examples described above, is thought to represent a subversion of the normal mechanisms of growth regulation.[18] However, several cellular proto-oncogenes and genes that code for particular mitogens, apparently are expressed during normal embrylogical development [reviewed in reference #101]. For example, TGF-α was shown to be expressed in mouse embryos (see above).[64,65] Similarly, five proto-oncogenes, c-myc, c-erb-A, c-

Ha-ras, c-src, and c-sis, also were found to be expressed at appreciable levels at discrete times during prenatal development of mouse embryos.[102] Expression of these genes, and the establishment of transiently-occurring, carefully-controlled autocrine loops of growth stimulation may represent necessary steps during development. However, the occurrence of these same (or similar) autocrine loops postnatally, by the inappropriate expression of one or more of these genes, may set the stage for clonal expansion and ultimately, tumorigenesis.[101]

Finally, recent evidence supports the existence of pathways of autocrine inhibition,[59] as well as autocrine stimulation. Transforming growth factor beta (TGF-β), while augmenting the growth-stimulatory effects of other mitogens in some cell systems, appears to function primarily as an antagonist of these mitogens, i.e. as an inhibitor of cell proliferation, for many cells in vivo [reviewed in references 18 and 104]. For example, TGF-β induces a reversible inhibition of growth in human prokeratinocytes,[106] but irreversible inhibition of cell division, by inducing terminal differentiation, in normal human bronchial epithelial cells.[105] Moreover, transformed versions of these same cell types appear to have lost their responsiveness to growth inhibition by TGF-β.[18,104,106] Thus, the loss of a normal inhibitory response, e.g. to TGF-β, could produce the same result overall, as the acquisition of an abnormal stimulatory response. In addition, the so-called tumor suppressor genes (also termed "anti-oncogenes" or "emerogenes"), may serve to further modulate the effects of the growth-promoting genes.[108] Fontana (A. Fontana, cited in reference #119) reported that patients with glioblastomas showed immunosupression, presumably as a result of high levels of a TGF-β-like protein found in these individuals. It is not yet clear whether the TGF-β-like material was tumor-derived; if so, the immunosuppressive effects of the material also may have played a role in tumor development.

To summarize, the regulation of cell proliferation and of cell differentiation appear to consist of a series of complex pathways, including both positive and negative growth signals, that interact in ways yet to be fully defined.[18] Additional work is required to demonstrate how the pathways become deregulated during cell transformation, and how this deregulation contributes to tumorigenicity in vivo, e.g. in the transformation of neuroglial cells into malignant gliomas.

GENETIC CHANGES ASSOCIATED WITH EXPOSURE TO IONIZING RADIATION

Data are not yet available to identify specific genetic alterations in the human malignant gliomas associated with exposure to ionizing radiation.[1,2] However, activation of oncogenes by ionizing radiation has been reported in several experimental systems. For example, Guerrero et al.[109] exposed mice 35-60 days old to 150 rads of gamma radiation once per week for 4 consecutive weeks, and observed the induction of thymoma tumors. The radiation-induced thymomas all exhibited activation of the c-Ki-ras gene. The activation of c-Ki-ras in these thymomas was later shown to result from a single base mutation within the first exon of the gene, resulting in a single amino acid substitution (glycine being replaced by aspartic acid) in the c-Ki-ras gene product of the tumor cells.[110] Sawey et al.[111] induced skin tumors by irradiation of four-week old rats with 800-1600 rads of 0.8 MeV electrons. Of 12 tumors analyzed by DNA transfection techniques and by Southern blotting, six showed an activation of c-Ki-ras, and ten showed c-myc amplification (up to 20-fold), with concomittant overexpression of c-myc mRNA (up to 6-fold). Moreover, none of the 12 tumors was negative for both c-Ki-ras and c-myc activation.[111] Borek et al.[112] isolated DNA from X-ray-transformed cells and used the DNA to transfect three recipient, indicator cell lines. The DNA transfections were accompanied by conversion of the recipient cells to the transformed phenotype, suggesting the radiation activation of transforming sequences. The identity of the transforming gene(s) could not be established. The c-myc gene also was implicated in radiation-induced in vitro cell transformation: Sorrentino et al.[113] found that mouse C3H 10T1/2 cells that were infected with a viral vector carrying the c-myc gene, showed a 15-20-fold enhancement of cell transformation by 600 rads of gamma rays. Thus, the constituitive expression of c-myc appeared to contribute to radiation-induced cell transformation.

The results above suggest that ionizing radiation activated both known, and as yet unidentified, cellular transforming genes. Radiation-induced point mutations, gene amplification, and partial gene deletion each was implicated as a potential activating lesion. Recent results suggest that the role of DNA deletions may be greater than heretofore appreciated. Molecular analysis of the mutations induced by ionizing radiation, at several different genetic loci in different lines of mammalian cells, revealed that the majority of these mutations consisted of large deletions [reviewed in reference #114]. In contrast, mutations induced by ultraviolet radiation [115] or by mutagenic chemicals [116] in the same experimental systems, produced either point or small deletion mutations. Moreover, ionizing radiations of different radiation quality, e.g. I-125 decay in the cellular DNA,[114] or external beam irradiation with alpha particles,[116] caused no appreciable changes in the frequency of large deletion mutations.

The genetic analyses above were performed primarily using the Southern blotting technique, which can detect only those deletions larger than about 500 base pairs.[114] Consequently, the potential existence of smaller deletions among the remaining mutations could not be determined. Pastink et al.[117] recently addressed this question by using base-sequence analysis to study the molecular nature of X-ray induced mutations at the "white" locus in the fruit fly Drosophila. They chose five mutants for detailed analysis from the group of radiation-induced mutants that did not show large deletion mutations (about 40 percent of the total). In all five mutants, the mutations were found to be small deletions of from 6 to 29 base pairs.[117]

Thus, the data from both Drosophila and from cultured mammalian cells showed that the mutations produced by ionizing radiation were predominantly deletions, or gene rearrangements such as translocations and inversions.[114, 117] The induction of these mutations illustrates the potential of ionizing radiation for oncogene activation, since gene deletions and rearrangements represent the putative activating lesions for several oncogenes (see above).

CONCLUSIONS AND IMPLICATIONS FOR HUMAN RADIATION EXPOSURE IN SPACE

Glioma tumor cells from different sources were found to exhibit a common, non-random pattern of genetic changes. For example, about 50 percent of human gliomas contained DM chromosomes, and a large percentage of the DMs carried multiple copies of the HER gene. The common pattern of changes suggests the involvement of a limited number of genes in the "spontaneous" development of gliomas. Whether similar alterations occur in radiation-induced gliomas remains to be seen. Nevertheless, alterations in the structure and the expression of these genes represent important new experimental endpoints that can be used to investigate the etiology of malignant gliomas, both "spontaneous" and radiation-associated.

A number of genetic changes associated with activation of oncogenes, and with expression of the transformed phenotype in indicator cells, are known to be produced by ionizing radiation. Partial gene deletion has been shown to distinguish a number of oncogenes from their cellular cognates (proto-oncogenes). Moreover, a large fraction of somatic, radiation-induced mutations appear to consist of deletions of all or part of a gene. Thus, ionizing radiation clearly has the potential for proto-oncogene activation.

The preponderance of deletions among the radiation-induced mutations is striking, but its significance is not understood. Most radiation-induced lesions in DNA are thought to be relatively localized spacially, involving a few adjacent base pairs.[122] The evidence for a non-random distribution or "clustering" of deletion breakpoints,[117, 120] has been interpreted to implicate genetic recombination in the formation of deletion mutations.[114, 117] A model for recombination, initiated by the formation of DNA double strand breaks, was proposed by Szostak et al.[123] Double strand breaks are known to be formed in DNA irradiated in vitro or in vivo.[124]

A linear induction of mutations was suggested by recent data obtained with X-irradiated human lymphoblasts in vitro. The linear dose-response was observed with radiation doses down to less than 10 rads, and no dose-rate dependence was evident over

the range of one rad/day to 30 rads/min.[125] Moreover, X-ray-
induced mutation of genes carried on non-required, "marker"
chromosomes, where the potential contribution of radiation-induced
cell lethality was expected to be minimized, yielded mutation
frequency estimates higher by up to a factor of 200 than those
estimates obtained with the more conventional techniques.[126]
Clearly, more information is required in order to: a) identify the
mechanisms of radiation-induced mutagenesis; b) better define the
actual mutation rate in human cells, as a function of dose rate,
radiation quality, and other factors; and c) establish the
contribution of radiation-induced mutations to the process of
cell transformation. X-ray-induced transformation, at least
in vitro, appears to consist of two steps, only one of
which exhibits characteristics consistent with a random mutation
event.[127]

Space explorers are certain to encounter a radiation
environment, the qualitative and quantitative nature of which is
discussed by others in These Proceedings. Moreover, it appears
that radiation exposure, however small, will result in the
induction of mutations in the exposed individuals, and that some
of the mutations will increase the probability of cell
transformation/tumorigenesis. Post-irradiation intervention, as a
potential means of moderating the effects of radiation exposure,
appears not to represent a practical tactic at this time, because
the mechanisms in the carcinogenic process remain unknown. At
present, the risks associated with radiation exposure in space can
be diminished, if not eliminated, by the use of appropriate
shielding to reduce the dose to exposed individuals.[128]

REFERENCES

1. K. A. Hardy, M. G. Yochmowitz, R. E. Cordts, D. H. Wood, Y. L. Salmon, and A. B. Cox, these proceedings.
2. G. V. Dalrymple, W. A. Nagle, A. J. Moss, Jr., J. C. Mitchell, K. A. Hardy, and H. Davis, these proceedings.
3. J. M. Bishop, Science, 235, 305, 1987.
4. I. B. Weinstein, J. Cell. Biochem., 33, 213, 1987.
5. J. W. Hopewell, D. N. Edwards, G. Wiernir, Br. J. Cancer, 34, 666, 1976.
6. P. L. Kornblith, M. D. Walker, and J. R. Cassady, Cancer, Principles and Practice of Oncology, V. T. DeVita et al., eds. (J. B. Lippincott Co., Philadelphia, 1985), p. 1437.
7. K. Jellinger, Acta Neurochir., 42, 5, 1978.
8. M. F. Rajewsky, Recent Results Cancer Res, 84, 63, 1983.
9. E. L. Jones, C. E. Searle, W. E. Smith, J. Pathol., 109, 123, 1973.
10. J. W. Beard, Viral Oncology, G. Klein, Ed. (Raven Press, N.Y., 1980), p. 55.
11. F. Deinhardt, ibid., p. 357.
12. W. T. London, et al., Science, 201, 1246, 1978.
13. D. D. Copeland, F. S. Vogel, D. D. Bigner, J. Neuropathol. Exp. Neurol., 34, 340, 1975.
14. D. D. Bigner and C. N. Pegram, Adv. Neurol., 15, 57, 1976.
15. H. Kalter, T. I. Mandybur, I. Ormsby, J. Warkany, Cancer Res., 40, 3973, 1980.
16. W. Schmahl, H. Kriegel, Teratogen Carcin. Mut., 5, 159, 1985.
17. J. F. Knowles, Int. J. Radiat. Biol., 41, 79, 1982.
18. C-H Heldin, C. Betsholtz, L. Claesson-Welsh, Biochim. et Biophys. Acta, 907, 219, 1987.
19. S. H. Bigner, et al., Cancer Genet. Cytogenet, 22, 121, 1986.
20. J. R. Shapiro, Semin-Oncol., 13, 4, 1986.
21. M. D. Waterfield, et al., Nature, 304, 35, 1983.
22. R. F. Doolittle, et al., Science, 221, 275, 1983.
23. K. Alitalo, et al., Biochim. Biophys. Acta, 907, 1, 1987.
24. B. Westermark, M. Nister, C. H. Heldin, Neurol. Clin., 3, 785, 1985.
25. M. B. Sporn and G. J. Todaro, N Eng J Med, 303, 878, 1980.
26. R. A. Weinberg, Science, 230, 770, 1985.
27. C. H. Heldin, B. Westermark, and A. Wasteson, Proc. Natl. Acad. Sci. USA, 78, 3664, 1981.
28. M. Nister, C. H. Heldin, A. Wasteson, and B. Westermark, Proc. Natl. Acad. Sci. USA, 81, 926, 1984.
29. P. F. Lens, B. Altena, R. Nusse, Mol. Cell Biol., 6, 3537, 1986.
30. R. I. Freshney, et al., Prog. Exp. Tumor Res., 29, 12, 1985.
31. R. Ballotti, et al., EMBO J, 6, 3633, 1987.
32. C. D. Stiles, Cancer Res., 45, 5215, 1985.
33. S. E. Lynch, J. C. Nixon, R. B. Colvin, H. Antoniades, Proc. Natl. Acad. Sci. USA, 84, 7696, 1987.

34. T. B. Barrett, et al., Proc. Natl. Acad. Sci. USA, 81, 6772, 1984.
35. J. Downward, et al., Nature, 307, 521, 1984.
36. T. Hunter, Nature, 311, 414, 1984.
37. H. Riedel, J. Schlessinger, and A. Ullrich, Science, 236, 197, 1987.
38. E. Chen, et al., Biochem. Soc. Symp., 52, 65, 1986.
39. A. Ullrich, et al., Nature, 313, 756, 1985.
40. L. Coussens, Nature, 320, 277, 1986.
41. T. Takeya and H. Hanafusa, Cell, 32, 881, 1983.
42. A. Ullrich, et al., Nature, 309, 418, 1984.
43. A. L. Schechter, et al., Nature, 312, 513, 1984.
44. C. I. Bargmann, M-C Hung, and R. A. Weinberg, Nature, 319, 226, 1986.
45. A. L. Schechter, et al., Science, 229, 976, 1985.
46. T. Yamamoto, et al., Nature, 319, 230, 1986.
47. C. I. Bargman, M-C Hung, and R. A. Weinberg, Cell, 45, 649, 1986.
48. L. Coussens, et al., Science, 230, 1132, 1985.
49. Y. Kokai, J. A. Cohen, J. A. Drebin, and M. I. Greene, Proc. Natl. Acad. Sci. USA, 84, 8498, 1987.
50. P. P. DiFiore, et al., Science, 237, 178, 1987.
51. R. M. Hudziak, J. Schlessinger, and A. Ullrich, Proc. Natl. Acad. Sci. USA, 84, 7159, 1987.
52. T. J. Velu, et al., Science, 238, 1408, 1987.
53. T. A. Libermann, et al., Cancer Res., 44, 753, 1984.
54. T. A. Libermann, et al., Nature, 313, 144, 1985.
55. A. J. Wong, et al., Proc. Natl. Acad. Sci. USA, 84, 6899, 1987.
56. C. K. Chou, et al., J. Biol. Chem, 262, 1842, 1987.
57. A. M. Honegger, et al., Mol. Cell Biol., 7, 4568, 1987.
58. B. Ozanne, C. S. Richards, F. Hendler, D. Burns, and B. Gusterson, J. Pathol., 149, 9, 1986.
59. M. B. Sporn and A. B. Roberts, Nature, 313, 745, 1985.
60. G. J. Todaro, D. C. Lee, N. R. Webb, T. M. Rose, J. P. Brown, Cancer Cells, Vol. 3, (Cold Spring Harbor Laboratory, New York, 1985), p. 51.
61. G. J. Todaro, C. Fryling, and J. E. DeLarco, Proc. Natl. Acad. Sci. USA, 77, 5258, 1980.
62. H. Marquardt, M. W. Hunkapiller, L. E. Hood, and G. J. Todaro, Science, 223, 1079, 1984.
63. J. Massaque, J. Biol. Chem, 258, 13614, 1983.
64. D. R. Twardzik, J. E. Ranchalis, and G. J. Todaro, Cancer Res., 42, 590, 1982.
65. D. R. Twardzik, Cancer Res, 45, 5413, 1985.
66. D. C. Lee, R. Rochford, G. J. Todaro, and L. P. Villarreal, Mol. Cell. Biol., 5, 3644, 1985.
67. D. C. Lee, T. M. Rose, N. R. Webb, and G. J. Todaro, Nature, 313, 489, 1985.
68. G. Carpenter and J. G. Zendegui, Exptl. Cell Res., 164, 1, 1986.

69. R. Derynck, et al., Cancer Res., 47, 707, 1987.
70. R. Derynck, et al., Cell, 38, 287, 1984.
71. J. Scott, et al., Science, 221, 236, 1983.
72. A. Gray, T. J. Dull, and A. Ullrich, Nature, 303, 722, 1983.
73. J. J. Smith, R. Derynck, and M. Korc, Proc. Natl. Acad. Sci. USA, 84, 7567, 1987
74. O. D. Laerum and M. F. Rajewsky, J. Natl. Cancer Inst., 55, 1177, 1975.
75. D. Schubert, et al., Nature, 249, 224, 1974.
76. R. T. Schimke, Cell, 37, 705, 1984.
77. G. R. Stark and G. M. Wahl, Ann. Rev. Biochem., 53, 447, 1984.
78. J. L. Hamlin, J. D. Milbrandt, N. H. Heintz, and J. C. Azizkham, Int. Rev. Cytol., 90, 31, 1984.
79. J. A. Rey, et al., Cancer Genet. Cytogenet., 29, 201, 1987.
80. J. A. Rey, M. J. Bello, J. M. deCampos, M. E. Kusak, and S. Moreno, Cancer Genet. Cytogenet., 29, 223, 1987.
81. A. A. Saadi, F. Latimer, M. Madercic, and T. Robbins, Cancer Genet. Cytogenet., 26, 127, 1987.
82. X. O. Breakefield and D. F. Stern, Trends Neurosci., 9, 150, 1986.
83. A. Eva, et al., Nature, 295, 116, 1982.
84. S. H. Bigner, et al., Cancer Genet. Cytogenet., 29, 165, 1987.
85. J. Trent, et al., Proc. Natl. Acad. Sci. USA, 83, 470, 1986.
86. K. W. Kinzler, et al., Science, 236, 70, 1987.
87. A. Johnsson, C. Betsholtz, C. H. Heldin, and B. Westermark, EMBO J., 5, 1535, 1986.
88. A. B. Schreiber, M. E. Winkler, and R. Derynck, Science, 232, 1250, 1986.
89. J. Folkman and M. Klagsbrun, Science, 235, 442, 1987.
90. P. Pantazis, P. G. Pelicci, R. Dalla-Favera, and H. N. Antoniades, Proc. Natl. Acad. Sci. USA, 82, 2404, 1985.
91. L. T. Williams, J. A. Escobedo, M. T. Keating, and S. R. Coughlin, J. Cell. Physiol. Supp., 5, 27, 1987.
92. B. D. Tong, et al., Nature, 328, 619, 1987.
93. C. H. Heldin and B. Westermark, J. Cell, Physiol. Supp. 5, 31, 1987.
94. N. A. Giese, K. C. Robbins, and S. A. Aaronson, Science, 236, 1315, 1987.
95. G. N. Gill, J. B. Santon, and P. J. Bertics, J. Cell. Physiol. Suppl., 5, 35, 1987.
96. P. P. DiFiore, et al., Cell, 51, 1063, 1987.
97. G. Carpenter, J. Cell Sci. Supp., 3, 1, 1985.
98. M. A. Anzano, A. B. Roberts, J. M. Smith, M. B. Sporn, and J. E. DeLarco, Proc. Natl. Acad. Sci. USA, 80, 6264, 1983.
99. A. Rosenthal, P. B. Lindquist, T. S. Bringman, D. V. Goeddel, and R. Derynck, Cell, 46, 301, 1986.
100. D. J. Giard, et al., J. Natl. Cancer Inst., 51, 1417, 1973.
101. R. I. Ohlsson and S. B. Pfeifer-Ohlsson, Exptl. Cell Res., 173, 1, 1987.

102. D. J. Slamon and M. J. Cline, Proc. Natl. Acad. Sci. USA, 81, 7141, 1984.
103. A. B. Roberts, C. A. Frolik, M. A. Anzano, and M. B. Sporn, Federation Proc., 42, 2621, 1983.
104. M. B. Sporn, A. B. Roberts, L. M. Wakefield, and R. K. Assoian, Science, 233, 532, 1986.
105. T. Masui, et al., Proc. Natl. Acad. Sci. USA, 83, 2438, 1986.
106. G. D. Shipley, et al., Cancer Res., 46, 2068, 1986.
107. H. L. Moses, et al., J. Cell. Physiol. Suppl. 5, 1, 1987.
108. G. Klein, Science, 238, 1539, 1987.
109. I. Guerrero, P. Calzada, A. Mayer, and A. Pellicer, Proc. Natl. Acad. Sci. USA, 81, 202, 1984.
110. I. Guerrero, A. Villasante, V. Corces, and A. Pellicer, Science, 225, 1159, 1984.
111. M. J. Sawey, A. T. Hood, F. J. Burns, and S. J. Garte, Mol. Cell. Biol., 7, 932, 1987.
112. C. Borek, A. Ong, and H. Mason, Proc. Natl. Acad. Sci. USA, 84, 794, 1987.
113. V. Sorrentino, V. Drozdoff, L. Zeita, and E. Fleissner, Proc. Natl. Acad. Sci. USA, 84, 4131, 1987.
114. R. A. Gibbs, J. Camakaris, G. S. Hodgson, and R. F. Martin, Int. J. Radiat. Biol., 51, 193, 1987.
115. L. F. Stankowski, Jr., and A. W. Hsie, Radiation Res., 105, 37, 1986.
116. J. Thacker, Mutation Res., 160, 267, 1986.
117. A. Pastink, C. Vreeken, A. P. Schalet, and JCJ Eeken, Mutation Res., 207, 23, 1988.
118. S. H. Bigner, et al., Cancer Res., 88, 405, 1988.
119. J. L. Marx, Science, 239, 257, 1988.
120. A. J. Grosovsky, E. A. Drobetsky, P. J. DeJong, and B. W. Glickman, Genetics, 113, 405, 1986.
121. A. C. Upton, Radiation Carcinogenesis: Epidemiology and Biological Significance, J. D. Boice, Jr. and J. F. Fraumeni, Jr., Eds., (Raven Press, New York, 1984), p. 9.
122. J. F. Ward, Radiation Res., 104, S-103, 1985.
123. J. W. Szostak, T. L. Orr-Weaver, R. J. Rothstein, and F. W. Stahl, Cell, 33, 25, 1983.
124. F. Hutchinson, Prog. Nuc. Acid Res. Mol. Biol., 32, 115, 1985.
125. A. J. Grosovsky and J. B. Little, Proc. Natl. Acad. Sci. USA, 82, 2092, 1985.
126. C. Waldren, L. Correll, M. A. Sognier, and T. T. Puck, Proc. Natl. Acad. Sci. USA, 83, 4839, 1986.
127. A. R. Kennedy, J. Cairns, and J. B. Little, Nature, 307, 85, 1984.
128. J. R. Letaw, R. Silberberg, and C. H. Tsao, Nature, 330, 709, 1987.

NEW ASTRONAUT RADIATION EXPOSURE LIMITS AND IMPLICATIONS OF PROPOSED CHANGES IN QUALITY FACTORS

D. S. Nachtwey - NASA Johnson Space Center, Houston, TX 77058
R. J. M. Fry - Oak Ridge National Lab., Oak Ridge, TN 36739

Astronauts are judged to be a type of radiation worker because of their technologically enhanced exposures to natural ionizing radiation in space. As radiation workers they are subject to a system of dose limitation. Since 1970, NASA has employed a set of space radiation exposure limits recommended by an Advisory Panel of the Space Science Board of the National Academy of Sciences--National Research Council (NAS/NRC). Recently, NASA asked the National Council on Radiation Protection and Measurements (NCRP) to re-evaluate the astronaut radiation exposure limits. NCRP Scientific Committee 75 in its soon-to-be-published report, "Guidance on Radiation Received in Space Activities," has recommended the new limits shown in the table.

NCRP RECOMMENDED ASTRONAUT RADIATION EXPOSURE LIMITS

	DEPTH (5 cm)	EYE (0.3 cm)	SKIN (0.01 cm)
30 DAY	0.25 Sv	1.0 Sv	1.5 Sv
ANNUAL	0.50	2.0	3.0
CAREER	1.0-4.0*	4.0	6.0

*The career depth-dose-equivalent limit is based on a maximum 3% lifetime risk of cancer mortality. The total dose-equivalent yielding this risk depends on sex and on age at the start of exposure. The limit is nearly equal to:

$$2.0 + 0.075 \,(\text{age} - 30) \text{ for males}$$
$$2.0 + 0.075 \,(\text{age} - 38) \text{ for females}$$

These new career limits represent a reduction from the 1970 NAS/NRC limits. However, eye and skin 30-day and annual limits are increased relative to the old limits. The rationale behind these new limits will be discussed.

The new limits are not expected to preclude any future missions such as staytimes of a year or more on Space Station, lunar base activities, or piloted missions to Mars; but the radiation exposures in such missions will, nonetheless, be major concerns. NASA is legally bound to maintain dose-equivalents "As Low As Reasonably Achievable" (the so-called ALARA principle), therefore, tradeoff studies of increased shielding versus cost of mass-to-orbit--to determine what is "reasonable"-- will be a continuing part of any manned program. Knowledge about Galactic Cosmic Ray (GCR) fluences, transport, and fragmentation processes in spacecraft and astronaut bodies must be improved to support such tradeoff studies.

In addition to physical uncertainties, there are uncertainties in the carcinogenic risks associated with GCR heavy ions and other high LET radiations. High LET radiation is more carcinogenic per unit of absorbed dose than is low LET radiation. In current terrestrial radiological health practice, a quality factor (Q) is applied to the dose in gray (Gy)* to obtain the dose-equivalent in sievert (Sv)*. Recently ICRP and NCRP have recommended increasing the value of Q by factors of 2 to 5 for some radiations. Increasing Q could lead to the exposure limits being exceeded in some anticipated missions.

*1 Gy = 100 rad; 1 Sv = 100 rem.

PROMOTION OF A NEW RADIOPROTECTIVE ANTIOXIDATIVE AGENT

J. Matsubara, A. Ikeda and T. Kinoshita

Department of Epidemiology, Faculty of Medicine
The University of Tokyo
Hongo 7-3-1, Bunkyoku , Tokyo, 113 Japan

ABSTRACT

Radioprotective effects of pre- and post- irradiation treatments giving zinc or manganese, or one of biological response modifiers(BRMs) ,or subjecting mice tp stressful conditions on the whole bodily irradiated mice were investigated with respect to MT induction.

The normal level of MT in murine liver is as low as 1-2 nano mole/g tissue, but the level can be elevated to several or even twenty times after the administration of a heavy metal or a certain BRM or both. A stress, e.g. dermal excision also induced hepatic MT synthesis.

These pretreatments to induce hepatic MT gave a strong tolerance against lethal damage to mice. The effect of combined use of BRM with manganese or zinc was more prominent. Since manganese has less adverse effects and easily eliminated from organs, the spot administration of even after the irradiation was considerasbly effective. Mice administered manganese menifested a quicker regaining in the number of leucocytes and erythrocyted than control mice .

These investigations suggested us a possibility that man can promote their body defense mechanism through the induction of MT in the organs by a certain stimulative measure e.g. the administration of a certtain less harmful metal or chemicals , physical exercises or controling various factors in our environment. Since the induction of MT by taking advantage of body's own biological protection mechanism is versatile with almost no adverse effects, it may opens new vistas of radiation protection in the future.

INTRODUCTION

Radiation protection is of vital importance because the radiation exposure and resulting hazard limit the extent of human activity in space. Until now radiatoion protection problem is mainly considered by health physicists who appreciate their successful history to control exotic radiation exposure and to develop various administrative measures during this half century[1]. However, the promotion of effective and safe medicinal radioprotectants remains far from sufficient. We have tried a positive consideration about body defense mechanism which we human are possessing, and performed some devices to stimulate biological protection mechanism in mice in order to promote tolerance in the body against a lethal dose of radiation damages.

MT is widely accepted as heavy metal detoxicating substance and is said to be ubiquitous protein in search of physiological function[2]. However, we regard MT a sort of emergency protein which can work against oxidative invasion due to exposure to radiation, oxidants or toxic substances, and various loads of physical, chemical or biological stressers because of the following evidences : (1) The normal level of MT in murine liver is as low as 1 - 2 nano mole /g tissue , but the level can be elevated to several or even twenty times after the administration of a heavy metal or exposure to radiation, cold, or bacterial infection, surgical injury or physical exercises [3,4]. (2) A remarkable radioresistance was found not only in metal (Mn 20mg/kg or Cd 3 mg/kg) salt administered mice but also in 2x2 cm^2 dorsal skin excised mice to stimulate zinc turnover inducing MT synthesis in the liver[5]. and (3) A purified Zu-Cu-thionein can work as electron donor similarly as glutathione with glutathione peroxidase in vitro in order to reduce hydroperoxide[6].

Recently obtained results that manganese salt was radioprotective even administered to mice after the irradiation of a lethal dose and it seemed not only reduce the damages but also to help recovery process from radiation injury and another evidence that the radioprotective action of a immunomodulator e.g. IL-1 or OK432 was intensified by the simultaneous administration of zinc or manganese, suggested us the coordinating mechanism of body protection aginst chemical environmental factors e.g.heavy metals or oxidants as well as biological ones e.g. microorganisms.

MATERIALS AND METHODS

ICR: Jcl male mice of 7 weeks old were subjected to the pre- or post-irradiative treatment, e.g. the administration(ip. or sc. injection) of Mn(20mg/kg) or Zn (5mg/kg) salt ,or some of immuno-modulators, e.g. IL-1(40 μg/kg) or OK432(15mg/kg) alone or a certain plant extract of traditional oriental medicine alone ,or together with zinc or manganese salt, or 2X2 cm^2 portion of dorsal skin excision. Mice were irradiated whole bodily with X ray of 6.5 or 7.5 Gy. Groups of five mice each with different pre- or post- treatment were used for MT assays, according to the Cd-hemoglobin affinity method (Onosaka et al.[7] and all the groups of each 16 mice irradiated were kept 30 days under observation for the survival study. In the meantime several microliter blood was taken from codal vein periodically and numbers of blood corpuscles were counted by a Microcell counting appratus. Details are written in our previous papers 3),5).

RESULTS

In the above pretreated mice increased MT synthesis, 8 to 20 fold of the control level was observed (see Table 1).These mice demonstrated striking radioresistance against lethal damage from a single dose of 6 - 8 Gy Xray as shown in Fig. 1 .

The post -irradiative administration of manganese salt also manifested the reduction of mortality of the mouse as shown in Fig.2. The mice without manganese administration showed 20% survival for 6.7 Gy irradiation and mice administered 20 mg Mn/kg 24 h prior to irradiation showed 100 % survival. While mice administered Mn immediately after irradiation(0day) , 0.25day, and 1 day after irradiation showed 80 % , 70 % and 55 % survivals respectively. The administration of manganese salt at 3 days post-irradiation or more was not effective but harnful.

Resuls from periodical leucocytes and erythrocytes countings with mice 20 mg Mn/kg administered 24 h prior to 6.3 Gy irradiation are presented on Fig. 3. Thogh the irradiation causes severe reduction of leucocytes and erythrocytes, however, mice administered manganese could regain those counts on 15th day of postirradiation which was the most critical time for bone marrow death caused by the semilethal dose of X rays. These results suggest the administration of manganese could help a certain process of radiation recovery.

DISCUSSION

Living organisms are not only exposed to natural radiation but to various oxidative substances as O_3 and NOx in the environment. They also have a mechanism to produce peroxideswithin the cell at phagocytosis. On the other hand, organisms have subtle protecting mechanisms against these harmful substances by scavenging free-radicals and oxidants against oxidative invasions. Fig.4 is illustrated to clarify the possible sites of protection by reviewing the protection mechanism against free-radical-mediated cell injury.

Indirect process of radiation damage in organic cells is regarded as follows: excitation of molecules by electro-magnetic waves causes the production of free radicals such as O_2 , OH , LOO producing lipid peroxides in biomembranes, and which will develop various episodes of biohazards. Normally naturally existing scavenge systems as superoxide dismutase, catalase and glutathione peroxidase systems work to quench these oxidized substances. The present author(JM) hypothesized that MT is involved not only in free-radical scavenging process(site A in Fig. 4) but also in glutathione peroxidase system(site B in Fig.4 [6]).

Since the nomal level of MT in the liver is not high but the synthesis of MT is readily induced by various stimulative factor e.g. the administration of metal or a certain immunomodulator-like drug . The treatment with manganese salts manifested an increase of tolerance against radiation as well as a promotion of recovery from damages in hematopoietic cells . Another result that the spleenal colony stimulation in irradiated animals was higher in metal administered mice in which hepatic MT was higher(our unpublished observation). According to Ebadi et al.[8] the level of MT is higher in the physiological condition in the brain where the level of glutathione is lower. Thus the synthesis of MT in the liver seems to be in accordance with a detoxication mechanism as well as immunostimulative mechanism in the liver though we need further evidences. We deem the study on the physiological function of MT provides broad aspects of the coordinating mechanism of chemical protection and immunological protection in the body though we have still many unknowns about what is the coordination .

Our results came out after nearly ten years lasting interdisciprinary approach [9,10]. Since the human body is an integration of all sort of biological and physico-chemical activities which maintain a strong homeostasis against various stressers, a protective measure should be performed with a comprehensive manner, not by a prescription of a chemical agent only. WR2721 is really harmful chemical which destroys our calcium homeostasis in the body [11]. While induction of metallothionein by taking advantage of body's own biological protection mechanism, e.g. administration of manganese or zinc salt in addition to a proper immunomodulator or oriental traditional remedies made from a certain plant extracts[12], is convenient and versatile with almost no adverse effect, it may open a new vista of radiation protection in the future.

References

1. Twenty fifth Hanford Life Sciences Symposium(Richland Oct/1986): Health Physics, Special Issue, , July (1988)
2. J.H.R.Kägi and M.Nordberg, Metallothionein Experientia Suppl. 34, Birkhäuser (1979)
3. J.Matsubara, et al., Environ. Res. 43, 66 (1987)
4. S.H.Oh et al. , Amer. J. Physiol. 234 (3), E282 (1978)
5. J. Matsubara, et al. Radiat. Res. 111 , 267 (1987)
6. J. Matsubara,In ` Metallothionein II' (J.H.R.Kagi and Y.Kojima Eds) Experientia Suppl. 52,603 Birkhäuser (1987)
7. S. Onosaka et al., Eisei Kagaku, 29,221 (1983)
8. M. Ebadi, Biol. Trace Elem. Res. 11,101 (1986)
9. J. Matsubara, et al. Environ. Res. 40, 525 (1986)
10. J. Matsubara, et al. Environ. Res. 41, 558 (1986)
11. Symposium on Perspective in Radioprotection : Pharmac. Ther. Special Issue , , June (1988)
12. J. Matsubara and S. Yonesawa, papaer in preparation

Table 1 The survival rate(%) of irradiated mice in various groups with varied pretreatments and levels of metallothionein in the liver.

Study(A)

Pretreatment	Survival Rate after 6.7Gy irradiation(%)	Concentration of Metallothionein in the liver(nano mol/g tissue)
Saline s.c.	25	4.0±0.15
Cd(3mgCd/kg) s.c.	94	31 ±6.3
Mn(10mgMn/kg) s.c.	84	19 ±2.5
Skin-exicised($2 \times 2cm^2$)	94	27 ±5.4

Study(B)

Pretreatment	Survival Rate after 6.5Gy irradiation(%)	Concentration of Metallothionein in the liver(nano mol/g tissue)
Saline i.p.	20	2.1±0.6
OK-432(5KE/mouse) i.p.	70	14 ±2.2
Lentinan(5mg/kg) i.p.	25	6.0±0.6
IL-1(40 μg/kg) i.p.	90	11 ±0.5
PSK(50mg/kg) i.p.	10	5.7±0.5
IL-1 + Zn i.p.	94	64 + 13
OK432 + Zn i.p.	90	54 + 2.5

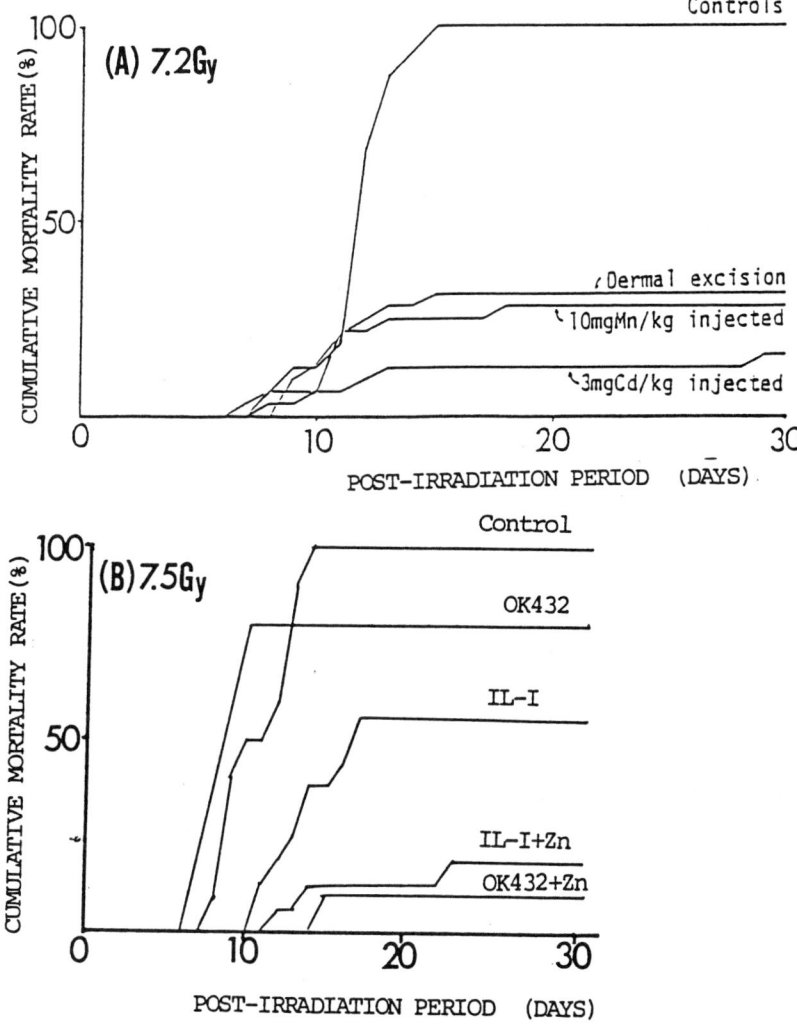

Fig. 1 Daily changes in cumulative mortality of mice during 30 days post irradiation ((A) 7.2 Gy, (B) 7.5 Gy single dose).

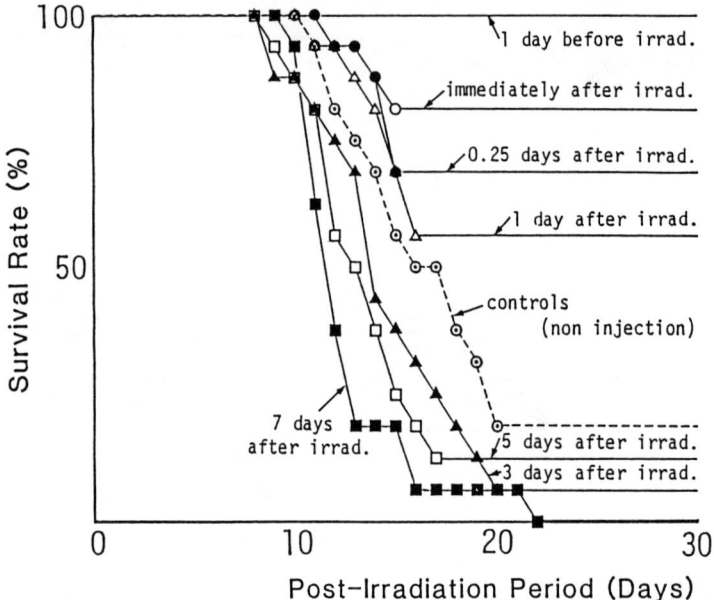

Fig. 2 Survival rates of mice during 30 days of 6.7 Gy post-irradiation according to the timing of $MnCl_2$ (20 mg/kg) s.c. injection

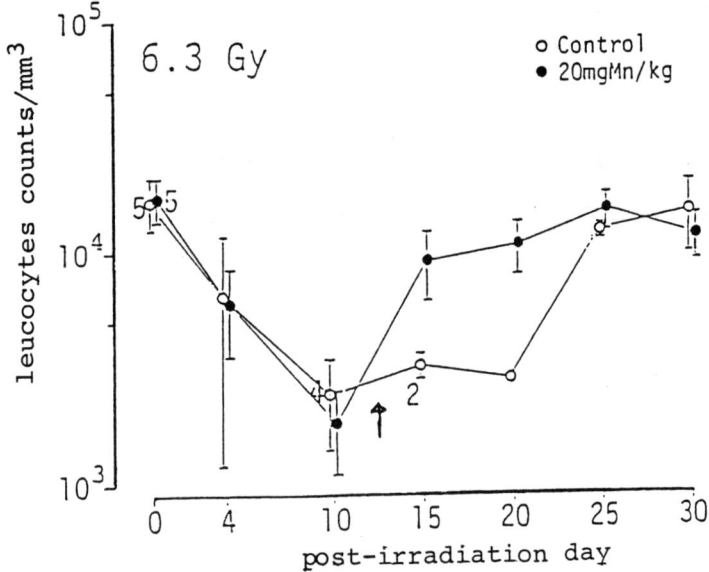

Fig. 3 Shifts of leucocytes counts of manganese administered mice during 30 days post-irradiation

Fig. 4 MECHANISM OF CHEMICAL PROTECTION
AGAINST ENVIRONMENTAL FACTORS
— Protection against Free radical-mediated Cell-injury —

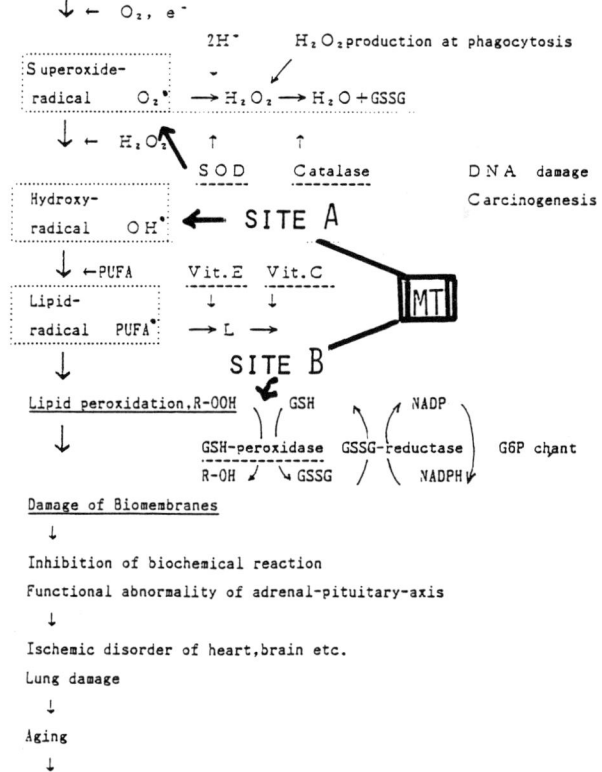

VII. Future Needs and Strategies

SPACE STATION: INFRASTRUCTURE FOR RADIATION
MEASUREMENTS IN LOW EARTH ORBIT

B. D. Meredith
NASA Langley Research Center, Hampton, VA. 23665

ABSTRACT

The National Aeronautics and Space Administration has recently begun the design and development phase of the Space Station Program. International negotiations with the Europeans, Japanese and Canadians are progressing and they will soon join the United States in the development to complete plans for a cooperative, international space endeavor. The Space Station will provide a permanently manned facility in low earth orbit to support scientific research, technology development and commercial activities. In addition, free-flying platforms will be developed to carry a variety of user payloads in polar orbit, in co-orbit with the station and eventually into geosynchronous orbit. The Station and platforms represent an infrastructure of research facilities in space which could be utilized for measurements of naturally occuring radiation and secondary emissions over extended periods of time. High energy radiation experiments at the Station/platforms might vary in objective from an intent to improve shielding or sensor technology to the scientific purpose of characterizing the radiation environment for improved model accuracy. This paper describes the Space Station Program content, schedule and approach for inputting user requirements into the design process. Conference participants can then assess to what extent this future capability in space matches their future research initiatives.

INTRODUCTION

With the advent of an orbiting Space Station, human activities in earth orbit will no longer be constrained by the one week nominal stay-time afforded by the Space Shuttle. The International Space Station, to be developed by the U.S. in cooperation with the Europeans, Canadians and Japanese, will provide a permanent human presence in space. To the scientific or commercial users of the Station, "human presence" represents a valuable resource in orbit, enabling crew interaction with payloads for operation, servicing and/or contingencies. It also makes possible certain benefits which, if no longer unique to human capability, are most effectively done by people; i.e, interpretation of and response to unexpected opportunities.

© 1989 American Institute of Physics

In addition to crew involvement, the Space Station will supply other essential resources for science, technology and commercial payloads. These include pressurized volume within laboratory modules, structure for mounting payloads externally, electrical power, thermal management and control, data services and servicing support. Standard, instrument racks within the Space Station lab modules will be available for accommodation of payloads and control equipment. The laboratories will support a variety of activities including microgravity research, materials processing, life science research and technology experimentation. Instrumentation will no doubt be needed within the modules to monitor the internal radiation environment for the health and welfare of the crew, equipment and research samples. Since the use of Space Station "common" modules is an integral part of current scenarios devised for manned Mars and Lunar base initiatives, high energy particle interaction with the module walls and the overall radiation environment will be of particular interest to NASA. Payload attachment equipment will be developed by the program to provide structural, thermal, power and data interfaces to those payloads mounted on the Station truss structure external to the pressurized volume. Certain attached payloads will have viewing preferences for observing the Earth or pointing inertially toward the stars or sun. It is likely that sensor packages will be attached to the Station for characterizing the radiation environment at the operating orbital location of 250 nautical miles nominal altitude (463 km.) and 28.5 degree inclination.

The Space Station design will incorporate features which will allow it to evolve gracefully over the 30 year operational lifetime. It is anticipated that increased user demands on the Station will precipitate the need for incremental increases in resources such as electrical power, crew and lab volume. However, evolution involves more than growth of resources. There is likely to be a need in the future for added functional capability (e.g., the assembly and checkout of large Mars or Lunar vehicles) and for accommodation of advances in subsystem technologies. The results of current evolution planning within NASA will be reflected in the initial design of the Space Station so that users can look forward to increased utilization over the life of the facility.

In addition to the manned element, the Space Station Program also includes free-flying platforms which will be capable of carrying large or multiple payloads and will be designed for on-orbit servicability and growth. The polar platforms are intended to support a number of relatively small payloads most of which operate in a synergistic fashion to observe the Earth and its atmosphere across a wide spectrum of wavelengths. Polar platforms will be designed for ascending and descending nodal crossings and will be capable of operating over an altitude range of 500 to 900 km. The co-orbiting platforms will fly at the same inclination as the core Station, but can vary in altitude from 463 to 1000 km. Typical payloads for the co-orbiters consist of large astronomical telescopes or commercial materials processing facilities. The materials production platforms would fly in formation with the Space Station for periodic crew tending but apart from the Station to maintain needed microgravity over long durations of time. A programmatic decision was recently made to defer co-orbiting platforms from the baseline program to the subsequent phases. Plans for the evolutionary phases of the Space Station program also include the development of platforms for carrying science payloads to geosynchronous orbit. Any of the three types of platforms could support instrument packages which measure the background radiation at their particular orbit locations.

The remainder of this paper will be devoted to describing the Space Station program background, schedule, configuration and mechanism for inserting payload requirements into the design process. Potential users can then judge the applicability of this capability in space to their research needs.

PROGRAM BACKGROUND AND SCHEDULE

Table I summarizes the key milestones for the Space Station definition and development phases.

Table I Space Station definition and development schedule

Definition Milestones:		Development Milestones:	
contract start	April '85	contract start	Dec. '87
baseline adopted	May '86	PDR	Jan. '89
CETF revision	Sept. '86	CDR	Aug. '90
complete defn. contracts	Jan. '87	first element launch	March '94
program cost review	Jan. '87	permanently manned	Early '96
revised baseline	April '87	phase I complete	Early '97

Phase B or the definition phase of the Space Station Program began in April of 1985 when NASA initiated eight study contracts for the purpose of defining systems, technical approaches and configurations for the Station and platforms. Efforts on the part of NASA and the contractors resulted in the adoption of a baseline configuration for the Space Station in May of 1986. After some modifications by the Critical Evaluation Task Force (CETF) the following September to primarily reduce astronaut EVA time associated with station assembly, the "dual keel" configuration became the program reference and principal output of the Phase B process. Key features of this configuration are described in the subsequent section of this paper.

In early 1987, a comprehensive assessment of program cost resulted in the decision to phase the Space Station capability in order to bring initial costs more in-line with budget realities. This resulted in the adoption of a revised baseline for phase I of the program. The updated baseline deferred certain capabilities (co-orbiting platforms for example) to phase II and reduced the amount of structure available to attach payloads on Space Station. The revised station baseline configuration (phase I) was then provided by NASA as the reference for contractors involved in the design and development activities or Phase C/D.

Letter contracts for Phase C/D were signed in December '87, by NASA and the prime contractors- McDonnell Douglas Astronautics, Boeing Aerospace, General Electric Space Division and Rockwell's Rocketdyne. Preliminary and critical design reviews are currently scheduled for January, 1989 and August, 1990 respectively. Space Station will be delivered to earth orbit in a piece-by-piece fashion via the Space Transportation System (STS) and possibly some expendable launch vehicles and assembled by the crew with robotic assistance. The first element of this assembly process is scheduled for launch in March of 1994. The current plan calls for Station to be ready for permanent human occupation by early 1996 and phase I should be completed one year later. The evolutionary design of the Space Station will allow additional capability and new technology to be accommodated in phase II or any subsequent phases of the development.

CONFIGURATIONS AND CAPABILITY

The "dual keel" configuration was adopted as the Space Station baseline during the definition phase of the program. Figure 1 shows that configuration with the CETF refinements.

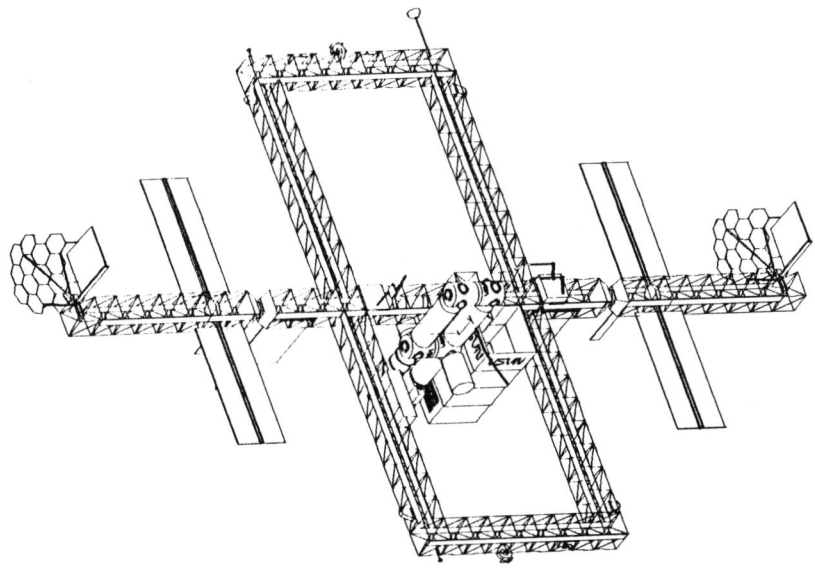

Fig. 1. Space Station configuration from the definition phase.

This configuration gets its name from the large, twin, vertical keels which would be utilized for attached payload structural support. The keels are separated by upper and lower booms which are prime real estate for astronomical and earth observing payloads. Each of the Station booms and keels are constructed from "building blocks" of 5 meter truss structures or bays. The keels are 105 meters in height and are spaced 35 meters apart. This spacing allows for growth in pressurized modules and new facilities and provides room for the on-orbit assembly of large space structures.

The largest structural element is the transverse boom which spans 145 meters and supports the power generation systems, thermal radiators, an unpressurized bay for servicing free-flying satellites, and the lab and habitation modules. Power generation is accomplished by a hybrid system composed of conventional photovoltaics and a new space technology; solar dynamics. This approach utilizes large dish-shaped mirrors to concentrate the solar energy and heat engines to drive electrical power generators. For this particular configuration, a total of 87.5 kilowatts of power is produced, of which, 50 kw. would be available for user purposes. In the initial module pattern, there are four primary modules interconnected by pressurized resource nodes. The United States will supply a habitation module which supports the baseline crew of 8 and a laboratory module. The Japanese will provide an experimental lab module with an attached, external pallet and the Europeans will provide another lab. The Europeans also plan to develop a polar platform that will complement the U.S. platform. A mobile manipulator system, the Mobile Servicing Center (MSC), will serve as a crane to assist in the assembly of the Station and when operated in conjunction with the Flight Telerobotic Servicer (FTS) can perform dexterous functions such as servicing of attached payloads insitu. The manipulator arm will be furnished by Canada and the U.S. will develop the mobile base and FTS.

In an effort to reduce development cost, the Space Station Program was redirected toward the concept of phasing station capability and, as a result, a revised baseline was defined and adopted in April of 1987. The phase I, baseline configuration is shown in figure 2.

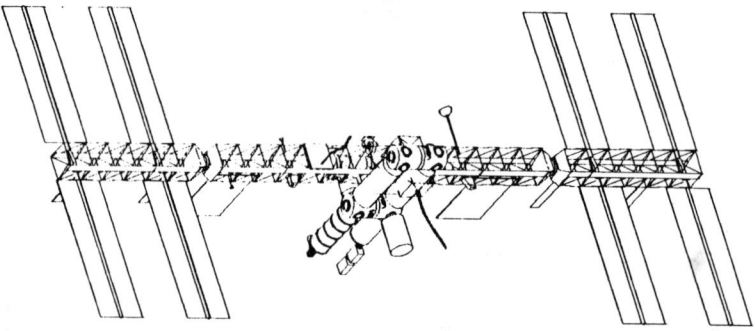

Fig. 2. Revised baseline

The most obvious impact of the revised baseline is the removal of the keels and upper and lower booms from the phase I configuration. This translates to reduced area for attaching user payloads to Space Station, at least in the initial phase. Also deferred to the subsequent phase of the program is the satellite servicing bay and the solar dynamic power technology. The total power generation capability of the phase I Station will be 75 kilowatts and about 45 kw. will be available to the users. The module type and pattern remain unchanged from the prior baseline and an initial crew of 8 is still planned. The development of a U.S. polar platform was also retained in the baseline program. The MSC will have somewhat reduced capability in that it will be unable to make plane changes or turn corners on the Station.

The phase I design will incorporate the necessary provisions or "scars" to allow the keels and booms to be added on-orbit in the subsequent development phase along with the addition of other deferred capability. Ultimately, the evolution of Space Station will be driven by user needs and by overall U.S. policy and objectives in space.

INSERTION OF USER REQUIREMENTS

With the phase C/D contractors currently onboard, NASA and the international partners can begin the detailed design task for an orbiting Space Station. Experimenters who wish to become users of the Space Station and/or polar platforms should take appropriate steps to ensure that their requirements are considered in the design process. While no single, unique approach is recognized for inserting user/mission requirements into the program, the road map illustrated in figure 3 offers one generalized solution.

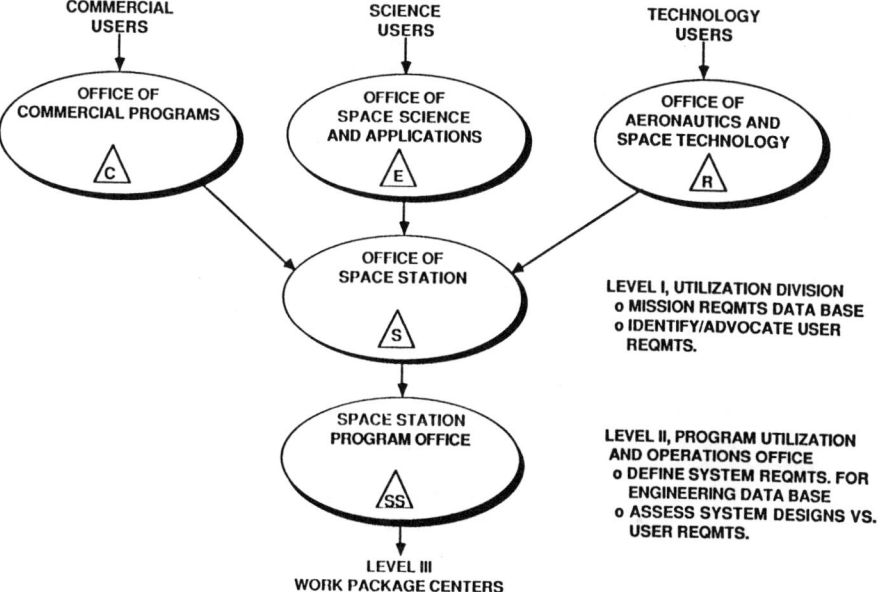

Fig. 3. User roadmap to the Space Station Program

The principal point-of-entry for potential users is the program offices at NASA Headquarters. They interface with the Office of Space Station, Code S at Headquarters, to identify and recommend missions and payload sets for the Station and platforms. The discipline of the proposed mission determines which office is the appropriate contact. Users whose mission is to develop advanced space technology, for example, high energy radiation sensors or shielding material, would interface with the Office of Aeronautics and Space Technology, Code R. Scientific endeavors, such as the characterization of the radiation environment in polar orbit, are represented by the Office of

Space Science and Applications, Code E. Commercial interests are represented by the Office of Commercial Programs, Code C. Commercial missions include, for example, the growth of semiconductor crystals in the zero-G space environment.

Within Code S or Level I of the Space Station Program management structure, the Utiliization Division interacts with users to identify mission requirements and periodically updates the Space Station Mission Requirements Data Base (MRDB) with new or refined data entries. The Level I organization is responsible for overall policy and advocates mission requirements to the Space Station Program Office or Level II. Level II provides the detailed, day-to-day program management and is physically located in Reston, VA. The Program Utilization and Operations Office of Level II is responsible for deriving system requirements from the mission data and assessing the capability of proposed designs to meet user requirements. Each of the four, Level III development centers then utilizes the Engineering Data Base of system requirements within their individual projects.

CONCLUDING REMARKS

In the mid-1990's, the International Space Station will afford experimenters an orbital facility in space for scientific investigations and commercial ventures. The Space Station infrastructure will offer an opportunity for high energy radiation research in polar and equatorial orbits and, in subsequent phases, in geosynchronous orbit. Space Station will provide essential resources, internal pressurized and external unpressurized mounting locations and crew interaction for payload operations, servicing and contingencies. Since the program is beginning the design and development phase, experimenters desiring to utilize this future capability should take appropriate measures outlined in the body of this paper.

SCIENTIFIC CONSIDERATIONS IN THE DESIGN OF THE MARS OBSERVER GAMMA-RAY SPECTROMETER

J. R. Arnold
University of California, La Jolla, CA 92093

W. V. Boynton
University of Arizona, Tucson, AZ 85721

P. Englert
San Jose State University, San Jose, CA 95192-0101

W. C. Feldman
Los Alamos National Laboratory, Los Alamos, NM 87545

A. E. Metzger
Jet Propulsion Laboratory, Pasadena, CA 91109

R. C. Reedy
Los Alamos National Laboratory, Los Alamos, NM 87545

S. W. Squyres
Cornell University, Ithaca, NY 14853

J. I. Trombka
Goddard Space Flight Center, Greenbelt, MD 20771

H. Wanke
Max-Planck-Institut fur Chemie, Mainz, Fed. Rep. Germany

ABSTRACT

Cosmic-ray primary and secondary particles induce characteristic gamma-ray and neutron emissions from condensed bodies in our solar system. These characteristic emissions can be used to obtain qualitative and quantitative elemental analyses of planetary surfaces from orbital altitudes. Remote sensing gamma-ray spectroscopy has been successfully used to obtain elemental composition of the Moon and Mars during United States Apollo 15 and 16 missions and the Soviet Luna and Mars missions. A remote sensing gamma-ray and neutron spectrometer will be included aboard the United States Mars Observer Mission. If proper care is not taken in the design of the spectrometer and choice of materials in the construction of the detector system and spacecraft, the sensitivity of these remote sensing spectrometers can be greatly degraded. A discussion of these design and material selection problems is presented.

© 1989 American Institute of Physics

I. INTRODUCTION

Remote sensing gamma ray spectrometers have been successfully flown on a number of USA and Soviet planetary missions. Gamma-ray spectrometers flown aboard Apollos 15 and 16 were used to obtain maps of the chemical composition of a number of key elements for about 20% of the lunar surface[1,2]. A NaI(Tl) detector was used. One of the instruments scheduled to fly on the Mars Observer mission in 1992 is a gamma ray spectrometer (GRS) utilizing a high purity germanium detector, hp(Ge), and also having the capability to detect low-energy neutrons.

By detecting the gamma rays and neutrons emitted from Mars, the abundances and stratigraphy of a number of elements will be determined along with results related to Martian volatiles (H_2O and CO_2) and climate. Relatively strong gamma ray lines are made by the decay of the naturally radioactive elements (K, Th and U) and by cosmic-ray interactions (mainly the inelastic scatter or capture of neutrons)[3,4]. Many of these gamma ray lines can be used to determine elemental abundances of all major and many minor elements in the top few tens of centimeters of the Martian surface with an areal resolution of the order of the spacecraft altitude, which will be ~360 km. Both the gamma rays and neutrons emitted from Mars will be sensitive indicators of the presence of hydrogen (water) in, or carbon-dioxide frost on, the Martian surface[5,6]. There are a number of sources of background that, if not either minimized or understood, would substantially degrade our ability to determine the elemental composition of a planetary surface from an analysis of the spectrum measured by a remote sensing gamma-ray spectrometer.

II. BACKGROUND COMPONENTS

In order to illustrate the various background components, let us look at the results obtained from the analysis of the Apollo Gamma Ray Spectrometer measurements. The first major background component to be considered is that due to charged particles in the cosmic rays interacting with the gamma-ray detector. Table 1 shows the ratio of the count rates in various energy intervals for gamma-ray interactions to those of gamma rays plus particle interactions. An active charged-particle anti-coincidence shield was used to significantly reduce if not completely eliminate the charged particle background. The basis for operation of such a shield is as follows. The gamma ray detector (high Z material) is surrounded by a plastic scintillator (low Z material). Charged particles will produce detectable interactions in both the plastic scintillator and gamma ray detector while the gamma rays will produce detectable interactions in the gamma ray detector but have a low probability of detection in the plastic scintillator. In this manner, coincidence pulse rejection is used to eliminate charged particle events in the measured gamma-ray pulse height spectrum. Thus, measurements with the shield disabled represents a mixed gamma-ray and charged particle measurement, while measurement with the shield enabled represents the comparatively pure gamma-ray component. As can be seen, the charged particle background dominates the count rate above 2 MeV.

Table 1

The ratio of gamma-ray interactions (anti-coincidence shield enabled) to that of gamma ray plus charged-particle interactions (shield disabled) as a function of energy measured during the Apollo 15 mission in orbit around the moon.

Energy (MeV)	Enabled $cm^{-2}s^{-1}MeV^{-1}$	Disabled $cm^{-2}s^{-1}MeV^{-1}$	Disabled/Enabled
.75	.3	.3	1.0
1.0	.18	.23	1.3
2.0	.08	.13	1.6
5.0	.0083	.06	7.2
10.1	.002	.05	25.
20.0	.00075	.051	68.

To illustrate the importance of the active shield, consider the gamma-ray pulse height spectrum obtained during Apollo 16, shown in Figure 1. The strongest characteristic gamma ray emission lines for such elements as O, Fe, Ti, Al are found at energies above 3 MeV. These lines would be much harder to detect without the use of an active charged-particle anti-coincidence shield.

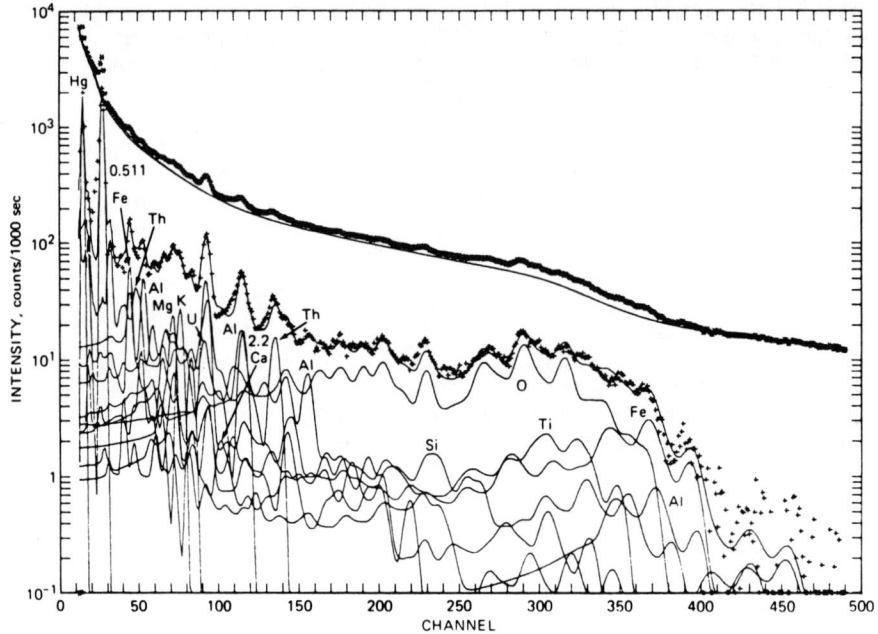

Figure 1

The observed pulse-height spectrum (*, top curve), the spectrum due to the continuum of photons from the Moon (top line), and the net discrete line gamma-ray pulsed height spectrum (+) obtained with the Gamma-Ray Apollo Remote Sensing spectrometer during the Apollo 16 mission and the calculated contributions (solid lines) to this spectrum by individual elelments in the moon. The energy scale is 19.3 keV/channel.

Figure 2 shows the actual gamma-ray pulse height spectrum measured in lunar orbit and the background total. The various major background components are also indicated in Figure 2. The major background components shown are[1,7]:

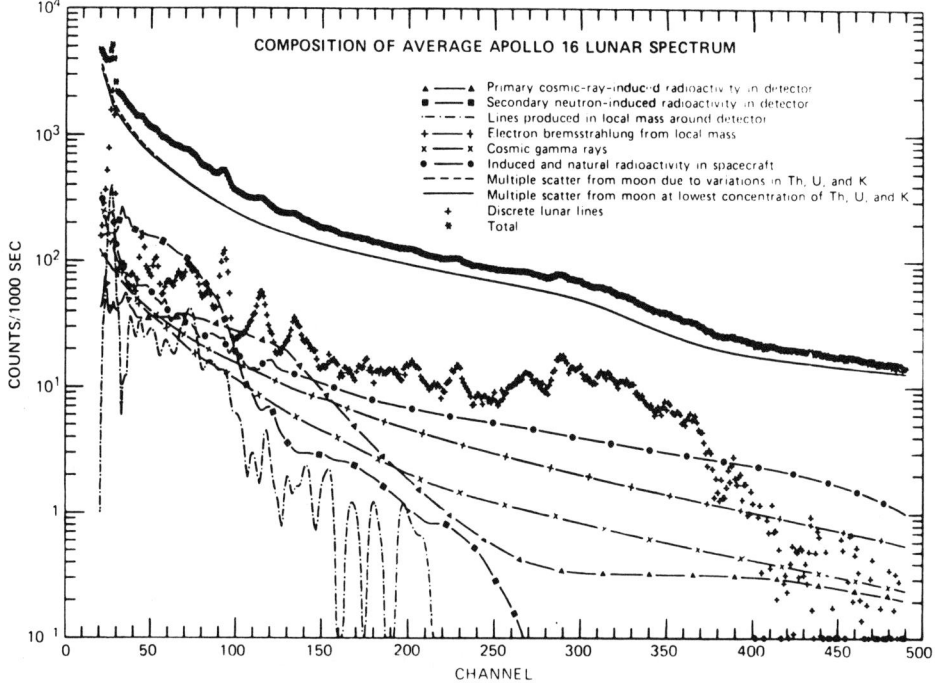

Figure 2

Total gamma-ray emission pulse height spectrum (*, top points), the net lunar gamma-ray spectrum (+), and the components of the background as measured by the Apollo Gamma-Ray Remote Sensing Spectrometer in lunar orbit. The energy scale is 19.3 keV/channel.

1. Gamma-ray continuum from the surface. Since a major portion of this background component can be attributed to multiple scatter in the surface, the shape of the continuum can vary over the planetary surface, due for example to variations in concentration of the naturally radioactive species K, Th, U. This variation was seen quite distinctly on the Moon.

2. Spacecraft gamma-ray emission. These emissions can be attributed to natural radioactivity and to cosmic-ray primary, cosmic-ray secondary, and surface leakage particle induced activation. Prior to the launching of Apollos 15 and 16, detailed surveys of the spacecraft were carried out to determine its radiation cleanliness[8]. Table 2 lists a number of sources located during the survey.

Table 2
Known Radiation on Apollo 15 and 16

Radioisotope	Activity	Identification and Location
Sources Always Present		
Potassium - 40	0.7 microcuries	EPS and ECS Radiators Thermal Paint
Potassium - 40	1.5 microcuries	KOH Electrolyte-Pyro and Re-entry Batteries
Potassium - 40	16 microcuries	KOH Electrolyte-Fuel Cells
Potassium - 40	2.1 microcuries	LM-type Battery in SM
Potassium - 40	0.003 microcuries	Mass Spectrometer Thermal Paint
Thorium - 232	5.8 microcuries	Mapping Camera Lens
Thorium - 232	microcurie range	Guidance System Heat Sinks in CM
Mercury - 203	0.1 microcuries	Gamma Ray Spectrometer
Iron - 55	1.0 microcurie	X-Ray Spectrometer
Polonium - 210	5.0 microcuries	Alpha Particle Spectrometer
Source Jettisoned Shortly After Lift-Off		
Uranium - 238	0.1 curie	Launch Escape System Ballast Plates
Source Jettisoned in Lunar Orbit		
Polonium - 238	1.0 microcurie	Subsatellite Particle Detector (Apollo 15 only)
Sources on LM Descent Stage		
Plutonium - 238	$40(10)^3$ curies	Radioisotope Thermoelectric Generator on LM
Promethium - 147	200 millicuries	Landing Point Designator Paint-LM
Potassium - 40	9.2 microcuries	Five Batteries

Sources on LM Ascent Stage

Tritium	14.7 curies	Portable Life Support System
Promethium - 147	21.3 curies	Radioluminescent Discs in Lunar Module
Promethium - 147	0.2 curies	Self-Luminious Switch Tips in LM
Potassium - 40	3.7 microcuries	Two Batteries

In order to reduce the background due to gamma-ray emission from the spacecraft, the Apollo gamma-ray detector systems were placed on booms and extended away from the spacecraft. The criteria used to determine boom length (8m) on the Apollo spacecraft was to extend the detector away from the spacecraft such that the solid angle subtended by the boom and the spacecraft be at most one tenth of that subtended by the surface of the moon (at the detector). The spacecraft contribution due to cosmic-ray interaction and natural radioactivity for the lunar case was determined by measuring the gamma-ray emission as a function of distance, at intermediate and full extension of the boom during trans-lunar flight. These measurements were repeated in lunar orbit and the effects of the particle leakage from the lunar surface was thus determined.

3. Emissions from materials surrounding the detector. Contributions from the natural radioactivity and induced activity from local mass were inferred in the case of the Apollo measurements as a residual after all other components were subtracted from the pulse height spectrum . Great care was taken in constructing the Apollo gamma-ray spectrometer to reduce both naturally radioactive materials such as K-40 in the glass of phototubes and other materials that might significantly interfere with the measurement of the characteristic surface gamma-ray emission spectrum. A more comprehensive program of control has been developed for the Mars Observer Gamma Ray Spectrometer and will be discussed later.

4. Gamma-ray emission from the gamma-ray detector. Detector activation due to cosmic-ray primary and secondary and surface leakage particles can produce both continuous and discrete-line features in the measured pulse height spectra. The shapes of the background components are time dependent. This is due to the fact that the exciting flux changes with time

(e.g., trans-Mars, Mars orbit, solar flares) and the important activated species have ranges of half-lives from fractions of seconds to many years. Interaction cross-sections and detector response functions must be determined for all the component nuclear species produced must be determined, and calculational methods for predicting these backgrounds as a function of time and nature of the incident flux must be developed. Such a calculation system was developed for the Apollo program[9] and is now being extended for application to the Mars Observer Gamma Ray Spectrometer program. Accelerator experiments in both Europe and the United States have been performed to study activation in detector systems using such detector materials as NaI, CsI, hpGe, and BGO. Further studies will be carried out by the Mars Observer Gamma Ray Spectrometer Flight Investigation Team at accelerators in France and Switzerland. Emphasis will be placed on the studies of high purity Ge and BGO. Spaceflight experiments carried out on Apollo 17 and Apollo-Soyuz were used to verify the calculation system. Results of the expected response of a NaI detector in earth orbit, and the activation due to cosmic rays and the time dependence due to passing through the trapped radiation belts (i.e., the South American Anomaly)[9] are shown in Figure 3.

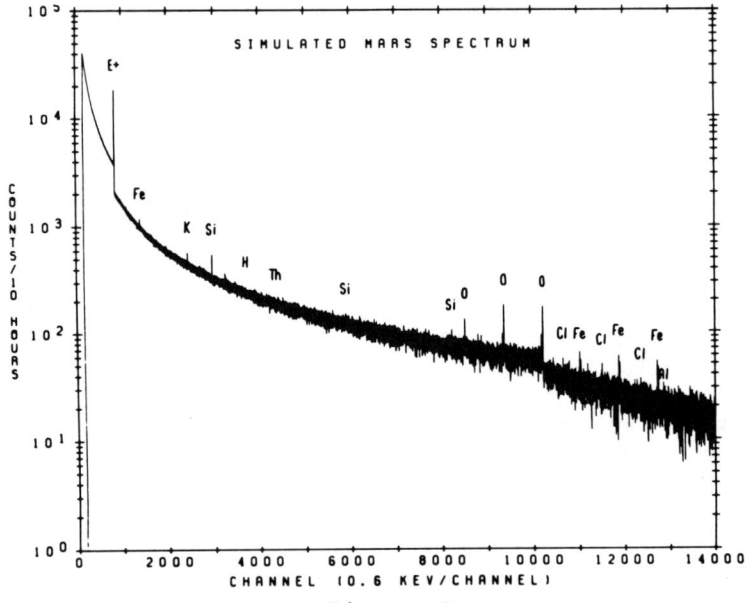

Figure 3

Time dependent expected response of a Na(Tl) detector in Earth orbit due to activation by cosmic ray primaries and secondaries and to trapped radiation.

5) The cosmic gamma ray background. This component is constant in time and small relative to gamma ray emission from a planetary surface. Detailed measurements of this component were carried out during the Ranger 3 and Apollo 15 and 16 missions[7,10]. The origin of this emission is still unknown. Subsequent balloon and satellite flights have confirmed the Apollo results[11].

6) Jovian electron bremsstrahlung emission. High-energy electrons interacting in the materials surrounding the detector produce a bremsstrahlung X-ray/gamma-ray spectrum. This is a continuum and is seen at all energies of interest in remote gamma ray spectrometers. This flux varies with time and space (i.e., in the solar system) and therefore becomes rather difficult to predict unless simultaneous electron measurements are made. This effect was first noticed as a difference in magnitude between the total lunar spectra measured for Apollo 15 and Apollo 16. This difference was traced to the change in magnitude of the high-energy electron flux near the Moon, which is believed to be of Jovian origin[7].

III. BACKGROUND REDUCTION PROGRAM FOR THE MARS OBSERVER GAMM RAY SPECTROMETER (GRS)

A more detailed consideration of the program for the background reduction for the Mars Observer Gamma Ray Spectrometer will be given below. All instruments on the Mars Observer spacecraft are subject to constraints in regard to money, mass, power, and data rates, thereby restricting their designs. However, an equally important constraint for the Mars Observer GRS is that the spacecraft, other instruments, and the GRS itself cause minimal interfere with the planetary gamma rays and neutrons of interest. It must be remembered that the Apollo results were obtained with a NaI(Tl) detector while the Mars Observer utilizes a hpGe detector. The Mars emission spectrum expected with a hpGe detector is shown in Figure 4[12]. The improved spectral resolution should allow us to obtain elemental analysis for more elements as compared with the Apollo system. This in a sense then further complicates the material control problem since we hope to be able to measure the abundances of many more elements. Secondly, the Mars Observer GRS will be used to obtain global elemental composition maps, thus spectra must be determined over small spatial surface elements. From the experience obtained on Apollo, it is predicted that about one half hour of data accumulation will be required over a particular surface element in order to obtain enough statistical accuracy to perform detailed elemental analyses. If one looks at orbital characteristics and considers elements near the Martian equator, this will require summing up data over about a Martian year (about two Earth years). Over this period of time, the background contribution will change. Furthermore, the Martian particle leakage spectrum will vary as a function of spatial position in Mars orbit because of chemical variation in surface composition. This is particularly affected by major changes in the hydrogen content because of soil-to-ice compositional changes. Thus it is important to minimize the background contributions where possible and be able to predict the background variation with time and orbital position.

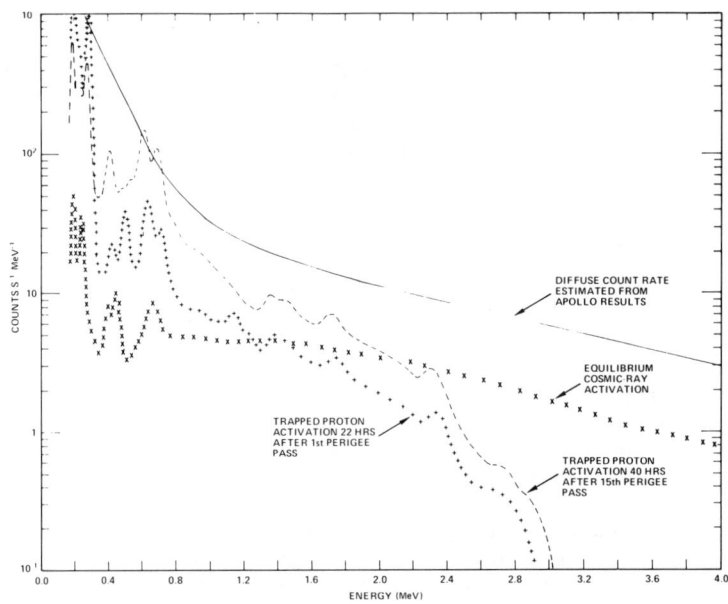

Figure 4

Predicted Mars surface emission spectrum using a hpGe detector of the size being considered for the Mars Observer GRS system[12].

With these factors in mind we look at the Mars Observer GRS design and the plans for controlling and for understanding and predicting the various background components.

A radiation survey of the Mars Observer spacecraft will be conducted to minimize unexpected natural and artificial radioactivities, and limits of radiations from the spacecraft that reach the GRS have been included in the mission's science requirements document. It has been requested that certain elements that are excited by thermal neutrons, such as gadolinium

and samarium, not be used if possible and that an inventory of
such materials and their amounts and positions in the spacecraft
be reported.

The GRS will be mounted on a six meter boom to reduce backgrounds
from the spacecraft and other instruments. The detector is
separated from the major electronics components. The electronic
components are packaged in the Central Electronic Assembly (CEA).
By mounting the CEA on the spacecraft the background from local
matter can be further reduced.

A cross section of the Mars Observer GRS detector design is shown
in Figure 5. The gamma ray detector will be hpGe cooled by a
passive radiator and surrounded by a plastic anticoincidence
shield to reject pulses produced in the hpGe by charged cosmic-ray
particles. By including several weight percent of boron in the
plastic scintillator, the anticoincidence shield can also be used
to detect low energy (thermal and epithermal) neutrons[13]. The
orbital velocity of the Mars Observer spacecraft, 3.4 km/s, is
faster than that of a thermal neutron, so the neutron counts in
the front, back, and side faces of the GRS anticoincidence shield
can be used as a Doppler filter to distinguish between thermal and
epithermal neutrons from Mars[13,14]. Energetic (~1 MeV) neutrons
can be detected by the irregular-shaped Ge inelastic scattering
peaks (which have long tails at energies higher than the inelastic
gamma ray energy) that they produce in the spectra[15].

While the Mars Observer GRS Flight Investigation Team will have
little control over the materials used in the spacecraft and other
instruments, strong restraints have been imposed on the GRS
instrument contractor relative to the materials that may be used
in construction. Testing for unintentional radioactivities and
elemental contamination in all materials used in the GRS
construction also is being carried out. An example of this
control program is presented to illustrate the approach. While
beryllium would be a good choice for building the instrument
(light and very few cosmic-ray induced gamma rays), the risk of
radioactive contaminations in most commercially available
materials, which was observed in samples of beryllium, was
initially found to be too great to permit their use. Graphite
epoxy is being seriously considered for some parts of the GRS
structure, but we discovered an unacceptably high concentration of
chlorine in epoxy samples. Chlorine may be an important element
in Martian duricrust, and chlorine also produces background lines
in important spectral regions. Control was then imposed on the
graphite epoxy materials and chlorine-free materials are now
available. Similar efforts for obtaining Be with low activity
have been implemented, and it is believed that some materials may
soon be commercially available.

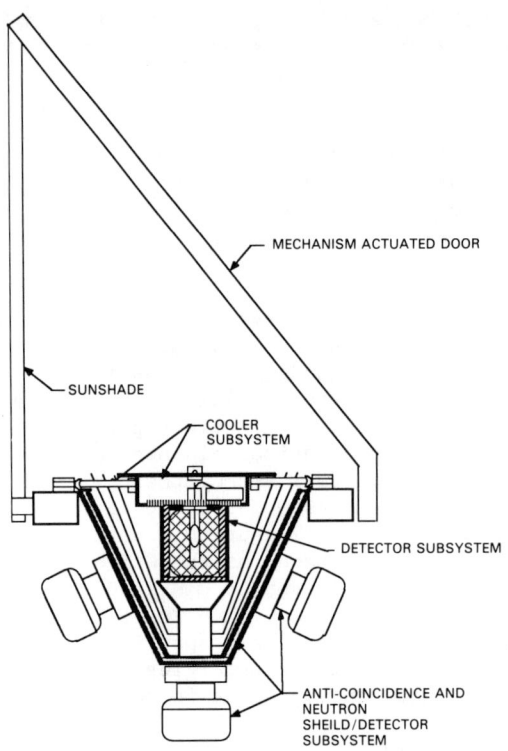

Figure 5

Cross section view of the preliminary design of the Mars Observer GRS.

An attempt has been made to limit in the GRS the amounts of almost all elements of interest in the Martian surface to levels that produce less background than that coming from the spacecraft or from the Martian surface. In order to quantify this effect of various chemical elements' gamma-ray emission as a function of position relative to the GRS detector, tables of acceptable mass divided by the square of the distance of that material from the hpGe detector have been compiled for several key elements, (Table 3). These limits were derived by calculating the expected gamma-ray flux at the detector for a number of important geochemical components in the Martian surface and comparing these fluxes with those expected from detector or spacecraft construction materials.

Table 3, Elements of Scientific Interest Permitted Near Detector

The following table gives the amount of each element of scientific interest that may be used in the vicinity of the detector without significant impact to the science. The limit is expressed as a function of mass of the element (in grams) divided by the square of the distance (in centimeters) from the element to the center of the GRS gamma-ray detector.

ELEMENT	m/r^2 LIMIT (g/cm^2)	ELEMENT	m/r^2 LIMIT (g/cm^2)
C	19.4	Mn	0.01
N	0.21	Fe	0.08
Na	0.09	Ni	0.0025
Mg	0.05	Cd	3.2 E-5
Al	0.61	La	0.00016
Si	0.31	Sm	3.0 E-5
S	0.09	Gd	2.3 E-6
Cl	0.003	Lu	7.0 E-7
K	0.0008	Th	5.0 E-7
Ca	0.06	U	3.0 E-8
Cr	0.014		

These tables indicate which materials in the GRS would contribute most significantly to our backgrounds. One of the most important scientific considerations to date was the decision whether to choose aluminum or titanium for the the Ge crystal housing and for other structural materials in the GRS instrument. It was decided that aluminum was a more desirable element to detect in the Martian surface than was titanium. Aluminum is a major rock-forming element whose abundance must be known to calculate absolute abundances of other elements. Moreover, it is a diagnostic element for the presence of aluminosilicates, such as feldspars, that could be important components of the ancient Martian cratered highlands. Thus it was decided that the GRS instrument would be built with titanium and graphite epoxy instead of aluminum .

Finally, accelerator tests are being planned to study detector and material activation problems. Studies using accelerators with protons of both ~GeV and 600 MeV energies are planned for the next few years. Further programs for predicting the induced activity in hpGe are under development. Many of these programs are extensions of those programs that were developed for the analysis of the Apollo remote sensing gamma-ray spectrometer data[16].

ACKNOWLEDGEMENTS

The majority of this work was supported by the Solar System Exploration Division of NASA Headquarters under RTOP's 838-59-50 and 157-03-50. The activities at the Los Alamos National Laboratory was performed under the auspices of the U.S. Department of Energy. L. G. Evans of CSC and D. M. Drake of LANL made major contributions to this work.

Bibliography

1. M.J. Bielefeld, R.C. Reedy, A.E. Metzger, J.I. Trombka, and J.R. Arnold, Proceedings of the 7th Lunar Science Conference (Pergamon, N.Y., 1976), p. 2661.

2. M.I. Etchegaray-Ramirez, A.E. Metzger, E.L. Haines, and B.R. Hawke, J. Geophys. Res. 88, A529 (1983).

3. R.C. Reedy, J.R. Arnold, and J.I. Trombka, J. Geophys. Res. 78, 5847 (1973).

4. R.C. Reedy, Proceedings of the 9th Lunar and Planetary Science Conference (Pergamon, N.Y., 1978), p. 2961.

5. L.G. Evans, and S.W. Squyres, J. Geophys. Res., 92, 9153, (1987).

6. D.M. Drake, W.C. Feldman, and B.M. Jakosky, J. Geophys. Res., 93, 6353, (1988).

7. J.I. Trombka, C.S. Dyer, L.G. Evans, M.J. Bielefeld, S.M. Seltzer, and A.E. Metzger, Ap.J., 212, 925, (1977).

8. A.E. Metzger, and J.I. Trombka, Proc. of the Nat. Symp. on National and Manmade Radiation in Space, E.A. Warman (Ed.), NASA TM X-2440, p. 714, (1972).

9. C.S. Dyer, J.I. Trombka, S.M. Seltzer, and L.G. Evans, Nucl. Instr. and Meth., 173, 585, (1980).

10. A.E. Metzger, M.A. van Dilla, E.C. Anderson, and J.R. Arnold, Nucleonics, 20, No. 10, 64, (1962).

11. C.E. Fichtel, and J.I. Trombka, Gamma Ray Astrophysics: New Insight Into the Universe, NASA SP-453, (1981).

12. L.G. Evans private communication.

13. D.M. Drake, W.C. Feldman, and C. Harlbut, Nucl. Instr. and Meth, A247, 576, (1986).

14. W.C. Feldman, and D.M. Drake, Nucl. Instr and Meth., 1986, A245, 182, (1986).

15. J. Bruckner, H. Wanke, and R.C. Reedy, J. Geophys. Res., 92, E603, (1987).

16. J.I. Trombka and C.E. Fichtel, Physics Reports, 97, No. 4, (1987).

PARTICLE BACKGROUND EFFECTS FOR HUBBLE SPACE TELESCOPE (HST) AND THE LYMAN FAR ULTRAVIOLET SPECTROSCOPIC EXPLORER

B. E. Woodgate and W. B. Fowler NASA/Goddard Space Flight Center, Greenbelt, MD 20771

ABSTRACT

Requirements on orbital background counting rates are described for the UV and visible detectors on the planned second generation instrument Space Telescope Imaging Spectrograph (STIS) and for the proposed Lyman Far Ultraviolet Spectroscopic Explorer.

Expected background rates and shielding requirements are analyzed for detector types in both instruments for low earth orbits, and for high orbits in the case of Lyman.

Requirements

The HST is currently planned to be launched in 1989 into a 560 km high 28.5^0 inclination orbit. A second generation instrument, STIS, is being designed to be inserted in orbit into the spacecraft several years later. It will contain photon-counting array detectors for use in the UV and Charge Coupled Device (CCD) arrays for use in the visible and near-IR.

The Lyman Far Ultraviolet Spectroscopic Explorer is a mission currently undergoing a Phase A Study by the European Space Agency, and it has been accepted (January 1988) for a Phase A Study by NASA under the Explorer program. It is expected to contain photon-counting array detectors for the UV and either photon-counting or CCD arrays for star tracker detectors. The orbits being considered are either a low orbit similar to the HST, or a highly elliptical 48 hour period orbit with 120,000 km apogee by 1000 km perigee. We will refer to the latter as the Lyman elliptical orbit.

For both STIS and Lyman, many of the key scientific observations will be accomplished by long exposures, up to 10 hours. For photon-limited sensitivity at all brightness levels, we require that the background be negligible.

A goal of 1 count pixel^{-1} hour^{-1} would accomplish this for detectors and types of background event where one event is equivalent to one detected photon, such as photon-counters for all events, and CCDs with single photon events, such as phosphorescence.

However, for galactic cosmic rays passing through CCDs, about 1000 electrons will be released by a minimum ionizing particle in a 7 micron layer of silicon typical of a thinned CCD, rendering each pixel affected unuseable for the exposure. For readout noise levels near or below 10 electrons rms, these events frequently affect multiple pixels. If the probability of a pixel being affected during the exposure time is low, then the events can be voted out using multiple exposures, as is done for ground-based astronomical observations. With this procedure in mind, we select a background event rate goal of 10^{-2} events pixel^{-1} hour^{-1} for cosmic rays through CCDs.

Direct Energetic Particle Background

We first consider the prompt effect of particles outside of the trapped particle belts, where astronomical observations will occur. The dominant penetrating particles are galactic cosmic ray protons, and their energy spectrum[1] is shown in Figure 1. The curve labelled 'free space protons' is applicable to the Lyman elliptical orbit. The curve for a 463 km circular orbit at 32° inclination to the equator may be used as an approximation for the HST orbit, bearing in mind that the fluxes are variable, for example with solar activity.

Above the belts with no geomagnetic shielding, the peak proton rate is 2×10^{-3} cm^{-2}s^{-1} MeV^{-1}, with a mean energy of 500 MeV and a FWHM about 10^3 MeV, so that the total rate on a detector surface is 2 cm^{-2}s^{-1}, for Lyman.

Below the belts, with geomagnetic shielding, the peak proton rate is 10^{-5} cm^{-2}s^{-1} MeV^{-1}, with a mean energy of 8×10^3 MeV and a FWHM about 10^4 MeV, and a total rate on a detector surface of about 0.1 cm^{-2}s^{-1}, for HST. Alternatively, this rate may be read from Figure 2 for low earth orbit[2]. The proton rate for protons at 28.5° is about 0.3 cm^{-2}s^{-1}, depending on solar activity. We adopt a nomimal value of 0.2 cm^{-2}s^{-1} for subsequent calculations. For a photon-counting array detector with 25 micron square pixels, this results in a rate of 4×10^{-3} pixel^{-1} hour^{-1}, a factor 250 better than the count rate goal for photon counters.

Even for Lyman, without geomagnetic shielding, the background rate is 4×10^{-2} pixel^{-1} hour^{-1}, a factor 25 better than required.

A further large background reduction factor occurs for the Multi-Anode Microchannel Array (MAMA)[3] photon-counting detector, which is selected for STIS and is a candidate for Lyman. The MAMA logic will reject oblique events which excite more than four pixels, or eight microchannel pores, simultaneously within the few hundred nanosecond time resolution. For pores with a length to diameter ratio about 120, particles will be rejected and not counted, except those arriving within 4° of the axis of the pores, so that an angular anti-coincidence factor of approximately 1000 will result for isotropic particles.

For the CCDs the situation is much less favorable. As integrating detectors there is no anti-coincidence rejection. The proton rate of 4×10^{-3} pixel^{-1} hour^{-1} calculated above for HST, is only a factor 2.5 better than our requirement. This is just acceptable and means that for HST, thinned CCDs are useable for long exposures provided that multiple exposures and voting in the data analysis is used.

However, above the trapped radiation belts, as for the Lyman elliptical orbit one in every 25 pixels will contain a hit of about 1000 electrons in a one hour exposure, which will cause a large number of coincident hits even with multiple exposures, so that a CCD is unsuitable for long exposures outside the geomagnetic shielding. The assumption of a 7 micron thickness was taken as typical of a thinned CCD. For a thick CCD, the amplitude of the deposited change will be much higher, and the number of neighboring pixels affected would be much higher.

Effects of Trapped Particles

First, we assume the detectors and windows will be shielded from trapped electrons by 8 to 10 mm thick aluminum. Figure 3, taken from Viehmann, et. al.[4], shows that this shielding will reduce the trapped electron flux well below the trapped proton flux level. It will result in a proton spectrum at the detector peaking in the 50-100 MeV range with a flux about 2×10^3 $cm^{-2}s^{-1}$ maximum in the South Atlantic Anomaly (SAA) at the HST orbit height. This is a factor 10^4 higher than the geomagnetically shielded cosmic ray rate outside the SAA for the HST orbit. If we assume isotropic particles, this would result in a background rate of 4×10^{-2} pixel^{-1} hour^{-1} for the MAMA photon counters from direct protons if we include the angular anti-coincidence rejection factor of 1000. It may, therefore, be well worthwhile to try to operate these detectors and observe throughout the SAA passage. However, it will be necessary to check the effects of Cerenkov radiation and fluorescence.

For the Lyman elliptical orbit, according to models by Daly and Tranquille[5], the spacecraft will go through a dense region of the trapped particles, with a proton flux above 15 MeV of about 2×10^4 protons $cm^{-2}s^{-1}$. Then the peak direct proton background rate for a MAMA detector would be 0.4 pixel^{-1} hour^{-1}, or an average of 0.06 pixel^{-1} hour^{-1} if the passage through the belt peak takes 10 minutes.

For the CCDs, the charge deposited by the trapped protons would be very large, particularly since the 50-100 MeV protons would deposit about ten times more charge into a given thickness of silicon than a minimum ionizing particle, and virtually all pixels would be affected. Therefore, a CCD must be rendered inactive during trapped radiation belt passage, and cleared thoroughly after exit before beginning exposures.

Cerenkov Radiation in the Trapped Particle Belts

We now estimate the effect of Cerenkov radiation in detector windows in the trapped particle belts. Discussions of window effects, such as Cerenkov, fluorescence and phosphorescence, will not affect Lyman, since it will not use windows. Figure 4, developed from shielding studies by SAI[6,7] and particle radiation levels from Stassinopoulos[8], shows the spectrum of protons at the peak of the SAA for the HST orbit, with and without shielding. The proton flux above the energy threshold which can produce Cerenkov radiation in LiF or MgF_2 windows is approximately 200 $cm^{-2}s^{-1}$. Figure 5, from Viehmann and Eubanks[9] shows the number of protons produced per energetic proton as a function of long wavelength cut-off, for both Cerenkov and fluorescence.

For a MAMA detector with a CsI photocathode and a MgF_2 window 3mm thick, each proton will produce about 200 Cerenkov photons in the accepted wavelength range, radiating into the Cerenkov cone angle. The photon angles will be randomly distributed around the rim of the cone for a given energy and refractive index. Over the wavelength range 1150-1700Å covered by CsI the refractive index of MgF_2 varies from 2 to 1.5. For a proton energy of 300 MeV the cone half-angle will vary from zero to 45°, according to Jelley[10], so that the photon angular distribution from a single proton fills the the 45° cone, over about 2 steradians.

The 2048x2048 pixel MAMA detector is divided into four independently functioning quadrants. An independent detector quadrant subtends 0.1 steradians at the window, so that 1/20 of the photons will enter the detector, or on an average three detected photons per event for a cathode efficiency of 0.3. The probability of only one photon being detected is about 0.13. Other events will be rejected by mutual anti-coincidence. For a window area of 70 cm^2, and a detector quadrant area of 6 cm^2, the Cerenkov flux at SAA peak for STIS is 400 $cm^{-2}s^{-1}$ or .003 counts $pixel^{-1}s^{-1}$. For an SAA passage lasting about 10 minutes, about 2 counts $pixel^{-1}$ would be accumulated. The rate for a Cs_2Te photocathode is very similar since it has about a third of the quantum efficiency of CsI, but the number of Cerenkov photons produced per proton is approximately three times higher within its larger wavelength range. Rates in the CCD bandpasses are irrelevant since CCDs are unuseable in the SAA, as shown above.

Fluorescence in the Trapped Particle Belts

The next effect investigated is the fluorescence induced in detector windows during passage through the trapped particle belts. Here, the entire range of proton energies contributes to the window fluorescence, so the shielding around the window and detector is important. Figure 4 shows the effect on the proton spectrum of the shielding used on the Goddard High Resolution Spectrograph (GHRS), a first-generation instrument on the HST, and we assume a similar spectrum for STIS. Then the input peak proton flux in the SAA after shielding is 3000 $cm^{-2}s^{-1}$.

From Figure 5, for a CsI photocathode near a MgF_2 window 3 mm thick, each proton will produce about 30 fluorescent photons in the wavelength range, emitted into 4π steradians. Smith, Becher, Fowler and Flemming[11,12], find experimentally an average fluorescent yield of 2.5 counts/40 MeV proton with a 10% Q.E. CsTe photocathode on a 2mm MgF_2 window (into 2π steradians). This corresponds to 40 photons/proton into 2π steradians for a 3mm window for a 3000Å cut-off, in agreement with Figure 5. For the STIS MAMA detectors, with each independent quadrant of the detector subtending about 0.1 steradian, and with a quantum efficiency of 0.3, on average only one fluorescent photon will be detected per quadrant for every 12 incident protons, so that anti-coincidence rejection between fluorescent photons will be negligible. In a minority of cases, anti-coincidence rejection will occur between the photon passing through the detector and a fluorescent photon. For a window area of 70 cm^2, and a detector quadrant area of 6 cm^2, the detected fluorescent flux at SAA peak for STIS is 3000 cm^{-2}s^{-1} or 0.02 counts pixel^{-1}s^{-1}. For an SAA passage lasting about 10 minutes, about 12 counts pixel^{-1} would be accumulated. This is higher than our limit for observing faint sources, but useful observations of brighter objects could be made through the SAA. Using a windowless detector would eliminate this problem.

For the STIS Cs$_2$Te detector, in addition to the window from which the detector quadrant subtends 0.1 steradians, another window may hold the cathode which, therefore, subtends 2π steradians at the detector. In this case any fluorescent photons will occur in anti-coincidence with the proton through the detector, and not be counted. These fluorescence photons would also further suppress counts from protons passing normal to the microchannel plate face. With a 0.1 quantum efficiency cathode, and 45 fluorescent photons per photon emitted into 2π, an average of 4.5 photons would interact. The probability of zero photons interacting is about 1%, so that the detected proton rate would be decreased by a further factor 100.

Phosphorescence

We now estimate the effect of detector window residual phosphorescence after emergence from the trapped particle belts. Measurements have been made[4,13] on phosphorescence induced by particles, for several samples of MgF$_2$ and LiF and the results are highly variable from sample to sample. Testing and selection on each window for fluorescence and phosphorescence would be necessary if the expected levels are a concern.

For a good quality sample of MgF$_2$ viewed by a Cs$_2$Te photocathode, Viehmann found ratios of phosphorescence to fluorescence of 1.2×10^{-4}, 4×10^{-5} and 2.2×10^{-5} for 1 min., 10 min. and 90 min. respectively after particle irradiation. It will generally take about 10 minutes to exit from the SAA maximum to the edge of the belt, so for observations beginning outside the belt, using our fluorescence calculations above, the maximum counting rate from phosphorescence is 8×10^{-7} counts pixel^{-1}s^{-1}, or 3×10^{-3} counts pixel^{-1} hour^{-1}, which is negligible.

After emergence from the particle belts, with appropriate frame readouts to remove any residual image, the CCDs will be useable. From Fig. 5, extrapolating the fluorescence yield curve to 10,000Å, and using a quantum efficiency of 0.7, the detected phosphorescence is a factor 50 more than in the UV, but will still only accumulate about 0.15 counts pixel^{-1} hour^{-1}. This is small, but care should be taken to avoid orders of magnitude increases by incautious selection of window material.

Fluorescence Outside the Trapped Particle Belts

Since we find that for a windowed detector, the fluorescence dominates over the direct proton background inside the trapped particle belts, we should review the situation for the cosmic rays outside the trapped particle belts. The cosmic rays will similarly cause fluorescence in a window, so we may use a similar ratio of fluorescence to direct protons, since the geometric situation at the detector is similar. Then the counting rate is 8×10^{-3} pixel^{-1} hour^{-1}; for the photon counter, and 0.4 pixel^{-1} hour^{-1} factor for the CCD at the HST orbit. For the CCD with a window in front in the elliptical Lyman orbit, the fluorescent background would be 4 pixel hour^{-1}.

Cerenkov Radiation Outside the Trapped Particle Belts

Outside the trapped particle belts, each cosmic ray is sufficiently energetic to produce Cerenkov radiation in a detector window. With a typical energy of 500 MeV at the HST orbit, the Cerenkov cone half-angle will be in the range 40° to 60°, so that the photons from a single proton interaction will be distributed over an annulus subtending 2 steradians, similar to the trapped proton case. For the MAMA photon counter with its anti-coincidence, the probability of only one photon being detected is again about 0.13. With a cosmic ray proton flux of 0.2 cm^{-2}s^{-1} at the HST orbit, the Cerenkov background rate is 5×10^{-3} counts pixel^{-1} hour^{-1} for the MAMA photon counter.

For the CCDs on HST, anticoincidence rejection will not occur between the detected photons of the Cerenkov shower. For the worst case for Cerenkov background, the CCD could have a 3mm thick MgF$_2$ window and be sensitive to the UV range with a quantum efficiency of 0.7. From Figure 5, each proton would produce about 600 Cerenkov photons. With a detector geometry similar to the photon counter, with a 6 cm^2 detector quadrant subtending 0.1 sterandians at a 70 cm^2 window, the average detected Cerenkov flux is 0.18 photoelectrons pixel^{-1} hour^{-1}.

Conclusions

The numerical results are summarized in the Table. We assume approximately 10 mm of aluminum shielding to exclude electrons.

Outside the Trapped Particle Belts

Photon counters have negligible backgrounds from direct protons, fluorescence, Cerenkov and phosphorescence.

CCDs have a significant background from direct protons in low orbit, which can be vetoed by using multiple exposures. In high orbit, above the geomagnetic shielding, the background rate would limit their use to short exposures of a few minutes. This would still allow the use of thinned CCDs for star tracking in high orbit, provided care was taken not to mistake a cosmic ray hit for the star being tracked.

Inside the Trapped Particle Belts

CCDs will be unuseable for astronomy in the particle belts.

Photon counters (not including those with phosphors and CCDs for readout, which have not been evaluated), may have a detectable but low accumulated background through the particle belt passage, due to fluorescence in a window, or due to direct protons if no anti-coincidence is used. They would still be useful for observations of all but the faintest objects or spectra.

A windowless photon counter would not be subject to fluorescence, Cerenkov or phosphorescence backgrounds.

A windowless MAMA detector contains angular anti-coincidence in its readout logic against direct protons, and should have negligible background throughout particle belt passage.

Table I. Summary of Background Results

Results are expressed in counts pixel^{-1} hour^{-1} for 25x25μ pixels.
All effects are due to protons. Electrons are assumed shielded out.

	Low orbit (HST) 560 km circular			Elliptical Orbit (Lyman/FUSE) 120,000 x 1000 km		
	UV MAMA Photon-Counter 3mm MgF$_2$ window		CCD 3mm MgF$_2$ window	UV MAMA Photon-Counter Windowless		CCD 3mm MgF$_2$ window
OUTSIDE PARTICLE BELTS						
Direct Protons	4x10^{-6}		4x10^{-3} (Note B)	4x10^{-5}		4x10^{-2} (Note B)
Cerenkov	5x10^{-3}		0.2	0		2
Fluorescence	8x10^{-3}		0.4	0		4
Phosphorescence	3x10^{-3}		0.15	0		1
WITHIN PARTICLE BELTS						
Direct Protons (Note A)	7x10^{-3}		--	0.06		--
Cerenkov	2		--	0		--
Fluorescence	12		--	0		--

A. Background is accumulated over 10 minutes within the hour, representing a particle belt passage.

B. Each event will produce about 1000 electrons charge in a thinned (7μ) CCD.

REFERENCES

1. M. O. Burrell and J. J. Wright, The Estimation of Galactic Cosmic Ray Penetration Dose Rates, NASA TN D-6600 (1972).
2. J. W. Haffner, Radiation and Shielding in Space (Academic Press, N.Y., 1967), p. 289.
3. J. G. Timothy, Opt. Eng. $\underline{24}$, 1066, 1985.
4. W. Viehmann, A. G. Eubanks, G. F. Pieper, and J. H. Bredekamp, Appl. Opt. $\underline{14}$, 2112, 1975.
5. E. J. Daly and C. Tranquille, Lyman Radiation Environments, European Space Technology Center Document WMA/EJD/LYM/5, 1987.
6. T. W. Armstrong, B. J. Colborn, C. W. Hill, J. A. Nabor, and D. C. Shreve, Study of Particle Radiation and Shielding for the High Resolution Spectrograph Aboard the Space Telescope, Phase I Report SAI-134-79-933 LJ, 1979.
7. C. W. Hill and T. W. Armstrong, Study of Particle Radiation and Shielding for the High Resolution Spectrograph Aboard the Space Telescope, Phase II and Final Report SAI 134-81-144 LJ, 1980.
8. E. G. Stassinopoulos, The ST Environment: Expected Radiation Levels, NASA/GSFC Report X-601-78-30, 1978.
9. W. Viehmann and A. G. Eubanks, Noise Limitations of Multiplier Phototubes in the Radiation Environment of Space, NASA TN D-8147.
10. J. V. Jelley, Cerenkov Radiation (Pergammon Press, N.Y., 1958), p. 281.
11. L. C. Smith, J. Becher, W. B. Fowler, K. Flemming, SPIE $\underline{279}$, 156 (1981).
12. L. C. Smith, J. Becher, W. B. Fowler, K. Flemming, NASA Technical Paper 1852 (1981).
13. W. Viehmann, private communication, (1980).

FIGURE CAPTIONS

Figure 1 - Low orbit differential energy spectra for cosmic ray protons.

Figure 2 - Galactic flux rate as a function of geomagnetic latitude.

Figure 3 - Calculated electron and proton-induced count rates for peak flux densities at 550 km altitude as a function of effective shielding, normalized to 1 mm thick window thickness and fluorescence efficiency of one count MeV^{-1} 2πsr^{-1}. Inset: Relative variation of peak flux densities of electrons (E>0.5 MeV) and protons (E>5 MeV) at the center of the SAA with orbital altitude.

Figure 4 - Spectrum of protons at the peak of the SAA for the approximate HST orbit with and without shielding. Also included is the calculated flux at the digicon windows of the Goddard High Resolution Spectrograph (GHRS).

Figure 5 - Number of Cerenkov photons and fluorescence photons produced by a relativistic electron or proton in 1 mm thick windows of various materials as a function of photocathode cutoff wavelength.

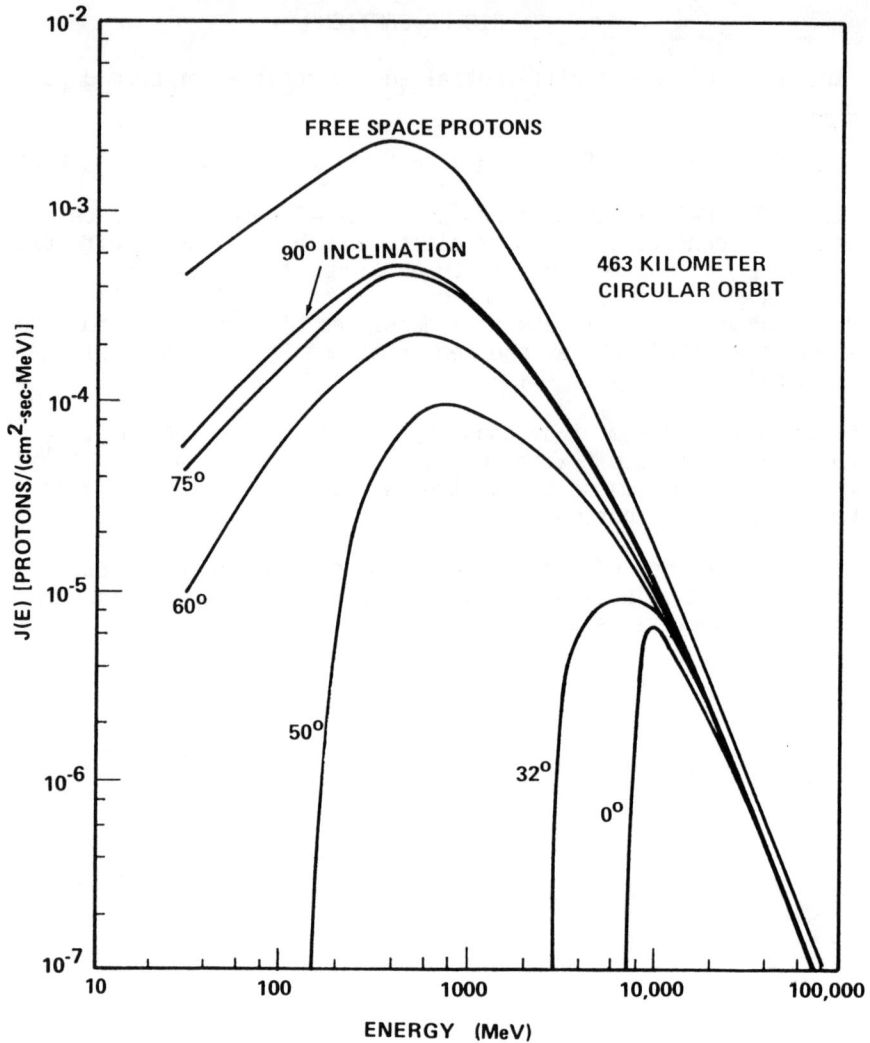

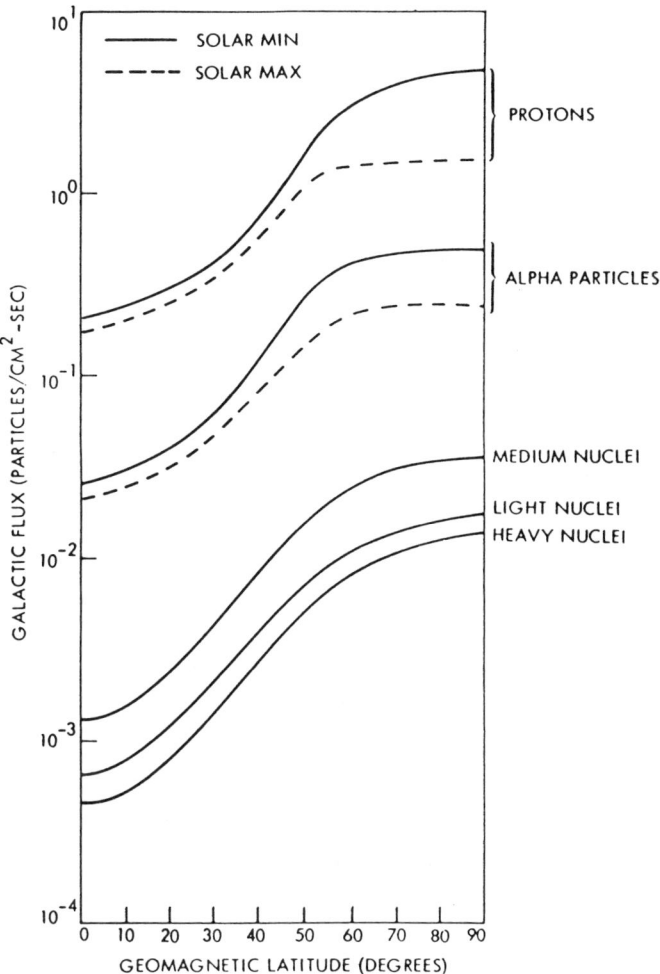

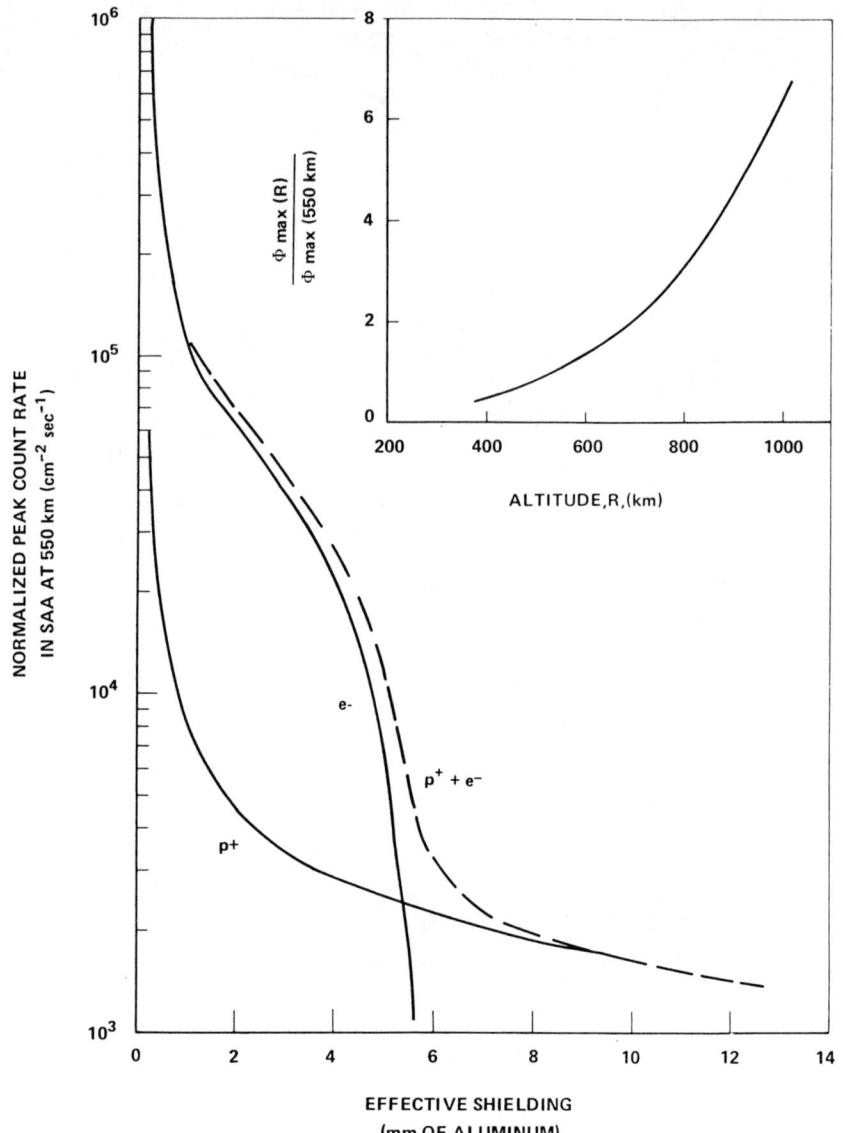

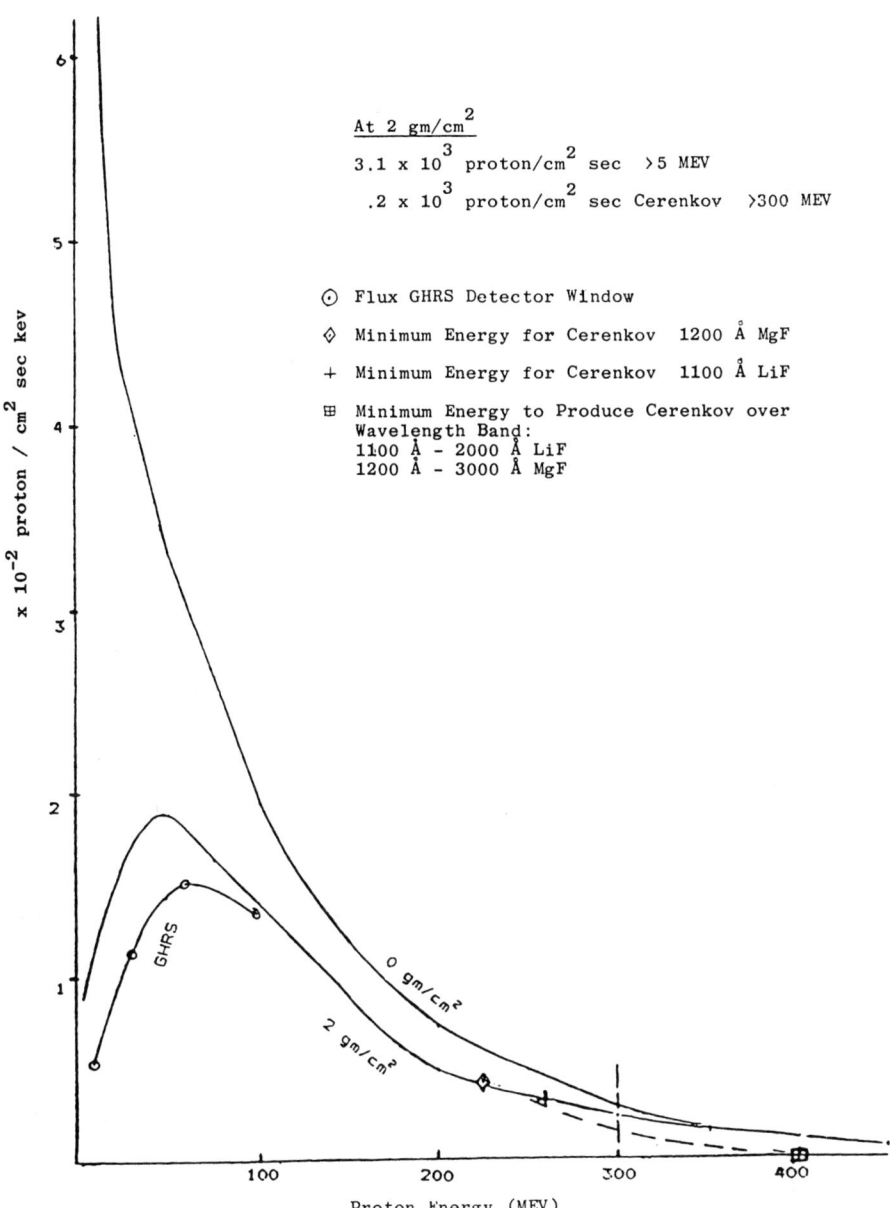

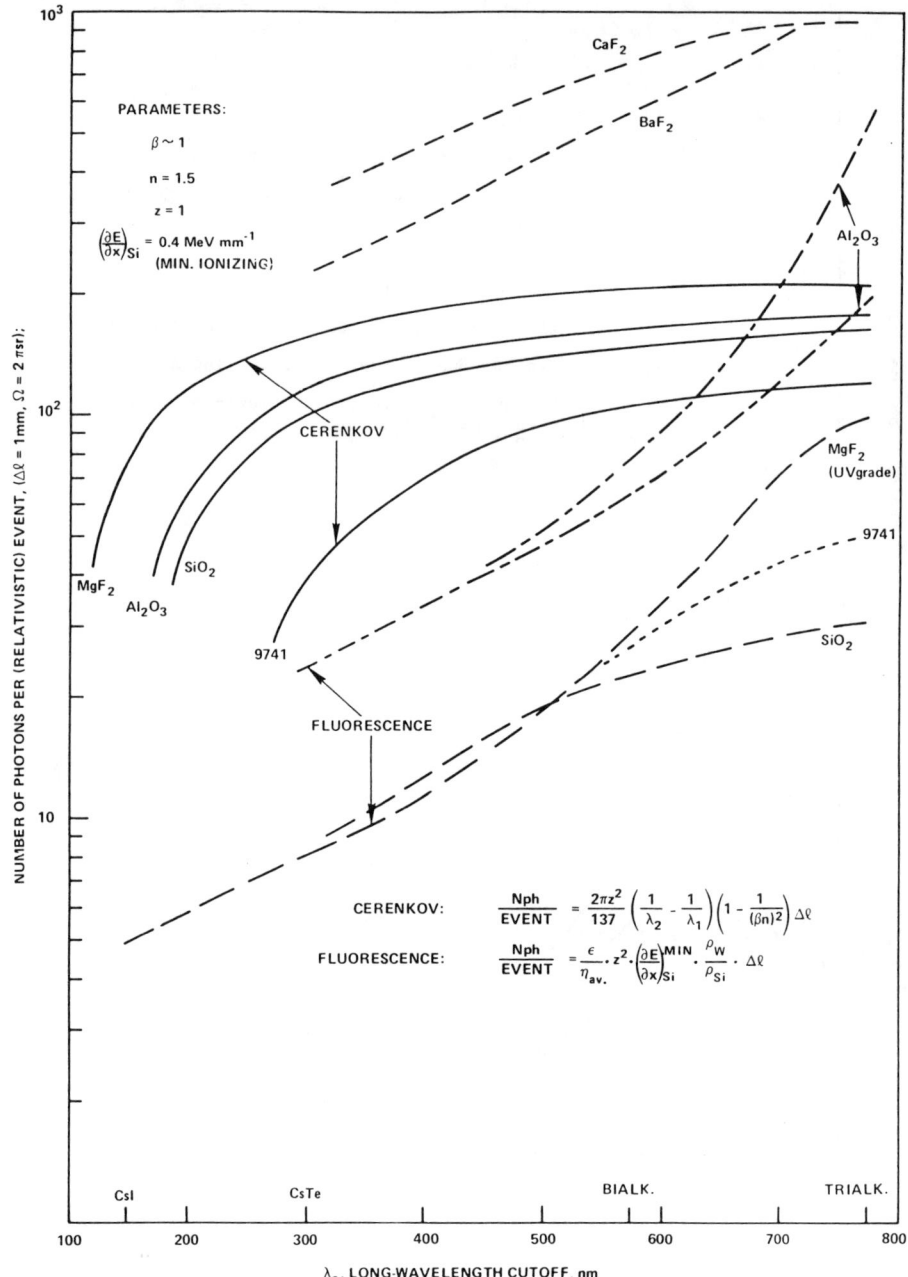

RADIATION ENVIRONMENT EVALUATION FOR ESA PROJECTS

E. J. Daly

ESA/ESTEC, 2200AG Noordwijk, The Netherlands

ABSTRACT

The effects of the natural space radiation environment on ESA projects are discussed and methods and tools used by ESA for evaluating this environment and its effects are described. Various computer-based methods have been adopted, developed and applied to a variety of problems. Environment models are discussed, together with dose and upset-rate computation. Projects under consideration include astronomy and astrophysics missions, manned spacecraft and polar platforms. In the light of these activities, areas of concern requiring further attention are identified.

INTRODUCTION

The successful execution of space programs depends on proper evaluation of the hostile natural space environment. Apart from radiation, this environment is composed of atomic oxygen, various types of plasma environment, meteoroids and debris, UV, and thermal environments. The radiation environment encountered by spacecraft varies considerably in character and causes a wide range of problems for spacecraft and missions. This environment is composed of a variety of energetic particles whose energies range from keV to GeV and beyond, including geomagnetically trapped electrons and protons, galactic cosmic-ray ions and solar-flare particles, together with their associated secondary interaction products. The effects of these particles on space missions are similarly varied; they give rise to degradation of electronic components and materials, single-event interactions in electronics, interference with detector operation and damage to biological systems. These resultant *effects* clearly determine which aspects of the environment are of interest; for example, magnitudes of particle fluxes are important for determining radiation dose and damage, although for space-borne astronomical detectors the concern may be the time for which certain flux and/or energy thresholds are exceeded, thereby causing interference.

ENVIRONMENT

The radiation environment of a spacecraft is composed of:
- Trapped Radiation - energetic particles trapped in the Earth's (or some other planet's) magnetic field - the *radiation belts*;
- Solar flares - these produce energetic protons and other ions with energies

© 1989 American Institute of Physics

of 10's or 100's MeV. These events are sporadic and extremely varied in type, peak particle fluxes, duration and heavy ion population;
• Cosmic rays - low fluxes of energetic heavy ions up to TeV energies and beyond, including ions of all elements in the periodic table;
• Others - secondary radiations, induced radioactivity etc.

The Earth's radiation belts (Figure 1) consist mainly of electrons of up to a few MeV and protons of up to several hundred MeV energies. Since these belts are linked to the geomagnetic field, locations are most conveniently described in terms of geomagnetic L and B coordinates, where L is the equatorial radius of the geomagnetic field-line in idealized dipole space and B is the field strength. Spatially, the 'outer' electron belt extends beyond the geostationary orbit to about 10 Earth radii (R_E) and is brought to low altitude by geomagnetic field lines at high latitude, forming the *polar horns*. The trapped proton belt is confined to lower altitudes. An offset and tilt of the geomagnetic axis with respect to the Earth's rotation axis brings the belts to low altitude over the South Atlantic, creating the *South Atlantic Anomaly*. It is clear, therefore, that most spacecraft encounter some part of the trapped radiation environment.

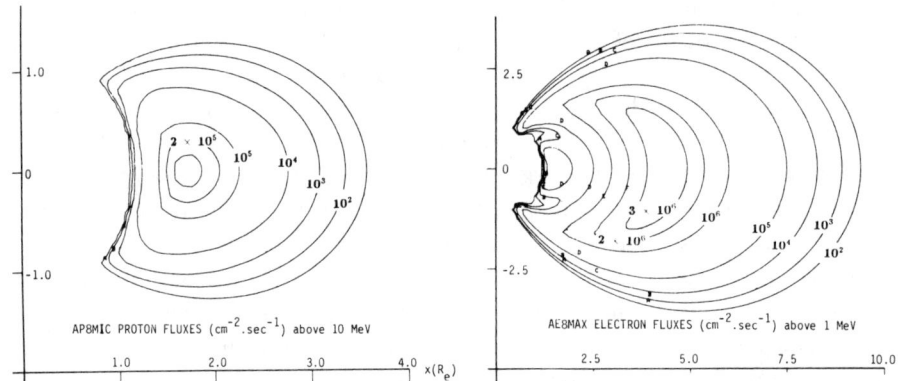

Figure 1: The Earth's radiation belts in idealized dipole space, according to the AP8 and AE8 models[1,2]. Average omnidirectional integral fluxes above energy thresholds are shown.

Low Earth Orbits (LEO) of low inclination, < 30° and < 500km altitude, encounter energetic protons and electrons on passes through the South Atlantic Anomaly, but not necessarily on every orbit. Protons are the most significant component of this environment, having a 'hard' spectrum extending to hundreds of MeV. Energetic particle fluxes in LEO are strongly anisotropic, with an east-west asymmetry and particle motions nearly normal to the field, both consequences of atmospheric absorption[3]. Shuttle operations will lead to many payloads operating in the LEO environment and most American and European manned activities in the next decades will take place there, including missions involving the Space Station, co-orbiting platforms and extensive EVA activities.

Polar orbits, normally 500km-1000km altitude, ~98° heliosynchronous, encounter energetic electrons in the polar horns and protons and electrons on passes through the South Atlantic Anomaly. These orbits are very valuable for Earth observation, having good solar aspect and ground coverage properties. Polar platforms, including Earth resources, meteorology and geophysical missions will make use of this orbit. Eventually, manned missions may also be performed here.

High altitude orbits can pass through the most severe parts of the radiation belts. Geostationary orbits lie beyond the centre of the outer electron belt but still experience a severe electron environment; energetic proton fluxes are low at these altitudes. This orbit will clearly remain very popular. Eccentric orbits are also valuable; for astronomy missions they can provide long periods of clear observing time and good ground-station coverage, and they can visit important parts of geospace.

Solar-flare events giving rise to energetic particle fluxes are unpredictable on timescales longer than a few days and are very varied in their characteristics. Although they last for only a matter of days, they can contribute significant integrated particle fluences and doses in that time as a result of very high particle fluxes. Figure 2, from McGuire[4], shows the record of significant energetic proton flare events from which their association with the solar cycle and their variability are apparent. Apart from energetic protons, flares produce large fluxes of heavy ions.

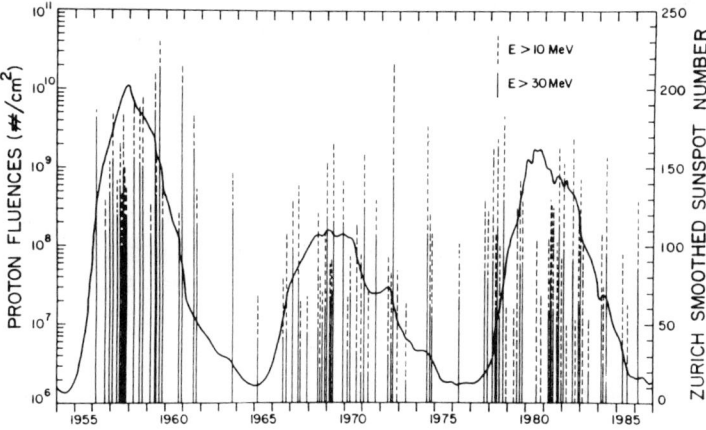

Figure 2: Solar-flare proton fluences (vertical lines) superimposed on the sunspot cycle (continuous curve).

Galactic Cosmic Rays (GCR's) are a continuous component of the radiation environment. Fluxes of these particles are low but, because they include heavy ions of elements such as iron, they cause intense ionization as they pass through matter, are difficult to shield against, and therefore constitute a significant hazard.

Orbit type has an effect on the solar-flare and cosmic-ray environments encountered by spacecraft. The particles constituting these environments originate outside the magnetosphere and the geomagnetic field has a strong influence on their fluxes within the magnetospheric cavity. Bending of particle trajectories by the geomagnetic field results in *geomagnetic shielding* of locations from some of these particles. Geomagnetic shielding is weak at high altitude and over the poles, so geostationary, eccentric and polar orbits are at least partially exposed to solar-flare and cosmic-ray particle fluxes. LEO orbits are normally well shielded by the geomagnetic field from flare particles but energetic GCR particles can be encountered.

At a given location in the field there will be minimum cut-off energies necessary for ions to penetrate to that point. Störmer's theory gives a cut-off rigidity, P_c, for particle arrival at a point, depending on the point's geomagnetic R, λ polar coordinates and the angle of ion arrival from east, γ [5]:

$$P_c = \frac{M \cos^4 \lambda}{R^2 \left[1 + (1 - \cos^3 \lambda \cos \gamma)^{1/2}\right]^2} \quad (1)$$

(Rigidity is related to particle energy by $P = (A/Z)(E^2 + 1862E)/10^3$, where P is in GV and E is in MeV). M is the normalized dipole moment of the Earth. From Equation (1), it can be seen that cosmic-rays penetrate the geomagnetic field more easily from the west ($\gamma=\pi$) than from the east ($\gamma=0$). The R, λ coordinates can be computed from B and L according to the method of Roberts[6]. For vertical arrival, the expression simplifies to

$$P_c \simeq 16/L^2 \text{ since } \gamma = \pi/2 \text{ and } R = L\cos^2 \lambda \quad (2)$$

An approximate value of 16 for the constant is used to fit with observed *effective* cut-offs, accounting for penumbral arrival found in more complete treatments[7]. Magnetospheric disturbances, which often follow solar-flares, result in a lowering of cutoff; this has been described by Adams et al.[5] as:

$$\Delta P/P_0 = .54 \exp(-P_0/2.9 \text{GV}) \quad (3)$$

The effect of arrival direction and magnetospheric state on proton cut-off energy is seen in Figure 3.

EFFECTS AND THEIR ASSESSMENT

Spacecraft electronics can be significantly degraded by exposure to ionizing radiation. Component damage, single-event upset ('bit-flips', SEU) and

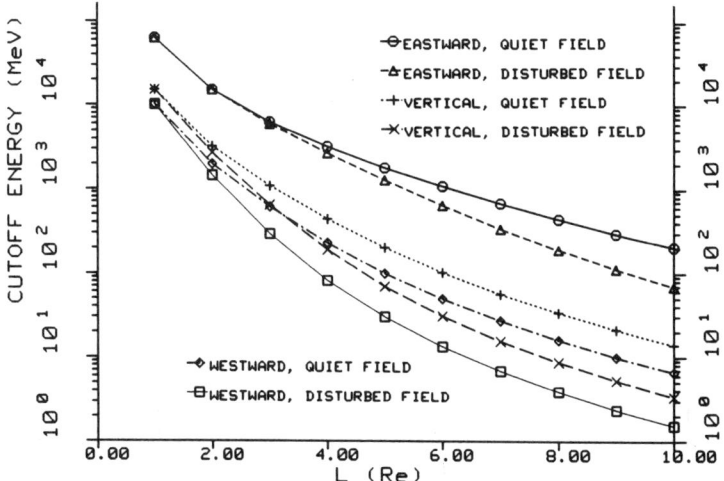

Figure 3: Geomagnetic shielding: proton arrival at the geomagnetic equator from different directions under quiet and disturbed geomagnetic conditions.

'latchup' can occur. SEU occurs with the passage of a highly-ionizing particle, such as a cosmic-ray ion, leading to the generation of enough free charge in the sensitive volume of a chip to 'flip' the bit. Latchup occurs when such ionization creates a parasitic current path, potentially destroying the device. Assessment criteria used to evaluate these problems include absorbed dose, single-event-upset rate and latchup probability, although latchup is normally difficult to quantify.

Radiobiological problems in space are well appreciated, but knowledge about some of the unique effects of the *space* radiation environment is still lacking. Clearly, serious effects can result from radiation exposure during flare events. However, late effects from prolonged low-level exposure and the effects of energetic heavy ions remain uncertain[8]. In order to quantify the environment for evaluating radiobiological effects, *dose-equivalent* is used. This is related to the physical absorbed dose by Quality Factors and Dose Modifying Factors which describe the effectiveness of the radiation in producing damage in biological systems. Uncertainties in these factors reflect uncertainties in the evaluation of effects. Dose-equivalent may be a poor way of quantifying the radiobiological effects of HZE cosmic-ray ions[8] and the use of HZE fluence or some 'hit-rate' parameter may be more appropriate. Ultimately, *risk* is the parameter to be evaluated[9].

Instrument operation can be significantly impaired by the radiation background; primary particles or particles resulting from interactions in matter can 'hit' a detector element and be misinterpreted by the instrument. Among possible interactions leading to false signals are bremsstrahlung, nuclear-reaction products, neutron-capture gamma-rays and Cerenkov radiation in optical components. To evaluate problems in this area, it is usually necessary to know the primary or product particle fluxes at the detector, and to derive a hit-rate. Flux and energy thresholds may need to be considered, depending on shielding characteristics and/or detector sensitivity. Also of some importance is the fraction of the time for which the background level prohibits the normal function of the instrument, such as around an orbit passing through the radiation belts, or as a result of solar-flare events.

Solar cell performance and cover-glass optical properties can be degraded by exposure to radiation. For these problems, the traditional assessment criterion has been the so-called 'damage-equivalent fluence' of mono-energetic electrons and protons, a quantity which attempts to relate the space environment to a ground-test environment causing the same degradation by means of damage coefficients[10].

Optical, electrical and physical properties of materials can be significantly affected by the space radiation environment; the quantities absorbed dose and particle fluence are used for assessment.

The evaluation of solar-flare effects depends on the particular problem under consideration. For example, a study of observation time loss on an astronomy mission, due to to flare interruption, involves analysis of the duration of flare fluxes above a threshold energy, determined by detector shielding, and above a background flux level[11]. This is clearly different from dose evaluation, which normally considers the magnitudes of integrated fluxes. A flare which is not significant in terms of fluence or dose can nonetheless be significant in terms of duration. This kind of problem necessitates re-evaluation of raw flare data, such as from the IMP-series satellites.

ESA PROJECTS AND THE RADIATION ENVIRONMENT

Past ESA space activities have included a variety of scientific and applications satellites and, in cooperation with NASA, the Spacelab manned project. Below is a partial summary of past ESA (and some European national) projects.

<u>Science</u>:- ESRO-2,4 ; HEOS-1,2 ; TD-1 ; Cos-B ; Geos-1,2 ; Spacelab* ; ISEE*; IUE† ; Exosat ; Giotto ; instruments on Salyut-7 and Mir space stations‡ (IRAS: national US/UK/NL)

<u>Manned</u>:- Spacelab*

<u>Communications Satellites</u>:-
OTS ; ECS-1,2,4 ; Marecs-A,B (+ national programs)

<u>Earth Observation/Meteorology</u>:- Meteosat-1, 2 - Geostationary Meteorological (+ French national: Spot-1 Earth Observation)

* with NASA.   † with NASA and UKSERC

In the past, ESA projects have been influenced by the radiation environment in a number of ways. Clearly, knowledge of deleterious radiation effects have impacted on parts selection so that in general, advanced components could not be flown. An example of a severe problem is the extensive parts revision which has been necessary on the Ulysses spacecraft as a result of latchup sensitivity uncovered late in its development.

In-flight dosimetry on ESA spacecraft has shown a much milder geostationary environment than expected. Integrating pMOS RADFET dosimeters[12] have flown on the OTS, GEOS-2 and METEOSAT satellites and, as seen in Figure 4, register doses well below values predicted on the basis of the AE8 trapped electron model[13].

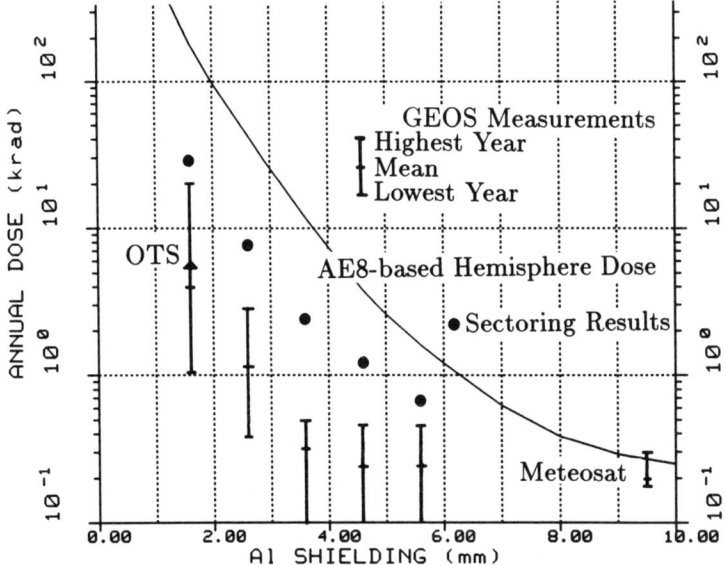

Figure 4: Geostationary orbit dosimetry results.

Radiation background has been seen in telescopes, leading to interruption of operations and inteference with observations. The affected satellites include:

Exosat, in a $30R_E$ apogee, 70° orbit, implemented instrument switch-offs below 70000km to avoid radiation-belt background. Solar-flare particle events interrupted observations and, as expected, the instruments experienced solar-cycle-modulated cosmic-ray background[14] (Figure 5).

Cos-B, in a $16R_E$ apogee, 90° orbit saw a modulated CR background (Figure 5), but also observed flaring in a detector which was not obviously correlated with any solar or magnetospheric activity[15].

IUE, in a low-inclination, moderately eccentric 5.2-$8R_E$ orbit encountered outer zone electrons. This was reflected in local-time and solar-cycle dependence of the background monitor output[16].

ESA are participants in X-ray astronomy experiments flying on the Mir space station. Mir's orbit is roughly circular at about 350km with an inclination of about 52°. The GSPC instrument registers an enhanced background at higher latitudes. This doesn't fit with trapped electron polar horn crossings but fits very well with the expected increase in cosmic-ray fluxes at high latitude due to weakening geomagnetic shielding[17].

IRAS, a US-UK-Netherlands satellite in polar orbit, encountered the anticipated radiation background on South Atlantic and polar horn crossings. An on-board veto system was provided to prevent particle-induced signals from registering as IR data; this it did very effectively[18].

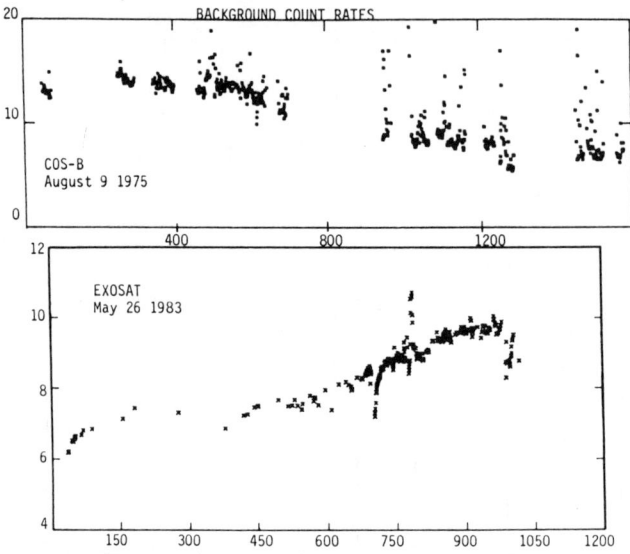

Figure 5: Radiation Background: Cos-B and Exosat cosmic-ray background as a function of mission day. Solar cycle variations are clear.

As the following list of projects under study and development shows, ESA anticipates a considerable expansion in its activities up to the year 2000.

Science[19]:-
Hipparcos (astrometry) ; Space Telescope* ; Ulysses* ; ISO (IR astronomy) ; GRO-Comptel* (γ astronomy) ; Cluster + SOHO = ISTP* (solar-terrestrial) ; Eureca (Shuttle-launced microgravity platform) ; *cornerstone missions*: XMM (X-ray astronomy) ; FIRST (far-IR astronomy) ; CNSR * (cometary) ;
phase A: Lyman (UV astronomy) ; Quasat (VLBI radio-astronomy) ; Cassini Titan Probe* ; *competing studies*: GRASP (γ astronomy); Vesta ‡ (asteroid probe) ; Caesar (cometary) ; *others*: Giotto resurrection ; Mir payloads†.

Manned:- Spacelab* ; Columbus Pressurized Module for ISS* ; MTFF: Man-Tended Free Flyer platform ; Hermes 'mini-shuttle' ; EVA capability.

Communications Satellites:- Olympus ; ECS/Marecs family ; DRS ; (+ national programs).

Earth Observation and Meteorology:-
Meteosat Operational Program and extension ; ERS-1,2 (μ-wave remote-sensing) ; SS Polar Platform ; (French national: Spot-2,3..).

* with NASA. † with USSR ‡ with USSR and CNES

The radiation environment is likely to become an even more important consideration for future projects. The pressure to use advanced VLSI components, lightweight solar cells, new materials etc., is mounting as the enhancement to scientific and operational returns from missions which make use of such technologies is recognised.

More sensitive detectors are planned for space-borne telescopes where radiation background and damage problems may impose operational limits.

Extended manned activities are planned including long stays on the ISS, visits to MTFF, Hermes operations and EVA activities. Visits to polar orbits are not presently feasible, but visits to 50° - 60° latitudes may be undertaken.

Mission planning, involving consideration of such factors as mass budgets and ground-station coverage, may lead to trade-offs being made which result in choices of orbits with radiation problems.

Internal charging of dielectric materials can occur as a result of energetic electron penetration and charge build-up. This can lead to discharges, disrupting spacecraft electronics. In the past, this has mainly been a worry for missions to the giant planets which have particularly hostile magnetospheres, but should not be overlooked for Earth-orbiting spacecraft. The NASA/USAF CRRES spacecraft will carry an experiment designed to evaluate the problem in Earth-orbit[20]. Planetary missions may be included in future ESA plans.

TOOLS FOR RADIATION ENVIRONMENT ANALYSES

Given the multitude of possible radiation-related problems which can occur and the fact that virtually every spacecraft has to consider the radiation environment to some degree, tools are clearly needed in order to evaluate the environment and its effects on future missions. The tools can be divided into *environment* models, describing the basic space radiation environment and its spatial and temporal characteristics, and *interaction* tools which include consideration of radiation transport through matter and which yield some quantity which can be related to the effect of concern (e.g. dose).

Magnetic field models are needed for trapped particle, solar flare and GCR environment modeling since, in the former, particle morphologies are described in terms of location in idealized geomagnetic dipole space, and the latter two need the field in order to include geomagnetic shielding effects. The Earth's offset, tilted and distorted quasi-dipolar field is described by numerical field models such as the IGRF.

The UNIRAD system (Figure 6) is used by ESA for routine mission analyses. It provides mission fluxes, flare fluences, doses and damage-equivalent fluences. It consists of trapped particle models, flare models, an orbit generator, geomagnetic field models, a code for computing doses in simple geometries and a code for computing solar cell damage-equivalent fluences. The orbit generator and geomagnetic field model programs provide geomagnetic B/B_0 and L coordinates at points on a user-defined orbit which the TREP program uses to access the NSSDC trapped particle models, AE8 and AP8. These models provide time-averaged, integral omnidirectional electron and proton fluxes as functions of particle energy and the geomagnetic coordinates L and B/B_0. TREP uses these fluxes to compute fluxes as a function of orbit time and orbital-average spectra. The models are simplifications, inasmuch as short-term variations, diurnal variations and pitch-angle distributions are not treated.

The old AE4 outer zone electron model included two interesting features not present in later models[21]. The first was a statistical model giving standard deviations, and the second, a model for local time flux modulation; both models being functions of E and L. These have been extended and can be applied to the AE8 model by TREP. There is no pitch-angle dependence in the AE8 and AP8 models and none has yet been implemented in TREP.

Use of a geomagnetic field model requires definition of an epoch and this has lead to problems in the past when the field model epoch was set to times incompatible with the epoch used in setting up the particle models[22].

Solar-flare particle events, because of their unpredictability and large variability in magnitude, duration and spectral characteristics, have to be treated statistically[23]. Burrell (see ref. 23) developed a modified Poisson statistic to describe the probability of a number of events n occurring during a time t, based on a previously observed frequency of N during time T:

$$p(n,t;\ N,T) = \frac{(n+N)!(t/T)^n}{n!N!(1+t/T)^{N+n+1}} \quad (4)$$

Flare characteristics are often based on the worst event observed over the last two solar cycles (20 and 21), which occurred in August 1972 (see Figure 2). This *anomalously large* (AL'72) dominated all other events of the past 25 years in terms of peak flux, integrated flux (fluence) and duration.

King[23] examined IMP energetic proton data for a number of flare events and, treating the AL'72 flare separately, produced a log-normal distribution of *ordinary* flare fluences as a function of proton energy and an exponential fit to the AL'72 integral fluence spectrum:

$$J(>E) = 7.9 \times 10^9 \exp\left\{\frac{30-E}{26.5}\right\} \quad (5)$$

with E in MeV and J in protons/cm^2

Models of mean, worst-case and anomalous solar-flare proton spectra based on King's work are used in TREP, together with the Burrell distribution (Equation 4). Depending on the mission duration, a number of flares are predicted and the resulting total fluence spectrum is produced.

Geomagnetic shielding of flare protons is computed on the basis of the trajectory in B, L space. Stassinopoulos and King[24] reported a model which has total cutoff at L=5; it is assumed that no protons can penetrate to lower values. Computing the geomagnetic cut-off for vertically arriving protons, we find that this model corresponds to a quiet magnetosphere vertical cutoff model excluding protons of E<200MeV from L<5 R$_E$ (see Figure 3). This model is adequate for most cases and is included in TREP.

Figure 3 shows that protons of lower energy can penetrate below L=5 with other arrival directions, especially in a disturbed magnetosphere where the geomagnetic shielding is weakened. For westward arrival at the L=5 geomagnetic equator in a disturbed magnetosphere, the energy cut-off could be as low 30MeV. An arrival-direction-dependent model for flare proton cutoffs throughout B, L space can be used by TREP if required.

UNIRAD includes the SHIELDOSE code[25] for dose computation in simple aluminum geometries. Problems in modeling the interactions of radiation with matter have been well studied for many years and there is no shortage of applicable computer codes and interaction data (cross-sections etc.). The problem is often in deciding which one to use. Shielding (attenuation) characteristics, the production of secondaries, scattering, and so on, have to be considered. It may also be important to consider geometry in some cases. When performing such radiation analyses, it is necessary to define requirements so that the tools fit the job, avoiding 'over-kill' or inadequate methods. For example, it makes little sense to use a complex-geometry multi-particle Monte-Carlo code such as CERN's GEANT for routine analyses of radiation doses in satellite subsystems, although it is very useful for other problems; the choice is often between efficient engineering tools having simplified physics and research tools with a more complete physical treatment.

Transport and absorbed dose can be computed using Monte-Carlo codes but often dose look-up tables based on Monte-Carlo runs with mono-energetic particle populations can be used to give doses in simple geometries for arbitrary input spectra. This is the procedure used by SHIELDOSE. Monte-Carlo codes follow the transport of populations of particles explicitly, including energy loss, scattering and the generation of secondary particles. Condensed-history Monte-Carlo techniques are normally used where the results of small quasi-continuous steps in the particle transport are followed, rather than the vast number of individual interactions. The SHIELDOSE data set is based on slab-geometry ETRAN Monte-Carlo runs and a conversion formula is used to give spherical geometry results. Doses due to electrons, electron-induced bremsstrahlung and protons are computed.

Continuous-slowing-down or straight-ahead treatments can also be applied, for example with the CHARGE code[26]. Here, particle scattering is not treated explicitly but through analytical corrections.

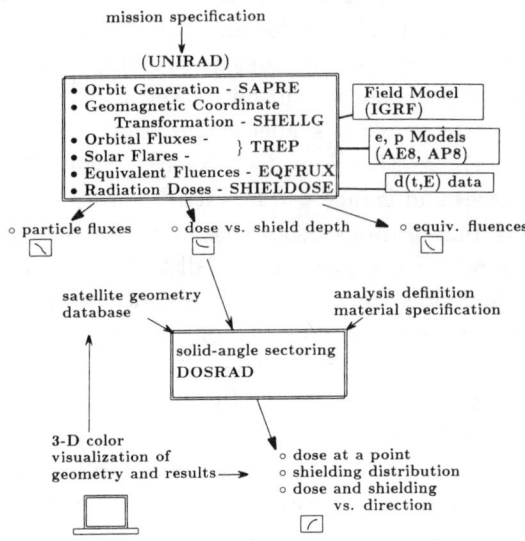

Figure 6: ESABASE/RADIATION Architecture.

For routine 'engineering-level' evaluation of more complex shielding geometries, the ESABASE/Radiation is tool is used. ESABASE is a *system-level design tool* for performing geometry-related analyses, and features: a geometry description language; visualization tools, including gateways to EUCLID, PATRAN, GEOMOD etc.; an hierarchical data structure which can also handle variable spacecraft geometries; and analysis modules. Analyses include Radiation, Electrostatic-charging, Contamination, Plume-impingement, Atomic-oxygen, Mass-properties, Thermal-analysis, Perturbation, Pointing and Occultation.

ESABASE/Radiation includes the UNIRAD suite of programs described above and, for geometrical analysis, uses the 'sectoring' approximation with a generalized satellite geometry database. Its architecture is shown in Figure 6. Rays are traced in all directions from the point of interest, the encountered aluminum-equivalent thickness t_i is computed for each ray, the dose is derived from a sphere-geometry dose-depth curve and a summation performed over all directions:

$$\sum_{i=1}^{n} d(t_i) \frac{\Omega_i}{4\pi} \qquad (6)$$

where $d(t)$ represents the dose-depth curve and Ω_i the elemental solid angle around ray i. This procedure is approximate, in that it ignores electron angular scattering and bremsstrahlung angular effects, but is good for ion transport where attenuation is in a straight line.

Among its outputs, the code produces shielding distributions and directional dose contributions. It has been applied to studies of ISS elements, astronaut EVA and the GEOS dosimetry package as shown in Figure 7.

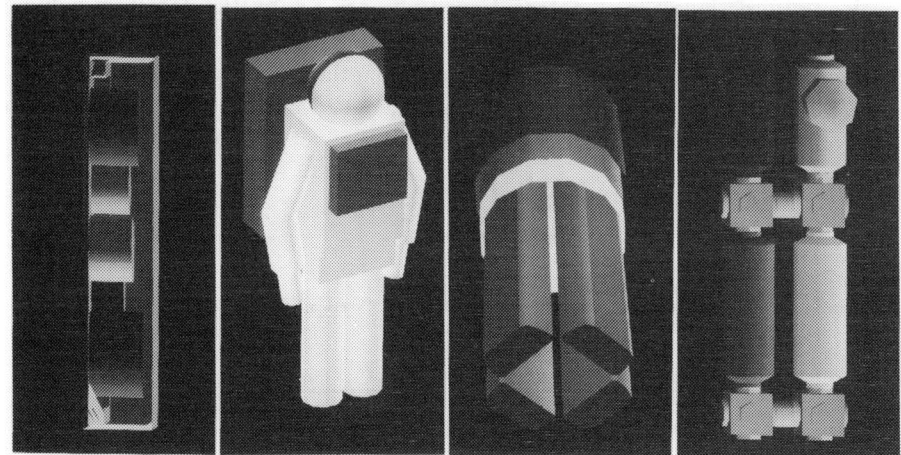

Figure 7: Examples of Geometries Analyzed by ESABASE/Radiation.

Damage-equivalent fluences for evaluating solar cell degradation are computed with the EQFRUX program[10]. Damage coefficients D(E) derived from cell testing are included. and these are used to relate test beam exposure to the cover-glass-shielded space radiation environment.

Cosmic-ray environments have been comprehensively modeled by Adams and co-workers at NRL[5], based on satellite data for cosmic-ray and flare ion fluxes. They developed the CREME suite of programs[27] which include these models, together with upset/hit rate computation. The environments available from CREME include solar-cycle modulated GCR fluxes for most ions, 10%-worst-case GCR fluxes, anomalous component fluxes and anomalous, worst-case and ordinary flare fluxes including mean or worst-case composition. Earth shadowing and transport through shielding material are also considered. For treating geomagnetic shielding, CREME computes an orbital attenuation function on the basis of Shea and Smart's world map of vertical cut-off rigidities. Energy and LET spectra are produced, the latter being composites of the fluxes of all ions as functions of ion stopping powers in silicon.

To compute an upset or hit rate for an electronic device or a detector from the predicted fluxes, device characteristics must be specified, particularly the size of the sensitive volume and the *critical charge* in the volume which results in upset (bit-flip) or registers as a 'count'. The rate is found by integrating over the composite differential LET spectrum, $\phi(L)$, and the path-length distribution for the sensitive volume, $p(l)$.

$$U = S/4 \int_{E_c/L_{\max}}^{l_{\max}} p(l) \int_{E_c/l}^{L_{\max}} \phi(L)\, dL\, dl \;\rightarrow\; \Phi S/4 \text{ in a very sensitive detector} \tag{7}$$

where S is the total surface area of the sensitive volume and Φ is the integral omnidirectional flux. Integration limits are set by the sensitive volume dimensions and the critical energy, E_c; $E_c/L_{\max}$ is the shortest path capable of supporting upset, $l_{\max}$ is the maximum pathlength, E_c/l is the minimum particle LET necessary to cause upset on a pathlength l and $L_{\max}$ is the maximum LET of the spectrum. Figure 8 shows an overview of CREME.

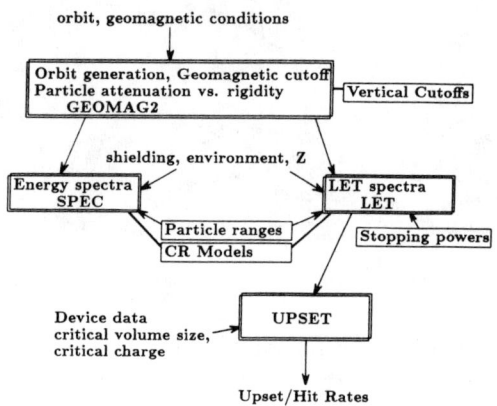

Figure 8: CREME Architecture.

Ion-beam tests can establish the threshold LET, defining the critical charge ($E_c = L_c \times depth$), and saturation cross-section, giving information about the sensitive volume. However, all the upsetting 'bits' may not be the same, and often no sharp threshold in LET is apparent. In these cases, an integral over the measured response curve may be performed[28]. The upset rate computation is sensitive to the value chosen for this critical charge and derived from the LET threshold because, whereas particle fluxes fall rapidly with increasing LET, the sensitivity increases rapidly.

As indicated previously, quantification of the radiation environment for biological effects requires computations of dose-equivalents which require, in addition to the absorbed dose, knowledge of Quality Factors and dose-modifying factors. Quality Factors for high-LET (stopping-power) radiations are sometimes LET-dependent and so transmitted spectra are needed in such cases.

Other environments sometimes have to be considered, including induced radioactivity through activation of materials[29], the radiation produced by on-board sources (e.g. thermoelectric generators) and the environments about other planets.

For special studies, Monte-Carlo codes are available. The ETRAN derivatives (ETRAN, TIGER, CYLTRAN, ITS) are available for coupled electron-photon transport problems.

Another interesting code is GEANT, developed at CERN. This is a state-of-the-art complex-geometry multi-particle code covering a wide energy range and an extensive range of particles and physical processes. It is interfaced with the GHEISHA hadron code and the EGS-4 electron-photon code. This allows analysis of special problems, for example, production of neutron-capture photons followed by Compton scatter and energy deposit in correlated detecting elements, a process possible on the GRO/Comptel instrument.

CONCLUSIONS AND DISCUSSION

Although it is clear from the above that there are many tools available for analysing the space radiation environment, some weaknesses are apparent.

The trapped particle models in general use are mainly based on data acquired before 1971 and so are around 20 years old. Ways in which these models could be improved include the obviously prudent inclusion of more recent data and implementation of local-time, pitch-angle and statistical models, possibly including modeling of long- and short-term temporal variations.

DMSP data from polar orbit[30] imply a less severe electron environment than predicted. Averaged SCATHA[31] and NTS-2[32] data are about a factor 2 below predictions, which seems a reasonable level of agreement, but these data, and those from ATS[33], show clearly the very large variability in outer zone electron fluxes over a number of different timescales. NTS[34] and ATS data show the 27-day periodicity in these fluxes and GOES data[35] shows the diurnal effects very clearly. The CRRES spacecraft, due to fly in the early 90's, will be instrumented to provide much valuable environment data for use in updating radiation belt models. However, existing data sets, including from some of the above satellites, could be used to create interim improved models. Vital though it is, CRRES cannot provide total spatial and temporal coverage, and radiation environment measurement should become a more routine 'housekeeping' function on-board as many future spacecraft as possible.

The flare record for the last solar cycle[4] should allow updating of models of flare characteristics and statistics.

For astronomy missions, anticoincidence techniques and on-board particle detectors, allowing real-time corrective action, can be very useful. The latter measure removes a lot of the reliance on background predictions from the averaged and simplified models.

Much work remains to be done in the area of radiobiological effects where the special nature of the space radiation environment means there is a lack of truly relevant exposure data.

It is important that the necessary updating and development of models and predictive tools, together with environment and effects measurements, keep pace with the expected expansion of space activities world-wide.

ACKNOWLEDGEMENTS

I am grateful to Len Adams for providing the RADFET dosimetry results, to Cecil Tranquille for the Cos-B and MIR/GSPC background information, to Tony Peacock for Exosat and Cos-B data and to Günther Thörner for the sector analysis results for the Geos dosimeter.

REFERENCES

1. Sawyer D.M. and Vette J.I., 'AP8 Trapped Proton Environment For Solar Maximum and Solar Minimum', NSSDC 76-06, NASA-GSFC (1976).
2. Vette J.I. et al., AE8 model for inner and outer zone electrons at solar maximum and solar minimum: unpublished electron model replacing AE5, AE6 and AEI7 (see NSSDC reports 72-06, 72-13 (AE-4), 72-10 (AE5) and 76-04 (AE6)).
3. Hess W.N., 'The Radiation Belt and the Magnetosphere', Blaisdell Publ. Co. (1968).
4. McGuire R.,'A Characterization of the Interplanetary Environment: Proton and Ion Fluences and Spectra 1-100 MeV/Nucleon 1972-1986', these proceedings.
5. Adams J.H., Jr, Silberberg R. and Tsao C.H., 'Cosmic Ray Effects on Microelectronics, Part I: The Near-Earth Particle Environment', NRL Memorandum Report 4506 (1981).
6. Roberts C.S., 'Coordinates for the Study of Particles Trapped in the Earth's Magnetic Field: A Method of Converting from B, L to R, λ Coordinates', J. Geophys. Res. 69, 5089 (1964).
7. Smart D.F. and Shea M.A., 'A Study of the Effectiveness of the McIllwain Coordinates in Estimating Cosmic-Ray Vertical Cutoff Rigidities', J. Geophys. Res. 72, 3447 (1967).
8. Bücker H. and Facius R., 'Radiation Protection Problems for the Space Station and Approaches to their Mitigation', Adv. Space Res. 6, 305 (1986).
9. Sinclair W.K., 'Radiation Protection Standards in Space', Adv. Space Res. 6, 335 (1986).
10. Tada H.Y., Carter J.R., Anspaugh B.E. and Downing R.G., 'Solar Cell Radiation Handbook, 3rd edition', NASA JPL 82-69 (1982).
11. Daly E., 'Solar Flare Proton-Induced Background Counts in the ISO IR Detectors', ESTEC Internal Note WM/EJD-10-84.
12. Holmes-Siedle A., Adams L., Marsden S. and Pauly B., 'Calibration and Flight Testing of a Low-Field pMOS Dosimeter', IEEE Trans. Nuc. Sci. NS-32, 4425 (1985).
13. Thörner G., ESTEC, personal communication (ESTEC/WMA/GT/GEO/1, 1987)
14. Peacock A., ESTEC, personal communication
15. Tranquille C., ESTEC, personal communication, (Technical note ESTEC/WMA/CT-88)
16. Imhoff, C.L. 'Long Term Trends in IUE Radiation Background', NASA IUE Newsletter 27, 9 (1985).

17. Tranquille C. and Daly E., 'Correlation Between Background X-Rays Measured by the Sirene II Instrument and Terrestrial Radiation Environment Models', ESTEC/WMA/CT/MIR/3 (1988).

18. Long E.C. and Langford D., 'Test Results and In-Orbit Operation of the Infrared Astronomical Satellite Circumvention Circuit', Proc. SPIE, 264 (1984).

19. 'Report on the Scientific Satellites of the European Space Agency', ESA SP-1090 (1987).

20. Coakley P.G., Treadway M.J. and Robinson P.A., 'Low Flux Laboratory Test of the Internal Discharge Monitor (IDM) Experiment Intended for CRRES', IEEE Trans. Nuc. Sci. NS-32, 4066 (1985).

21. Singley G.W. and Vette J.I., 'The AE-4 Model of the Outer Radiation Zone Electron Environment', NSSDC 72-06 (1976).

22. Konradi A., Hardy A.C. and Atwell W., 'Radiation Environment Models and the Atmospheric Cutoff', J. Spacecraft 24, 284 (1987).

23. King J.H., 'Solar Proton Fluences for 1977-1983 Space Missions', J. Spacecraft 11, 401 (June 1974).

24. Stassinopoulos E.G. and King J.H., 'Empirical Solar Proton Model For Orbiting Spacecraft Applications', IEEE Trans. on Aerosp. and Elect. Systems AES-10, 442 (1973).

25. Seltzer S.M., 'Electron, Electron-Bremsstrahlung and Proton Depth-Dose Data for Space-Shielding Applications', IEEE Trans. Nucl. Sci. NS-26, 4896 (Dec. 1979).

26. Yucker W.R. and Lilley J.R., 'CHARGE Code for Space Radiation Shielding Analysis', McDonnell Douglas Report DAC-62231 (1969).

27. Adams J.H., Jr., 'Cosmic Ray Effects on Microelectronics, Part IV', NRL Memorandum Report 5901 (1986).

28. Harboe-Sorensen R., Adams L., Daly E.J., Sansoe C., Mapper D. and Sanderson T.K., 'The SEU Risk Assessment of Z80A, 8086 and 80C86 Microprocessors Intended for Use in a Low Altitude Polar Orbit', IEEE Trans. Nucl. Sci. NS-33, 1626 (1986).

29. Dyer C.S., 'Radioactivity Induced in Gamma Ray Spectrometers', these proceedings.

30. Gussenhoven M.S., 'New Radiation Measurements at Low-Altitudes', these proceedings.

31. Mullen E.G. and Gussenhoven M.S., 'SCATHA Environmental Atlas', AFGL-TR-83-0002 (1983).

32. Pruett R.G., 'Comparison of DMSP and NTS-2 Dosimeter Measurements With Predictions', J. Spacecraft 17, 270 (1980).

33. Paulikas G.A. and Blake J.B., 'Effects of the Solar Wind on Magnetospheric Dynamics: Energetic Electrons at the Synchronous Orbit', Aerospace Corporation Report ATR-79(7642)-1, NASA-CR-159948 (1978).

34. Blake J.B, 'Magnetospheric Radiation Environment in a 12-Hour Circular Orbit', J. Spacecraft 18, 477 (1981).

35. 'Solar Geophysical Data Prompt Reports', issued weekly by NOAA/SESC, Boulder, Co.

AIP Conference Proceedings

		L.C. Number	ISBN
No. 1	Feedback and Dynamic Control of Plasmas – 1970	70-141596	0-88318-100-2
No. 2	Particles and Fields – 1971 (Rochester)	71-184662	0-88318-101-0
No. 3	Thermal Expansion – 1971 (Corning)	72-76970	0-88318-102-9
No. 4	Superconductivity in d- and f-Band Metals (Rochester, 1971)	74-18879	0-88318-103-7
No. 5	Magnetism and Magnetic Materials – 1971 (2 parts) (Chicago)	59-2468	0-88318-104-5
No. 6	Particle Physics (Irvine, 1971)	72-81239	0-88318-105-3
No. 7	Exploring the History of Nuclear Physics – 1972	72-81883	0-88318-106-1
No. 8	Experimental Meson Spectroscopy –1972	72-88226	0-88318-107-X
No. 9	Cyclotrons – 1972 (Vancouver)	72-92798	0-88318-108-8
No. 10	Magnetism and Magnetic Materials – 1972	72-623469	0-88318-109-6
No. 11	Transport Phenomena – 1973 (Brown University Conference)	73-80682	0-88318-110-X
No. 12	Experiments on High Energy Particle Collisions – 1973 (Vanderbilt Conference)	73-81705	0-88318-111-8
No. 13	π-π Scattering – 1973 (Tallahassee Conference)	73-81704	0-88318-112-6
No. 14	Particles and Fields – 1973 (APS/DPF Berkeley)	73-91923	0-88318-113-4
No. 15	High Energy Collisions – 1973 (Stony Brook)	73-92324	0-88318-114-2
No. 16	Causality and Physical Theories (Wayne State University, 1973)	73-93420	0-88318-115-0
No. 17	Thermal Expansion – 1973 (Lake of the Ozarks)	73-94415	0-88318-116-9
No. 18	Magnetism and Magnetic Materials – 1973 (2 parts) (Boston)	59-2468	0-88318-117-7
No. 19	Physics and the Energy Problem – 1974 (APS Chicago)	73-94416	0-88318-118-5
No. 20	Tetrahedrally Bonded Amorphous Semiconductors (Yorktown Heights, 1974)	74-80145	0-88318-119-3
No. 21	Experimental Meson Spectroscopy – 1974 (Boston)	74-82628	0-88318-120-7
No. 22	Neutrinos – 1974 (Philadelphia)	74-82413	0-88318-121-5
No. 23	Particles and Fields – 1974 (APS/DPF Williamsburg)	74-27575	0-88318-122-3
No. 24	Magnetism and Magnetic Materials – 1974 (20th Annual Conference, San Francisco)	75-2647	0-88318-123-1
No. 25	Efficient Use of Energy (The APS Studies on the Technical Aspects of the More Efficient Use of Energy)	75-18227	0-88318-124-X

No. 26	High-Energy Physics and Nuclear Structure – 1975 (Santa Fe and Los Alamos)	75-26411	0-88318-125-8
No. 27	Topics in Statistical Mechanics and Biophysics: A Memorial to Julius L. Jackson (Wayne State University, 1975)	75-36309	0-88318-126-6
No. 28	Physics and Our World: A Symposium in Honor of Victor F. Weisskopf (M.I.T., 1974)	76-7207	0-88318-127-4
No. 29	Magnetism and Magnetic Materials – 1975 (21st Annual Conference, Philadelphia)	76-10931	0-88318-128-2
No. 30	Particle Searches and Discoveries – 1976 (Vanderbilt Conference)	76-19949	0-88318-129-0
No. 31	Structure and Excitations of Amorphous Solids (Williamsburg, VA, 1976)	76-22279	0-88318-130-4
No. 32	Materials Technology – 1976 (APS New York Meeting)	76-27967	0-88318-131-2
No. 33	Meson-Nuclear Physics – 1976 (Carnegie-Mellon Conference)	76-26811	0-88318-132-0
No. 34	Magnetism and Magnetic Materials – 1976 (Joint MMM-Intermag Conference, Pittsburgh)	76-47106	0-88318-133-9
No. 35	High Energy Physics with Polarized Beams and Targets (Argonne, 1976)	76-50181	0-88318-134-7
No. 36	Momentum Wave Functions – 1976 (Indiana University)	77-82145	0-88318-135-5
No. 37	Weak Interaction Physics – 1977 (Indiana University)	77-83344	0-88318-136-3
No. 38	Workshop on New Directions in Mossbauer Spectroscopy (Argonne, 1977)	77-90635	0-88318-137-1
No. 39	Physics Careers, Employment and Education (Penn State, 1977)	77-94053	0-88318-138-X
No. 40	Electrical Transport and Optical Properties of Inhomogeneous Media (Ohio State University, 1977)	78-54319	0-88318-139-8
No. 41	Nucleon-Nucleon Interactions – 1977 (Vancouver)	78-54249	0-88318-140-1
No. 42	Higher Energy Polarized Proton Beams (Ann Arbor, 1977)	78-55682	0-88318-141-X
No. 43	Particles and Fields – 1977 (APS/DPF, Argonne)	78-55683	0-88318-142-8
No. 44	Future Trends in Superconductive Electronics (Charlottesville, 1978)	77-9240	0-88318-143-6
No. 45	New Results in High Energy Physics – 1978 (Vanderbilt Conference)	78-67196	0-88318-144-4
No. 46	Topics in Nonlinear Dynamics (La Jolla Institute)	78-57870	0-88318-145-2
No. 47	Clustering Aspects of Nuclear Structure and Nuclear Reactions (Winnipeg, 1978)	78-64942	0-88318-146-0
No. 48	Current Trends in the Theory of Fields (Tallahassee, 1978)	78-72948	0-88318-147-9

No. 49	Cosmic Rays and Particle Physics – 1978 (Bartol Conference)	79-50489	0-88318-148-7
No. 50	Laser-Solid Interactions and Laser Processing – 1978 (Boston)	79-51564	0-88318-149-5
No. 51	High Energy Physics with Polarized Beams and Polarized Targets (Argonne, 1978)	79-64565	0-88318-150-9
No. 52	Long-Distance Neutrino Detection – 1978 (C.L. Cowan Memorial Symposium)	79-52078	0-88318-151-7
No. 53	Modulated Structures – 1979 (Kailua Kona, Hawaii)	79-53846	0-88318-152-5
No. 54	Meson-Nuclear Physics – 1979 (Houston)	79-53978	0-88318-153-3
No. 55	Quantum Chromodynamics (La Jolla, 1978)	79-54969	0-88318-154-1
No. 56	Particle Acceleration Mechanisms in Astrophysics (La Jolla, 1979)	79-55844	0-88318-155-X
No. 57	Nonlinear Dynamics and the Beam-Beam Interaction (Brookhaven, 1979)	79-57341	0-88318-156-8
No. 58	Inhomogeneous Superconductors – 1979 (Berkeley Springs, W.V.)	79-57620	0-88318-157-6
No. 59	Particles and Fields – 1979 (APS/DPF Montreal)	80-66631	0-88318-158-4
No. 60	History of the ZGS (Argonne, 1979)	80-67694	0-88318-159-2
No. 61	Aspects of the Kinetics and Dynamics of Surface Reactions (La Jolla Institute, 1979)	80-68004	0-88318-160-6
No. 62	High Energy e^+e^- Interactions (Vanderbilt, 1980)	80-53377	0-88318-161-4
No. 63	Supernovae Spectra (La Jolla, 1980)	80-70019	0-88318-162-2
No. 64	Laboratory EXAFS Facilities – 1980 (Univ. of Washington)	80-70579	0-88318-163-0
No. 65	Optics in Four Dimensions – 1980 (ICO, Ensenada)	80-70771	0-88318-164-9
No. 66	Physics in the Automotive Industry – 1980 (APS/AAPT Topical Conference)	80-70987	0-88318-165-7
No. 67	Experimental Meson Spectroscopy – 1980 (Sixth International Conference, Brookhaven)	80-71123	0-88318-166-5
No. 68	High Energy Physics – 1980 (XX International Conference, Madison)	81-65032	0-88318-167-3
No. 69	Polarization Phenomena in Nuclear Physics – 1980 (Fifth International Symposium, Santa Fe)	81-65107	0-88318-168-1
No. 70	Chemistry and Physics of Coal Utilization – 1980 (APS, Morgantown)	81-65106	0-88318-169-X
No. 71	Group Theory and its Applications in Physics – 1980 (Latin American School of Physics, Mexico City)	81-66132	0-88318-170-3
No. 72	Weak Interactions as a Probe of Unification (Virginia Polytechnic Institute – 1980)	81-67184	0-88318-171-1
No. 73	Tetrahedrally Bonded Amorphous Semiconductors (Carefree, Arizona, 1981)	81-67419	0-88318-172-X

No. 74	Perturbative Quantum Chromodynamics (Tallahassee, 1981)	81-70372	0-88318-173-8
No. 75	Low Energy X-Ray Diagnostics – 1981 (Monterey)	81-69841	0-88318-174-6
No. 76	Nonlinear Properties of Internal Waves (La Jolla Institute, 1981)	81-71062	0-88318-175-4
No. 77	Gamma Ray Transients and Related Astrophysical Phenomena (La Jolla Institute, 1981)	81-71543	0-88318-176-2
No. 78	Shock Waves in Condensed Matter – 1981 (Menlo Park)	82-70014	0-88318-177-0
No. 79	Pion Production and Absorption in Nuclei – 1981 (Indiana University Cyclotron Facility)	82-70678	0-88318-178-9
No. 80	Polarized Proton Ion Sources (Ann Arbor, 1981)	82-71025	0-88318-179-7
No. 81	Particles and Fields –1981: Testing the Standard Model (APS/DPF, Santa Cruz)	82-71156	0-88318-180-0
No. 82	Interpretation of Climate and Photochemical Models, Ozone and Temperature Measurements (La Jolla Institute, 1981)	82-71345	0-88318-181-9
No. 83	The Galactic Center (Cal. Inst. of Tech., 1982)	82-71635	0-88318-182-7
No. 84	Physics in the Steel Industry (APS/AISI, Lehigh University, 1981)	82-72033	0-88318-183-5
No. 85	Proton-Antiproton Collider Physics –1981 (Madison, Wisconsin)	82-72141	0-88318-184-3
No. 86	Momentum Wave Functions – 1982 (Adelaide, Australia)	82-72375	0-88318-185-1
No. 87	Physics of High Energy Particle Accelerators (Fermilab Summer School, 1981)	82-72421	0-88318-186-X
No. 88	Mathematical Methods in Hydrodynamics and Integrability in Dynamical Systems (La Jolla Institute, 1981)	82-72462	0-88318-187-8
No. 89	Neutron Scattering – 1981 (Argonne National Laboratory)	82-73094	0-88318-188-6
No. 90	Laser Techniques for Extreme Ultraviolt Spectroscopy (Boulder, 1982)	82-73205	0-88318-189-4
No. 91	Laser Acceleration of Particles (Los Alamos, 1982)	82-73361	0-88318-190-8
No. 92	The State of Particle Accelerators and High Energy Physics (Fermilab, 1981)	82-73861	0-88318-191-6
No. 93	Novel Results in Particle Physics (Vanderbilt, 1982)	82-73954	0-88318-192-4
No. 94	X-Ray and Atomic Inner-Shell Physics – 1982 (International Conference, U. of Oregon)	82-74075	0-88318-193-2
No. 95	High Energy Spin Physics – 1982 (Brookhaven National Laboratory)	83-70154	0-88318-194-0
No. 96	Science Underground (Los Alamos, 1982)	83-70377	0-88318-195-9

No. 97	The Interaction Between Medium Energy Nucleons in Nuclei – 1982 (Indiana University)	83-70649	0-88318-196-7
No. 98	Particles and Fields – 1982 (APS/DPF University of Maryland)	83-70807	0-88318-197-5
No. 99	Neutrino Mass and Gauge Structure of Weak Interactions (Telemark, 1982)	83-71072	0-88318-198-3
No. 100	Excimer Lasers – 1983 (OSA, Lake Tahoe, Nevada)	83-71437	0-88318-199-1
No. 101	Positron-Electron Pairs in Astrophysics (Goddard Space Flight Center, 1983)	83-71926	0-88318-200-9
No. 102	Intense Medium Energy Sources of Strangeness (UC-Sant Cruz, 1983)	83-72261	0-88318-201-7
No. 103	Quantum Fluids and Solids – 1983 (Sanibel Island, Florida)	83-72440	0-88318-202-5
No. 104	Physics, Technology and the Nuclear Arms Race (APS Baltimore –1983)	83-72533	0-88318-203-3
No. 105	Physics of High Energy Particle Accelerators (SLAC Summer School, 1982)	83-72986	0-88318-304-8
No. 106	Predictability of Fluid Motions (La Jolla Institute, 1983)	83-73641	0-88318-305-6
No. 107	Physics and Chemistry of Porous Media (Schlumberger-Doll Research, 1983)	83-73640	0-88318-306-4
No. 108	The Time Projection Chamber (TRIUMF, Vancouver, 1983)	83-83445	0-88318-307-2
No. 109	Random Walks and Their Applications in the Physical and Biological Sciences (NBS/La Jolla Institute, 1982)	84-70208	0-88318-308-0
No. 110	Hadron Substructure in Nuclear Physics (Indiana University, 1983)	84-70165	0-88318-309-9
No. 111	Production and Neutralization of Negative Ions and Beams (3rd Int'l Symposium, Brookhaven, 1983)	84-70379	0-88318-310-2
No. 112	Particles and Fields – 1983 (APS/DPF, Blacksburg, VA)	84-70378	0-88318-311-0
No. 113	Experimental Meson Spectroscopy – 1983 (Seventh International Conference, Brookhaven)	84-70910	0-88318-312-9
No. 114	Low Energy Tests of Conservation Laws in Particle Physics (Blacksburg, VA, 1983)	84-71157	0-88318-313-7
No. 115	High Energy Transients in Astrophysics (Santa Cruz, CA, 1983)	84-71205	0-88318-314-5
No. 116	Problems in Unification and Supergravity (La Jolla Institute, 1983)	84-71246	0-88318-315-3
No. 117	Polarized Proton Ion Sources (TRIUMF, Vancouver, 1983)	84-71235	0-88318-316-1

No. 118	Free Electron Generation of Extreme Ultraviolet Coherent Radiation (Brookhaven/OSA, 1983)	84-71539	0-88318-317-X
No. 119	Laser Techniques in the Extreme Ultraviolet (OSA, Boulder, Colorado, 1984)	84-72128	0-88318-318-8
No. 120	Optical Effects in Amorphous Semiconductors (Snowbird, Utah, 1984)	84-72419	0-88318-319-6
No. 121	High Energy e^+e^- Interactions (Vanderbilt, 1984)	84-72632	0-88318-320-X
No. 122	The Physics of VLSI (Xerox, Palo Alto, 1984)	84-72729	0-88318-321-8
No. 123	Intersections Between Particle and Nuclear Physics (Steamboat Springs, 1984)	84-72790	0-88318-322-6
No. 124	Neutron-Nucleus Collisions – A Probe of Nuclear Structure (Burr Oak State Park - 1984)	84-73216	0-88318-323-4
No. 125	Capture Gamma-Ray Spectroscopy and Related Topics – 1984 (Internat. Symposium, Knoxville)	84-73303	0-88318-324-2
No. 126	Solar Neutrinos and Neutrino Astronomy (Homestake, 1984)	84-63143	0-88318-325-0
No. 127	Physics of High Energy Particle Accelerators (BNL/SUNY Summer School, 1983)	85-70057	0-88318-326-9
No. 128	Nuclear Physics with Stored, Cooled Beams (McCormick's Creek State Park, Indiana, 1984)	85-71167	0-88318-327-7
No. 129	Radiofrequency Plasma Heating (Sixth Topical Conference, Callaway Gardens, GA, 1985)	85-48027	0-88318-328-5
No. 130	Laser Acceleration of Particles (Malibu, California, 1985)	85-48028	0-88318-329-3
No. 131	Workshop on Polarized ^{3}He Beams and Targets (Princeton, New Jersey, 1984)	85-48026	0-88318-330-7
No. 132	Hadron Spectroscopy–1985 (International Conference, Univ. of Maryland)	85-72537	0-88318-331-5
No. 133	Hadronic Probes and Nuclear Interactions (Arizona State University, 1985)	85-72638	0-88318-332-3
No. 134	The State of High Energy Physics (BNL/SUNY Summer School, 1983)	85-73170	0-88318-333-1
No. 135	Energy Sources: Conservation and Renewables (APS, Washington, DC, 1985)	85-73019	0-88318-334-X
No. 136	Atomic Theory Workshop on Relativistic and QED Effects in Heavy Atoms	85-73790	0-88318-335-8
No. 137	Polymer-Flow Interaction (La Jolla Institute, 1985)	85-73915	0-88318-336-6
No. 138	Frontiers in Electronic Materials and Processing (Houston, TX, 1985)	86-70108	0-88318-337-4
No. 139	High-Current, High-Brightness, and High-Duty Factor Ion Injectors (La Jolla Institute, 1985)	86-70245	0-88318-338-2

No. 140	Boron-Rich Solids (Albuquerque, NM, 1985)	86-70246	0-88318-339-0
No. 141	Gamma-Ray Bursts (Stanford, CA, 1984)	86-70761	0-88318-340-4
No. 142	Nuclear Structure at High Spin, Excitation, and Momentum Transfer (Indiana University, 1985)	86-70837	0-88318-341-2
No. 143	Mexican School of Particles and Fields (Oaxtepec, México, 1984)	86-81187	0-88318-342-0
No. 144	Magnetospheric Phenomena in Astrophysics (Los Alamos, 1984)	86-71149	0-88318-343-9
No. 145	Polarized Beams at SSC & Polarized Antiprotons (Ann Arbor, MI & Bodega Bay, CA, 1985)	86-71343	0-88318-344-7
No. 146	Advances in Laser Science–I (Dallas, TX, 1985)	86-71536	0-88318-345-5
No. 147	Short Wavelength Coherent Radiation: Generation and Applications (Monterey, CA, 1986)	86-71674	0-88318-346-3
No. 148	Space Colonization: Technology and The Liberal Arts (Geneva, NY, 1985)	86-71675	0-88318-347-1
No. 149	Physics and Chemistry of Protective Coatings (Universal City, CA, 1985)	86-72019	0-88318-348-X
No. 150	Intersections Between Particle and Nuclear Physics (Lake Louise, Canada, 1986)	86-72018	0-88318-349-8
No. 151	Neural Networks for Computing (Snowbird, UT, 1986)	86-72481	0-88318-351-X
No. 152	Heavy Ion Inertial Fusion (Washington, DC, 1986)	86-73185	0-88318-352-8
No. 153	Physics of Particle Accelerators (SLAC Summer School, 1985) (Fermilab Summer School, 1984)	87-70103	0-88318-353-6
No. 154	Physics and Chemistry of Porous Media—II (Ridge Field, CT, 1986)	83-73640	0-88318-354-4
No. 155	The Galactic Center: Proceedings of the Symposium Honoring C. H. Townes (Berkeley, CA, 1986)	86-73186	0-88318-355-2
No. 156	Advanced Accelerator Concepts (Madison, WI, 1986)	87-70635	0-88318-358-0
No. 157	Stability of Amorphous Silicon Alloy Materials and Devices (Palo Alto, CA, 1987)	87-70990	0-88318-359-9
No. 158	Production and Neutralization of Negative Ions and Beams (Brookhaven, NY, 1986)	87-71695	0-88318-358-7

No. 159	Applications of Radio-Frequency Power to Plasma: Seventh Topical Conference (Kissimmee, FL, 1987)	87-71812	0-88318-359-5
No. 160	Advances in Laser Science–II (Seattle, WA, 1986)	87-71962	0-88318-360-9
No. 161	Electron Scattering in Nuclear and Particle Science: In Commemoration of the 35th Anniversary of the Lyman-Hanson-Scott Experiment (Urbana, IL, 1986)	87-72403	0-88318-361-7
No. 162	Few-Body Systems and Multiparticle Dynamics (Crystal City, VA, 1987)	87-72594	0-88318-362-5
No. 163	Pion–Nucleus Physics: Future Directions and New Facilities at LAMPF (Los Alamos, NM, 1987)	87-72961	0-88318-363-3
No. 164	Nuclei Far from Stability: Fifth International Conference (Rosseau Lake, ON, 1987)	87-73214	0-88318-364-1
No. 165	Thin Film Processing and Characterization of High-Temperature Superconductors	87-73420	0-88318-365-X
No. 166	Photovoltaic Safety (Denver, CO, 1988)	88-42854	0-88318-366-8
No. 167	Deposition and Growth: Limits for Microelectronics (Anaheim, CA, 1987)	88-71432	0-88318-367-6
No. 168	Atomic Processes in Plasmas (Santa Fe, NM, 1987)	88-71273	0-88318-368-4
No. 169	Modern Physics in America: A Michelson-Morley Centennial Symposium (Cleveland, OH, 1987)	88-71348	0-88318-369-2
No. 170	Nuclear Spectroscopy of Astrophysical Sources (Washington, D.C., 1987)	88-71625	0-88318-370-6
No. 171	Vacuum Design of Advanced and Compact Synchrotron Light Sources (Upton, NY, 1988)	88-71824	0-88318-371-4
No. 172	Advances in Laser Science–III: Proceedings of the International Laser Science Conference (Atlantic City, NJ, 1987)	88-71879	0-88318-372-2
No. 173	Cooperative Networks in Physics Education (Oaxtepec, Mexico 1987)	88-72091	0-88318-373-0
No. 174	Radio Wave Scattering in the Interstellar Medium (San Diego, CA 1988)	88-72092	0-88318-374-9
No. 175	Non-neutral Plasma Physics (Washington, DC 1988)	88-72275	0-88318-375-7

No. 176	Intersections Between Particle and Nuclear Physics (Third International Conference) (Rockport, ME 1988)	88-62535	0-88318-376-5
No. 177	Linear Accelerator and Beam Optics Codes (La Jolla, CA 1988)	88-46074	0-88318-377-3
No. 178	Nuclear Arms Technologies in the 1990s (Washington, DC 1988)	88-83262	0-88318-378-1
No. 179	The Michelson Era in American Science: 1870–1930 (Cleveland, OH 1987)	88-83369	0-88318-379-X
No. 180	Frontiers in Science: International Symposium (Urbana, IL 1987)	88-83526	0-88318-380-3
No. 181	Muon-Catalyzed Fusion (Sanibel Island, FL 1988)	88-83636	0-88318-381-1
No. 182	High T_c Superconducting Thin Films, Devices, and Application (Atlanta, GA 1988)	88-03947	0-88318-382-X
No. 183	Cosmic Abundances of Matter (Minneapolis, MN 1988)	89-80147	0-88318-383-8
No. 184	Physics of Particle Accelerators (Ithaca, NY 1988)	87-07208	0-88318-384-6
No. 185	Glueballs, Hybrids, and Exotic Hadrons (Upton, NY 1988)	89-83513	0-88318-385-4